AF378868

An Engaging Approach to the Science and Engineering of Materials

Khaled Morsi

An Engaging Approach to the Science and Engineering of Materials

The SPHERE

 Springer

Khaled Morsi
Department of Mechanical Engineering
San Diego State University
San Diego, CA, USA

ISBN 978-3-032-06230-7   ISBN 978-3-032-06231-4   (eBook)
https://doi.org/10.1007/978-3-032-06231-4

This Springer imprint is published by the registered company Springer Nature Switzerland AG
The registered company address is: Gewerbestrasse 11, 6330 Cham, Switzerland

If disposing of this product, please recycle the paper.

Preface

Today, we live in a unique age, with smartphones, iPads, and computers being an integral and deeply rooted part of our lives. We are surrounded by many examples of materials used in products that directly and indirectly impact us. Airplanes, cruise ships, space shuttles, wind turbines, and cars are but a few of these examples. Materials science and engineering is a field rich in concepts and often in need of an effective communication approach. One of the main issues is how we engage the student and entice him or her to explore, learn, and benefit from this critical science. This book, *An Engaging Approach to the Science and Engineering of Materials: The SPHERE*, attempts to instill the topic in the subconscious of students, helping them live, breathe, and appreciate this critical science that is affecting us in many prominent and non-obvious ways.

Our primary approach is different from traditional approaches. Here, knowledge is delivered to students within an engaging discovery-based storyline. *An Engaging Approach to the Science and Engineering of Materials: The SPHERE* strategically provides students with the knowledge needed for an introductory course in materials science and engineering. The Sphere is a time capsule that plays a pivotal role in the book's storyline. The spherical shape is also unique; planets, the moon, the sun, atoms, and white blood cells can all be approximated as spheres. Moreover, most nanoparticles are spherical. In sports, we find baseballs, tennis balls, soccer balls, basketballs, billiard balls, etc. More importantly, the spherical geometry subtly alludes to the nature of materials, always preferring the lowest energy configuration. The spherical geometry delivers this as it provides the least surface area to volume ratio of all shapes, thus minimizing surface energy.

The book covers conventional materials science topics as well as topics of current and future importance. As such, a whole chapter is devoted to nanomaterials, signifying another unique feature of the present book. The book's most unique feature is the constant dialogue between the professor and the student, through which subtle details of materials and historical facts are revealed, together with clarifications of typical student misconceptions.

Instructor Resources

An instructor's solutions manual and PowerPoints are also provided online to the instructor.

San Diego, USA   Khaled Morsi

Acknowledgments Writing this book has been a humbling experience. The more one learns, the more one acknowledges one's ignorance of the abundance of knowledge out there. My ultimate debt and gratitude are to God.

I am grateful to my late parents, Tawheeda Nadim and Mohamed Morsi, who dedicated their lives to raising me from the day I was born. I am also thankful for my wife, Amira, and my children, Yusuf, Nadeen, and Lana, who are the sunshine of my life. I also thank my sister, Prof. Rasha Morsi, for her unwavering love and support.

Life has many trials and tribulations. It would have been all the more difficult without the support of family and friends. Indeed, I owe them a great deal.

On the academic front, I am genuinely indebted to current and previous professors and scholars, who profoundly influenced me through their work and guidance. These include (but are not limited to) Profs. M. Ashby, W. Callister, B. Derby, A.S. El-Gizawy, R. German, J. Gordon, and R. A. Higgins. Special thanks to all my past and present graduate and undergraduate research students, members of the Advanced Materials Processing Laboratory.

Knowledge is cumulative in nature; consequently, our current understanding would not have been possible without the remarkable efforts and contributions of past and present researchers and educators. Despite my shortcomings, I humbly hope that this effort of mine will benefit people now and for years to come.

Competing Interests The author has no competing interests to declare that are relevant to the content of this manuscript.

About This Book

The storyline revolves around an undergraduate university student, **Alex**, who stumbles across an incredible find, which turns out to be a **time capsule (The Sphere)**, that contains some of the most important or notorious materials man has ever encountered. A central fictional character in the book is **Prof. Bainite C. Crystaloff (BCC)**, who happens to be Alex's neighbor and learns from Alex about this find. They both set out on a journey of discovery to identify the materials in The Sphere. Through investigative discovery, Alex learns all the fundamental knowledge needed to ascertain the identity of each material.

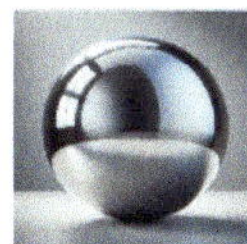
The Sphere

A time capsule containing seven concentric layers of unidentified material, representing materials of great importance in the history of mankind. Alex attempts to reveal the identity of each layer after learning relevant knowledge from Prof. Crystaloff's textbook, **"As a *Matter* of Fact"**. Now and then, the sphere symbol appears, and discussions ensue between Alex and Professor Crystaloff regarding the mysterious layers. These will be identified as *grey boxed italics* within the "As a Matter of Fact" text.

Black italics are used for their conversations when they are unrelated to the Sphere. This way, the text representing As a Matter of Fact will be identified as black standard non-italic font.

Book Characters

The book revolves around conversations between Alex and Prof. Crystaloff, a materials science and engineering Professor at the local university.

Alex (a gender-neutral name) is a male or female undergraduate student who discovers a time capsule (The Sphere) and pursues a journey of learning and discovery to identify the materials placed within it.

Professor Bainite C. Crystaloff (BCC), Alex's neighbor and mentor.

The professor has several roles in the book:

1. To help with the investigation process and advise Alex.
2. To communicate ethical, historical, and philosophical aspects when encountered in the book.
3. To share his experiences and lifelong lessons with Alex.
4. Professor Crystaloff constantly directs Alex to his book, **As a *Matter* of Fact**, supposedly a traditional materials science textbook he had recently published. Hence, the present book (The Sphere) integrates "As a *Matter* of Fact" within its pages to provide a holistic learning experience. Essentially, As a *Matter* of Fact, provides the traditional textbook learning style packaged within a novel storyline framework (The Sphere).
5. To quiz Alex as new knowledge is acquired.
6. To dialogue with Alex throughout the book, for example, when misconceptions or questions arise. Hence, students' general misconceptions are identified and addressed.
7. As a highly eccentric character, Prof. Crystaloff draws unorthodox and imaginative analogies to explain why materials behave as they do. This approach attempts to bring the readers into the Psyche of materials, for the lack of a better word.
8. Although students acquire knowledge within a university setting, they typically are unaware of the environment in which professors work and their duties. Alex is exposed to some day-to-day activities in a professor's life through discussions with the professor.

The book contains example problems and their solutions in chapters to reinforce specific concepts. Additional problems are also listed at the back of each chapter.

AI Statement

ChatGPT 4.0 has only been used to generate some images, such as the image of the Sphere, Alex, Prof. Crystaloff, and others, throughout the book. The latter are identified as ChatGPT 4.0 images in their figure captions.

Data and Information Statement

The data, information and graphs presented in this book are for educational purposes only. For any design considerations, the reader is advised to refer to more specialized design books.

Book Structure

The book is divided into seven parts, each containing one or more chapters. In most cases, a layer is identified at the end of each part.

Contents

Part V   Layer 6: Communication is Key

Part VI   Layer 1: Looks can be Deceiving

Part VII   Layer 7: This is Cutting Edge Technology

About the Author

Dr. Khaled Morsi is a Professor of Mechanical Engineering at the Department of Mechanical Engineering, San Diego State University, US. He earned his D.Phil. in 1996 from the Department of Materials, University of Oxford, UK, and has taught materials science and engineering for over 25 years. He has taught various topics, including Materials Science and Engineering, Nanomaterials, Powder Metallurgy, Materials, Design and Manufacturing, Advanced Materials Processing, and Thermodynamics. Dr. Morsi has received multiple awards for his teaching and mentorship of students. He also firmly believes in the importance of materials education and research as two sides of the same coin. As such, he has advised over 150 Ph.D., MS, and undergraduate students in materials research. His materials research focuses on the processing and properties of metals, intermetallics, nanomaterials, and their composites. Dr. Morsi has published extensively in the field and is in the top 1% most cited scientists in Materials according to the Stanford/Elsevier top 2% Scientists in Materials. He has served as Associate Editor for the *ASME Journal-Engineering Materials and Technology* (JEMT) and is on the editorial board of other materials journals. He also served as Chair of the Powder Materials Committee and the Advanced Characterization, Testing, and Simulation Committee at The Minerals, Metals and Materials Society (TMS).

Part I

The Sphere (Layer 2: The Search Begins)

"The Sphere"

As a highly inquisitive engineering student, Alex was constantly intrigued by the abundance of different types of natural and engineering materials and even more interested in finding new ways in which such materials can be used to benefit mankind. Then, one night, Alex dreamed of a metallic sphere with incredible sparkling and glamour hovering closely. Just before touching it, the alarm bell rang, and Alex woke up. The same dream recurred several times. Then, one spring morning, Alex got out of bed, drank the usual fragrant morning coffee, and stepped outside into the backyard to plant a small Guava tree. A tree that would add to the already exotic collection of fruit trees in the backyard. The sound and feeling of hitting a hard surface resounded as the digging started. After brushing off some of the dirt, a gleaming, almost sparkling surface appeared. The object was removed, only to discover the same Sphere appearing in all the dreams. The sphere appeared polished since it reflected Alex's face when looking at it. After obtaining confirmation from the proper authorities that the sphere was generally safe, it was determined that it was made up of two identical hollow hemispheres, each with what appears to be concentric material layers (numbered 2–7). The central cavity, however, contained a series of material blocks or samples left over for examination. All the samples were labeled with a number between 1 and 7, each supposedly representing the internal structure of the concentric material layer in the sphere with the same number. However, the materials making up the layers were not identified. Also, concentric layer 1 appeared to be missing.

(a)

(b)

The Sphere

(a) 3D model of sphere made up of two halves (hemispheres), (b) inside of one
     hemisphere showing what appears to be 6 concentric material layers numbered
     (2–7), with a central cavity containing samples representing material layers (1–7)

This sophisticated discovery takes Alex aback, then immediately remembers a next-
door neighbor, **Prof. Bainite C. Crystaloff (BCC)**, a Materials Science and Engineering Professor at the local University. Alex rushes to the Professor's doorstep and
rings the bell. His wife answers and informs Alex that the Professor is conducting
research in his lab at the University. With great eagerness to share this exciting find,
Alex arrives at the Lab. There he was leaning over a microscope to examine a material microstructure that his research students had just produced in his lab. Professor
Crystaloff turns his head with a welcoming smile.

What are you carrying, Alex? It seems heavy.

After being updated on what happened, the professor is also taken aback by this unusual find. He opens the sphere and examines its internal structure.

How unusual, I've never seen anything like it in my life.

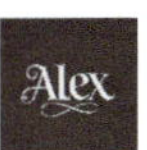

Can you help me investigate this object and use it to learn about the wondrous world of materials you keep talking to me about?

That's a splendid idea, Alex. Let's start on this adventure tomorrow. Get some rest; you will need it. Also, take my recently published book "As a Matter of Fact" with you. You will need it on your exciting journey. In it, you will learn about all types of materials and how their structures can affect their properties and use. This information may help you reveal the identities of those mysterious layers.

There is something a bit weird, though. I noticed seven samples, numbered 1–7, placed in the internal cavity of the sphere, but I can only see six layers. I don't see a layer 1.

Yes, I noticed this too. Let's start on layer 2 and work until we figure out this mysterious layer 1.

The Sphere is safely locked in a cabinet in the Professor's lab. On the way home, Alex became intrigued and curious to know more about this sphere; what was it made of? What are all these concentric layers? What are these blocks or samples? Why are there seven samples but six layers? Why was it buried in the backyard? Who brought it? So many questions. If only Alex had known about engineering materials, the answer to at least one of these questions could have been known. Who knows what that could lead to?

The Sphere
Layer 2

The Search Begins

The following day, the layer 2 sample (also called sample 2) is removed from the sphere's central cavity. The sample is tetragonal (square base and more extended height) and appears golden yellow.

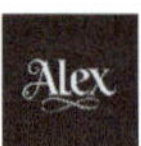

I like the color. Could it be gold, Prof.?

Come on, we've only just examined sample 2 using our naked eyes. Our eyes are incredible, but they have limitations. Did you know that we cannot resolve anything less than the diameter of a human hair? Can you see the atoms making up the sphere or its samples? Or how their atoms are arranged? Or even deduce their chemical compositions? We can see their colors, but let's not make any assumptions now. Remember, you are a scientific investigator; we only get excited after confirming our findings. How about this? You lead the scientific investigation and seek help from me whenever you need it, or I will jump in and provide you with helpful input when I can.

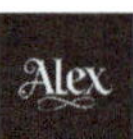

Ok then, as an investigator, I will start by weighing sample 2, measuring its dimensions, and calculating its volume and then density (since density is mass/volume). Once I know the density, I will compare it to the values of known materials and hopefully get lucky.

When we don't know anything about sample 2, any data is good but be cautious about how you interpret your findings. If you think you can find its density and compare it to the densities of known materials, then you are ignoring some potential pitfalls. How do you know that sample 2 does not contain internal porosity/holes or more than one type of atom? Both can influence the measured density and throw you off track. Not to mention that there are different materials out there that have the same density. For example, copper and nickel both have a density of 8.9 g/cm^3, which means that if you weigh a solid block of dimensions 1 cm $\times$ 1 cm $\times$ 1 cm and if it were all made of either nickel or copper, then it will always weigh 8.9 g either way. What if you measure the density and the above issues pop up? I love your enthusiasm, but for now, just read "As a Matter of Fact," starting with Chapter 1, Materials: An Integral Force Behind Civilizations and Technological Advancements. The more knowledge you acquire, the more likely you will succeed. Take this advice from someone who has sailed the oceans of knowledge and is still a humble seeker, even at my age.

Professor, I just measured the sample dimensions and mass, which enabled me to work out the density of this sample 2; it is 8.79 g/cm^3.

Congratulations on your first experiment and data point! Now, you need to read and build up your knowledge of materials.

As a 'Matter' of Fact

By Prof. Bainite C. Crystaloff (BCC)

Chapter 1
Materials: An Integral Force Behind Civilizations and Technological Advancements

1.1 Introduction

This chapter discusses the historical development of materials. It shows that our current time is indeed a special time in mankind's history regarding material development.

1.2 A Materials World

Our globe possesses riveting beauty and an abundance of riches, enough to sustain life with all its needs. These include water, food, materials for clothing, and all the prerequisites needed for technological advancements over time. The most fundamental starting point for all of these is the element, of which we have around 90 naturally occurring ones in the Earth's crust. Table 1.1 lists some of these elements and their approximate abundance. Important ones are also present in our oceans and atmosphere. For example, without the element oxygen, I would not have been able to write this book, nor would you have been able to read it. However, before we delve into our present-day world of materials, one must ultimately think of how our knowledge of materials developed over time. This is important to gain perspective and appreciation of the era in which we were born. Due to the cumbersome nature of such a vast topic, only a brief and rather selective treatment is given here.

It is not an overstatement to say that materials have been the force behind the advancements of civilizations throughout history. Let us start with the materials used by the first humans who inhabited this planet. Although dates of the periods may vary depending on geographic locations, archaeologists generally divide the Stone Age into the **Paleolithic period** (also called the Old Stone Age, ~ 2.5 million to ~ 8500 BCE) and the **Neolithic period** (New Stone Age, ~ 8500 to ~ 4500 BCE). This is then followed by the **Chalcolithic Period** (also

© The Author(s), under exclusive license to Springer Nature Switzerland AG 2026

K. Morsi, *An Engaging Approach to the Science and Engineering of Materials*,

https://doi.org/10.1007/978-3-032-06231-4_1

Table 1.1 Abundance of some of the elements in the Earth's crust

Element	Presence in the Earth's crust (wt.%)
Oxygen (O)	46
Silicon (Si)	28
Aluminum (Al)	8
Iron (Fe)	5.8
Calcium (Ca)	4
Sodium (Na)	2.4
Potassium (K)	2.3
Magnesium (Mg)	2.1
Titanium (Ti)	0.4
Hydrogen (H)	0.1
Phosphorous (P)	0.1
Manganese (Mn)	0.1
Copper (Cu)	0.0058
Lead (Pb)	0.0001
Tin (Sn)	0.00015
Silver (Ag)	0.000008
Platinum (Pt)	0.000004
Gold (Au)	0.0000004

Stephen L. Sass, The Substance of Civilization, Arcade Publishing, New York (2011)

known as the **Copper Age**, ~ 4500 to ~ 3200 BCE), the **Bronze Age** (~3200 to ~ 1200 BCE), and then the **Iron Age** (~ 1200 to ~ 500 BCE, although some may argue that we are still in the Iron Age). Figure 1.1 shows this general timeline.

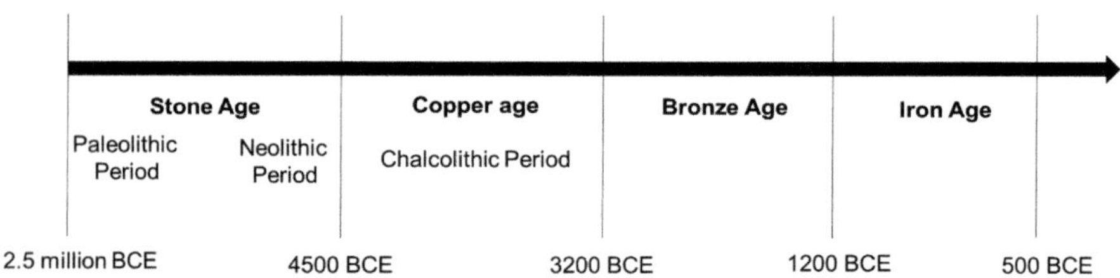

Fig. 1.1 Timeline (nonlinear) for the Stone, Copper, Bronze, and Iron Ages

1.3   The Stone Age (~ 2.5 Million to ~ 4500 BCE)

Although it is called the Stone Age, and stone was used, other materials, including wood, tree resin, and sinew (animal tendons), were used during this period. Two of the stones used were **obsidian** (a black glass) and **flint**. What is so special about these two materials is their ability to be shaped and sliced by **knapping,** in which a **hammer stone** is used to hammer out obsidian or flint rock flakes. This resulted in extremely sharp obsidian or flint rock objects that could be turned into tools. For example, a sharp flint piece could be attached to wood (using tree resin and/or sinew) to produce an axe, spear, or knife. Such artifacts were used in hunting but also as general tools. Since wood and other organic matter decay over time, it is possible to confirm the time of use of such ancient organic materials after discovering them preserved in protective environments, for example, trapped under ice for thousands of years. The melting of the ice allowed access to such artifacts. This led to the field of *Ice Patch Archeology*. Unlike glaciers, which are known to move over time (and in doing so, expose ancient artifacts to environmental decay), ice patches occur on leveled ground and have the advantage of being stagnant, thus giving them the ability to preserve buried artifacts in the exact location for thousands of years. Wooden spears, atlatls (spears thrown using a hurling device to reach further distances), and arrows were successfully excavated, showing incredible features such as the sinews and feathers used to produce weapons, such as arrows. Of course, once the ice melts, time becomes of the essence, and if these ancient, fragile artifacts were not immediately preserved, they too could deteriorate. Our present historical understanding, however, was greatly enabled through the development of radiocarbon dating, which was introduced by University of Chicago **Professor Willard Libby** in the late 1940s, for which he later (1960) received the Nobel Prize.

Another significant development during the Stone Age was using **clay** to form artifacts and their subsequent firing (heating) to convert them into hard pots (for storing food, grains, and drinks) and clay tablets, which preserved writing. By the way, clay tablets are much better preserved than papyrus. Changing moldable clay into hard artifacts through heating (firing) was possibly the first time mankind intentionally changed a material's structure, properties, and use. This is the basis of what we call materials engineering today. Examples of some of the materials used in the Stone Age are given in Fig. 1.2.

1.4   The Chalcolithic Period (The Copper Age, ~ 4500 to ~ 3200 BCE)

This was the period when copper became prominent, and its smelting from naturally occurring rocky minerals was introduced. Notably, relatively pure copper was also found before this period, typically in small amounts known as **native copper** (Fig. 1.3a). However, due to their generally low abundance, the artifacts

Fig. 1.2 Stones used in ancient times, **a** Obsidian, **b** Flint, and **c** Shaped flint on arrowhead. (Images generated using Chat GPT 4.0 by K. Morsi)

produced were typically restricted to making small items, such as jewelry. Sometimes, extremely large native copper was found; for example, in the Smithsonian Institution in Washington, D.C., lies a copper boulder weighing ~ 3000 lb. It was really after the introduction of smelting that the production of copper in large quantities became possible. A copper-rich ore, **malachite** ($Cu_2CO_3(OH)_2$) (Fig. 1.3b), was one of the minerals that could be smelted to produce copper. In fact, due to its beauty, it was possibly used as Jewelry. Heating malachite with charcoal (from burned wood) in a confined space (maybe a pit) generates carbon monoxide (CO) (from the reaction between charcoal and oxygen), which causes a reduction process to take place that frees up the copper and the temperatures generated (~ 1200 °C) were enough to melt copper (which has a melting point of 1083 °C). There it was the **smelting** of copper, which allowed the production of copper in large quantities enough to produce tools and weapons. Copper-arsenic (Cu–As) alloys were produced during this period.

Fig. 1.3 **a** Native copper nuggets, **b** Malachite stones. (images generated using Chat GPT 4.0 by K. Morsi)

1.5   The Bronze Age (~ 3200 to ~ 1200 BCE)

Although the Bronze Age is said to have started ~ 3200 BCE, at that time, the use of copper and copper-arsenic (Cu-As) was still dominant; only about a thousand years later did bronze take over. Unlike Cu-As, bronze is an alloy of copper and tin (Sn). The drive of our ancestors to use bronze and Sn may have been the depletion of As-containing rocks and the use of tin-containing rocks. However, in certain locations, such as in Ancient Egypt, archeological evidence suggests that the Ancient Egyptians added tin to copper intentionally. In any case, there is an inherent difference between As and Sn. When heated in air, arsenic converts to arsenic trioxide, which is volatile and easily sublimes, thus bypassing the liquid phase altogether. Arsenic trioxide gas is extremely dangerous and can attack the nervous system; people dealing with these alloys while hot (during smelting or subsequent hot working) can be crippled. The production of these volatile fumes also results in variability between the composition of batches and their overall properties. This is not the case for bronze, since the tin in bronze produces tin oxide that is not volatile like arsenic trioxide.

1.6   The Iron Age (~ 1200 to ~ 500 BCE)

Before the Iron Age, mankind was acquainted with iron through **iron meteorites** sent down to Earth from outer space as meteoroids (Fig. 1.4). Note that meteorites are meteoroids that survived the Earth's atmosphere and landed on Earth. Although large masses of iron meteorites have been found on Earth, a much larger occurrence is smaller pieces due to meteorites breaking up while entering the Earth's atmosphere. This made it difficult for our ancestors to make anything useful from celestial or meteoric iron on any large scale. Around 2000 years before iron smelting, the discovery of 9 tubular iron beads represents the oldest known artifacts made from iron (found in Egypt). The origin of that iron was determined to be iron meteorites. This was confirmed due to its significant nickel content and large grain sizes (more about that later). However, given that iron and steel are possibly the most impactful materials in human history, meteoric iron may have indicated the shape of things to come. It also served as an early example of alloying (meteoric iron typically contains more than 5 wt.% nickel as an alloying element).

Apart from meteoric iron, iron is the fourth most abundant element in the Earth's crust, typically in the form of minerals such as hematite (Fe_2O_3) and magnetite (Fe_3O_4). The Iron Age began only after the successful smelting of iron from these minerals. However, smelting iron is not as easy as smelting the previous lower melting point materials like bronze and copper. The main problem was that iron melted at 1538 °C, and the highest temperature achievable at the time was ~ 1200 °C. So, how did our ancestors smelt iron? The smelting process involved burning wood to produce charcoal (placed in a pit) and blowing air through clay pipes; the pits were also covered with clay. Charcoal reacts with oxygen to form carbon monoxide,

Fig. 1.4 Iron meteoroid heading to Earth. (Image generated using Chat GPT 4.0 by K. Morsi)

which reacts with iron oxide in the ores and reduces it to iron. The temperature was not high enough to melt iron, but it was enough to soften it considerably to make it feasible to squeeze it out through some hammering processes while hot. The resulting iron was known as **bloomery iron**. The iron from this process was soft and still needed to be hardened through subsequent processes, such as heating again in charcoal and eventually quenching and tempering. Logically, with the lack of any modern scientific principles at the time, there must have been a great deal of trial and error, and reproducibility must have been a significant issue due to the extremely high sensitivity of composition and properties to the manufacturing conditions as we know them today. Remarkably, we use these exact techniques today, albeit using more sophisticated equipment and precision built on scientific principles.

Other metals were also extracted throughout ancient times, especially those with relatively low melting points that were easily released from their compounds or structures during smelting. Gold was typically found in quartz (as veins) or free-standing nuggets. As one can imagine, elements can be found in abundance in one location, while in other cases, they can occur in very small quantities but spread over large geographic locations. Gold was very abundant in Egypt. The casket of Tutankhamun was made of solid gold (weighing an incredible 240 kg).

Apart from elements having ancient origins (e.g., carbon, iron, copper, sulfur, arsenic, silver, gold, tin, antimony, thallium, and lead), the earliest discovery of elements in modern history can be traced back to 1669, when phosphorus was first discovered. Figure 1.5 shows the timeline of the historical discoveries of some of today's most recognizable elements.

Early attempts to universally classify or arrange elements in some logical order were difficult but also resulted in important insights. It was not until 1869, when Russian chemist **Dmitri Mendeleev** generated the foundation of what is known today as the periodic table, that a consistent logical arrangement was put forward. His genius was such that he left gaps for yet-to-be-discovered elements, which were later inserted. In fact, in 1955, in recognition of Mendeleev's contributions, element 101 was named Mendelevium.

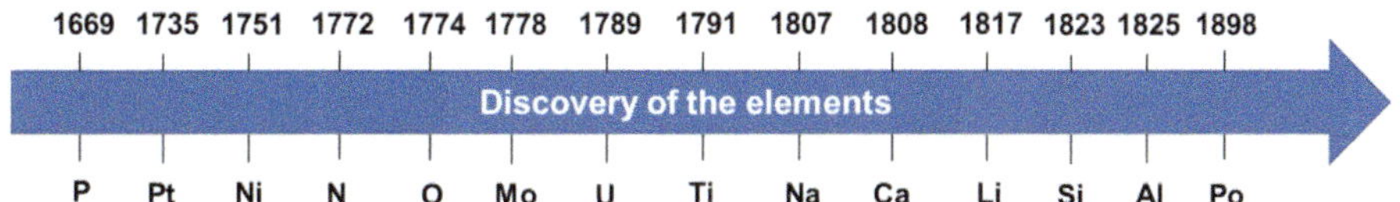

Fig. 1.5   Timeline (nonlinear) for the discovery of some of the elements (based on data from Mary Elvira Weeks and Henry M. Leicester, Discovery of the Elements, 7th Edition, Journal of Chemical Education, Easton, PA, USA (1968).)

1.7   Materials in Our Daily Lives

Today, ignoring how materials affect our daily lives is almost impossible. Just a glance at the endless products surrounding us, such as cars, airplanes, drones, bullet trains, cruise ships, solar cells, wind turbines, space shuttles, submarines, smartphones, and electronic devices (Fig. 1.6), we immediately realize the significance of materials in our lives today.

Materials Science is indeed a technology-*enabling* science. Without it, many engineering products would not exist today; in other words, they would not have "**material**ized" into anything useful.

Imagine if it were not for semiconductors, Silicon Valley would have been without the word *silicon*, and we can undoubtedly forget transistors coming onto the scene. Would computers be in our lives today without knowledge of electronic materials? Without biomaterials, what would have happened to all these patients needing knee, hip, or brain implants? Fig. 1.7a illustrates a hip implant, while Fig. 1.7b is an X-ray showing the implanted device. Hydrocephalus is a condition where excess cerebrospinal fluid builds up in the brain, for example, because of a congenital blockage somewhere in the brain. One of the ways to treat this condition is through the placement of a device called a shunt. Here, a hole is drilled in the skull, allowing the placement of a catheter, which provides an exit path for the fluid. These days, programmable valves are used to control the rate of fluid drainage. Figure 1.7c and d shows an image of the placement of a shunt device in a skull and an MRI showing the shunt catheter (white) reaching into the ventricle where the fluid had built up. It cannot be overstated that the development of biomaterials has made an extraordinary

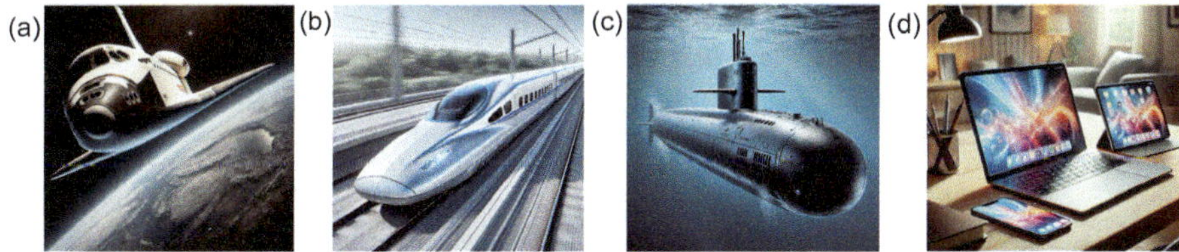

Fig. 1.6   a Materials used in a space shuttle may endure high levels of stress and severe conditions, **b** Bullet train (speeds of bullet trains may exceed 250 mph), **c** a submarine, **d** a variety of electronic devices available today. (images generated using Chat GPT 4.0 by K. Morsi)

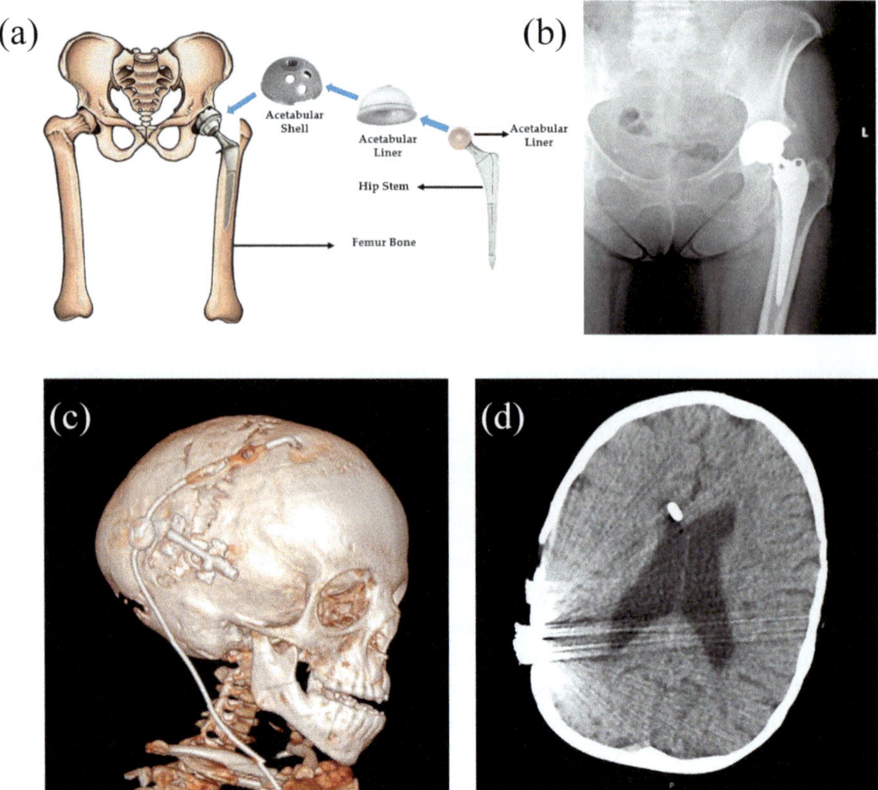

Fig. 1.7 **a** Illustration of an artificial hip implant (image from Soliman, M. M.; Chowdhury, M. E. H.; Islam, M. T.; Musharavati, F.; Nabil, M.; Hafizh, M.; Khandakar, A.; Mahmud, S.; Nezhad, E. Z.; Shuzan, M. N. I.; et al. A Review of Biomaterials and Associated Performance Metrics Analysis in Pre-Clinical Finite Element Model and in Implementation Stages for Total Hip Implant System. Polymers **2022**, 14, 4308. https://doi.org/10.3390/polym14204308), **b** X-ray of implanted hip implant (image from Berdini, M.; Procaccini, R.; Zanoli, G. F.; Faini, A.; Verdenelli, A.; Gigante, A. Influence of Femoral Stem Geometry on Total Hip Replacement: A Comparison of Clinical Outcomes of a Straight and an Anatomical Uncemented Stem. J. Clin. Med. **2024**, 13, 6459. https://doi.org/10.3390/jcm13216459), **c** Image of shunt placement in skull, **d** MRI of brain showing the tube from the shunt in white which helps drain the excess fluid in a hydrocephalus patient. **c** and **d** images from Ros, B.; Iglesias, S.; Linares, J.; Cerro, L.; Casado, J.; Arráez, M. A. Shunt Overdrainage: Reappraisal of the Syndrome and Proposal for an Integrative Model. J. Clin. Med. 2021, 10, 3620. https://doi.org/10.3390/jcm10163620). Images reproduced (**a** and **b**) and reproduced with modification (**c** and **d**) under the terms and conditions of Creative Commons Attribution (CC BY) license (https://creativecommons.org/licenses/by/4.0/)

difference in many people's lives. Such materials not only need to have the proper mechanical and physical properties to fulfill their tasks, but they also need to be biocompatible with the human body. They should not elicit any negative response from the body. Here, biocompatibility plays a significant role.

Without lightweight materials, such as aluminum, titanium, magnesium alloys, or polymer composites, would we have been able to have planes flying over enormous distances without stopping to refuel? Or very fast racing cars? Would we have been able to explore space or our deepest oceans without a deep knowledge and understanding of materials? What about skyscrapers? Here, steel-reinforced concrete plays an important role. Burj Khalifa in Dubai is currently the tallest building in the world, more than half a mile high. All these achievements were only possible due to an in-depth knowledge of engineering materials.

How about our exploration of features and structures smaller than the resolution of our naked eyes? Would microscopes have been possible without knowledge of lens materials and their optical properties? Indeed, most, if not all, of our advancements would have remained as ideas penciled in on paper without our knowledge of materials. By the way, both the pencil and the paper are materials. A pencil is made from a mixture of clay and graphite, with the latter comprising stacked graphene atomic layers, each consisting of carbon atoms strongly bonded in the form of 2D-hexagonal arrangements. However, weaker bonds exist between these parallel layers; thus, each layer can easily slide past the other upon applying a shear force. This provides the basis for using graphite as a pencil material that deposits markings on paper or as a high-temperature solid lubricant used in the hot working of metals and alloys. Due to the unique properties of individual graphene layers, researchers have produced them in standalone form and used them in more advanced applications, including as property-enhancing reinforcements in nanocomposites. It should be clear by now how significant and critical materials science is.

We are indeed very fortunate to live in an age when materials have erupted onto the scientific arena, mainly within the last one to two hundred years (a drop in the ocean of time), and are constantly impacting societies. Unlike the Stone, Bronze, or Iron Ages, today, we are overwhelmed by an enormous number of materials to choose from when we are faced with selecting a material for an engineering application. Most would say it's a nice problem to have. This is where world-renowned **Prof. Michael F. Ashby** of Cambridge University in the UK made one of his several incredible contributions to the field by developing a methodology by which engineers can select the best material for a particular application from potentially hundreds of thousands available to us today. This has led to the introduction of materials selection and process selection charts for choosing the best material and the best manufacturing process to make the final component. Such an approach helped advance the use of materials in engineering and other disciplines. Likewise, throughout this book, mention is made of the incredible contributions of many scientists who have shaped our current understanding of materials.

Around the world, there are continuous advances and developments in new materials, from metallic foams that can absorb impact energy or provide thermal management (open-cell foams) to nanostructured/nanoscale materials with extraordinary properties that can exceed by an order or orders of magnitude the properties of conventional materials such as strength, thermal conductivity, and stiffness, to name a few. Figure 1.8 illustrates such a material, multi-wall carbon nanotubes, with a

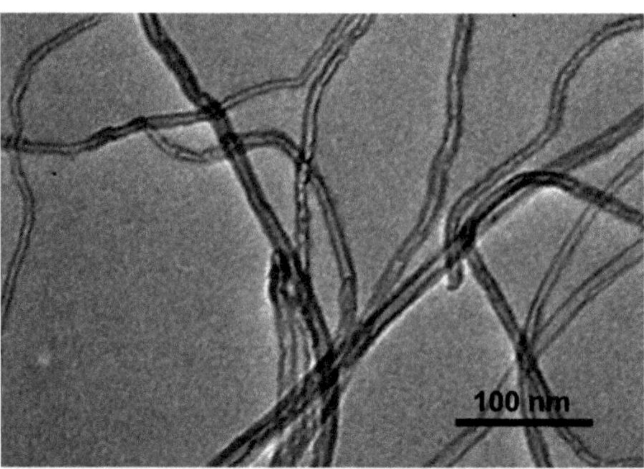

Fig. 1.8 Transmission electron micrograph of multi-wall carbon nanotubes, note the high aspect ratios (reproduced with modification from MDPI, M. Sorcia-Morales, F. C. Gómez-Merino, L. Sánchez-Segura, J. L.Spinoso-Castillo and J. J. Bello-Bello, Multi-Walled Carbon Nanotubes Improved Development during In Vitro Multiplication of Sugarcane (Saccharum spp.) in a Semi-Automated Bioreactor, Plants 2021, 10(10), 2015; https://doi.org/10.3390/plants10102015 reproduced and modified from MDPI under Creative Commons Attribution (CC BY) license (https://cre ativecommons.org/licenses/by/4.0/))

6–13 nm diameter and a 2.5–20 μm length. Carbon nanotubes can be 100 times stronger than steel and five times stiffer.

Smart materials have also been developed that remember the shape they were initially made in, so they can revert to it when needed. An example of these materials is NITINOL, Nickel Titanium Naval Ordinance Laboratory, first discovered by **William Beuhler**, who coined the name. This alloy is today used in numerous applications, such as springs, heart stents, plates for bone healing, antennas, actuators, and window knobs/locks used in greenhouses for climate control, among many other applications. By the way, the shape memory effect is not restricted to metals; some polymers display it. Back to Nitinol, due to another property it possesses, that is, super-elasticity, they are also used as eyeglass frames, which can be bent to large deflections yet return to their shape when the load is removed.

Professor, it seems incredible; imagine them even thinking about designing such a material.

Certainly! The person who discovered it and the people who later developed this material were very smart. However, its shape memory behavior was discovered totally by accident. It's a very interesting story. **William Beuhler**, *who worked at the US Naval Ordinance Laboratory (NOL), was investigating intermetallic materials that contained nickel and titanium. He had cast a few bars of the material, and after intentionally dropping one of the materials, he noticed a dull sound, like dropping a piece of lead. That particular bar had cooled much more than other samples after casting. However, when he dropped the other warmer bars, they produced a bell-like ringing sound. He knew that any change to the acoustic properties of a material could point to a change in its atomic structure, and this change seemed to depend on temperature.* **Raymond Wiley** *then joined him in studying the properties of this material further. Wiley, who worked in a different group at NOL, was more interested in the fatigue properties of materials. In 1961, Wiley was about to present research progress to management, so he brought a thin, long folded strip of nitinol, thinking he could display its great fatigue properties. The strip was shaped like an accordion so people could easily stretch and compress it as it was passed around the table during the meeting. The strip reached an associate director, who was also a pipe smoker; naturally, he became curious about how this so-called great strip would behave under heat. So, he exposed the strip to the flame of his lighter, and then the weirdest thing happened. The accordion-shaped thin strip stretched out on its own into its original shape before it was folded. This was the beginning of the life of a material that has significantly impacted our modern-day society. Today, very large space antennas can be crushed to occupy a smaller volume and not take up too much space in a space shuttle, and then, once in space, the heat from the sun would allow it to return to its initial stretched-out shape. You'd be excused for thinking that it's a thing of science fiction, but we are living through it, Alex!*

So, if materials science is an enabling science behind many of the present-day engineering products, then what is *Materials Engineering*? To say we engineer a material is to say we manipulate its makeup or internal structure to give it what we desire in terms of properties and performance. This can only happen if we understand the inner workings of materials—what makes one material strong and another weak? What makes one malleable while the other brittle? Why is one material conductive while another is an insulator? Along this train of thought, one can imagine many materials with various properties and attributes. However, our basic scientific understanding of causes and effects within materials enables us to manipulate the materials to improve performance.

It seems there are a lot of materials to choose from, Prof., not to mention our ability to manipulate them.

Alex, I believe there is no bad material; and only our lack of imagination prevents us from using a given material for a practical application(s). Materials come in different shapes, forms, and properties. Sometimes, you may encounter a material or property you think is useless. For example, a material that cannot conduct heat well and cannot be used as a heat sink for thermal management is considered useless for this application. However, another application can use the same material as a thermal insulator and save energy. A material prone to fracture, like a ceramic, may not make the best sword material, but would make great bulletproof armor. The fact that it cracks (and absorbs energy in doing so) upon the impact of a bullet is precisely why it is successful in such an application. So, we should never look down on a material just because it is weak or has a low melting point, as someone may come along and find a great use or application for it. On a different note, have you ever looked down from a plane at night? If you have, you must have noticed how beautifully decorated the Earth is with night lights, another impact of materials and electricity. Would the earth have looked the same at night if you were 'somehow' airborne thousands of years ago? Indeed, we are in special times.

1.8   The Golden Rules of Materials

Before proceeding, two golden rules in materials science must be mentioned.

Structure Affects Properties

It is well known that the **structure of a material generally affects its properties and behavior**. It, therefore, follows that if we can manipulate or control the structure of a material, then we have a doorway to controlling its properties and performance. The word structure can refer to entities of different length scales, ranging from angstroms to much larger sizes. On one extremely small scale, we could be talking about the electronic structure of an atom, while on another, we could be talking about the size of crystals or grains (microstructure) that make up countless engineering materials. On an even larger scale, we could talk about a large component containing cavities or cracks visible to the naked eye. Would we trust a machine component that contains a big crack to carry a load? I don't think so!

If given a block of pure metal, it would appear opaque. However, if we could see the interior of this block, we would know that it is made up of a large number (up to millions) of crystals stuck together. Such materials are **polycrystalline**, which makes up all metals and many ceramics. Inside each of these crystals (also called grains), we would see a beautifully ordered long-range arrangement of atoms. Although such atomic arrangements would be the same in every grain/crystal, they would be oriented or aligned differently, as seen in Fig. 1.9a for three grains identified by red, light blue, and dark blue colors. In essence, the grains or crystals would be randomly oriented to each other, and so would their atomic arrangements. As mentioned before, the structure could also refer to the size of these grains; as seen in Fig. 1.9, the grain size decreases from Fig. 1.9a–c. Such differences in structure can lead to differences in material properties and behavior. For example, the same metal with a smaller average grain size will be stronger than one with a larger average grain size. We also have processes by which we can make a whole material out of only one single grain or crystal. The material would then be called a **single crystal**. Si, as a semiconductor, is a notable example.

It is also possible to have engineering materials without long-range order for their atomic arrangements. These are called **non-crystalline materials** (or **amorphous materials**). The fact that their atoms are not arranged in an ordered manner means their structure is different from that of crystalline materials, and hence their properties will be different. Examples of amorphous materials include glass, many other polymers, and, recently, metallic glass. So, we need to remember that structure generally affects properties.

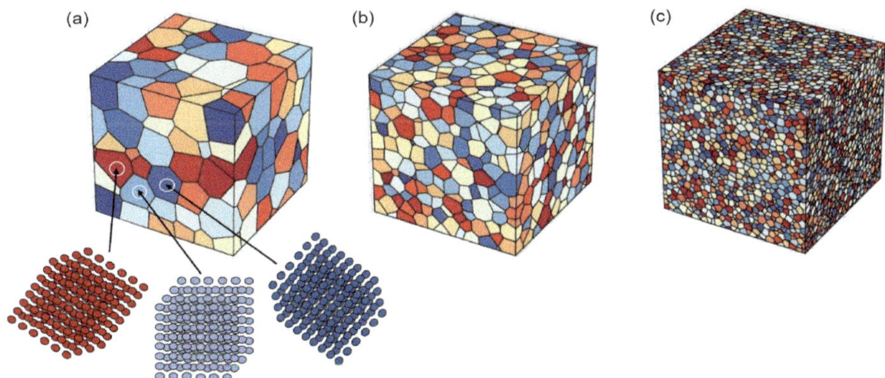

Fig. 1.9 Polycrystalline material showing; **a** large grains and atomic arrangement in neighboring grains with different orientations, **b** and **c** the same material as in (**a**) but with smaller and smaller grains. (Cubic grain structures generated using MicroStructPy, an open-source code. Copyright for MicroStructPy is held by Georgia Tech Research Corporation (Kenneth (Kip) Hart and Julian Rimoli)

One Goes up, and the Other Comes Down

Concerning the **strength and ductility**, if we can manipulate the structure of a metal or alloy to make it stronger, then it will become less ductile (i.e., more brittle). Conversely, if we change the structure to make it more ductile, the strength of the metal or alloy will be reduced. Hence, if strength goes up, ductility (the ability of a material to change shape permanently under a tensile load, for example, by stretching it) and malleability (the ability of a material to change shape permanently under a compressive load, for example, by forging or pressing) will go down, and vice versa.

With these golden rules in mind, let us now embark on a wondrous world of materials, a dynamic world with many beautiful arrangements, variations, and opportunities.

1.9 The Basic Classes of Materials

One basic way to look at engineering materials would be to divide them into three classes: **metals, ceramics,** and **polymers**, each with advantages and disadvantages. We form **composites** if we combine metals, ceramics, or polymers. It is also possible to divide materials into even more classes. However, in this introductory book, we will stick to simple basics. Part of our job as engineers is to determine which material would be the most suitable for a particular engineering application. For example, polymers have an incredible advantage in that they are typically lightweight compared with ceramics and metals. Hence, they could be great candidates for applications that require lightweight materials, such as components in race cars or aircrafts (typically with some addition of carbon or glass fibers to stiffen these components).

Many **polymers** are also relatively cheap to make. Guess why most toys are made of polymers (plastics) these days? But if they tell me to use a polymer in a high-temperature turbine that needs to operate at 1000 °C, I will tell them, think again. Most polymers would not survive temperatures above 600 °C, let alone 1000 °C. When we compare this temperature, however, to the melting point of a refractory metal such as tungsten (~ 3422 °C), steel (~ 1370 °C), or an alumina ceramic (2072 °C), we can see why polymers would not stand a chance to compete as an actual high-temperature engineering material. By that, I mean materials that can be used, for example, in jet engines or furnaces, whether as standalone components or coatings. At lower temperatures, however, we will find many applications for polymers; their generally poor electrical conductivity makes them great as electrical insulators. Guess which material we have covering our electric wires? An example would be a thermoplastic polymer such as polyethylene. This is not to say that today, we have not managed to engineer special polymers that we now call conductive polymers. We have!

On the other hand, **metals** can conduct electricity and heat in a way that would be the envy of most polymers and ceramics. Many metals are ductile and malleable, thus capable of being drawn, bent, and formed into different shapes. Yet, when we heat metals to high temperatures in air, they tend to oxidize. That is usually considered a big disadvantage. Additionally, they may not be optimum in wear-resistant applications.

Then come **ceramics**, typically compounds between a metal and a non-metal. Ceramics usually have high melting points, hardness, and wear resistance. A very popular ceramic is aluminum oxide (Al_2O_3), known as alumina. Unlike metals such as tungsten or titanium, we can heat alumina to high temperatures in air without worrying about oxidation. If it could talk, it would say, "I am already a stable oxide at that temperature; what *more can you do to me?*". Ceramics, however, suffer from severe brittleness and fast fracture. Unlike traditional ceramics, such as pottery, engineering ceramics have been engineered to be suitable for engineering applications. The brittleness of ceramics and their unreliability are issues only if we intend to load them in tension or bending. If, on the other hand, we apply a compressive force or pressure to a ceramic, it displays strengths an order of magnitude greater than when loaded in tension or bending. The reason for such a difference will be explained in a later chapter.

Therefore, each material class offers properties that can fulfill a particular application's needs but will not be the best material for everything we need. For example, we can have a ceramic that is hard and wear-resistant but at the same time brittle, fractures easily, and is electrically non-conductive. A metal can conduct electricity well and be tough and ductile, but may not possess the high wear resistance needed for a particular application. Combining a metal and a ceramic into one material produces a **composite** that can conduct electricity, provide wear resistance, and is not vulnerable to brittle fracture. The advantage of using composite materials is that we can access the best advantages of different classes of materials while minimizing their disadvantages. It's a win–win! Composites also give us unique control over properties and behavior.

Composite materials are typically named after the major constituent from which they are made (called the **matrix**). For example, aluminum disks containing 20 vol.% particles of silicon carbide (SiC) ceramic would be called aluminum matrix composites since aluminum is the major constituent. Aluminum would be the matrix, and the SiC would be called the **reinforcement**. Such materials have applications in brake disks and even horseshoes. Aluminum is lightweight and conducts heat well, dissipating frictional heat during operation. However, hard and wear-resistant SiC particles are added to prevent it from being worn easily over time. Along the same train of thought, we can have many different composites. Polymer matrix composites (e.g., epoxy reinforced with glass fibers or carbon fibers), ceramic matrix composites (tungsten carbide (WC)-cobalt (Co) composite, also referred to as a CerMet ("Cer" for ceramic and "Met" for metal)). WC–Co composites are used as cutting tools and as drill bits for oil drilling. The material is also one of the major triumphs of powder metallurgy. The ancient Egyptians were one of the earliest races to use composites. Straws and sand were mixed with mud to make mud bricks for building sturdy houses.

As mentioned, if we want to improve the wear resistance of a metal, we can add a ceramic to it. However, in what form will we bring the ceramic to the metal? We can do that by adding the ceramic in the form of particles (as in the case of the aluminum-SiC mentioned above), short or long fibers, or we can start with a porous

3D ceramic preform that is later infiltrated with the metal, etc. Let's say we added ceramic particles to reinforce the metal matrix. Then, the next question is, how much volume content do we add of the ceramic particles? If we add a low volume content of ceramics, we may improve wear resistance and strength, but not to the same level as when we add a higher volume content of the same ceramic particles. So, do we keep adding ceramic particles to strengthen our composite? At some point, we risk the ceramic particles becoming so many that they start touching/contacting each other throughout the composite volume (i.e., they percolate or interconnect). Once we have that interconnected brittle ceramic material network, guess what happens? Brittle behavior can become dominant—a crack can readily travel through the connected brittle ceramic regions.

Not only does the volume content of reinforcements affect the properties of the composite, but also the size and shape of the reinforcements. If the reinforcement size is below 100 nm, the composite produced is called a **nanocomposite**. This versatility in microstructural design results in highly tailorable materials in structure and, therefore, properties. These properties can be electrical, thermal, optical, mechanical, magnetic, etc. Hence, composites are incredibly versatile as engineering materials.

Materials research continues to expand the impact of materials; now, **nanomaterials** are often in the limelight. Figure 1.10 shows an example of recent research into self-sustaining power sources for wearable electronics. Here, we see aligned

Fig. 1.10 Aligned nitrogen-doped carbon nanotubes (CNT) for use as nano-electricity generators for wearable nanoelectronics. Image modified from Il'ina, M. V.; Soboleva, O. I.; Khubezov, S. A.; Smirnov, V. A.; Il'in, O. I. Study of Nitrogen-Doped Carbon Nanotubes for Creation of Piezoelectric Nanogenerator. J. Low Power Electron. Appl. **2023**, 13, 11. https://doi.org/10.3390/jlpea13010011. Reproduced and modified under the terms and conditions of the Creative Commons Attribution (CC BY) license (https://creativecommons.org/licenses/by/4.0/)

nanotubes that have been doped with nitrogen. These nanomaterials display a piezo-electric effect that can be utilized in wearable electronics. Here, slight deformations resulting from daily movements can generate electricity, which can power wearable electronics. The future promises more and more advancements through a growing portfolio of materials.

 *Alex, students like you enrich my life as a professor and keep me going. As you continue reading As a Matter of Fact, I want you to grasp the fundamentals. I have tried to make them easy to understand. Once you achieve this, believe me, Alex, understanding how a material behaves under any circumstance will become second nature. I often tell my students to put themselves in the shoes of the material and see how they would react. You will see that many times, materials will behave like you would. Don't tell my colleagues this; otherwise, they may accuse me of being crazy. I want you to live the material.*

Problems

1.1.   List the three most abundant elements in our Earth's crust, the most abundant first.
1.2.   Give an example of a stone used in the Stone Age, and explain how it was used.
1.3.   Were there other materials used in the Stone Age? If so, which ones?
1.4.   What is the name of a spear-hurling weapon used during the Stone Age?
1.5.   What is the process of knapping?
1.6.   What is meant by native copper?
1.7.   Describe the process of smelting as used in the Copper Age.
1.8.   What is Bronze?
1.9.   When were oxygen, aluminum, and silicon discovered?
1.10.   What is a shape memory alloy? Give an example of one.
1.11.   What are the advantages of metals, ceramics, and polymers?
1.12.   What are the disadvantages of metals, ceramics, and polymers?
1.13.   What is meant by a matrix in a composite?
1.14.   In a situation where strengthening a metal can be achieved by adding ceramic particles. Is there a limit to the amount of ceramics you can add? Or not? Explain.
1.15.   What is the difference between ductility and malleability?
1.16.   When could you classify a material as a nanocomposite?
1.17.   What will happen to its ductility if a metal is strengthened by some means?
1.18.   How would it be possible to make a polymer stiffer?
1.19.   What is the definition of a nanomaterial?
1.20.   What is a polycrystalline material?
1.21.   What is a single crystal?
1.22.   What is a CerMet, and can you give an example of such a material?

Chapter 2
Atoms and Atomic Bonding

2.1 Introduction

This chapter introduces atoms and atomic bonding concepts, including primary and secondary bonding.

2.2 Introduction to the Structure of Atoms

Many of the elements we know today were discovered during the past 350 years. It is common knowledge that all matter comprises atoms as building blocks; this includes the air we breathe, the water we drink, and the glass into which we place the water. Although the concept of an atom was hypothesized even as far back as the time of the ancient Greeks, our current understanding of the atom started to mature less than 200 years ago. The electron was the first subatomic particle to be discovered, owing to the combined work of **Michael Faraday** and **Sir William Crookes** after 1833. The first indication of the presence of positively charged **protons** was realized in 1886 by **Eugene Goldstein**, who was using a cathode ray tube at the time, and the name proton was suggested by **Rutherford** in 1907. It later became obvious that the mass of an atom was greater than the sum of the masses of protons and electrons. Hence, **neutrons** were eventually discovered, which made up that remaining mass.

Neutrons are subatomic particles that do not contribute to the charge of an atom and are hence electrically neutral. They, however, significantly contribute to the mass of an atom, as they weigh almost the same as protons. A neutron or a proton can be referred to as a **nucleon**. In 1913, the **Bohr model** described an atom with **electrons** revolving in *designated* energy shells or orbits (e.g., K, L, M, N) around a nucleus that contained protons and neutrons. In essence, the model promoted the concept of *quantized* energies for electrons, and only certain energy levels were allowed for

K. Morsi, *An Engaging Approach to the Science and Engineering of Materials*,
https://doi.org/10.1007/978-3-032-06231-4_2

electrons to possess in an atom. Later, Bohr was awarded the Nobel Prize for his work. Figure 2.1 shows Bohr and Einstein in 1925.

Figure 2.2 shows a schematic (not to scale) of the Bohr model for the lithium (Li) atom, which has three protons and four neutrons in its nucleus. While the protons are positively charged (each with a charge of 1.6×10^{-19} C), the three negatively charged electrons (also 1.6×10^{-19} C each) balance the charges and make the atom electrically neutral. It is important to note that the Bohr model represents a simplified view of the atom compared to what we know today. However, it is still helpful in explaining certain behaviors.

Regarding electrons, if we assign a zero potential energy (E) for an electron placed at an infinite distance away from the nucleus, then any electron that is closer to the nucleus than infinity will have a relatively lower potential energy, i.e., a negative E

Fig. 2.1 Bohr and Einstein in 1925. Paul Ehrenfest, public domain image

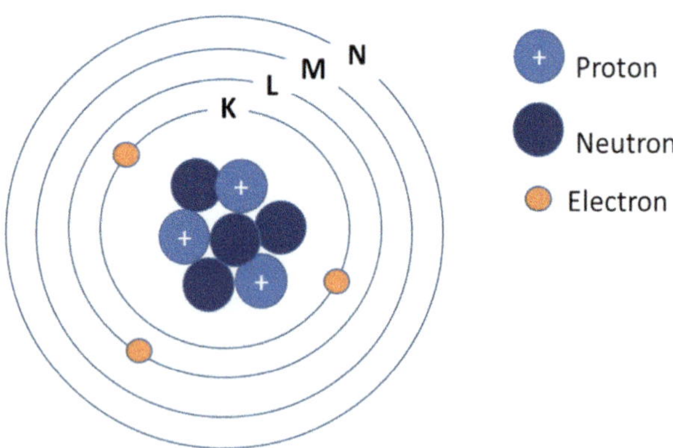

Fig. 2.2 Schematic of Bohr model for a lithium (Li) atom

value. It follows that the outermost electrons in the Bohr model will have a specific potential energy value, which will be negative. As we get closer to the nucleus, the potential energy of electrons in the nearer locations will become more and more negative (i.e., lower and lower). In another way, this means that the energy of electrons increases as we go from inner to outer electron energy shells. Hence, with reference to Fig. 2.2, the energy of electrons in the K shell will be lower than those in the L shell, which are lower than those in the M shell, which are lower than those in the N shell.

The outermost (**valence**) electrons are essential in atomic bonding and chemical reactions. They may leave their atom and join another atom from a different element, join other unbound electrons in an electron cloud, or even be shared with other atoms. Since each atom is initially electrically neutral, the electrical charge nature of the atom can change as a result of bonding. For example, if an electron is released from atom A and joins atom B, atom A will now have one less negatively charged electron, leaving the atom with one excess positively charged proton. Hence, the atom becomes a positively charged ion (called a **cation**). Meanwhile, atom B, which just received an extra electron, will have an excess negative charge and become a negatively charged ion (also called an **anion**). An atom that releases an electron to bond with another atom is termed **electropositive**, while the atom that receives an electron to bond with another atom is termed **electronegative**. Electronegativity is a measure of the ability of the atom to attract electrons for bonding. The **Pauling electronegativity scale** ranks the electronegativities of elements ranging from a little less than unity to a maximum value of 4. As such, nonmetals are found to have higher electronegativities than metals. For example, fluorine and sodium have electronegativities of 4 and 0.9, respectively. The electronegativities of many elements are presented in Fig. 2.3, which follows the element order typically found in the periodic table (discussed later). As can be seen, electronegativity generally increases from left to right and from bottom to top in the table. Electronegativity can significantly influence atomic bonding (as will be shown later).

One question that may arise is if an atom receives or loses an electron, would its size be affected? The simple answer is yes. The difference between the size of an atom and an ion depends on a couple of things. If the ion is a cation (positively charged ion), i.e., the atom had previously lost an electron, it would typically be smaller than its parent atom. The nucleus will have more of a net positive charge (since one electron has left), which exerts a greater pull on the remaining electrons. Meanwhile, an anion (a negatively charged ion) has gained an electron, making it larger than its parent atom. It is also important to note that even ions of the same element can have different sizes depending on the bonding type they undergo with other atoms. For example, the radius of a sodium ion in an ionic compound (e.g., a salt) is 0.98 Å, whereas in a sodium metal (metallic bonding) is 1.85 Å. More about bonding types later.

The number of electrons occupying a shell orbital is limited to only two, known as the **Pauli exclusion principle.** Also, if two electrons are in the same orbital, they must have opposite spins.

Electronegativities

1	2	3	4	5	6	7	8	9	10	11	12	13	14	15	16	17
H 2.1																
Li 1.0	Be 1.5											B 2.0	C 2.5	N 3.0	O 3.5	F 4.0
Na 0.9	Mg 1.2											Al 1.5	Si 1.8	P 2.1	S 2.5	Cl 3.0
K 0.8	Ca 1.0	Sc 1.3	Ti 1.5	V 1.6	Cr 1.6	Mn 1.5	Fe 1.8	Co 1.8	Ni 1.8	Cu 1.9	Zn 1.6	Ga 1.6	Ge 1.8	As 2.0	Se 2.4	Br 2.8
Rb 0.8	Sr 1.0	Y 1.2	Zr 1.4	Nb 1.6	Mo 1.8	Tc 1.9	Ru 2.2	Rh 2.2	Pd 2.2	Ag 1.9	Cd 1.7	In 1.7	Sn 1.8	Sb 1.9	Te 2.1	I 2.5
Cs 0.7	Ba 0.9	La 1.1	Hf 1.3	Ta 1.5	W 1.7	Re 1.9	Os 2.2	Ir 2.2	Pt 2.2	Au 2.4	Hg 1.9	Tl 1.8	Pb 1.8	Bi 1.9	Po 2.0	At 2.2
Fr 0.7	Ra 0.9															

Fig. 2.3 Electronegativities of a large number of elements

Table 2.1 Electric charges and masses of subatomic particles

Subatomic particle	Charge (c)	Mass (kg)
Electron	-1.6×10^{-19}	9.106×10^{-31}
Proton	$+1.6 \times 10^{-19}$	1.6725×10^{-27}
Neutron	0	1.675×10^{-27}

It is imperative to know that different elements have different numbers of protons, neutrons, and electrons in their atoms. To quantify these differences, we can define several terms. The **atomic number** is defined as the number of protons in the nucleus. The **mass number** is the *number* of protons and neutrons in the nucleus, and the **atomic mass** is the *mass* of protons and neutrons.

Professor, I know that atoms also contain electrons, but why aren't they included along with protons and neutrons to determine the atomic mass?

*You are theoretically correct in that they should be included! But when we consider the **mass of an electron** (~ 9.1×10^{-31} kg) and compare it to the **mass of a proton** and that of a **neutron** (~ 1.6725×10^{-27} kg and ~ 1.675×10^{-27} kg, respectively), we immediately realize that an electron is approximately 1836 times lighter than either of them. This makes electrons insignificant from the mass perspective.*

Table 2.1 summarizes the charges and masses of important subatomic particles.

As for the atomic number, it's just a number (a simple number *count* of protons) without units, whereas for the atomic mass, we need to assign a unit to it. Here, there are two approaches. The first is to calculate the total mass of one mole of atoms (i.e., 6.023×10^{23} atoms (which is **Avogadro's number**)) and present atomic mass in units of grams per mol (i.e., g/mol). The other approach uses the **unified atomic mass unit (u)**. Let's consider the carbon atom (^{12}C), which contains 6 protons and 6 neutrons. Note that $6 + 6 = 12$, which, as mentioned, is the mass *number*, appears as the superscript in ^{12}C. A unified atomic mass unit is defined as the "experimentally" determined mass of the ^{12}C atom divided by 12. Twelve is the number of nucleons (i.e., protons and neutrons). This makes one *unified atomic mass unit* equal to 1.66 $\times 10^{-27}$ kg, as shown in Eq. 2.1. Interestingly, if we add the masses of protons and neutrons within the carbon-12 atom, the mass of one nucleon will be slightly heavier than the experimentally determined value 1.66×10^{-27} kg, representing one unified atomic mass unit. The discrepancy is due to the **nuclear binding energy** (the energy that binds the constituents of the nucleus together). Recall, the nucleus contains positively charged protons and neutral neutrons. Naturally, one would expect the protons to repel each other in a way to increase their separation, but they don't separate! This is where the nuclear binding energy comes in; in other words, something is keeping them together. According to the conservation of energy principle, this energy must come from somewhere. Following Einstein's famous equation, $\mathbf{E = mc^2}$, the atom's

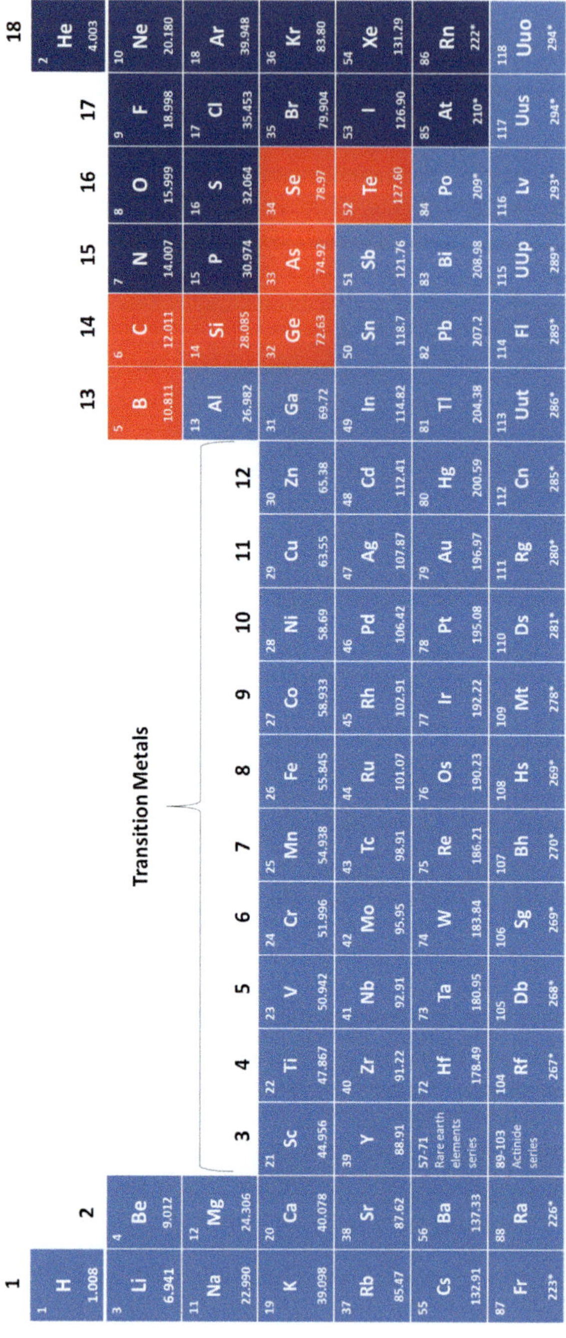

Fig. 2.4 The periodic table. (* = only the relative atomic mass of the most abundant or stable isotope is given)

mass will be reduced by this amount of nuclear binding energy. This is also sometimes referred to as the **mass defect**. That is why the definition of the unified atomic mass unit uses the *experimentally* determined mass of the ^{12}C atom, which includes the effect of the nuclear binding energy.

$$1u = \frac{\text{mass of } ^{12}\text{C atom}}{12} = 1.66 \times 10^{-27}\text{kg} \tag{2.1}$$

1 unified atomic mass unit (u) = experimentally determined mass of one nucleon.

Since 1 u was defined by dividing the experimentally determined mass of ^{12}C, then Carbon-12 must have an atomic mass of 12 u. Masses of all the elements can now be determined using u as the unit for mass.

What we have learned so far is that different elements have different numbers of protons, neutrons, and electrons, and as such, they have different atomic numbers and atomic masses.

All elements are organized in the famous periodic table, as shown in Fig. 2.4. The table organizes the elements in order of their atomic numbers, starting with hydrogen, which has an atomic number of 1. The columns are **groups** (numbered 1–18), and the rows are called **periods**. As seen in the periodic table, each element is placed in a box along with some important numbers. The light blue boxes are metals, the dark blue are nonmetals, and the red boxes represent elements between metals and nonmetals, sometimes referred to as **intermediate elements** or **metalloids**.

Let's look at one of the elements as an example; see Fig. 2.5 for titanium.

The number to the box's upper left in Fig. 2.5 is the atomic number, which means titanium has 22 protons in its nucleus. The number at the bottom is the relative atomic mass, sometimes called the **atomic *weight*** (not mass).

Professor, why is it called the relative atomic mass and not simply atomic mass, and why is it that the relative atomic mass of most of the elements in the periodic table is not a whole number? For example, titanium has a relative atomic mass of 47.867, chlorine's is 35.45, etc.

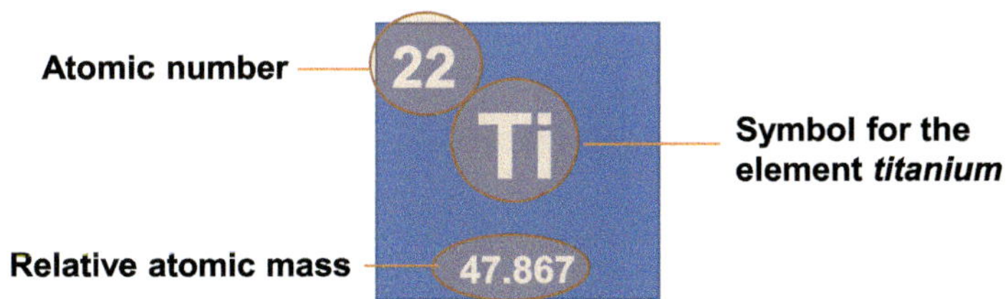

Fig. 2.5 Designation of elements within the periodic table

Great catch, Alex! We had initially made a simplified assumption: that every atom in an element had the same number of protons and neutrons, which would make the mass of all individual atoms the same. However, it turns out that even for the same element, despite all its atoms having the same number of protons, they may differ in the number of neutrons. In that case, we call these different atoms **isotopes** *of the same element and describe the element as being* **isotopic.** *The* **relative atomic mass** *is simply a weighted average of the atomic masses of the different isotopes in the element, considering their relative abundance.*

It turns out that many elements have isotopes. Three hydrogen isotopes are shown in Fig. 2.6: Hydrogen-1, Hydrogen-2, and hydrogen-3 (1H, 2H, and 3H; the superscript is the mass number). The most abundant isotope is ^{1}H (~ 99.9844%) (also called **Protium**), with only one proton and <u>no</u> neutrons. ^{2}H (**Deuterium**, abundance ~ 0.0156%) and ^{3}H (**Tritium**, amounting to only trace amounts) contain one and two neutrons, respectively. Since they all have one proton, they appear in the same position in the periodic table, under Hydrogen. Likewise, ^{12}C, ^{13}C, and ^{14}C are also shown in Fig. 2.6. Since the unified atomic mass unit was derived from the mass of ^{12}C, if carbon only had one isotope containing 6 protons and 6 neutrons, we would have found its unified atomic mass to be 12 in the periodic table, but we see that it is 12.011. The reason is that the other carbon isotopes weigh more than ^{12}C (since they have more neutrons). Hence, when a weighted average mass of the different isotopes is calculated, it results in a relative atomic mass of 12.011, not 12. It should also be noted that, in general, some isotopes are stable, and some are not and instead decay over time.

There are a few points to remember about the periodic table, and these are listed below:

1. Most elements can be considered metals (identified as light blue in the table).
2. Nonmetals are identified as dark blue.
3. Red boxes identify elements with characteristics between metals and non-metals; these have also been referred to as metalloids.
4. Elements within a group (i.e., column) would generally have the same valence and similar chemical and physical properties.
5. The atomic radius generally decreases from left to right and increases from top to bottom.
6. Electronegativity generally increases from left to right and from bottom to top.
7. Elements in the first group (column 1) are called the **alkali metals**.
8. Elements in the second column (group 2) are **alkaline earth metals**.
9. Groups (columns) 3–12 are the transition metals.
10. In group 17, we find the halogens.
11. Group 18 are inert (or noble) gases.

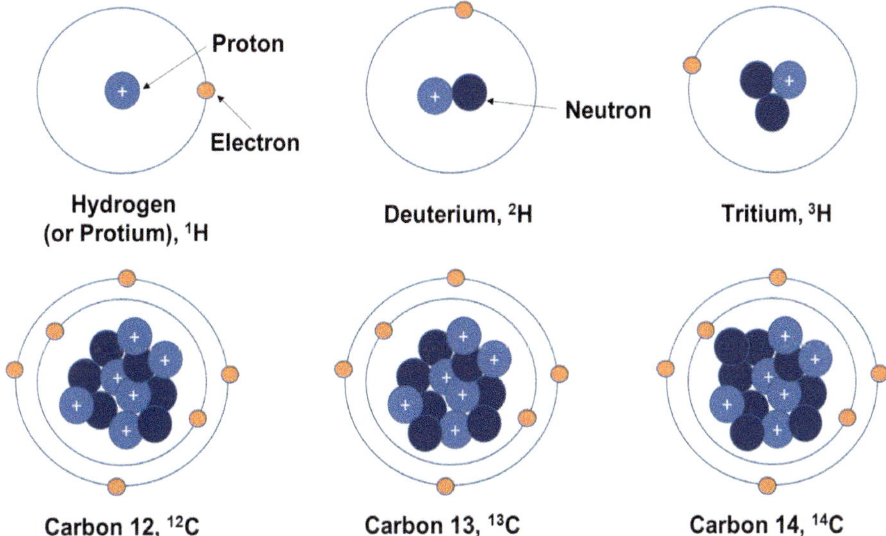

Fig. 2.6 Examples of isotopes of hydrogen and carbon. Hydrogen has no neutron, whereas deuterium and tritium have 1 and two neutrons, respectively. Also, carbon-12 has 6 neutrons, while carbon-13 and carbon-14 have 7 and 8 neutrons, respectively

Okay, let me quiz you, Alex! Give me an example of a transition metal, an alkali metal, an alkali earth metal, and an inert gas.

A transition metal would be iron, since it is in column 8 (i.e., between groups 3 and 12); an alkali metal would be sodium (in group 1), an alkaline earth metal would be magnesium (since it is in group 2), and an inert gas would be argon (group 18). I am wondering, though, why are alkali metals named as such?

Well done, Alex. You answered correctly!! Alkali metals are called as such because they form alkaline solutions with pH values greater than 7 when they react with water. Okay, so let me quiz you again. The most abundant isotopes of chlorine are ^{35}Cl and ^{37}Cl. If the atomic number of Cl is 17, how many neutrons do ^{35}Cl and ^{37}Cl have, respectively?

35 and 37 in ^{35}Cl and ^{37}Cl are the mass numbers, which refer to the total <u>number</u> of protons and neutrons. So, since the atomic number is only the number of protons, then for ^{35}Cl, the number of neutrons would be 35–17 = 18. Likewise, for ^{37}Cl, the number of neutrons would be 37–17 = 20.

Splendid, my friend, you've been paying attention!! Ok, what is the difference between atomic number and mass number? Don't look

The atomic number is the number of protons in the nucleus, whereas the mass number is the number of subatomic particles contributing to the atom's mass. It is the total <u>number</u> (not mass) of neutrons and protons

Fantastic! Now, go to Sect. 2.3, Introduction to Atomic Bonding in As a Matter of Fact, and we will talk again.

2.3   Introduction to Atomic Bonding

All solid materials are made of atoms that must be bonded together. Otherwise, they would drip through our fingers when we hold them. These bonds can be strong or weak, depending on the bonding type. For example, primary bonds such as metallic, covalent, and ionic are strong, resulting in materials with melting points typically between 1000 and 4000 K. Van der Waals bonding belongs to the weaker secondary bond category, resulting in materials with melting points generally between 100 and 500 K. The material classes typically have preferred bonding; for example, metals bond through metallic bonding, ceramics through covalent and ionic bonding, and polymers through covalent and van der Waals bonding.

Let us discuss these bonding types, starting with the primary bonds.

2.3.1   Primary Bonds

Covalent Bonding

This primary bond can be found in hydrogen (H_2), nitrogen (N_2), methane (CH_4), ceramics, and individual polymer molecules. In covalent bonding, the same or different atoms share valence electrons to complete their outermost shells. It's a collaboration of mutual benefit.

An example of covalent bonding is provided in Fig. 2.7, which represents a small section of the polyethylene polymer molecule. We can see that each carbon atom shares two outermost electrons with two other carbon atoms and its other two electrons with two hydrogen atoms. This way, the outermost electrons of each carbon atom increase from 4 to 8, filling the outermost shell. Likewise, each hydrogen atom

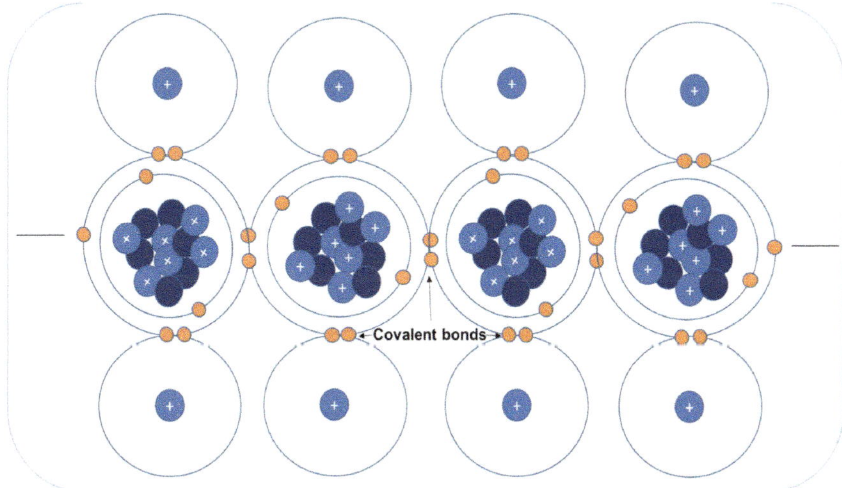

Fig. 2.7 Part of a polyethylene molecule, showing covalent bonding

receives a shared electron from the carbon atom, thus completing its outer shell with two electrons. This type of bonding is *directional*; one can easily see this in Fig. 2.7, where the electrons must be at a particular location to be shared with the intended atom. Covalently bonded materials will be strong and electrically insulating; an example is diamond.

Professor, polymers are covalently bonded. Knowing that the covalent bond is strong and hard to break, why are polymers comparatively soft and not as strong as metals?

Great question, Alex. While the bonding inside an individual polymer 'molecule' is covalent, the bonding that exists "between" individual molecules is weak secondary bonding. A material is as strong as its weakest link. If we were only to measure the strength of individual polymer molecules, they would make us proud because of their strength!

Covalent bonding also influences the electrical properties of the material. Both diamonds and polymers are electrically non-conducting; the reason, in both cases, is simply that the valence electrons are tied up in the covalent bonds and are, therefore, not free to travel. As will be seen later, secondary bonds do not even involve the exchange of electrons.

Consider a simplified model of two atoms brought closer together in space, and we can learn much about atomic interactions.

Figure 2.8 shows a schematic of the potential energy versus interatomic spacing between two atoms as they are brought closer together. We start by assigning zero potential energy of interaction to an interatomic spacing (between the two atoms) of infinity. As the atoms are brought closer together, the potential energy is reduced and becomes more negative (since, naturally, it will be lower than zero). This energy is attractive. However, as the atoms get very close, the outer negatively charged electrons of the two atoms start to overlap, resulting in a repulsive force and, hence, a repulsive potential energy, as seen in Fig. 2.8. If we add the contributions of both the **attractive (E_a)** and **repulsive potential energies (E_r)**, we obtain the **net potential energy (E_n)**, as can also be seen in Fig. 2.8. The relationship between attractive, repulsive, and net energies is described by Eqs. 2.3a and 2.3b:

$$E_n = E_a + E_r \tag{2.3a}$$

or

$$E_n = -\frac{A}{r^x} + \frac{B}{r^y} \tag{2.3b}$$

where the exponent x is less than y,, r is the interatomic spacing, and A and B are constants.

In Fig. 2.8, we see several things. First, the **net potential energy** decreases to a minimum value as the atoms are brought closer and closer together and then increases again (because of the repulsive energy brought about by the electrons). The lowest point in the net potential energy curve is what is referred to as the **bond energy** (also called the **binding energy**), which occurs at a unique interatomic spacing called the

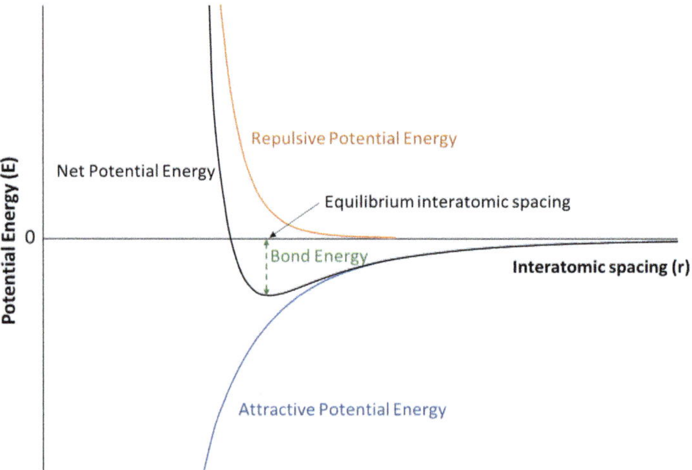

Fig. 2.8 A schematic (not to scale) showing the relation between the interatomic spacing and the attractive, repulsive, and net energies, as two atoms are brought closer together

equilibrium interatomic spacing. This is the spacing where atoms are considered to be at equilibrium. The atoms should stay at this spacing if we do not apply any external force or influence. Of course, there is no rest for the wicked; each atom is, in fact, in a state of constant vibration about an average position due to thermal energy. So, the equilibrium interatomic spacing should be viewed in this light as a time-averaged spacing. The magnitude of this bond energy has specific implications. It represents the energy needed to break the bond between the atoms; in other words, if we supply energy equal to this bond energy, we can expect the atoms to break away and the material to melt. One could initially imagine that a material with a high bond energy (i.e., high negative numerical value) would have a high melting point. This is certainly *generally* true for materials that bond similarly. For example, as the bond energy of covalently bonded material increases, so, in general, will their respective melting points. The same will be true within any bonding type. Note that although a material with a higher bond energy will generally have a higher melting point within any given bonding type, we cannot reach the same conclusion by comparing the bond energies of materials that are bonded differently. We cannot, therefore, compare one bond energy value from a metallically bonded material to one from an ionically bonded one, for example.

Table 2.2 lists specific bond energies for different bond types, and Table 2.3 shows bond energies for particular materials and substances according to their bonding types. Note that for van der Waals bonding, the bond energy refers to the intermolecular bond energies between covalently bonded substances. They are not the bond energies of the covalent bonds within the molecule but the weaker van der Waals bonds between molecules.

The interaction between atoms does not stop at the potential energies. Since the net force between atoms can be expressed as the differential of the net potential energy with respect to r (Eq. 2.2), we can also arrive at a similar expression for interatomic forces between the atoms:

Table 2.2 Bond energies of different bonds

Bond type	Bond energy (kJ/mol)
C–C	370
C=C	680
C≡C	890
C–H	435
C–N	305
F–F	160
H–H	435
O–Si	375
O–O	220
H–O	500

Source L. Van Vlak, Elements of Materials Science and Engineering, 6th edition, Addison Wesley Publishing company, (1989)

Table 2.3 Bond energies and melting points of different materials

Material/substance	Bond energy (kJ/mol)	Melting point (°C)
Van der Waals bonding		
Argon (Ar)	7.7	− 189 (69 kPa)
Methane (CH_4)	18	− 182
Chlorine (Cl_2)	31	− 101
Metallic bonding		
Aluminum (Al)	330	660
Tungsten (W)	850	3414
Covalent bonding		
Silicon (Si)	450	1410
Diamond (C)	713	> 3550

Source W. D. Callister Jr., D. G. Rethwisch, "Materials Science and Engineering, An Introduction (9E edition), Wiley, 2014

$$F_n = \frac{dE_n}{dr} \tag{2.2}$$

As with the potential energy curves, the net interatomic force versus interatomic spacing curve (Fig. 2.9) also has major implications. Although attractive and repulsive forces can operate simultaneously, especially at closer atomic distances, there will always be a net force considering the magnitudes of attractive and repulsive forces. The figure shows that when atoms are placed at a distance greater than the equilibrium interatomic spacing, a net attractive force returns them to the equilibrium interatomic spacing. However, a net repulsive force will result when the atoms are brought closer together at a separation smaller than the equilibrium interatomic spacing. Here, the magnitude of the repulsive force outweighs the attractive force. At this distance, the negatively charged electrons can now feel the effect of the negative charge of electrons from the other atom, resulting in a repulsive force that increases in magnitude as the atoms are brought closer and closer to each other. The result of these interactions is that the net interatomic force appears first to increase in attraction as the atoms are brought close, then decreases again as the atoms get very close, which means that the repulsive force is gaining dominance with further reduction in the interatomic spacing. This results in a net interatomic force that reduces to zero at the equilibrium interatomic spacing and becomes repulsive at smaller interatomic spacings.

By drawing a tangent to the curve at a force of zero (Fig. 2.10), we can calculate the slope of the tangent ($\Delta F/\Delta r$), which relates to the stiffness (S) of the material ($S = \Delta F/\Delta r$), by considering the bonds to be springs and applying Hooke's law. The figure also shows three possible slopes for three different materials. The largest slope, S_1, belongs to the material with the highest stiffness.

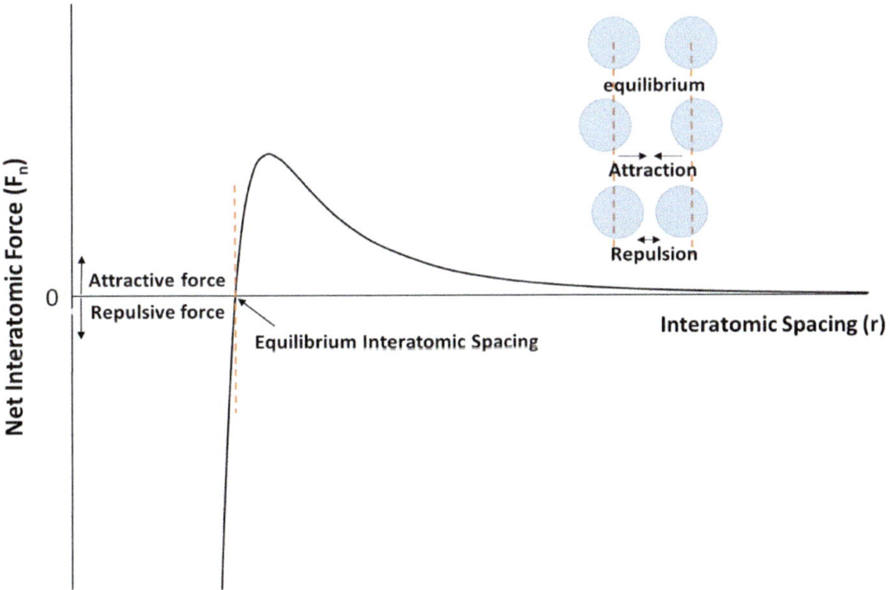

Fig. 2.9 Schematic (not to scale) of net force versus interatomic spacing curve. The distances between the drawn atoms are exaggerated for clarity, but when atoms are separated a net attractive force is generated, to pull them back to the equilibrium interatomic spacing position, and if the atoms are brought closer to each other, a net repulsive force is generated to push the atoms back to the equilibrium interatomic spacing position, as also seen in the atoms drawn in the figure

Metallic Bonding

Metals are found on the left-hand side of the periodic table. Their atoms typically have 1–3 electrons in their outermost energy shell, which gives them valences from 1 to 3. Metallic bonding is found in all metals. Here, the outermost electrons are released into a sea or cloud of electrons; by doing so, each atom becomes a positively charged ion, the charge of which will depend on how many electrons were in the outermost shell and released into the electron cloud.

Figure 2.11 shows a schematic of metallic bonding, where positively charged ions are somehow glued together by a negatively charged electron cloud. There are a couple of advantages to this type of bonding. First, since the electrons are delocalized and not bound by any ion, they are free to move, giving rise to materials with good electrical and thermal conductivities. Second, in metallic bonding, the bonds are non-directional, which means that the ions are free to pack and arrange themselves in whichever way they prefer. This leads to very high and efficient packing of ions in metallically bonded materials, which has major significance for specific properties (more about that later).

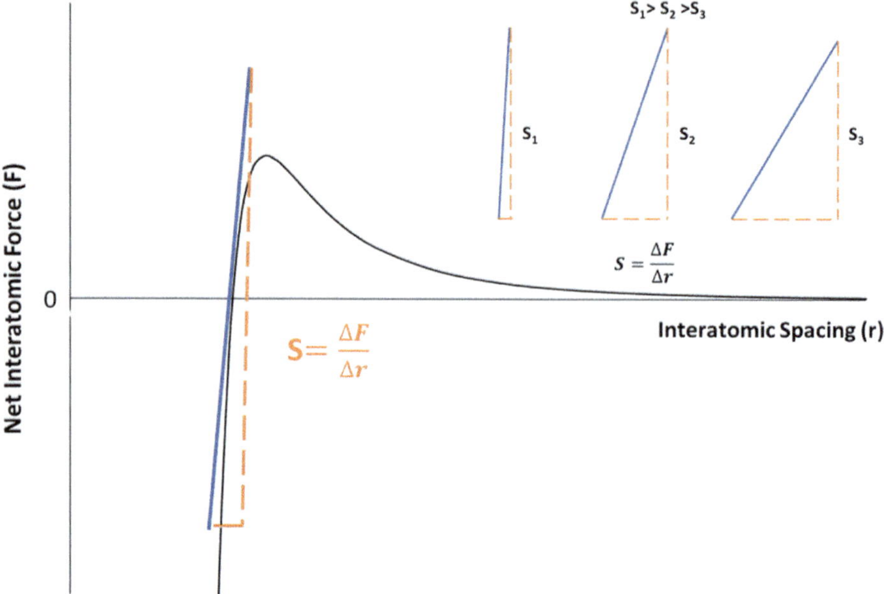

Fig. 2.10 Different Materials can have different slopes at $F_n = 0$

Fig. 2.11 Metallic bonding showing positively charged ions (cations) surrounded by a sea or cloud of electrons which act as the glue that holds them together

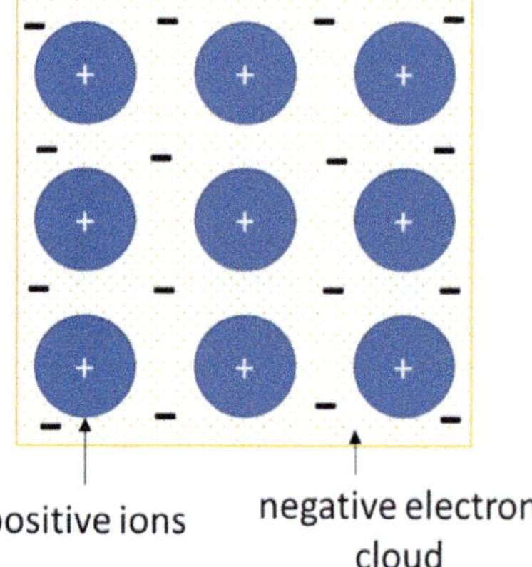

Ionic Bonding

This bonding often occurs in ceramics and typically occurs between a metal and a non-metal atom, in other words, between elements from the periodic table's left- and right-hand sides. An example would be sodium (Na) and chlorine (Cl) to form sodium chloride (NaCl), also known as table salt.

Sodium has one outermost electron, while chlorine has seven electrons in its outermost shell. If a sodium atom gives its outermost electron to a chlorine atom,

Fig. 2.12 Ionic bonding
showing cations and anions
bonded together

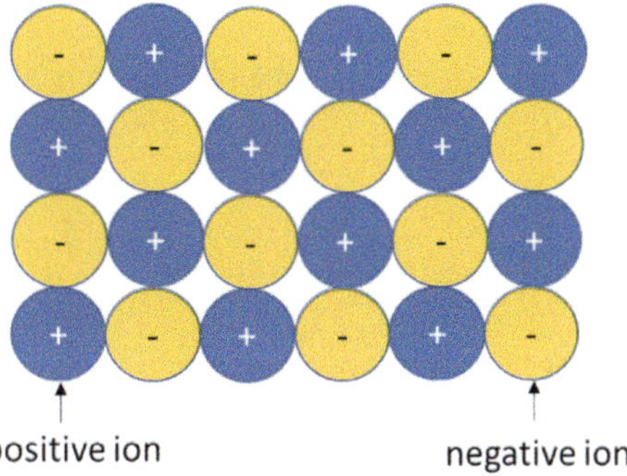

the sodium atom will then have a complete lower shell. When the chlorine receives
the sodium electron, it completes its outermost shell with eight electrons. In doing
so, sodium atoms become positive ions, while the chlorine atoms become negative
ions. Electrostatic (Coulombic) attraction ensues between the negative chlorine and
positive sodium ions, which forms the basis of ionic bonding. The bonds here are
also non-directional, with one condition: every charged ion is surrounded by ions
of the opposite charge, as seen in Fig. 2.12. Since the outermost (valence) electrons
are occupied in bonding, no free electrons are available to allow electricity to pass
through. Hence, ionic compounds are electrical insulators. However, when ionic
compounds melt, they can become conductive since the ions have been released
from their bonding and are free to move, by **ionic conduction.**

As mentioned, we can learn much about material behavior if we consider a simple
two-atom model, showing the potential energy and force interactions between two
atoms approaching each other in space. As these atoms are brought closer together, the
net potential energy decreases or, in other words, becomes more and more negative.
In the case of ionic compounds such as NaCl, when the atoms approach and become
close, a sodium ion can be formed if we supply 5.14 eV of energy to the sodium
atom, which in turn removes its outermost electron to produce a positively charged
Na ion (Na^+). When the electron joins the outermost shell of the Cl atom, 4.02 eV
is released from the atom, forming a negative ion, Cl^-. Hence, 5.14 eV of energy
was expended, and 4.02 eV was given back; this leaves 1.12 eV of net energy that is
done to form the Na^+ and Cl^- ions. The positive and negative ions tend to attract due
to Coulomb's law, thus generating an attractive energy; however, when the atoms
become very close, the outer electron clouds in each ion overlap, and a repulsive
energy ensues. The net energy can be expressed by Eq. 2.4.:

$$E_n = E_o - \frac{M.Q^2}{4\pi\epsilon_o r} + \frac{B}{r^n} \tag{2.4}$$

Here E_o is the net energy expended to form both the Na and Cl ions, also referred
to as the *ionization energy*, which is 1.12 eV, or 1.72×10^{-19} C, M is the Madelung
constant (1.75 for NaCl), $m = 8$, ε_o is the vacuum permittivity, 8.854×10^{-12} F/m, Q

is the charge (1 eV or 1.6 × 10^{-19} C), r = spacing between cation and anion (center to center) (for Na^+ and Cl^- (NaCl) it is 0.281 nm), $n = 8$ for NaCl, and B is given by Eq. 2.5:

$$B = \frac{Q^2 r_o^{n-1}}{4\pi n \varepsilon_o} \tag{2.5}$$

where r_o is the equilibrium interatomic (inter-ionic) spacing.

Figure 2.13 shows a schematic of the net potential energy versus interatomic spacing for an ionic compound. The ionization energy is clearly shown.

Using the rule of thumb presented in Table 2.4, we can roughly determine the dominant type of bonding in a compound from the difference in the electronegativities (EN) of the atoms making up the compound.

Table 2.5 applies the rule to H_2O, N_2, HCl, NaCl, and SiC.

It is important to note that more than one type of bonding can be found in a given material. For example, we know that polymers have strong covalent bonding acting

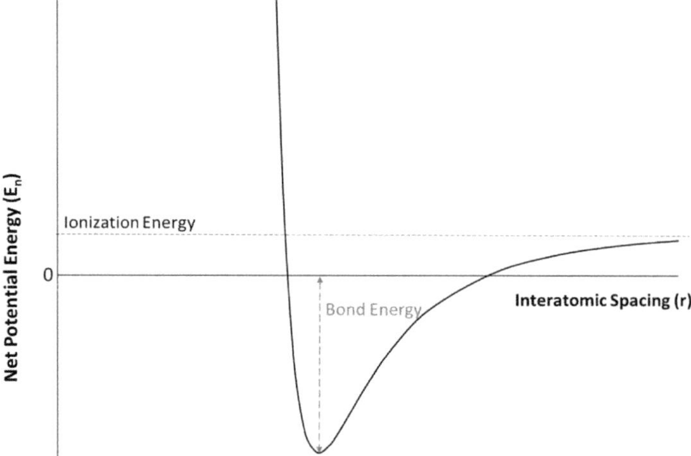

Fig. 2.13 Schematic (not to scale) of net potential energy vs. interatomic spacing in ionic bonding, showing the ionization energy (E_o) needed to produce two ions (positive and negative). (After M. F. Ashby and D. R. Jones, Engineering Materials 1 and introduction to their Properties and Applications, second edition, Butterworth and Heinemann, 1996.)

Table 2.4 Classification of dominant bonding in compounds according to differences in electronegativity between their constituent atoms (multiple sources)

Difference in electronegativity	Dominant bonding
Less than 0.5	Covalent
0.5–1.5	Polar covalent
Greater than 1.5	Ionic

Table 2.5 Examples of compounds and their dominant bonding according to Table 2.4 definitions

Compound	Electronegativities ()	Difference in electronegativities	Bonding type
H_2O	H (2.1), O (3.5)	1.4	Polar covalent
N_2	N (3.0)	0	Covalent
HCl	H (2.1), Cl (3.0)	0.9	Polar covalent
NaCl	Na (0.9), Cl (3.0)	2.1	Ionic
SiC	Si (1.8), C (2.5)	0.7	Polar covalent

along the C–C backbone of each polymer molecule and weak van der Waals bonding acting between molecules. In contrast, ceramics can have a mixture of covalent and ionic bonding.

A bond's ionic character (iC) can be estimated using Eq. 2.6, considering the electronegativities (EN) of the elements involved in the bonding.

$$iC = \left[1 - e^{\left(\frac{-(EN_1 - EN_2)^2}{4} \right)} \right] \times 100 \tag{2.6}$$

Figure 2.14 is a plot of the percent ionic character versus the difference in electronegativity (EN_1-EN_2) for the atoms of various compounds. An increase in the ionic bond character is associated with an increase in the electronegative difference between the atoms making up the compound or molecule.

2.3.2 Secondary Bonds

Secondary (or van der Waals) bonds are significantly weaker than primary bonds; they exist alone or in addition to primary bonds. Whereas primary bonds involve the transfer or utilization of outermost electrons in bonding, no electron transfer occurs in secondary bonding, as will be seen. Still, the molecules or atoms develop some temporary or permanent charged character, which aids in bonding with other charged atoms or molecules. Van der Waals bonding exists in all gases; a gas such as nitrogen (N_2) has a covalent bond between the two nitrogen atoms to produce the nitrogen molecule (N_2). However, nitrogen molecules can bond to each other through van der Waals bonding at a low temperature (− 198 °C) and atmospheric pressure, which results in its liquefication and the production of liquid nitrogen. We see van der Waals bonding also acting between polymer chains. Linear polymers like polyethylene comprise long-chained "spaghetti-like" polymer molecules. These polymer molecules have a backbone of C–C covalent bonds, making individual molecules quite strong. However, the bonding between polymer chains is the weaker van der Waals type; hence, the polymer strength and stiffness will be governed by

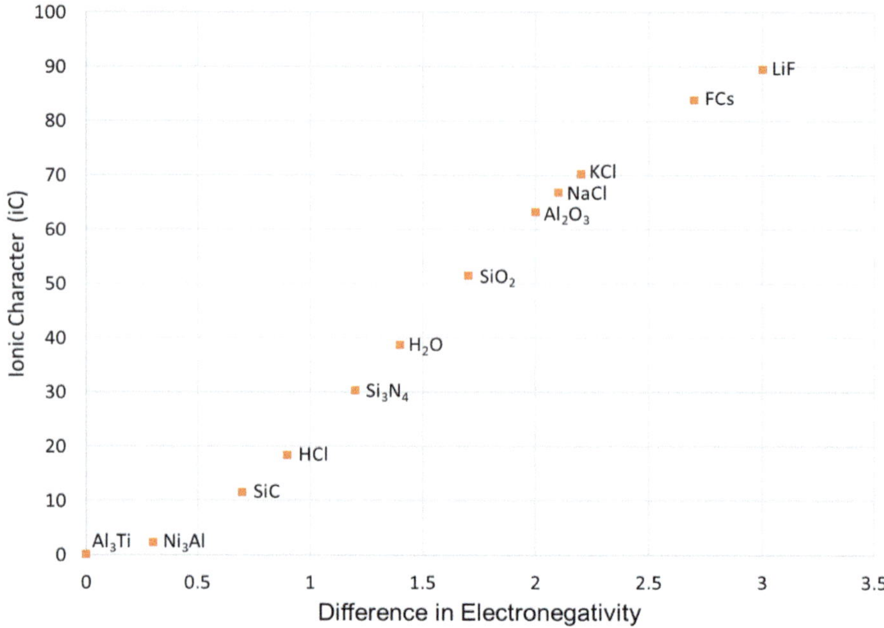

Fig. 2.14 Effect of difference in electronegativity on Ionic character of some compounds/molecules

its weakest link, the van der Waals bonds. Van der Waals bonds are also responsible for the bonding of water molecules through hydrogen bonding, as will be explained later. Without such bonds, water would have evaporated at subzero temperatures (not 100 °C), and we would not have been able to benefit from it the way we do today or even survive. Something to think about! As it happens, van der Waals bonding can be divided into four types: dispersion forces, permanent dipole–permanent dipole, permanent dipole–induced dipole, and hydrogen bonding, as shown in Fig. 2.15 and explained below.

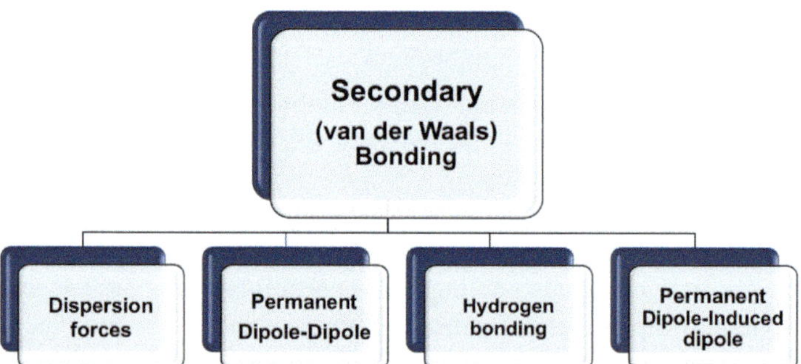

Fig. 2.15 Types of van der Waals bonding (after the classification of R.A. Higgins, Properties of Engineering Materials, Hodder and Stoughton, 1986)

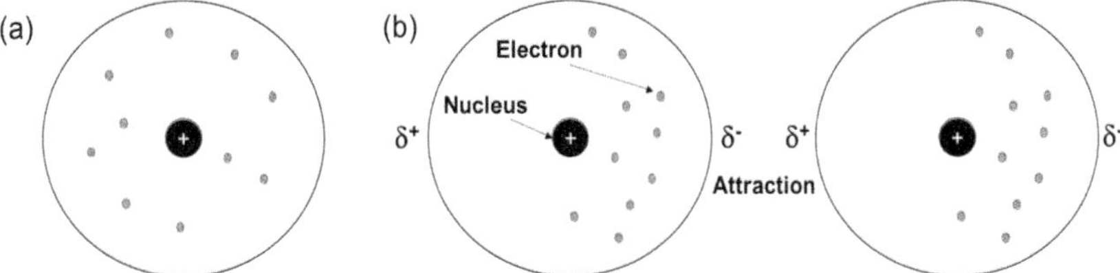

Fig. 2.16 **a** Neon atom with an even distribution of electrons averaged over time, **b** Neon atoms with instantaneous shift in the position of electrons resulting in instantaneous dipole moments. Negative regions of the atom attract the positive regions of other atoms (adapted from R.A. Higgins, Properties of Engineering Materials, Hodder and Stoughton, 1986, for the argon atom)

- **Secondary Bonding by Dispersion Forces**

This kind of bonding occurs in all atoms and molecules, including inert gas atoms and nonpolar (neutral) molecules. Let's closely examine the electron charges within atoms or molecules with respect to time. One will see that the electrons are constantly shifting locations but averaging out over time in a way that neutralizes the positive charges of the protons. This gives the general conclusion that the atom is electrically neutral. However, if one side of the molecule or atom suddenly has more electrons at any given instance than the other side, a negative charge will be instantaneously set up there. At the same time, another part of the atom or molecule will, therefore, have fewer electrons, thus allowing the proton's positive charge to dominate there and generate a slight positive charge in that location. This leads to what is called a **dipole moment**. A dipole moment is simply the product of the charge and the separation distance between the two charge centers (i.e., positive and negative charge centers). Then, the momentarily charged regions of such atoms will link up with regions of the opposite sign in other atoms, thereby enabling this secondary bonding. The momentary dipole can also induce or stimulate a dipole moment in a nearby uncharged atom, thus generating an opposite charge, resulting in coulombic attraction and bonding. Figure 2.16 illustrates the process.

- **Secondary Bonding by Permanent Dipole–Permanent Dipole**

Some molecules are what we call polar. An example would be HCl gas. Here, the chlorine and hydrogen atoms are covalently bonded. This means the hydrogen atom shares its sole electron with chlorine, and chlorine shares one of its seven outermost electrons with hydrogen. Since chlorine contains much more positively charged protons in its nucleus than hydrogen (1 proton), the positive charge in its nucleus provides a stronger pull on hydrogen's sole electron, which makes it move closer to it. This leaves the chlorine side of the HCl molecule with a slightly negative charge. At the same time, the hydrogen atom acquires a slightly positive charge (since it lost the complete influence of its electron in neutralizing the protons' positive charge). Another way of looking at it is that Cl is more electronegative than hydrogen, with electronegativities of 3 and 2.1, respectively. Hence, chlorine will attract electrons more than hydrogen. This process results in the HCl molecule having a permanent

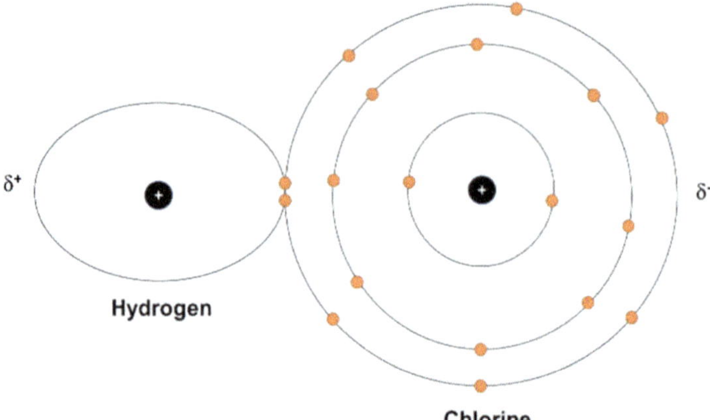

Fig. 2.17 Permanent dipole moment established in HCl gas molecule (multiple sources)

dipole moment, as shown in Fig. 2.17. The hydrogen atom in the molecule carries a slight positive charge, and the chlorine atom carries a slight negative charge. The HCl molecule will then bond with another HCl molecule where its positively charged hydrogen atom bonds with the negatively charged chlorine side of the other molecule, and vice versa.

- **Secondary Bonding by Hydrogen Bonding**

We often hear about the **hydrogen bridge**, a special case of the permanent dipole–permanent dipole case. However, it is a much stronger bond. We see this in ice (H_2O), where the electrons from each hydrogen atom are pulled closer toward the oxygen atom, leaving the nucleus of the hydrogen atoms relatively less shielded by its electron, thus establishing a permanent positive charge there. Likewise, a slight negative charge will be established in the oxygen atoms.

As such, the hydrogen atoms will bond with the oxygen atoms of other molecules, thus establishing bridges between the H_2O molecules (Fig. 2.18). Hydrogen bonding is not the only thing the atoms experience; we recall that all atoms and molecules experience vibrations due to thermal energy. As the temperature increases, these thermal vibrations can momentarily break hydrogen bonds. As the temperature is reduced, their effect is lessened, such that when ice is formed, hydrogen bonding becomes more dominant, and an open-type *structure* of bonded H_2O molecules is established. Since the structure is quite open, the density of ice is lower than that of water at freezing by about 9%. This simple fact has significant implications. If the density of ice is lower than water at freezing, then its volume will be higher. I'm sure some of us were in a hurry once to cool a beverage and place it in the freezer rather than the fridge to cool it quickly. Only to forget about it and come back later to find the vessel either broken or at least deformed (if it was made of a ductile material) due to the formation and volume expansion of ice.

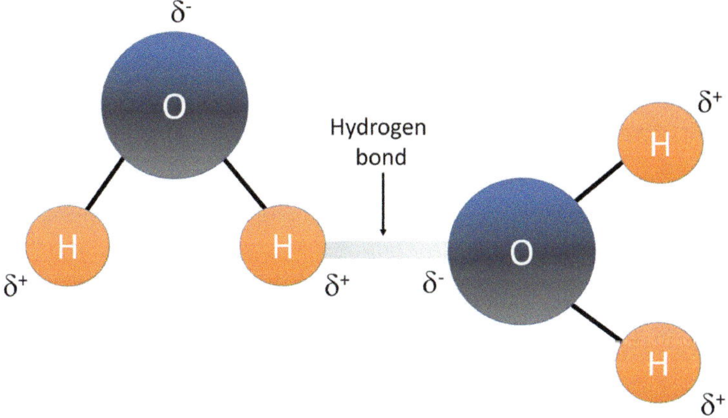

Fig. 2.18 Hydrogen bonding in H_2O (multiple sources)

Alex, we could take this knowledge and move on, or we could take it and reflect. Imagine if ice instead had a higher density than its liquid form (water), like any regular material would when it solidifies. In winter, when water freezes to produce a layer of ice on the surface of lakes, and oceans, that layer would sink rather than float because of its higher density, thus allowing it to be replaced by new water at the lake or ocean surface, which in turn also freezes and sinks and as the process is repeated, the lake or ocean would eventually freeze over. All lifeforms that the low-density ice layer would have protected would now die. Furthermore, icebergs would not even exist, for that matter. I would say this is something to think about.

- **Secondary Bonding by Permanent Dipole–Induced Dipole Moments**

Here, a polar molecule with a permanent dipole moment can induce a dipole moment in a closely positioned electrically neutral atom or molecule, resulting in electrostatic bonding. For example, the side of a polar molecule with a slight positive charge will induce a slight negative charge on an otherwise nearby neutral atom, resulting in Coulombic attraction.

As with primary bonds, Eq. (2.7) shows that the net potential energy of secondary (van der Waals) bonds consists of attractive energy (first term in the equation) and repulsive energy (second term in the equation) components.

$$E_n = -\frac{A}{r^6} + \frac{B}{r^n} \tag{2.7}$$

where n is typically ~ 12.

Let me quiz you, Alex. Do all atoms and molecules experience Van der Waals bonding?

Of course, since the dispersion forces are always in effect in all atoms and molecules. A momentary emergence of a slight charge on an atom or molecule is guaranteed. This can induce an opposite charge on another atom or molecule, leading to Coulombic attraction between the atoms or molecules, i.e., bonding

You've been paying attention. Roughly speaking, which materials do you expect to be electrically conductive? Metallically bonded, covalently bonded, or ionically bonded?

In general, metallically bonded materials, of course, since the electrons released into the sea or cloud of electrons are free to move. In ionic and covalent bonding, the electrons are tied up in the bonding and are not free to move.

I'm impressed!

Problems

2.1. What does electropositive and electronegative mean when referring to an element?
2.2. Define atomic number and mass number.
2.3. What is meant by saying that an element is isotopic? Can you give an example of such an element?
2.4. What is the difference between atomic mass and atomic weight?
2.5. What is nuclear binding energy, and how does it influence the atomic mass?
2.6. What is the Pauli exclusion principle?
2.7. Electrons are known to occupy energy shells. Which electrons have higher energy, those in the L or M shells?
2.8. What is the difference between protium and tritium?
2.9. State and discuss four van der Waals bonding types.
2.10. Where would you find alkaline earth metals in the periodic table?
2.11. Give two examples of primary bonds.
2.12. Which of the primary bonds are directional?
2.13. Compare the bond energies of primary bonds and secondary bonds.
2.14. Determine the ionic character of zirconia (ZrO_2).
2.15. The simplified two-sphere model has revealed important insights into the interactions between atoms, resulting in two primary plots. A force-interatomic

spacing plot and a potential energy-interatomic spacing plot. What important facts about a material can be deduced from these plots?

2.16. Sketch the net force versus interatomic spacing for two atoms of a metal on the same plot.

2.17. Sketch the attractive, repulsive, and net potential energy versus interatomic spacing for two atoms of an ionic compound on the same plot. Identify the ionization energy on the plot.

2.18. Why does aluminum have a lower melting point than tungsten?

2.19. Determine the dominant bonding type in oxygen (O_2) and chromium oxide (Cr_2O_3).

2.20. If the bond energy of a specific metal is higher than the bond energy of a particular ceramic, does this mean that the metal has a higher melting point than that ceramic?

2.21. The atomic number of carbon (C) is 6; determine the number of neutrons in ^{12}C, ^{13}C, and ^{14}C isotopes.

Chapter 3
Characterization of Materials

3.1 Introduction

This chapter introduces various characterization methods available to engineers and scientists for characterizing engineering materials. The chapter also introduces the structural aspects of materials.

3.2 Introduction to Visual Materials Characterization Methods

When presented with a block of material we know nothing about; we can first use one of our essential senses to learn more about it: eyesight. However, our eyesight is limited to only resolving or precisely seeing features approximately 0.1–0.2 mm (100–200 μm) and higher. Our eyesight will not do the job for anything smaller than that. This is problematic since most of the microstructural features of interest in materials lie below this resolution limit, hence the need to use other observation and imaging tools. Within this context, two issues become important. The first is to be able to examine the internal material features (microstructure or nanostructure), and the second is to be able to record and document the findings to communicate them effectively to the scientific and industrial communities.

Different microscopes are available to us today, and the choice of which one to use depends on the size of the features we are interested in observing. Regarding the ability to document and record our findings, today, cameras and image recording tools with very high resolutions are now abundantly available for use with any microscope. The word structure, as it relates to a material, can be classified according to the size of the entities of interest. For example, a **nanostructure** is any structure with a dimension between 1 and 100 nm. A **nanomaterial** is often defined as a material having a size between 1 and 100 nm in at least one dimension. This would make

K. Morsi, *An Engaging Approach to the Science and Engineering of Materials*,
https://doi.org/10.1007/978-3-032-06231-4_3

a coating with a thickness of 50 nm classified as a nano-coat, even if it covers an area as large as 100 cm^2. As explained later, most engineering materials consist of millions of minute crystals (also called grains) strongly attached to each other. These crystals are typically (but not always) between 1 and 100 μm. When we characterize these crystals, whether we are interested in their shape or size or even tiny microscale pores in the material, we are said to be characterizing the material's **microstructure**. Likewise, following the previous definition of nanomaterials, the material would be said to be **nanostructured** if the crystals are between 1 and 100 nm. Any structure of a material having a size above 100 μm is referred to as a **macrostructure**. Of course, it is understood that some critical material structural features also exist below 1 nm; for example, the structure of an atom with its protons, electrons, and neutrons is significantly smaller than 1 nm. Atoms may tend to form **ordered atomic arrangements**, but over very short atomic distances, typically between 0.1 and 1 nm, we call **short-range atomic order**. Such features are present in molten metal or water, where only the H_2O molecule is an ordered structure in which oxygen is bonded to two hydrogen atoms. However, the billions and billions of water molecules in water do not arrange themselves in any particular order over large distances, hence the term short-range atomic order. On the other hand, **long-range atomic order** refers to atomic order over larger distances, for example, 10 nm to centimeters. Examples of long-range atomic order can be found inside a 100 μm aluminum powder particle, where aluminum atoms are arranged in a periodic 3D ordered arrangement, or even in a 2 cm single crystal of silicon, where silicon atoms are also arranged in a long-range periodic manner.

Since a general rule in materials science dictates that structure affects the properties and behavior of materials, we must characterize and evaluate these structures. This characterization can be visual, mechanical, electrical, magnetic, etc. In this chapter, we are mainly concerned with visual characterization methods, which allow us to visualize the structure of our materials either through a conventional microscope or more sophisticated equipment. Figure 3.1 shows the structural scales in materials that can be characterized. Note, microstructures with grain sizes between 100nm and 1 micrometer are referred to as ultrafine-grained microstructures.

So far, we have stated that the macrostructure of a material can be observed with the naked eye. The development of microscopes has been key to our understanding of materials and their structures. There are two important features to consider when we talk about microscopy. One is **magnification**, and the other is **spatial resolution**.

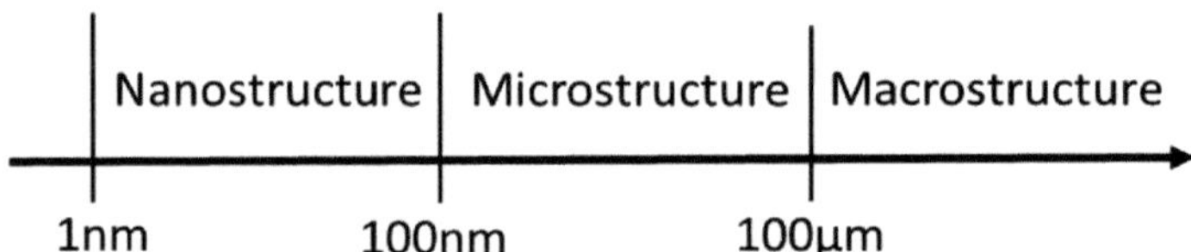

Fig. 3.1 Designation of different relevant structures (by scale) encountered in visual materials characterization (note, atomic scale is 10^{-10} m or 0.1 nm and microstructures with grain sizes between 100 nm and 1 micrometer are referred to as ultrafine-grained microstructures)

The spatial resolution *is the minimum distance that can be resolved or seen between two lines*. Depending on the lens/aperture system, one can theoretically magnify an image to a very high degree; however, as we keep magnifying to higher and higher levels, there will come a point where we will not be able to resolve finer features simply by increasing the magnification. This limit is imposed by the spatial resolution of the microscope used to image the material.

As previously mentioned, the resolution of our naked eye is about 0.1–0.2 mm or 100–200 μm. The resolution of microscopes depends on the wavelength of the illuminating source, roughly half of it. In light microscopes, since the minimum wavelength of visible light is around 400 nm, the resolution is about 200 nm or 0.2 μm. The maximum magnification that can be used without losing resolution is called the **useful magnification** and is given by Eq. 3.1:

$$\text{Useful Magnification} = \frac{\text{Resolution of human eye}}{\text{Resolution of the microscope}} \tag{3.1}$$

Applying Eq. 3.1 to light microscopes gives a useful magnification of 1000x (i.e., 200 μm/0.2 μm). Any attempt to exceed this useful magnification will not result in properly observing and resolving smaller features.

This limits the ability of light microscopy to characterize material features at extremely small size scales visually. For example, nanofeatures (1 nm–100 nm) are out of the question for a light microscope simply because of the limited wavelengths of the illumination source (visible light), i.e., 380–760 nm. For higher magnifications, we would need an illumination source capable of a much smaller resolution limit (which necessitates it having much shorter wavelengths than visible light). This is possible if we use electrons as the illuminating source instead of visible light. **Louis de Broglie** first theorized the wave-like characteristics of electrons and their having much shorter wavelengths than those found in visible light in 1925. Independently, an incredible avalanche of discoveries and innovations, including electron lenses and the first-ever electron microscope that produced the first-ever image, was published in 1932 by **Knoll and Ruska**, who also proposed the name electron microscope. For the first time, electrons were used to image materials. Ruska took over 50 years to be recognized for this outstanding achievement with a Nobel Prize in 1986 (only 2 years before he passed). Based on this principle, the **transmission electron microscope (TEM)** was first invented, followed by the **scanning electron microscope (SEM)**, both of which are very important characterization tools in materials science. The wavelengths used in the TEM are smaller than those in the SEM, which are smaller than those in the light microscope that uses visible light. It follows that the spatial resolution of the SEM is around 1 nm while that of the TEM is around 0.1 nm. The SEM and TEM can resolve smaller material features compared to the light microscope (0.2 μm resolution). While electrons are the illumination source in both SEM and TEM, their wavelength can be varied by varying what is known as the accelerating voltage (more about this later).

Ok, Alex, what is the useful magnification of an SEM?

Since the SEM's resolution limit is 1 nm, the useful magnification is $0.2 \times 10^{-3}/1 \times 10^{-9} = 200{,}000\times$, i.e., 200 times more than that of the optical microscope.

3.2.1 Material Preparation for Microscopic Observation

The material sample must be sectioned centrally for microstructural investigations to reveal the internal structure. Sectioning could be carried out using a diamond wheel (for ceramics and hard materials), abrasive wheel, hacksaw, bandsaw, etc. In each case, the resulting sectioned surface quality may be different. Even if the surface looks flat to the naked eye after sectioning, observation at higher magnification under the powerful lens of a microscope will reveal scratches and surface irregularities/ roughness, making it extremely difficult to see the material microstructure. Hence, a series of grinding papers of increasing grit numbers are used sequentially to grind the sample surface. Note that the larger the grit number, the smaller the hard particles embedded in the grinding paper; hence, the finer the surface finish on the sample. Table 3.1 shows the relation between grit number and equivalent particle size. These hard particles scratch the sample surface and progressively remove irregularities and larger scratches previously found on the surface. The smaller grit number paper contains larger embedded hard particles that remove larger amounts of material from the sample surface but leave scratches of their own behind.

Alex, did you know that although they are called sandpapers, they are not actually sand (SiO_2) particles embedded in these papers? The hard particles can be alumina (Al_2O_3) or silicon carbide (SiC) for example. A better name would be grinding paper

If the sectioned sample is relatively large and easily handled, then the operator would start by simply placing the sample's surface on the grinding paper attached to a rotating disk. If, however, the sectioned sample is small and not easily handled, it would need to be first mounted in epoxy resin or Bakelite. To place it in epoxy resin, the sample is first placed at the bottom of a plastic mold, with the surface to be examined facing downward, as shown in Fig. 3.2. Epoxy resin is then mixed with a hardener and poured over the sample. After the epoxy resin sets, the epoxy containing the sample is removed, as seen in Fig. 3.2, and placed on a rotating disk with the sample surface facing the disk surface. As mentioned, the rotating disk allows the placement of grinding paper on its top surface, as seen in the figure. Typically, we could start with a 220-grit size, which can remove relatively large irregular features from the sample surface. However, these same particles will leave

Table 3.1 Grit number and equivalent particle size (Buehler.com)

Grit number	Particle size (μm)
60	268
80	188
100	148
120	116
180	78
220	66
240	51.8
280	42.3
320	34.3
360	27.3
400	22.1
500	18.2
600	14.5
800	12.2
1000	9.2
1200	6.5

scratch marks on the sample surface. Hence, after finishing with the 220-grit paper, the same procedure is repeated using higher and higher numbered grit papers, e.g., 320, 400, and 600. After the final grinding stage, the sample surface may appear polished; you may even see your face in it. However, in reality, there would still be scratches on the surface that one can only see under the lens of a microscope. To remove these scratches, **polishing** operations need to be conducted. These would typically require a soft polishing pad placed on the polishing/grinding wheel; then, a 15 μm diamond particle suspension is sprayed onto the pad. However, before moving to the polishing stage, the sample is first placed in water in an ultrasonic cleaner, where the vibration and ultrasonic energy are used to remove any attached particles on the sample's surface left over from the previous grinding operation. If not removed, they may fall on the polishing pad and contaminate it (if larger than 15 μm in size) and will keep generating larger scratches on the sample. After rinsing the sample with deionized water, the polishing process is started using the 15 μm diamond polishing pad. Following the 15 μm polishing stage, the sample is ultrasonically cleaned again, placed on a 6 μm polishing pad, and polished. The process is continued using pads containing smaller and smaller diamond particles, for example, 3 μm then 1 μm, with ultrasonic cleaning in between each polishing stage. Typically, the final polishing stage makes use of alumina nanoparticles.

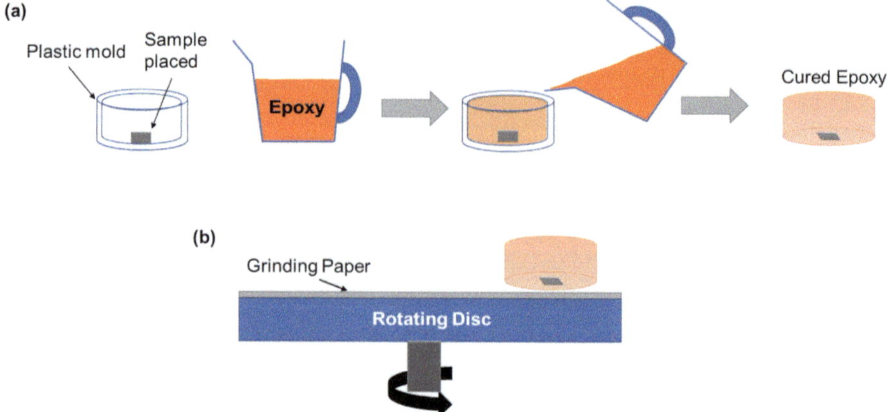

Fig. 3.2 **a** Process for placing sample in epoxy mold, **b** Epoxy helps hold sample for grinding and subsequent polishing operations

3.2.2 *Microscopic Characterization*

In this section, we discuss different types of microscopes and discuss their applicability in materials characterization, considering their magnification ranges and resolution limits.

- **The Stereomicroscope**

Stereomicroscopes generally have relatively low magnifications, typically up to 100 ×, and often much lower. However, recent research-grade microscopes can exceed 100 × magnification. Stereomicroscopes have an advantage in their depth of field, typically in tens of micrometers, with depth of field decreasing with an increase in magnification. They usually have large working distances between the lens and the sample; this permits the placement of large samples. It is also possible to obtain 3D images using stereo microscopy. Spatial resolutions can be around 1.5 μm. An example of a stereomicroscope is shown in Fig. 3.3a. Figure 3.3b is an image of an aluminum alloy foam taken with a stereomicroscope. Note the scale bar and the depth of field. Stereomicroscopy is, however, limited in its magnification. A more powerful microscope would be needed to see smaller things, such as microstructural features like fine grain sizes.

- **The Metallurgical Light Microscope**

These microscopes can be *upright*, where the sample is placed on the microscope stage and viewed from the top using a lens, or *inverted*, where the surface to be viewed is placed upside down and viewed from the bottom through an opening in the stage using a lens. The maximum magnification is typically 1000x, and the resolution is 0.2 μm. This microscope has a limited depth of field. Hence, a pore in the microstructure would appear like a black spot. Specifically, the depth of field

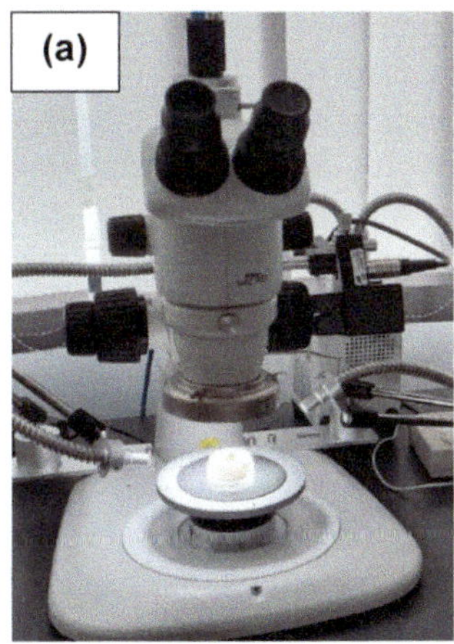

Fig. 3.3 **a** Photograph of a stereomicroscope. Image from Târtea, D. A.; Manolea, H. O.; Ionescu, M.; Gîngu, O.; Amărăscu, M. O.; Popescu, A. M.; Mercuţ, V.; Popescu, S. M. A Microscopy Evaluation of Emergence Profile Surfaces of Dental Custom CAD-CAM Implant Abutments and Dental Implant Stock Abutments. J. Pers. Med. 2024, 14, 699. https://doi.org/10.3390/jpm140 70699. Reproduced with modification under Creative Commons Attribution (CC BY) license (https://creativecommons.org/licenses/by/4.0/). **b** image of aluminum alloy foam taken by a stereo microscope, image from Costanza, G.; Del Ferraro, A.; Tata, M. E. Experimental Set-Up of the Production Process and Mechanical Characterization of Metal Foams Manufactured by Lost-PLA Technique with Different Cell Morphology. Metals 2022, 12, 1385. https://doi.org/10.3390/met120 81385, reproduced with modification under the terms and conditions of the Creative Commons Attribution (CC BY) license (https://creativecommons.org/licenses/by/4.0/)

depends on the magnification; for example, at $5 \times$ magnification, it is $15 \ \mu$m, while at $1000 \times$ magnification, it is $0.2 \ \mu$m. A rendering of a light microscope is shown in Fig. 3.4.

Fig. 3.4 Rendering of a light microscope (image generated using ChatGPT 4.0 by K. Morsi)

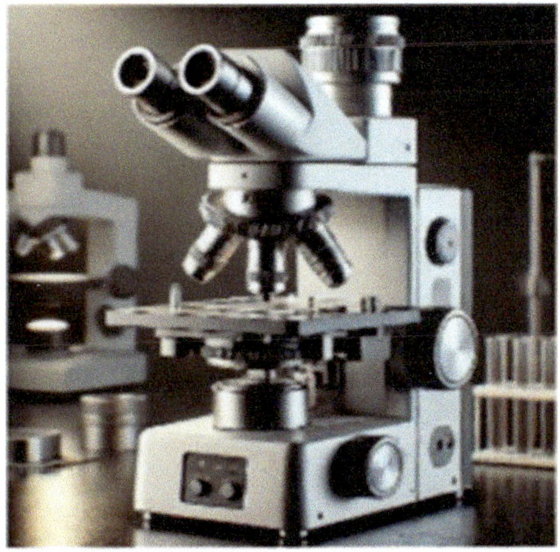

- **The Scanning Electron Microscope (SEM)**

Within the field of materials science, this microscope is invaluable. Although considerably more expensive than a regular light microscope, it provides much more information. While light microscopes use visible light as the illumination source (with a wavelength ranging from 380 to 760 nm) to observe microstructural features, they cannot resolve features smaller than half the wavelength of visible light, i.e., 400 nm/2 or ~ 0.2 μm. An illumination source with a shorter wavelength is needed to see smaller features. Electrons fulfill this requirement; they can be generated using a field emission or thermionic gun, while a bias voltage is applied to create an electron beam within a column placed under a very high vacuum. The electrons then pass through electromagnetic lenses, apertures, and scanning coils to scan the electron beam over the sample surface and ultimately generate an image. The accelerating voltage influences the wavelength of the electrons; for example, an electron beam with a wavelength of ~ 2.51 pm is produced at a 200 keV accelerating voltage, and even lower wavelengths are possible at higher accelerating voltages. It should be mentioned that the SEM typically uses accelerating voltages between 5 and 30 kV, and useful magnifications can be up to 200,000 X. We cannot see electrons with our naked eyes, so the image produced is a processed/reconstructed image in greyscale. Figure 3.5. shows an example of an SEM and the sample stage where the sample is placed, together with images that can be produced from such a characterization tool. Note here the depth of field, and magnifications possible using an SEM.

The approximate relationship between electron wavelength and accelerating voltage is depicted in the **de Broglie equation** (ignoring relativistic effects) as presented in Eq. 3.2:

$$\lambda(nm) = \frac{1.23}{\sqrt{V_{acc}}} \tag{3.2}$$

The short wavelengths give electron microscopes an excellent advantage in seeing smaller and smaller features.

Ok, what range of wavelengths can the SEM typically generate?

Since SEMs have accelerating voltages between 5 and 30 kV, using Eq. 3.2, the wavelengths of electrons can range from 0.017 nm to 0.007 nm, respectively. Hence, at the higher accelerating voltage, we expect to see the lowest wavelength in the range, allowing us to resolve smaller things. However, after reading up on this, I found that some sources use units of keV instead of kV for the accelerating voltage. Why is that?

First, a great answer. As for your question, whenever the accelerating voltage is given in kV, it only refers to the voltage. However, sometimes sources refer to the energy acquired by the electrons at the point of impact with the surface; this energy is given in keV. The higher the voltage, the higher the energy. So, the actual values or magnitudes of the accelerating voltage will not change here. This means that a 5 kV accelerating voltage can also be represented as 5 keV. Just remember that one refers to the actual voltage while the other refers to the energy of the electrons.

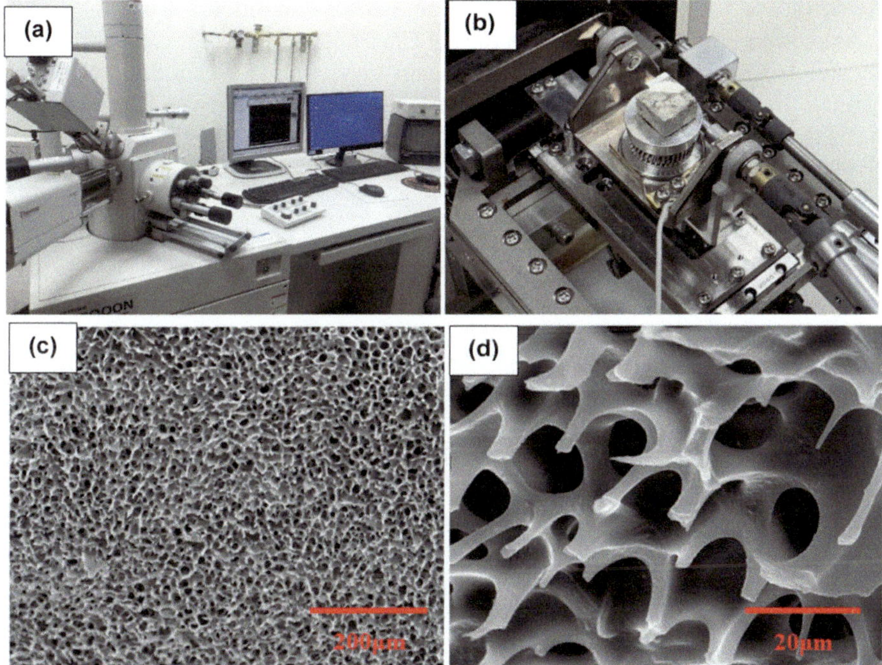

Fig. 3.5 **a** Scanning electron microscope, **b** sample placed inside microscope on sample stage (from Imtiaz, T.; Ahmed, A.; Hossain, M. S.; Faysal, M. Microstructure Analysis and Strength Characterization of Recycled Base and Sub-Base Materials Using Scanning Electron Microscope. Infrastructures 2020, 5, 70. https://doi.org/10.3390/infrastructures5090070es), **c** low magnification micrograph and **d** a higher magnification micrograph showing depth of field (from Qiu, L.; Zhang, Y.; Long, X.; Ye, Z.; Qu, Z.; Yang, X.; Wang, C. Scanning Electron Microscopy Investigation for Monitoring the Emulsion Deteriorative Process and Its Applications in Site-Directed Reaction with Paper Fabric. Molecules 2021, 26, 6471. https://doi.org/10.3390/molecules26216471). All figures are reproduced with modification under the terms and conditions of the Creative Commons Attribution (CC BY) license (http://creativecommons.org/licenses/by/4.0/)

The wavelengths generated in the SEM allow for much smaller resolution limits than the light microscope, i.e., 1 nm compared with 200 nm for a light microscope. Moreover, the depth of field is superior in the SEM, meaning we can see the contours of pores, whereas, in light microscopy, they may appear as spots due to the small depth of field. At $10 \times$ magnification, the depth of field in an SEM can be in millimeters, depending on the accelerating voltage.

SEMs not only produce crisp images at high magnification, with excellent resolution and depth of field, but the electron beam–material surface interaction can generate important information unmatched by light microscopes.

Figure 3.6 shows a schematic of an interaction volume generated when the high-energy electron beam hits the sample's surface; it indicates electrons traveling deeper under the surface (typically a few micrometers). When interacting with the sample atoms, the electrons experience both elastic and inelastic scattering. Elastic scattering refers to scattering where the electrons do not lose any significant energy.

As shown in the figure, different types of electrons and even X-rays emerge from the sample's surface upon the electron beam impacting it. **Secondary electrons** originate from the atoms of the actual sample being viewed, which get knocked out by the incoming electrons from the electron beam. Some energy will be lost in the event since knocking out an electron from its shell and generating an electron hole in its place require energy. Hence, secondary electrons are said to be generated by *inelastic* scattering. A secondary electron detector in the SEM picks up these electrons. They can provide excellent topographical information and images with

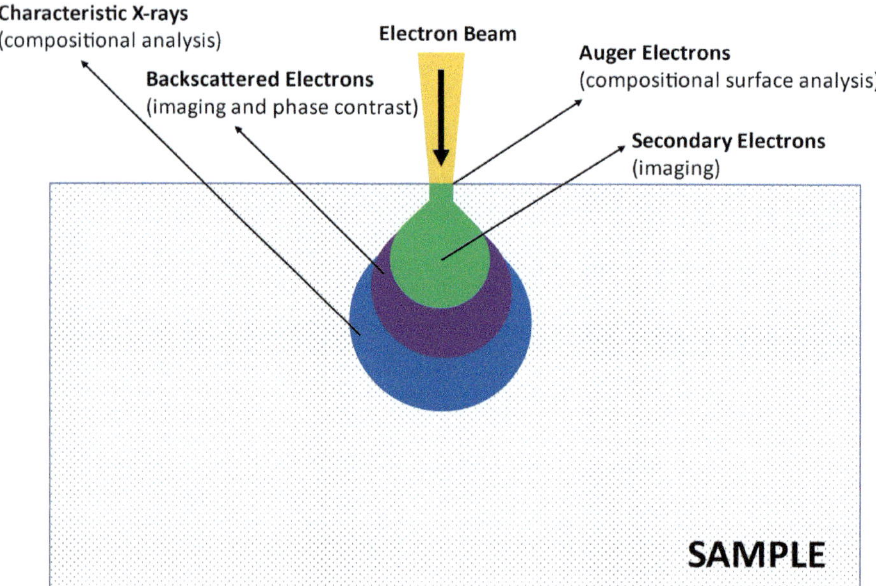

Fig. 3.6 Schematic of electron beam-material interaction, showing the interaction volume (multiple sources)

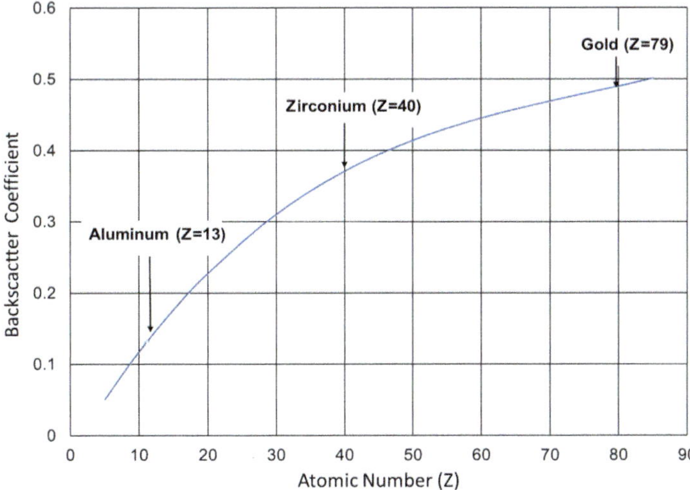

Fig. 3.7 Dependence of backscatter coefficient on atomic number. The higher the backscatter coefficient the brighter the image appears

great depth of field that can capture protruding or intruding surface features such as cavities or micropillars.

The **backscattered electrons** originate from the primary electron beam that gets elastically scattered back upon interaction with the material. The backscattered electron detector picks up such electrons. Now, we may wonder, how do the detectors know which electron is backscattered and which is secondary? The distinction comes from the energy of these electrons; backscattered electrons possess higher energy than secondary electrons. Note, backscattered electrons occur due to *elastic* scattering, whereas secondary electrons occur through *inelastic* scattering, where more energy is lost.

Here, the images produced are pretty revealing. It turns out that the backscattered electron signal coming out of the sample is highly dependent on the sample's atomic number; the higher the atomic number, the more backscattered electrons are. This dependence is seen in Fig. 3.7, where the backscatter coefficient (number of backscattered electrons divided by the total number of incoming electrons) increases with an increase in atomic number. A consequence of this effect is that within a microstructure, if there are two regions of different atomic numbers, the one with the higher atomic number will appear brighter since it will backscatter more electrons and register in the detector as brighter. This could be useful in identifying regions in the microstructure and even whether a microstructure is chemically homogenous after a reaction. For example, in Fig. 3.8, we see an image of a microstructure showing layers of Ni–Al intermetallic phases of varying compositions taken in the backscattered mode of the SEM. The image shows light gray and progressively darker gray layers. As shown in the periodic table, aluminum has an atomic number of 13, while Ni has an atomic number of 28. Since the Ni atom has a higher atomic number, it will backscatter relatively more electrons than aluminum and appear brighter. This means that the shade of each intermetallic compound layer will depend on the amount of aluminum and nickel atoms it contains. So, we see clearly that the layers going from right to left gradually include more and more aluminum atoms and fewer nickel atoms. That is why they become darker and darker.

The other important feature of electron microscopes comes from the **characteristic X-rays** that emerge from the sample when the incoming electrons from the electron beam strike it. These X-rays are emitted because the incoming electrons remove electrons from lower energy shells of the sample's atoms. When that happens, every ejected electron leaves behind an electron hole. Electrons from a higher energy shell drop and fill the hole and acquire a lower energy. Since energy is conserved, the difference in energy is given off in the form of X-rays having specific energies. These X-rays are unique to the element in question, like a fingerprint. Hence, the analysis of this data results in the ability to measure the percentage of elements within a sample and its composition. This is typically referred to as **X-ray microanalysis**

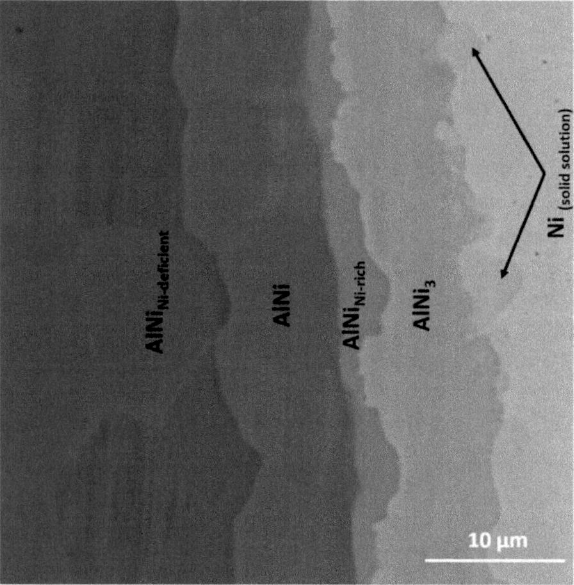

Fig. 3.8 Backscattered electron micrograph of Ni–Al intermetallic layers. It is clear that the shade of layers becomes darker from right to left. This is because the intermetallic phase composition increases in the lower atomic number element aluminum from right to left. The image shows the power of backscattered electron microscopy in revealing compositional inhomogeneities in a material's microstructure. Image from Kwiecien, I.; Bobrowski, P.; Wierzbicka-Miernik, A.; Litynska-Dobrzynska, L.; Wojewoda-Budka, J. Growth Kinetics of the Selected Intermetallic Phases in Ni/Al/Ni System with Various Nickel Substrate Microstructure. Nanomaterials 2019, 9, 134. https://doi.org/10.3390/nano9020134. Image reproduced under the terms and conditions of the Creative Commons Attribution (CC BY) license (http://creativecommons.org/licenses/by/4.0/)

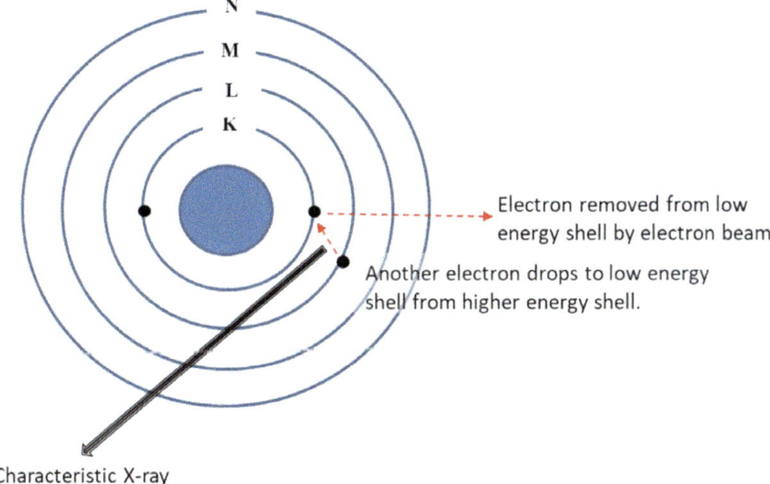

Fig. 3.9 Schematic of Bohr model showing generation of X-rays used in EDS

or **energy-dispersive X-ray spectroscopy** (abbreviated as EDX, EDXS, or EDS). Figure 3.9 illustrates the process of X-ray generation.

At surface regions where the electron beam meets the sample (Fig. 3.6), **Auger electrons** are ejected. Here, when incoming electrons from the electron beam strike an atom and remove an electron from the first innermost shell, an electron hole is left behind, as in EDS. An electron of higher energy will fill that hole and, in the process, attain a lower energy; however, in this case, the difference in energy results in the emission of an outer electron (called an **Auger electron**). The electron's kinetic energy is measured via **Auger electron spectroscopy (AES)** and used to determine the composition. AES can be used to determine elements with atomic numbers greater than 2 (i.e., higher than the atomic number of helium), which means that very light elements can be detected. This contrasts with EDS, which is generally limited to measuring elements with atomic numbers of 11 (i.e., sodium) and greater. In some variations, it can detect elements as light as boron (atomic number 5).

- **The Transmission Electron Microscope (TEM)**

Despite the great utility of the SEM, it generally uses accelerating voltages between 5 and 30 kV, which limits its resolution to about 1 nm. Increasing the accelerating voltage would result in a further decline in the electron wavelength to values much below the size of an atom. Using Eq. 3.2, at 100 keV, the wavelength would be 0.004 nm, with the diameter of an atom being around 0.3 nm. One would think that increasing the accelerating voltage in SEMs to extremely high values would result in being able to see even smaller and smaller features. However, there is a limit, as the observed material can be damaged at very high accelerating voltages, which can occur for polymeric and organic materials. On the other hand, TEMs typically operate at accelerating voltages between 60 and 300 kV. The higher accelerating voltages achieve two things. First, allow the electron beam to transmit through the sample. Here, the sample must be very thin, less than 100 nm. Special equipment is used to achieve this thinning operation. The second reason for high accelerating voltages is to reduce the electron wavelength and achieve better spatial resolution. Although using Eq. 3.2 would suggest extremely low wavelengths and correspondingly very low resolution limits, such predictions are limited by imperfect electron lenses in present-day microscopes. This results in a practical spatial resolution of around 0.1 nm. It, however, makes it feasible for TEMs to see atoms, as shown in Fig. 3.10 for a platinum nanoparticle taken with a **high-resolution transmission electron microscope (HRTEM)**. In the image, the crystallographic orientation of the faces of the nanoparticle is shown (more about this later). Note the 1 nm scale bar in the figure, which can be used to gauge the diameter of the atoms. It should be mentioned that since the electron beam goes through the thin sample to produce the image, we are not looking at individual atoms but a column of atoms running perpendicular to the plane of the image. The TEM can also have a scanning feature in what is known as a **scanning transmission electron microscope (STEM)**.

TEMs are invaluable for nanomaterials, as they can observe atomic defects, atomic arrangements, and more. However, TEMs have disadvantages, including their high cost and the incredibly painstaking task of preparing a sample less than 100 nm thick for conventional materials.

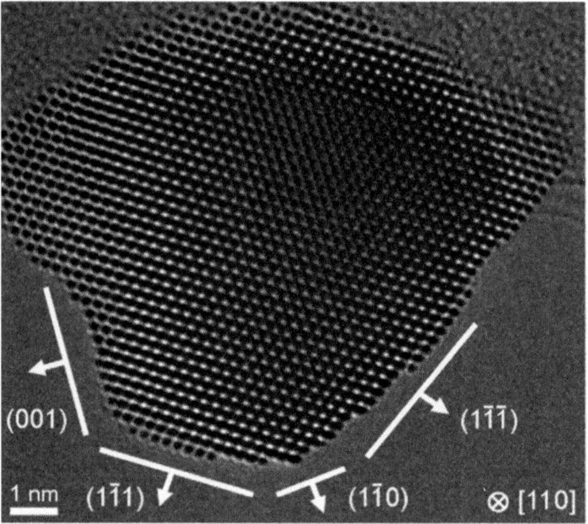

Fig. 3.10 An image of a platinum nanoparticle taken with a high-resolution transmission electron microscope, atomic positions shown, together with the crystallographic orientation of faces of the particle. Fig. 3 in Schneider et al., Atomic surface diffusion on Pt nanoparticles quantified by high-resolution transmission electron microscopy, Micron 63 (2014) 52–56. https://doi.org/10.1016/j.micron.2013.12.011. reproduced with permission from Elsevier

- **The Scanning Tunneling Microscope (STM)**

Invented in 1981 in IBM labs by **Gerd Binning and Heine Rohrer**, the scanning tunneling microscope significantly impacted nanotechnology. The microscope is built on the principle that if a conductive tip (also called a probe) is brought closer to a conductive surface (using piezoelectric materials) in a vacuum and under bias voltage, at some point, electrons will start jumping from the surface of the sample toward the electrically positively biased tip. The x–y–z stage is dependent on piezoelectric materials. These materials change dimensions on the application of a voltage, and on the flip side, when they are deformed, they generate a voltage. This affords a guiding system for the tip with very high nanoscale precision. The tip is atomically sharp because, at the atomic level, there will always be a region on the tip surface where an atom would be protruding. It turns out that the tunneling current produced by the conducting electrons is exponentially dependent on the separation distance between the tip and the surface of the sample. This provides incredible control over the separation distance. Two of the modes of operation are to move the tip closer to the sample surface until a current of a particular current intensity is produced. Then, the current is kept constant as the tip is scanned across the surface. For the current to remain constant, the distance between the tip and the surface must stay constant as the tip is scanned. The tip will rise and fall with the sample's surface topography. The other way is to scan the sample, observe the current's intensity, and calculate the distance. In both cases, an atomic scale image of the surface can be generated. Figure 3.11a shows a highly simplified schematic of the scanning tunneling microscope, in which a tunneling current is observed flowing from the sample surface through to the tip. Figure 3.11b, using the coarse scanning mode of STM, shows three graphene flakes that have been exfoliated (i.e., separated) from graphite. Figure 3.11c shows the measurements of the step from one graphene flake to the other, which is 0.35 nm (i.e., 3.5 Å).

Using the finer scanning mode, Fig. 3.11d shows the atomic structure of graphene, while Fig. 3.11e shows the atomic structure of graphite. The disadvantage of STM lies in requiring very clean and conductive sample surfaces to image. In 1986, **Gerd Binning** and **Heine Rohrer** received the Nobel prize for their contribution.

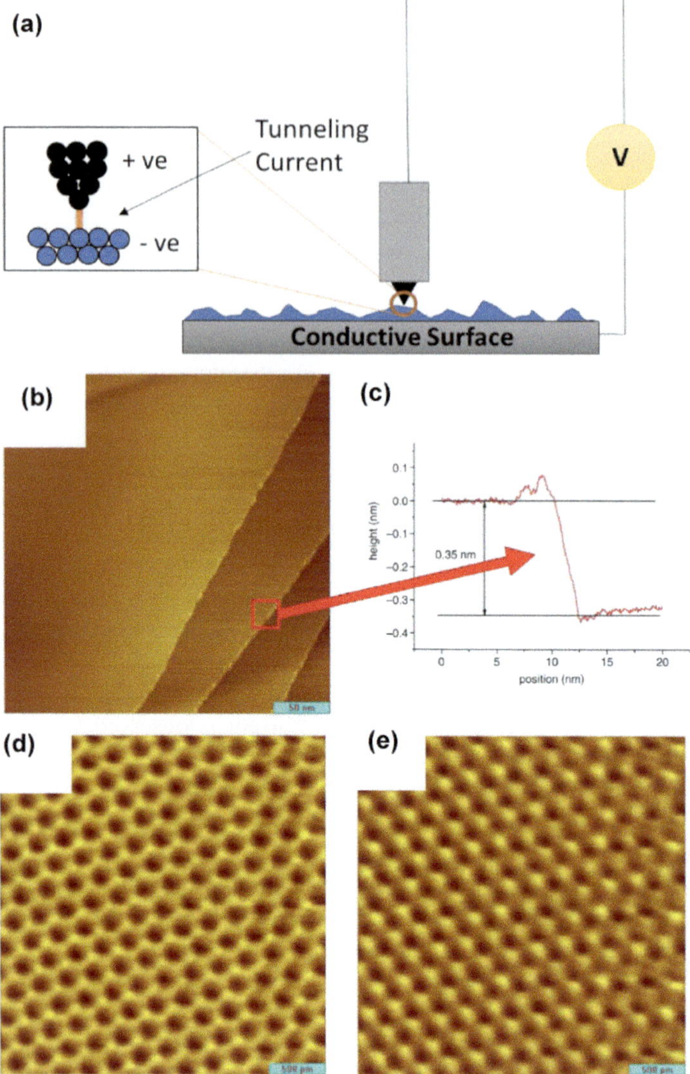

Fig. 3.11 **a** Schematic of scanning tunneling microscope (multiple sources), **b** coarse STM scan showing three graphene flakes on the surface of graphite, **c** displacement profile of tip showing the height of the step between two graphene layers, **d** fine scan showing the atomic structure of graphene flake, **e** fine scan showing atomic structure of graphite. **b–e** is modified from Fig. 1 in Adina Luican, Guohong Li, Eva Y. Andrei, Scanning tunneling microscopy and spectroscopy of graphene layers on graphite, Solid State Communications 149 (2009) 1151–1156. https://doi.org/10.1016/j.ssc.2009.02.059 (reproduced and modified with permission from Elsevier)

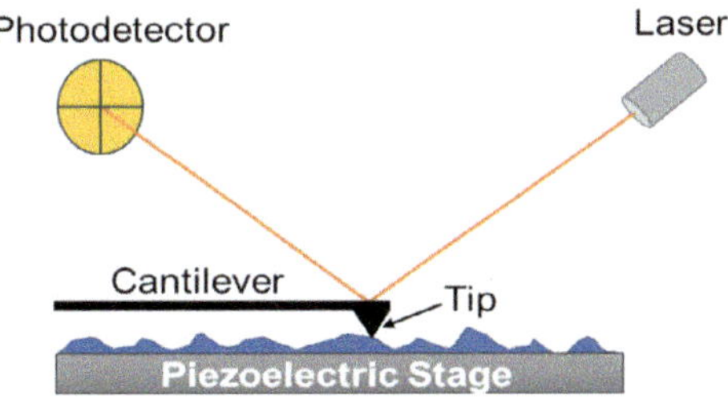

Fig. 3.12 Schematic of an atomic force microscope (multiple sources)

- **The Atomic Force Microscope (AFM)**

The atomic force microscope (AFM) was invented in 1986 by **Binning, Quate, and Gerber**. Unlike STM, the AFM can characterize conducting and non-conducting materials. The microscope (schematically shown in Fig. 3.12) uses a cantilever with an atomically sharp tip at its end that is scanned over the surface of a sample. Here, the microscope measures the displacement of the cantilever using light from a small diode laser and photodetector system. The force between the tip and the surface can also be measured and yield important mechanical property data, such as stiffness. AFMs can operate in a contact, tapping contact, or contactless mode. In contact mode, the deflection is measured as the probe scans over the sample's surface in a raster-type pattern while keeping the force constant. This mode helps obtain mechanical property data. Sometimes, the probe could get stuck in adsorbed moisture on the surface, also tip and surface damage are possible. The tapping mode can be used to reduce these problems. In the contactless mode, the probe is brought close to the sample surface without touching it; if it is within 1–10 nm from the surface, van der Waals attractive forces can be detected at this distance. The microscope can be used for several critical applications, including imaging and determining mechanical and electrical properties. It can also be used to manipulate the surface.

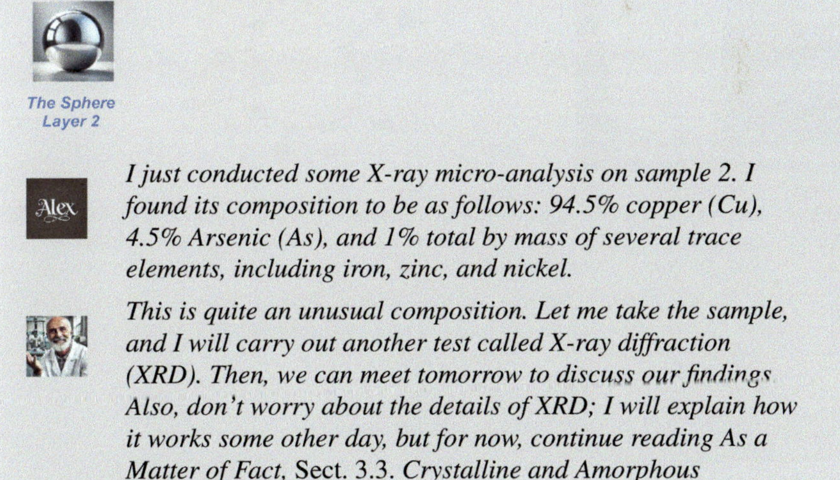

*The Sphere
Layer 2*

I just conducted some X-ray micro-analysis on sample 2. I found its composition to be as follows: 94.5% copper (Cu), 4.5% Arsenic (As), and 1% total by mass of several trace elements, including iron, zinc, and nickel.

This is quite an unusual composition. Let me take the sample, and I will carry out another test called X-ray diffraction (XRD). Then, we can meet tomorrow to discuss our findings. Also, don't worry about the details of XRD; I will explain how it works some other day, but for now, continue reading As a Matter of Fact, Sect. 3.3. Crystalline and Amorphous Materials.

3.3   Crystalline and Amorphous Materials

As mentioned, one of the simplest ways to classify materials is to divide them into metals, ceramics, polymers, and composites. Another way to classify materials is to divide them into crystalline and amorphous materials.

Amorphous materials are those materials in which their atoms are not arranged in any periodic order over long atomic distances, say 10 nm or more. This means that if we zoom into a material and observe how atoms are arranged together in one location. Then, we move to a different location, and the atomic arrangement will differ in both cases. A good example of an amorphous or non-crystalline material would be glass, which can be referred to as a highly viscous liquid where the atomic arrangement is not ordered. Figure 3.13a shows a glass vase, and Fig. 3.13b shows a 2D representation of amorphous silica (SiO_2) glass, showing the absence of long-range atomic order (i.e., the atomic structure is different from one part of the figure to the other). Molten metals are also amorphous.

Crystalline materials, conversely, are materials in which their atoms are arranged in some long-range periodic order. Figure 3.14a shows an image of an aluminum can, while Fig. 3.14b shows a 2D representation of long-range atomic order in aluminum. Who would have thought that aluminum would be called crystalline while a glass "crystal" vase would not?

Examples of **crystalline materials** include all metals, many ceramics, and some polymers. However, how we define crystallinity for polymers significantly differs from how we do for metals and ceramics, as explained later.

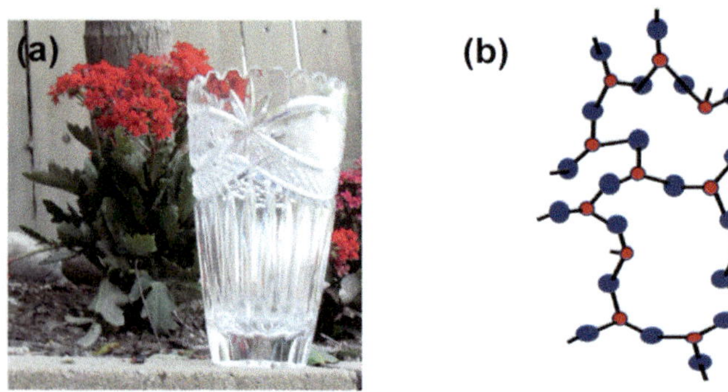

Fig. 3.13   **a** Glass crystal vase that is not crystalline (photography *courtesy of Yusuf Morsi*), **b** Silica (SiO_2) amorphous structure (2D representation), note the lack of long-range atomic order

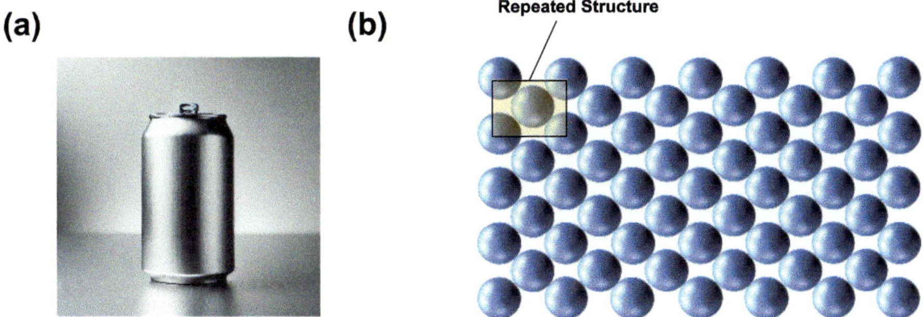

Fig. 3.14 **a** Aluminum can (generated using ChatGPT 4.0 by K. Morsi), **b** 2D illustrative example of long-range atomic order

Coming back to metals and ceramics, crystalline materials can be **polycrystalline** or **single crystals**. Most engineering alloys and ceramics are what we refer to as polycrystalline materials. If I hand you a block of metal, and if it is polycrystalline, that block will consist of very small 3D crystals or grains that are strongly bonded to each other. Figure 3.15 shows a 2D representation of four grains or crystals bonded together. If you zoomed inside any of these crystals, you would see a beautifully ordered arrangement of atoms. Such an ordered atomic arrangement can be imagined as being produced by repeating one single unit cell (atomic arrangement) in 3D to construct the long-range atomic order or pattern within each crystal. For an elemental metal, such as aluminum or gold, although each crystal or grain will have the same geometric arrangement of atoms, they will differ in the orientation or the alignment of this arrangement from one crystal (or grain) to another. Each crystal will be randomly orientated to the other. As shown in Fig. 3.15, in each grain (A, B, C, and D), the atoms are aligned in a particular direction (despite having the same 3D long-range geometric order). Where the grains meet is what we call a **grain boundary**.

Fig. 3.15 2D schematic of four grains, all with the same internal atomic arrangement but oriented randomly to one another

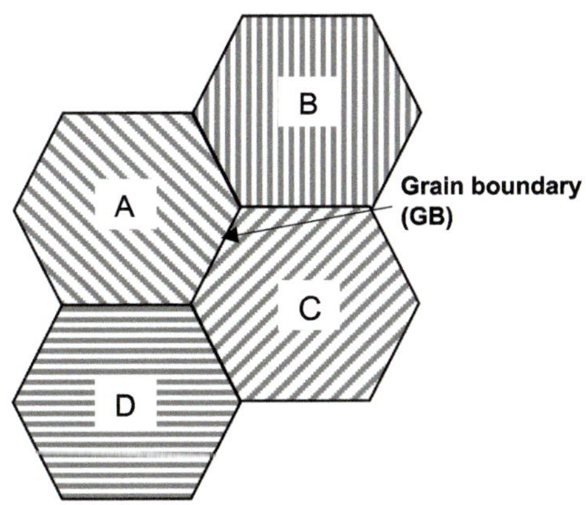

Fig. 3.16 Schematic
showing a grain boundary
between two parts of grains.
Inside each grain atoms are
seen with long-range order,
but inside the grain boundary
a random arrangement and
poor packing of atoms is
seen

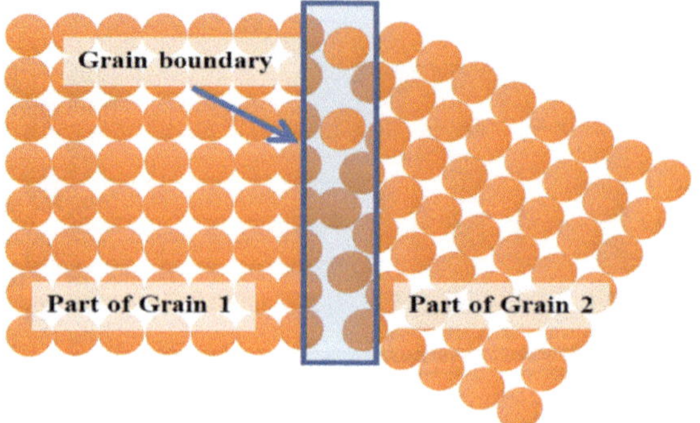

Although the atoms are arranged in an ordered fashion inside the grain, at the interface between the grains, you can imagine that the atoms from neighboring grains do not align perfectly when they meet. If they did, they would join to become one larger grain with the same 3D long-range atomic order. However, since the grains are randomly orientated, when they do meet, the atoms at the interface are randomly positioned and not well packed, unlike those within the grains, as shown in Fig. 3.16. The arrangement of atoms at grain boundaries gives the grain boundaries special characteristics that we will discuss later.

3.3.1 Insights into Polycrystalline Materials

The atoms inside the grain of a metal are very well packed. Each atom is surrounded by and bonded to many other atoms. We say the bonding of these atoms is satisfied. Now, considering the atoms at the grain boundaries (Fig. 3.16), we can see that the situation is different. The atoms there are not surrounded by the same number of atoms as the ones inside the grain (i.e., they are not packed as well), so their bonding is not as satisfied.

Not satisfied means these atoms want to bond but can't; this will make them more reactive and energetic than the ones inside the grain. So, are we saying that the grain boundaries possess energy because of that? Absolutely, yes! We refer to this as **grain boundary energy**, typically on the order of 0.5 J/m^2. As you can imagine, the more misorientation between grains 1 and 2 in Fig. 3.16, the less packed the atoms at the grain boundaries, and the less satisfied the bonding of these atoms. This makes these atoms and the grain boundary more energetic and reactive. If the misalignment between two connecting grains is small (say < 15°), we would call the boundary a **low-angle grain boundary**, and it would have energy but not as much as when the two grains are more out of alignment (> 15°). In this latter case, the boundary would be called a **high-angle grain boundary**.

So, how does this information help us? Well, if the material is made up of these 3D crystals or grains held together (very strongly, I might add) at grain boundaries, it may be prudent to find out what the average size of these grains is. Couldn't these grains be smaller in one material than another? Also, didn't we say that structure affects properties? A material with a small average grain size will behave differently from the same material with a larger grain size. Hence, we need to be able to measure the size of these grains. But before we can measure the grain size, we need to see the grain boundaries (i.e., the borders of these grains). They will not simply show up for us under a light microscope; hence, we cannot see where a grain ends and another one starts.

Here, we take advantage of the fact that the atoms at grain boundaries are more energetic and reactive than those inside the grains. That means they can be easily removed, for example, by heat or chemical attack. You see if we manage to remove the atoms at grain boundaries, then we will leave a groove in their place. It's as if we generated a microscopic trench around each grain, which reveals the boundaries of each grain and, hence, the grain size. The process of removing grain boundary atoms is called etching. **Etching** can be carried out chemically (**chemical etching**) by dipping the polished sample in an etchant of prescribed composition or by **thermal etching**, which typically is carried out by heating the polished sample to a temperature ~ 100 °C lower than the temperature used to manufacture it in the first place.

3.3.2   Grain Size Analysis

At this point, we need to think about how to measure a material's grain size. Typically, we would section the sample, then grind and polish the sectioned surface, and, in the process, remove scratches or irregular features on the surface. Afterward, if the material has no holes or pores, we will see a plain barren surface under the light microscope, with no sign of a grain or a grain boundary. Then, we take the sample and dip it into a chemical etchant for a while. The sample is then removed, washed, and dried. When we look under the microscope, we will see grooves where the grain *boundaries* are, through which we can determine the grain size. Figure 3.17 (side view) shows how the light rays are deflected away at the grooves of the grain boundaries and, hence, cannot reach the microscope lens, making them appear black in the image. While other light rays reflected back to the lens from the sample's surface will appear bright. We need to be careful, though, in our interpretation. Remember, we sectioned the 3D arrangement of grains, i.e., the image we are looking at (in Fig. 3.17 (top view)) is only a cross section of the sample showing a 2D representation of the grains. Typically, when the average size of the grains is measured in 2D, it is multiplied by a factor to convert it to a **3D representation of grain size**. Although the factor can vary, an approximate factor of 1.6 (sometimes referred to as the **Mendelson factor**) is typically quoted. This means that whatever grain size is measured from a 2D image needs to be multiplied by 1.6 to give the 3D grain size. However, let's first see how to measure the grain size from an image of the 2D cross section.

Published ASTM (American Society for Testing and Materials) standards explain how to measure grain size according to the **linear intercept method**. When we take a photographic image (at a particular magnification) of the etched microstructure,

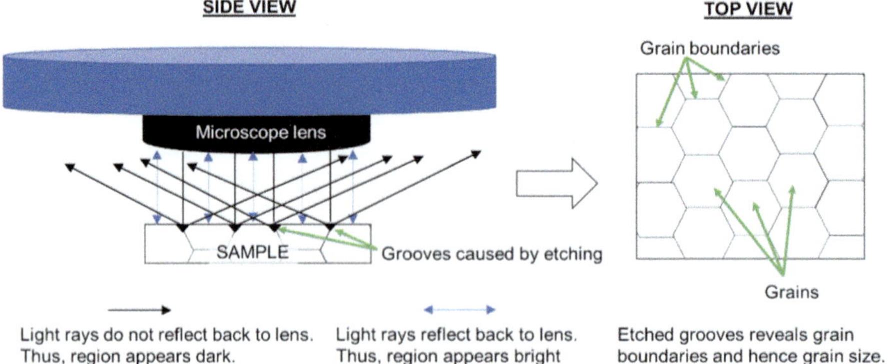

Fig. 3.17 Microscope observing polished and etched sample surface, showing how light rays are deflected at etched grain boundaries (grooves) thus revealing grain boundaries and grains

the image produced is called a **micrograph**. We then draw random lines on the micrograph and measure their total lengths (considering the magnification used to take it first).

There are two ways to approach this, which have their own rules. Using lines of known length drawn randomly within the planar 2D plane of the microstructure, we can count two things:

1. **Grain Intercept Method**

When a line passes through a *grain*, it is said that there is a line segment that lies on the grain. Here, we count the number of "grains" the line completely crosses. If the line starts or ends inside a grain, each will be counted as ½ of an intercept. Hence, the line in Fig. 3.18a has a total of $0.5 + 1 + 0.5 = 2$ intercepts. You could imagine that we would need the line or a series of randomly drawn lines to cross many grains to obtain accurate results. The ASTM E112–13 standard requires crossing at least 50 grains.

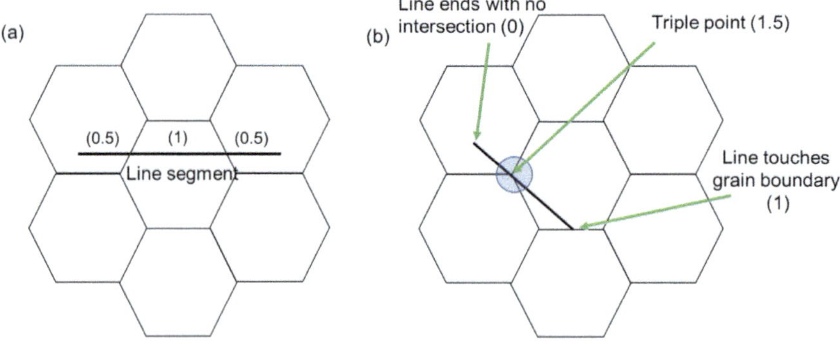

Fig. 3.18 Depiction of **a** Grain intercept method, **b** Grain boundary intercept method

2. Grain Boundary Intercept Method

Here, the line is drawn randomly in the same way, but this time we count how many times the line has been intersected by grain boundaries. The rules here are that if the line ends inside a grain, it does not count for anything, since a grain boundary does not intersect that line ending. If the line ends touching a grain boundary, then it would count as one intersection at that location. Also, if the line is tangent to a grain boundary, it counts as one intersection, while if it goes through a triple point junction, it will count as 1.5 intersections. Figure 3.18b clarifies these rules, in which we can count $0 + 1.5 + 1 = 2.5$ intersections.

Whether we are counting the number of grain intercepts (GI) or the grain boundary intercepts(GBI), the mean grain size is determined by dividing the actual length of the line (considering the magnification of the image) by either GI or GBI, as depicted in Eq. 3.3a and b given below:

$$\text{Mean grain size} \; = \; \frac{\text{total length of line}}{\text{total number of grain intercepts}} \tag{3.3a}$$

Or

$$\text{Mean grain size} = \frac{\text{total length of line}}{\text{total number of grain boundary intercepts}} \tag{3.3b}$$

Figure 3.19 shows a rough sketch of the microstructure of a hypothetical polycrystalline material after a supposedly typical etching procedure to reveal grain boundaries. Note that the length of the scale bar at the bottom right is 100 μm, which helps to judge the actual size of the microstructure. Also, if we compare the lengths of the randomly placed lines (all equal lengths) to the scale bar, the unmagnified real lengths of the lines can be determined. As such, the length of each line appears to be three times greater than the scale bar length of 100 μm. This means each line is 300 μm in length.

- **Grain Intercept Method**

> 7 horizontal lines (each $300\,\mu\text{m}$).
>
> Total line length $= 7 \times 300 = 2100\,\mu\text{m}$.
>
> Total number of intercepts $= 11 + 8 + 9 + 12 + 8 + 7 + 9 = 64$.
>
> Mean grain size $= 2100/64 = 32.81\,\mu\text{m}$.

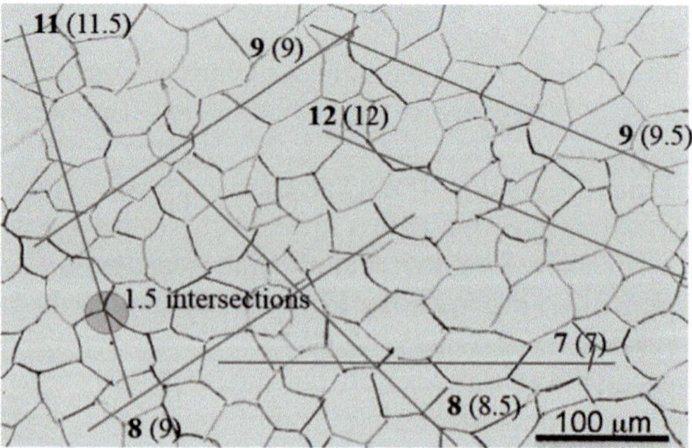

Fig. 3.19 Rough sketch of hypothetical polycrystalline microstructure, and randomly drawn lines of equal lengths used to measure the mean grain size, with bold numbers representing the number of grain intercepts, and numbers in parentheses represent number of grain boundary interceptions

- **Grain Boundary Intersection Method**

  7 horizontal lines (each $300\,\mu$m).

  Total line length $= 7 \times 300 = 2100\,\mu$m.

  Total number of intersections $= 11.5 + 9 + 12 + 9.5 + 7 + 8.5 + 9 = 66.5$.

  Mean grain size $= 2100/66.5 = 31.58\,\mu$m.

So, you see, both grain sizes are around 32 μm.

*The Sphere
Layer 2*

Alex

Professor, I polished sample 2 and etched it using a 50/50 distilled water and nitric acid solution. I found that several people had recommended this etchant recipe. I then measured the grain size using the linear intercept method and found that the average grain size was 220 μm. I also observed the presence of black spots in the microstructure, representing pores. More importantly, it shows unusual, irregular features in the microstructure.

Let me have a look. Ah, these features confirm the microstructure's dendritic nature. Dendrites are a telltale sign that the material was cast

What are dendrites, Prof.? Also, what is so special about the cast structure? How does it develop?

*A prevalent way of making metal structures is through casting. The metal is heated above its melting point and poured into a mold. If the mold has the shape of the final component, then after the metal solidifies into its intended shape, it will be called a metal **casting**. People often pour the molten metal into a simple cylindrical mold cavity so that, after it solidifies, it forms a simple cylindrical block of metal, an **ingot**. See Fig. 3.20 in As a Matter of Fact.*

As you explained, I can understand the importance of having a mold with a cavity shaped like the final component to be manufactured to produce a casting. However, I fail to understand what use a metal block, such as the ingot, would be.

It's not about the block itself but what you do with it. This ingot can now be processed further using different manufacturing processes. In other words, you can heat the ingot again, forge it, extrude it, or roll it into a different shape. Forging, extrusion, and rolling would then be called secondary manufacturing processes.

Look at Fig. 3.21 in As Matter of Fact, which shows an open die forging process schematic. Here, the ingot is first heated to a relatively high temperature (below its melting point), and a forging stroke is applied, as can be indicated by the white arrow. As the ingot is being forged, it tries to flow radially. However, the friction between the workpiece (i.e., ingot) and the die material resists radial material flow in that location. However, regions of the ingot away from the die/workpiece interface can easily flow (assuming the metal is malleable), giving rise to the bulging seen in the figure. This bulging, if not controlled, can lead to cracking. High-temperature lubricants such as graphite are used at the interface between the workpiece and die interface to reduce friction and bulging.

Now, let's return to the question of dendrites. Go to Sect. 3.4 on Casting and Dendrites in As a Matter of Fact, and you should find answers to your questions there.

3.4 Casting and Dendrites

Metals and alloys can be formed by casting (Fig. 3.20) or can be produced by thermomechanical processes such as forging (Fig. 3.21). With atoms and molecules, we often talk about **internal energy**, which consists of **kinetic** and **potential energy**. Potential energy depends very much on the bonding between atoms and molecules; the stronger the bond, the lower the potential energy, and the closer the atoms are. In the case of kinetic energy, however, it depends on different things depending on the state of matter. For a solid, it refers to the vibrational energy of the atoms. For liquids, it refers to the molecules' or atoms' vibrational energy, translational energy, and rotational energy. For gases, it relates to atoms' or molecules' translational and rotational energy. Figure 3.22 summarizes this.

Now, one could ask how the matter decides which state to occupy? Solid, liquid, or gas under a particular temperature and pressure? The answer depends on which state has the least energy under a specific temperature and pressure. Specifically, which one has the least **Gibbs free energy (G)**. Gibbs free energy refers to the energy (at constant temperature and pressure) in the system that is available to do work. It will want to do any process that the material can undergo to reduce its Gibbs free energy. This has significant implications for things like equilibrium, stable configurations, and states of matter.

Fig. 3.20 Molten metal poured into crucible (image generated using ChatGPT 4.0 by K. Morsi)

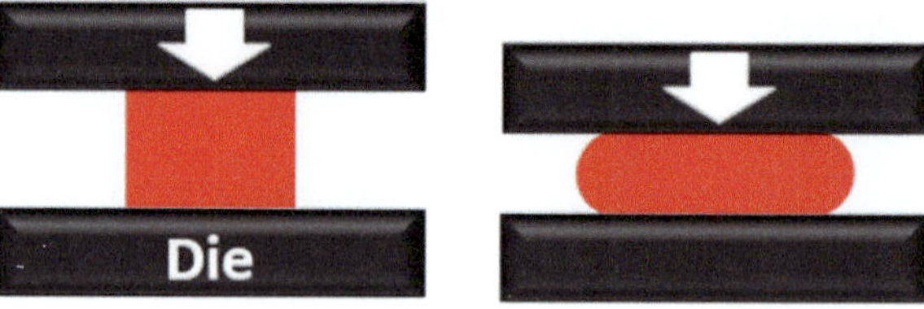

Fig. 3.21 Schematic of open die forging of a metal ingot

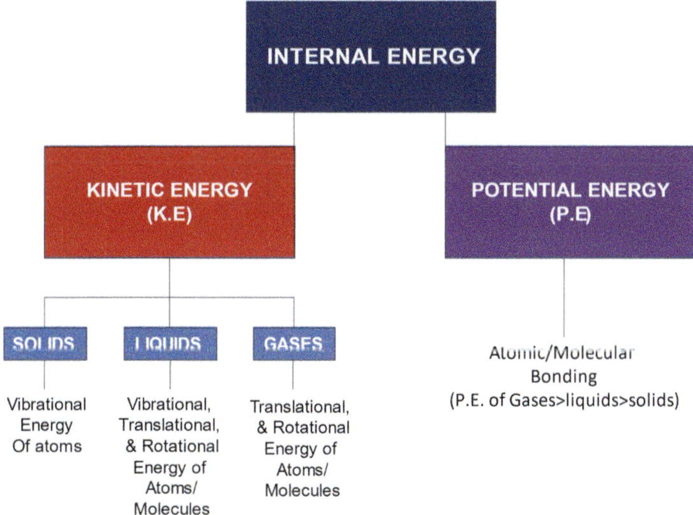

Fig. 3.22 Kinetic and potential energy origins in matter

The Gibbs free energy (G) for solids and liquids is defined by Eq. 3.4.

$$G = H - TS \tag{3.4}$$

where H is the enthalpy, which can be approximated to the Internal Energy (E) for solids and liquids (i.e., $H \cong E$), T is the temperature in Kelvins, and S is the entropy representing the degree of disorder or randomness. Gases have more disorder than liquids, and liquids have more disorder than solids. While internal energy is expected to increase with an increase in temperature due to an increase in thermal energy and potential energy, so does entropy, and as temperature increases, the "-TS" term in the equation becomes more and more dominant. As a result, the Gibbs free energy of matter changes with a temperature change. Figure 3.23 shows a schematic of the Gibbs free energy versus temperature for a molten metal (liquid) and its solid metal. As seen in both cases, the Gibbs free energy decreases with an increase in temperature; however, it does so at different rates. They meet at some point, coinciding with the melting or freezing point of the metal (T_m) (on the x-axis). At this point, the Gibbs free energy of molten and solid metals is equal. Hence, at T_m, both liquid and solid metals are stable and, therefore, can coexist in equilibrium. You can see this in real life when we heat a solid metal to its melting point; initially, the material is 100% solid, then instead of the temperature continuing to rise above T_m, it remains constant while the energy input is used to break some of the bonds in the solid and transforms the material there to liquid, and as time progresses, more liquid is formed until the material becomes 100% liquid, after which the temperature starts to rise again. During the initial period, when the temperature is constant, both

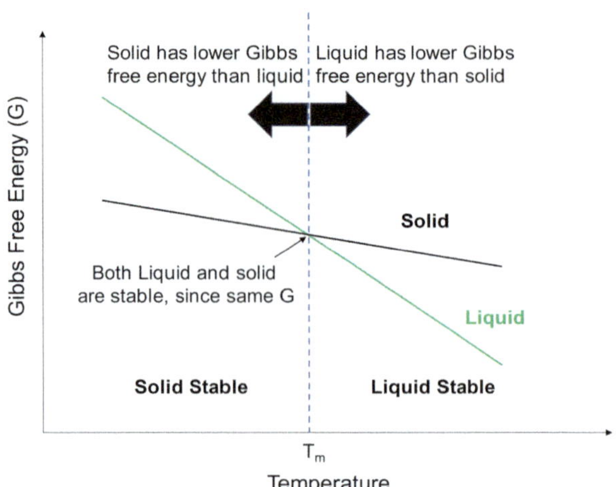

Fig. 3.23 Gibbs free energy versus temperature for a molten metal and a solid metal

liquid and solid exist and are said to be in equilibrium and are stable forms of matter. However, at temperatures below T_m, Fig. 3.23 shows that the Gibbs free energy of the solid is lower than that of the liquid. Hence, the solid becomes the favored and stable state (i.e., the one that would have the least energy). Likewise, above T_m, the liquid becomes the equilibrium state since it has a lower energy than the solid counterpart. That is how matter selects its state. You will see this energy-reducing phenomenon in many areas of materials, whether we are talking about grain growth during high-temperature annealing or when a material fractures. Remarkable: if only humans were as energy conscious as materials, what a wonderful world this would be.

Let's look more closely at matter in the gaseous state. When they vaporize, all metallic elements exist as single atoms in a gas, not molecules. In this state, atoms have considerable potential energy. Now, as energy is lost to the surroundings during cooling, the average kinetic energy of the atoms is lowered. This is only an average energy, since atoms can collide with each other and will, hence, individually possess different kinetic energies. So, when the average kinetic energy of the gas is reduced, some atoms (with lower energy than the average) will get closer to each other and will not be able to resist the van der Waals forces pulling them together. Thus, a liquid with a lower energy than the gas would eventually be formed. As the temperature is reduced further, at some point, the liquid solidifies as the primary bonds of the metal take over. Figure 3.24 shows the schematic of the atomic arrangements for metal gas, molten metal, and polycrystalline solid metal, displaying long-range atomic order and a grain boundary.

As the figure shows, potential energy (PE) decreases from gas to liquid to solid.

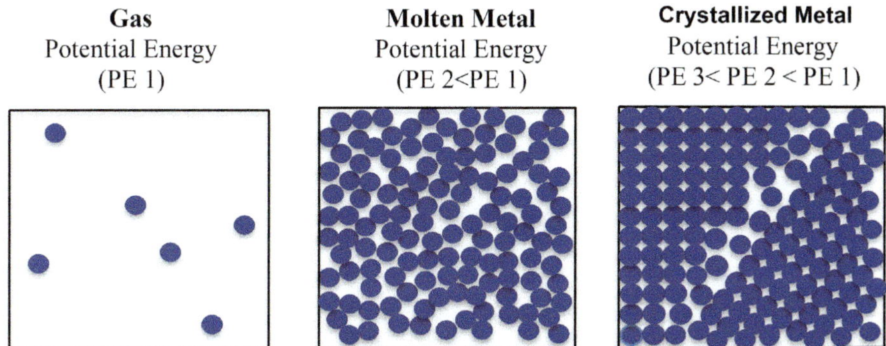

Fig. 3.24 Schematic of atoms, **a** in the gaseous state, **b** in the liquid (molten) state, **c** in the solid state (crystalline), showing a periodic arrangement of atoms separated by a grain boundary. PE 1, PE 2 and PE 3 are the potential energy of the gas, melt and crystallized solid respectively (based on multiple sources)

Okay, Alex. Aluminum's melting point is 660 °C, and its boiling point is 2467 °C. What do the boiling and melting points mean in terms of the Gibbs free energy?

At the boiling point, the Gibbs free energy of the metal gas and that of the molten metal are the same, and hence, they can co-exist in equilibrium at that temperature; however, if the temperature is increased beyond the boiling point, then the Gibbs free energy of the gas will be lower than that of the molten metal. Hence, matter prefers to be a gas rather than a liquid. The same can be concluded for the solid-to-liquid transformation at the melting point.

You seem confident, my friend, which is good. However, let me tell you something you did not know. If I asked you what the melting point of aluminum is, you would, in a heartbeat, tell me 660 °C, professor, and you would be right if you were referring to the melting point of aluminum at atmospheric pressure. Did you know that at high pressure, the melting point would increase? However, another more interesting fact is if we produced aluminum or any other material on the nano-scale, i.e., at sizes < 100 nm. Suddenly, the melting point of that material will depend on its particle size, decreasing with a decline in particle size. Crazy as it may seem, but the melting point of nanoscale aluminum can be room temperature.

- **Melting Point or Crystallization (Freezing)Temperature?**

As you heat a solid crystalline metal, there will come a temperature above which the solid metal becomes molten (i.e., an amorphous liquid). In other words, the atoms will gain thermal energy enough to break away from their primary bonds. This temperature is called the **melting point** of that metal. If the molten metal is cooled below its melting point, it crystallizes into a solid. Again, this means that the energy of the atoms is reduced because of the reduction in temperature, and the primary bonds start exerting their dominance again. That same temperature is now called the **crystallization (or freezing) temperature**. Hence, the same temperature has two different names depending on whether you are going from a crystalline solid to a molten metal (melting point) or from an amorphous molten metal to a crystalline solid (crystallization or freezing temperature).

Not all materials go through a crystallization temperature; in other words, not all amorphous molten materials become crystalline when they are cooled. Figure 3.25a is a cooling curve, showing exactly that: when the molten material cools, no distinct crystallization temperature is seen. In that case, as the material cools, the viscosity of the liquid will just become so high that the material will not flow like a liquid but will look and feel more like a solid. In this case, the solid material is hence termed **amorphous** since its atoms are not arranged in any long-range order. Metals can be produced today in the amorphous state through alloying and certain processing methods. These are called **metallic glasses**.

Why is there a reference to glass for these types of metals?

Since glass is amorphous, they named these metals as such. By the way, these metallic glasses have some extraordinary properties not observed by their crystalline counterpart.

Unlike amorphous materials that do not have a crystallization temperature, crystalline materials do. When a molten metal cools, it encounters a crystallization

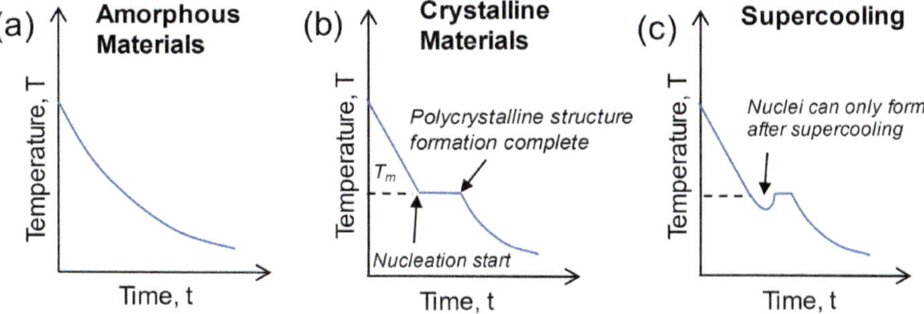

Fig. 3.25 Cooling curves for **a** an amorphous material that does not crystallize, **b** a metal that crystallizes, and **c** a metal that crystallizes after some supercooling (multiple sources)

temperature as it solidifies. The solidification process first involves forming many small nuclei that grow and eventually become grains. This means that while the material is a molten metal above the crystallization temperature, it is a polycrystalline material below it, e.g., Fig. 3.25b for an idealized solidification case. Hence, nucleation is a key factor in solidification.

There are two main types of nucleation processes: **homogeneous** and **heterogeneous nucleation**. In *homogeneous nucleation*, the molten metal needs to be cooled significantly below its crystallization temperature for nuclei to be thermodynamically able to form. Such a process is referred to as **supercooling** or **undercooling**. Before the formation of a given nucleus, several atoms from the molten metal form a **cluster** or an **embryo**, the growth of which leads to the formation of the nucleus. These clusters still form even in the molten metal (i.e., above the melting point), but they are highly unstable at these temperatures and break up, so they do not form a stable nucleus. It is important to note that for a nucleus to be stable, it needs to have a minimum size or radius (assuming, of course, a spherical nucleus geometry), which is called the **critical radius**. If a nucleus forms with a radius smaller than the critical radius, it would be unstable and tend to dissolve back into the molten liquid. Interestingly, a significant degree of supercooling is needed to produce a stable nucleus in homogeneous nucleation, Fig. 3.25c. For example, if a very pure molten nickel is cooled to 250 °C below its melting point, a stable nucleus will still not be able to form. As such, homogeneous nucleation is rarely encountered in real life, and instead, *heterogeneous* nucleation is the dominant mechanism. In *heterogeneous nucleation*, a stable nucleus is formed on preexisting impurities, intentionally added particles (called inoculants), or the surface of a casting mold wall. Since a surface is already present for the deposition of atoms, it reduces the energy needed to form a stable nucleus. Hence, the supercooling in heterogeneous nucleation is barely noticeable (e.g., ~ 1 °C). Figure 3.25b is a good *approximation* of the cooling curve for heterogeneous nucleation.

Alex, could you tell me the freezing temperature of pure water? At what temperature does pure water become ice?

That's easy: 0 °C.

Alex, I chose my words carefully. I asked about "pure" water. I agree that if you ask anyone, they will say 0 °C, but that is because water solidifies via *heterogeneous nucleation; you see, it's not highly pure, and there will always be impurities for ice to crystallize on. However, if you had purified the water to extreme levels, you would not see ice forming even if the temperature was -20 °C. You know, you will need a very large supercooling to achieve homogeneous nucleation of very pure ice. You may think this kind of thing is irrelevant, but it plays a significant role in whether or not we get rain. I will leave it to your inquisitive mind to research this one!*

As mentioned, nucleation involves a small collection of atoms bonding in the same basic atomic arrangement of the intended material. During casting, some impurities may inevitably be present. These impurities or the surface of the inner wall of the mold may provide good nucleation sites for heterogeneous nucleation.

Ideally, in casting, tiny seeds or nuclei form in many parts of the melt volume. As time progresses, the nuclei grow at the expense of the liquid, i.e., the atoms from the amorphous liquid start to condense on the nuclei and continue building the long-range atomic order on each nucleus. All nuclei will have the same atomic periodic arrangement. The only difference is that the atomic arrangement is oriented differently from one nucleus to the other; in fact, they are randomly oriented to each other. The nuclei grow continuously until all the liquid is consumed, and they grow to the extent where they start touching each other at grain boundaries. Hence, the initial nuclei have now become **grains** or **crystals** of polycrystalline materials. This is depicted in Fig. 3.26.

Understanding the solidification behavior of materials is important for processes such as casting and single-crystal manufacture.

Now, let us think about what happens when we cast molten metal. In fact, let's direct our focus on metals being cast into an ingot rather than a casting. Here, the mold plays a very important role. Molds can either be made of metals or a material that does not conduct heat as well, such as sand. The microstructure will differ depending on how well the mold conducts heat away from the molten metal when poured inside. Under high cooling rates, equilibrium conditions are not established. When considering something like this, we need to understand how solid nuclei grow within a solidifying melt under non-equilibrium conditions, i.e., solidification that is not too slow. When such a condition occurs, atoms are deposited, and each nucleus grows into cooler melt regions. As this happens, the formation of new solid material locally releases the heat of fusion and raises the temperature in that location. Hence, the continued growth of the nucleus must now occur in a different direction.

Specifically, a perpendicular direction to the initial growth into relatively cooler melt regions, and so on, such that branches start to form, and from these branches,

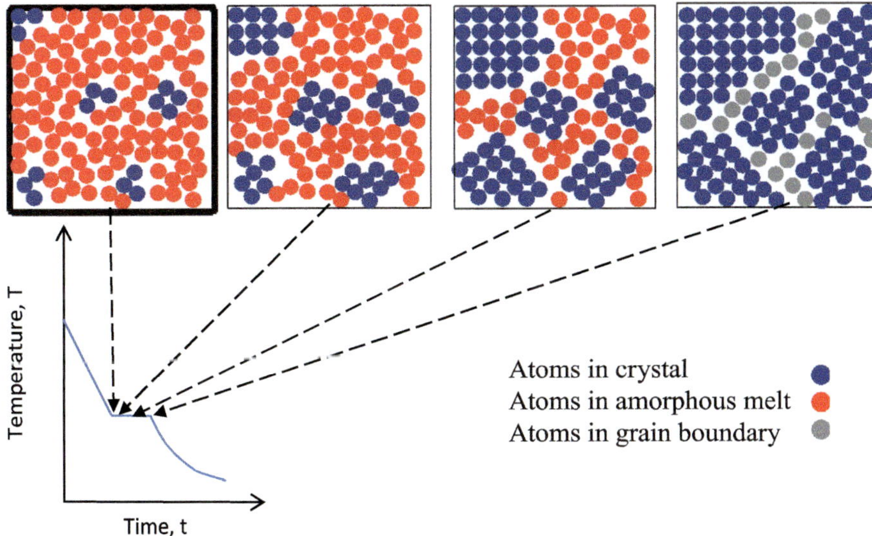

Fig. 3.26 Cooling curve showing amorphous atomic structure gradually changing to become crystalline. Dirt particles have not been included for clarity (multiple sources)

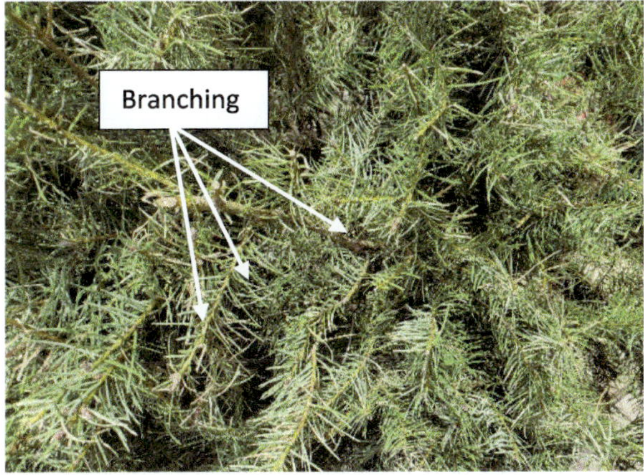

Fig. 3.27 Part of fir-tree showing secondary and tertiary branches (photography: K. Morsi)

other tertiary branches form, such that a fir-tree-like structure of the growing nucleus evolves, which is called a **dendrite**. The word is derived from the Greek dendron, which means a tree (hence, the structure resembles a trunk, branches, and twigs). Figure 3.27 shows a part of a fir-tree with secondary and tertiary branches.

Figure 3.28 shows how the growth of a dendrite evolves, with the formation of secondary and tertiary arms or branches. Of course, we can imagine that there would be many nuclei within the melt. When a given nucleus grows, there will come a point when the extending arms meet the arms of other growing nuclei. At this point, the arm stops its outward growth and thickens instead. The continuation of these processes leads to an equiaxed grain structure showing no signs of initial dendritic growth. However, sometimes, in casting processes, when there is a lack of liquid in some local regions of the solidifying melt, inter-dendritic pores develop due to shrinkage. These pores or shrinkage cavities give away the original dendritic structure. These are depicted in Fig. 3.28a and b.

Also, if an alloy is cast, impurity atoms may appear in relatively high concentrations on the outskirts of dendrites, again revealing the microstructure's dendritic origin.

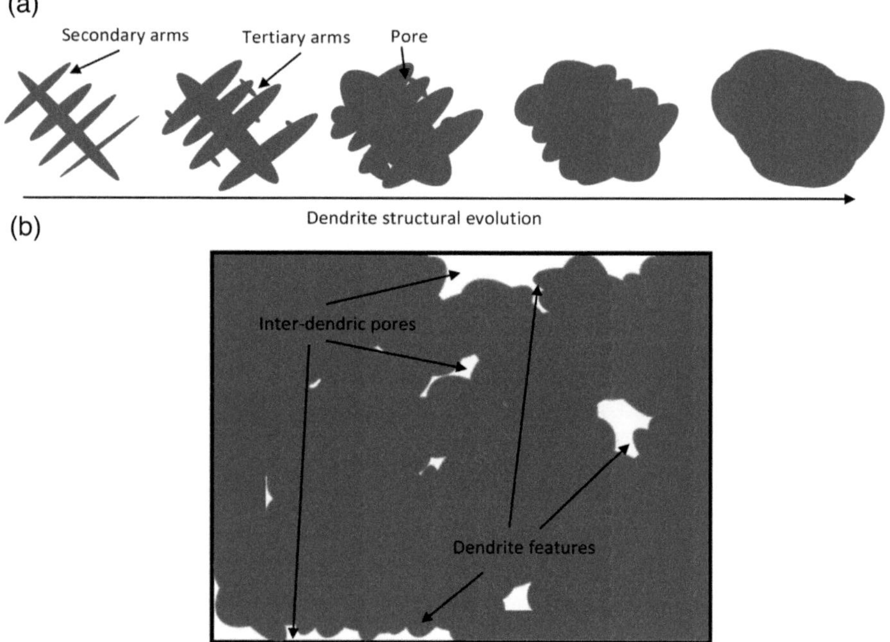

Fig. 3.28 **a** Schematic of dendrite growth showing secondary and tertiary arms, and **b** part of a microstructure showing evidence of dendritic growth, and inter-dendritic porosity

The Sphere
Layer 2 Identified

Professor, the microstructure in Fig. 3.28 looks very similar to that of the layer 2 sample. Do we need to conduct more tests?

Thanks, Alex. Let's see what information we have on layer 2 so far.

Density: ~ 8.87 g/cm^3

Composition: Copper (94.5%, Arsenic (4.5%), other elements including iron, zinc, and nickel (total 1%)

Microstructure: Dendritic origin

Grain size: 220 μm.

Structure: Crystalline.

Alex, your hard work paid off; we can now identify layer 2, but ensure you are seated first. **Sample 2** *is from a material not manufactured in our modern time; it is much, much older. The material is from the Chalcolithic period, or what is known as the Copper Age, the age that immediately preceded the Bronze Age. We are talking about a material over 5000 years old, Alex. You see, mankind has been dealing with metals for a long time; the smelting of ores has brought copper (Cu) to the forefront. The ores from which copper was smelted contained arsenic (As), which ended up in the copper product. As you can imagine, arsenic was not the healthiest choice; imagine the fumes during smelting. The replacement of As with tin (Sn) became prominent later on, giving rise to the Bronze Age. Bronze is a copper-tin alloy*

The compositional analysis using EDS showed that the alloy contained copper as the major element, but also 4.5% Arsenic. No one uses 'As' to that level (of 4.5%) today, which gave me a clue about its ancient origins. Moreover, the microstructure clearly showed the dendritic origin of the alloy, confirming that it was cast. They used sand/clay molds back then to cast artifacts, which imposed slow cooling rates on the molten metal as it solidified, so the grain size was large. Additionally, the X-ray Diffraction results confirmed that the material is crystalline

But why would this particular material be included in the Sphere? The only conclusion I could make is that during this Chalcolithic period, smelting began. Smelting was not only the reason for the Copper Age, but the general principles of this process would go on to produce the Bronze and the Iron Ages. What an impact!

This is fantastic. I can't believe it. You did mention the word "alloys," though; I've heard this word before, but what does it mean?"

Haven't you heard of 14-karat, 18-karat, or 21-karat gold (Fig. S3.1)? These are also called alloys. Alloys are when you mix two or more elements <u>on the atomic level</u>. In this case, 14-karat means you have 14/24 of the material weight made up of gold atoms; the rest could be another element, for example, copper atoms. People do that to ensure that the gold possesses strength; if the gold is 24-karat, i.e.,24/24, 100% of the material is gold, it will be soft and easily deformed if used as jewelry. People, therefore, add impurity atoms to gold to strengthen it

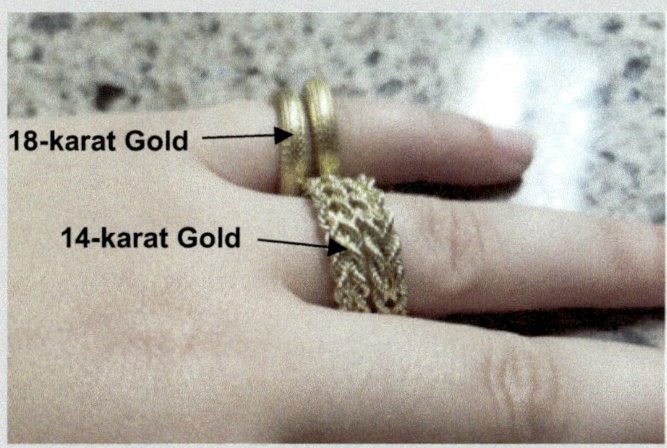

Fig. S3.1 18-karat and 14-karat rings (*photography: K.Morsi*)

But how does this work? Why does gold get stronger?

I will tell you all about this on another day; knowledge is vast, my young scientist; even with all my decades of experience, my knowledge is insignificant to the vast knowledge out there. But we must not let a day pass without becoming more knowledgeable than the day before. You see, we are climbing the mountain of knowledge; it will take hard work and patience to reach higher and higher heights. Only then can one benefit other seekers of knowledge and mankind in general (Fig. S3.2). Don't worry, I will tell you about it; your desire to learn is infectious!

Fig. S3.2 When you know more you can help others to climb the mountain of knowledge (*image generated using ChatGPT 4.0 by K.Morsi*)

Isn't it exciting how we found out the identity of this mysterious layer 2? Keep thinking about layer 3 until we meet again. I want you to maintain your voracious appetite for knowledge

Problems

3.1. The grain size of a material was determined to be 0.099 μm. Would we classify this material as nanostructured, microstructured, or macrostructured?

3.2. What is the difference between the optical (light) microscope and the stereomicroscope?

3.3. What is the difference between a scanning electron microscope and a transmission electron microscope?

3.4. Discuss the difference between secondary electrons and backscattered electrons within the context of scanning electron microscopy.

3.5. Determine the wavelength of electrons in an electron beam under an accelerating voltage of 10 keV.

3.6. What is the useful magnification of a stereomicroscope?

3.7. State the advantages of a scanning electron microscope.

3.8. What is the electron beam- material interaction volume (also sketch it)?

3.9. In what way can using the scanning electron microscope in the backscattered mode be useful?

3.10. Explain X-ray microanalysis and its limitations.

3.11. What are Auger electrons, and what are they used for?

3.12. With the aid of diagrams, discuss the difference between a tunneling electron microscope and an atomic force microscope.

3.13. When preparing a material surface for microstructural observation, do you grind the surface with progressively higher or lower grit sizes? And why?

3.14. If the misalignment at grain boundaries is 10°, would you classify this grain boundary as a high-angle or low-angle grain boundary?

3.15. Which one has more energy, a low-angle or high-angle grain boundary, and why?

3.16. Determine the average grain size of the austenitic steel microstructure shown in Fig. C3.1. using the linear intercept method (grain intercept approach).

3.17. Determine the average grain size of the austenitic steel microstructure shown in Fig. C3.1, using the linear intercept method (grain boundary intercept approach).

3.18. A grain size of 100 μm was measured from a 2D image of a microstructure. Since grains are 3D entities, what 3D grain diameter does the 2D 100 μm grain size represent?

3.19. Sketch the cooling curves (temperature–time curves) for a metal undergoing homogeneous nucleation and one undergoing heterogeneous nucleation during casting.

3.20. Sketch the cooling curve (from the melt) of an amorphous material. What noticeable feature is missing from it?

3.21. What is undercooling in solidification? Use plots as appropriate.

3.22. What are dendrites, and how do they develop?

Fig. C3.1 Microstructure of an austenitic steel (Image from: Dong Xu, Cheng Ji, Hongyang Zhao, Dongying Ju & Miaoyong Zhu, A new study on the growth behavior of austenite grains during heating processes, Scientific Reports, 7: 3968 (2017). https://doi.org/10.1038/s41 598-017-04371-8. Reproduced with modification under the terms and conditions of http://cre ativecommons.org/licenses/ by/4.0/)

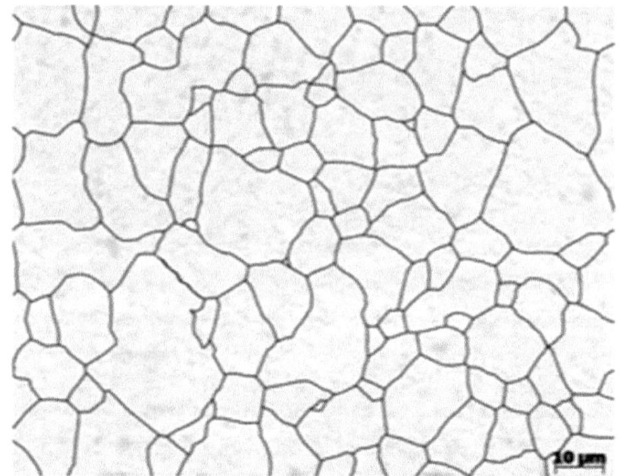

3.23. Above the melting point, which one has a lower Gibbs free energy, molten or solid metal?

3.24. What is an alloy?

3.25. At a temperature T, the Gibbs free energy of phase A is lower than that of phase B, which one is thermodynamically more favorable to form?

Part II
Layer 3: The Search Continues

Layer 3: The search continues.

Layer 3 seems interesting. It is silverish light gray, and after sectioning and polishing it, no pores are observed under the microscope. Density measurements of the sample place it around 2.85 g/cm^3.

It is not iron or a ferrous alloy (which has a much higher density, ~ 7.8 g/cm^3). The density makes me think it could be aluminum (since it is closer to the density of aluminum, 2.7 g/cm^3). Still, I feel I should be a better scientist than to hastily conclude so early. Who knows? Even if it was aluminum, couldn't it be an aluminum alloy? I guess more tests are needed.

I agree with you, Alex. Why don't you learn more about aluminum and its alloys before proceeding? If anything, even if the layer does not turn out to be aluminum, you would have learned something about one of the most important materials today. First, go to "As a Matter of Fact", but this time, Chap. 4: Atomic Arrangements. Learn about the repeated unit cell and what it can tell us; then, you can delve into lightweight metals and alloys (including aluminum) in the chapter right after it. Remember, build your knowledge bank slowly but surely!

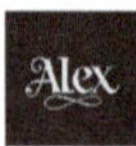

Sure, but before I go, Professor, is it aluminum or aluminium?

This one has an interesting story; I will tell it since you asked. You see, aluminum or aluminium as an atom has existed for millions of years as part of an ore called bauxite. It is the third most abundant element in the Earth's crust, behind silicon and oxygen, and just above iron. Additionally, it is the most abundant "metal" in the Earth's crust.
*The story of alumi**nium**, alumi**num**, or alum**ium** started with the British Scientist **Sir Humphry Davy**, who was trying to extract metals from their compounds, for example, aluminum from alumina (Al_2O_3), unsuccessfully, though, as he points out. In 1808, at the Royal Society, he presented "Electrochemical Researches of the Decomposition of the Earths." In his own words, Sir Davy explains," Had I been so fortunate as to have obtained more certain evidences on this subject, and to have procured the metallic substances I was in search of, I should have proposed for them the names silicium, **alumium**, zirconium, and glucium". You can see that here he refers to it for the first time as alum**ium**. He later changed it to aluminum; however, another scientist proposed the name aluminium during that time, so the division began. Now, the US and Canada are the aluminums, and the British and generally the rest of the world are the aluminiums. Funny story! Since we live in the United States, I will be using aluminum. The first to successfully extract aluminum cost-effectively were **Hall and Herout** in 1886, using an electrolysis process to extract aluminum effectively from one of its oxides, alumina (Al_2O_3). Alumina, however, also needed to be extracted cost-effectively from the Bauxite ore. A year or two after the extraction of aluminum, Bayer established a cost-effective process for extracting alumina from bauxite. Hence, the stage was set for the success of aluminum since it can now be produced cost-effectively. Did you know that before aluminum was produced cost-effectively, it was treated as a precious metal? It was said that **Napoleon III** used aluminum utensils and plates in state dinners while using silver and gold for less important guests. From the beginning of the twentieth century, life became good for aluminum; the Wright brothers used it in 1903 in their first biplane. In 1913, we had aluminum foil on the market in the US. Now, the metal is used in many applications, including power transmission lines, airplanes, laptops, iPhones, and iPads. With the present surge towards electric vehicles, aluminum is set to gain even more importance due to its lightweight. Ok, that's enough; now go to* Chap. 4 & 5 in As a Matter of Fact *and learn more about atomic arrangements and this exciting material called aluminum, and then we'll talk again. I must go to my lab and meet with my graduate students; we have a research paper due.*

Professor, I have always wondered what a professor's exact job is.

A professor can teach or carry out other activities on top of teaching. For example, tenure-track professors in the US are required to do two more things besides teaching. One is service, which means conducting university business, like participating in hiring committees for recruiting new professors or other committees dealing with the department, college, and the university. For example, they can participate in committees evaluating student scholarship applications. Conducting cutting-edge research is the third job requirement. Remember, without research, we would not have been able to write textbooks. This is why I have included stories about the research endeavors of some prominent historical figures who have contributed tremendously to our current understanding through their research. Nowadays, a professor typically needs to acquire research grants by proposing great research ideas. These grants are highly competitive to get, as you are competing with the best professors in the nation. It could also take months to write such grants, and even up to 6 months to hear back about the decision of the funding agency. Often, by the time you hear back, you will have forgotten what idea you proposed in the first place ☺. If you successfully acquire a grant, you hire graduate students and conduct the research. Then, you publish your results in prestigious scientific journals so that you can disseminate your research findings to the scientific community. So, the professor who teaches you is also a scholar in his or her field of research and may be very well-known internationally. Many undergraduates also engage in undergraduate research, which I also highly recommend. Alex, professors typically have a tremendous amount of knowledge that students need to take advantage of. I keep telling students not to waste office hours. Go and seek help and knowledge from your professor. He or she will be highly equipped to enlighten you! Unfortunately, most students don't take advantage of this. What a shame! I'm sure you would, though, right?

Of course, Professor. Why should I waste hours trying to understand a topic when a professor can give it to me on an aluminum platter ☺?

Good one Alex :)) By the way, offering office hours to our students is part of our job! Now that you know our job, I hope you will be compassionate and understanding toward us when we appear to be busy! For now, I have to leave and join my grad students. Talk to you later.

Chapter 4
Atomic Arrangements

4.1 Introduction

This chapter discusses atomic arrangements within crystalline materials. The basic science of crystallography is also discussed.

4.2 The Repeated Unit Cell

In the early 1900s, **Max von Laue** discovered that crystals could diffract X-rays, and like visible light, X-rays also had a wave nature. He received the Nobel Prize in physics in 1914. This discovery was quickly followed in 1913 by the pioneering work of the **Braggs** (father and son), which allowed the determination of crystal structures of crystalline materials. The Braggs received a Nobel Prize in 1915. Even the helical structure of DNA was determined much later through X-ray crystallography, landing a Nobel Prize in 1962, one of many for the field of X-ray crystallography, including those for 2D graphene, fullerenes, and quasicrystals.

Thanks to these groundbreaking efforts, we know that crystalline materials such as metals and many ceramics contain ordered long-range 3D arrangements of atoms, called lattices. Each lattice is constructed from a repeated building block of atomic arrangement called a **unit cell**.

A great deal can be learned from studying these unit cells. However, first we need to be able to visualize them, which can be done in two ways. One way is to draw the atoms within the unit cell as they should typically appear. In this case, we would call the model a **hard-sphere model** of the unit cell (e.g., Fig. 4.1a for a cubic unit cell). Note that the **actual unit cell** boundaries would be drawn with its corners at the center of the corner atoms of the unit cell. This means that most of the volume of each corner atom in Fig. 4.1a would reside outside the actual unit cell. Yet, such a hard-sphere unit cell model is helpful in several ways. First, it reveals which atoms

K. Morsi, *An Engaging Approach to the Science and Engineering of Materials*,
https://doi.org/10.1007/978-3-032-06231-4_4

are touching each other. Second, it can show the total number of atoms that fall within the boundaries of the unit cell; therefore, we could also determine the total volume of atoms residing inside the unit cell boundaries. This information can be useful, for example, it can help us determine the theoretical density of a material, which we will discuss later. Figures 4.1, 4.2 and 4.3 represent simple cubic, body centered cubic and face centered cubic unit cells.

Although the hard-sphere model is useful, it still has some deficiencies. For example, viewing all the atoms in the unit cell is impossible. For example, frontal atoms can block the view of the atoms behind them. Moreover, if we try to draw an internal vector within the volume of the unit cell, it becomes impossible to see. To overcome these issues, we redraw the unit cell with all the atoms reduced in size,

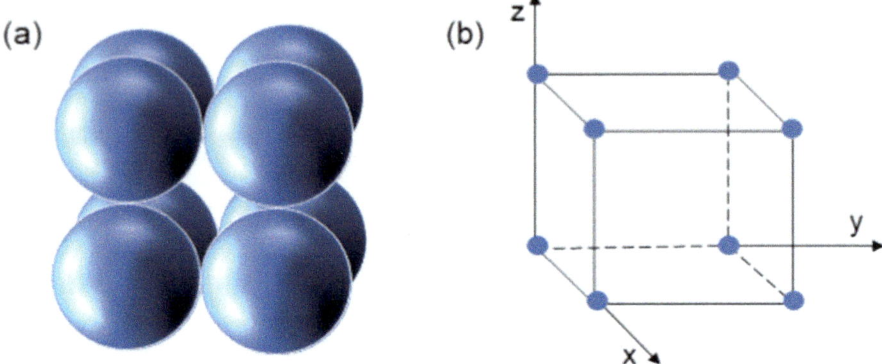

Fig. 4.1   Illustration of **a** Hard-sphere model (**imagine the unit cell being drawn as a cube with corners at the center of the corner atoms**) and **b** Reduced-sphere model of **simple (or primitive) cubic** unit Cell

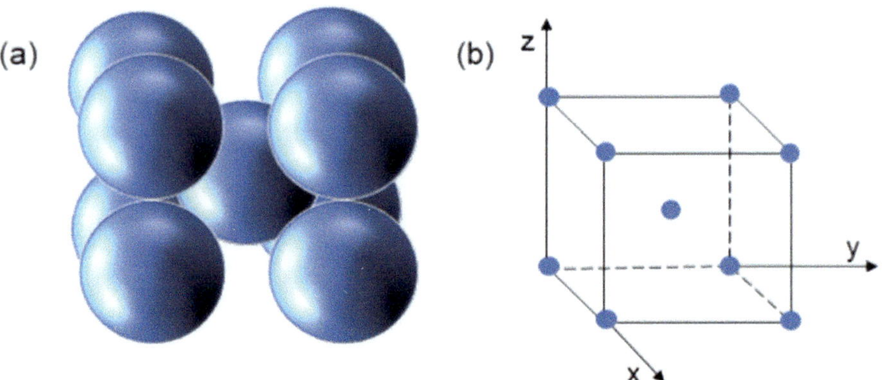

Fig. 4.2   Illustration of **a** Hard-sphere model (**imagine the unit cell being drawn as a cube with corners at the center of the corner atoms**) and **b** Reduced-sphere model of **body-centered cubic** unit cell

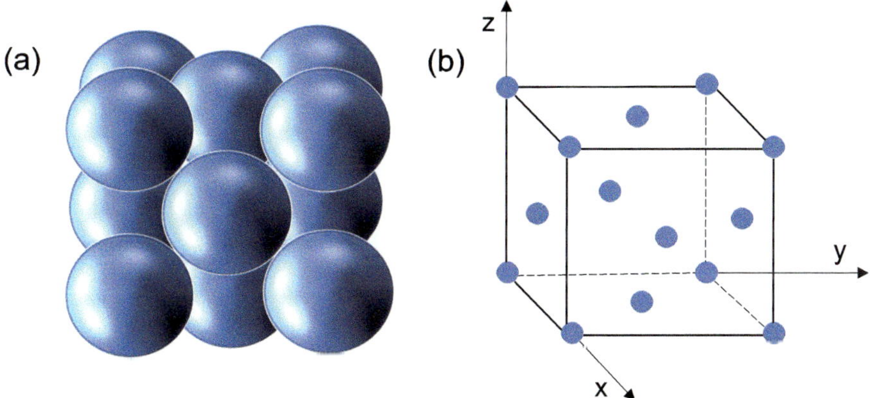

Fig. 4.3 Illustration of **a** Hard-sphere model (**imagine the unit cell being drawn as a cube with corners at the center of the corner atoms**) and **b** Reduced-sphere model of **face-centered cubic** unit cell

leading to the **reduced-sphere model**, which, although unrealistic regarding atomic (ionic) size, makes things easy to visualize, e.g., Fig. 4.1b.

Although unit cells do not have to be cubes, one of the simplest is a **simple** (or primitive) cubic unit cell. As seen in Fig. 4.1a, an atom is centered on each of its eight corners. The only element that has this crystal structure is polonium (Po). *Marie Curie* received the Nobel Prize in chemistry (1911) for extracting this radioactive element, in addition to radium. The simple cubic crystal structure can only generate poor atomic packing; hence, metals do not favor it.

If, in addition to the eight corner atoms, another atom was placed in the cube's center, it would undoubtedly allow for better packing of atoms for a cubic volume. In that case, the unit cell would be called a **body-centered cubic unit cell**, or **BCC** (Fig. 4.2). Instead of placing a central atom in the simple cubic unit cell, if we placed six atoms (one in the center of each cube's six faces); we would have even better packing than the simple cubic and the BCC unit cells. This new unit cell would be called a **face-centered cubic** unit cell, or **FCC** (Fig. 4.3).

Since all these unit cells are still cubic, the simple cubic, FCC, and BCC belong to the **cubic crystal system**. The definition of a cube is that it has equal sides and an angle between faces that is always 90°. If any of these change, we cannot call the unit cell a cube anymore, and the crystal system it belongs to would be anything other than the cubic crystal system. For example, if only the height changes and becomes longer than the rest of the sides, the unit cell would be called a **tetragonal unit cell** and belong to the **tetragonal crystal system**.

Along the same train of thought, if we consider the number of ways we could change the sides or the angles between faces, we will end up with **seven** basic crystal systems; one of them is the cubic. These seven crystal systems are displayed below (note: sides and angles that describe each crystal system are referred to as **lattice parameters**):

1. **The Cubic Crystal System:** All sides are equal, and all angles between faces are 90°.

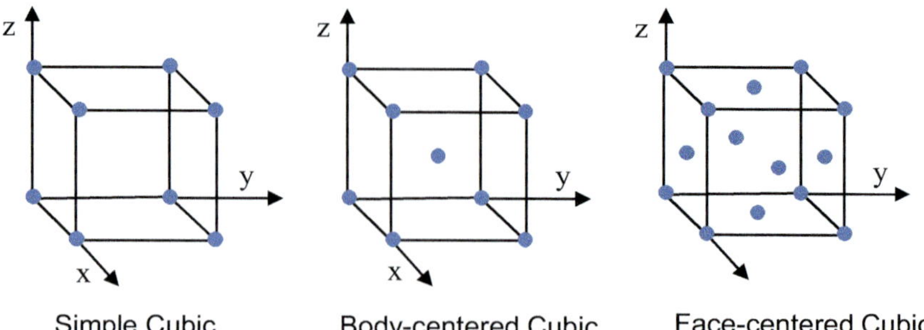

Simple Cubic   Body-centered Cubic   Face-centered Cubic

2. **The Tetragonal Crystal System**: Angles between faces are all 90°, $a = b$ and $c \neq a$ or b.

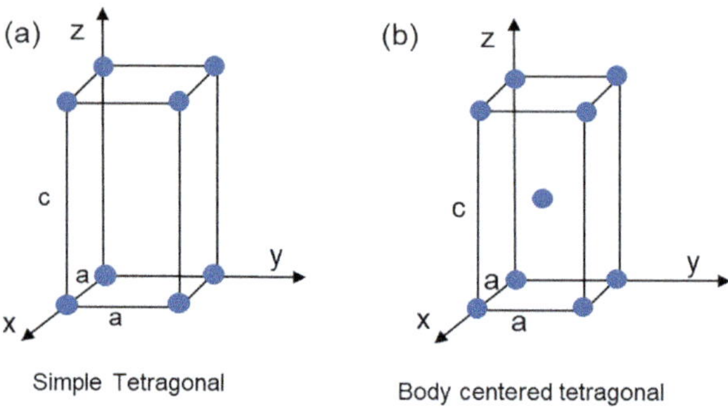

Simple Tetragonal   Body centered tetragonal

3. **Hexagonal Crystal System**: Sides on basal planes (i.e., top and bottom faces) are equal, angles between sides on basal planes = 120°, $c/a = 1.63$.

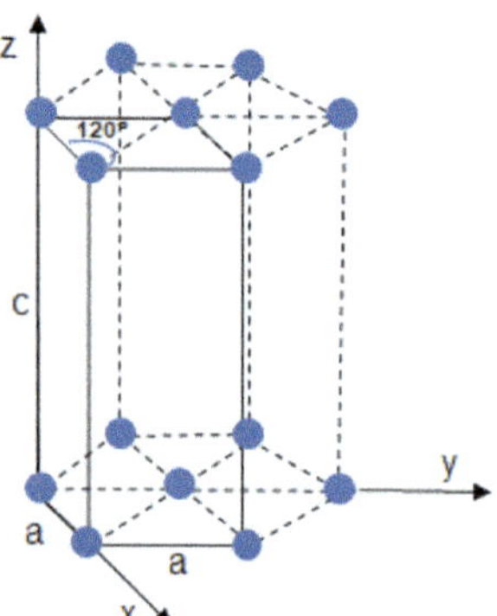

4. **Monoclinic Crystal System**: $a \neq b \neq c$, $\alpha = \beta = 90°$, $\gamma \neq 90°$.

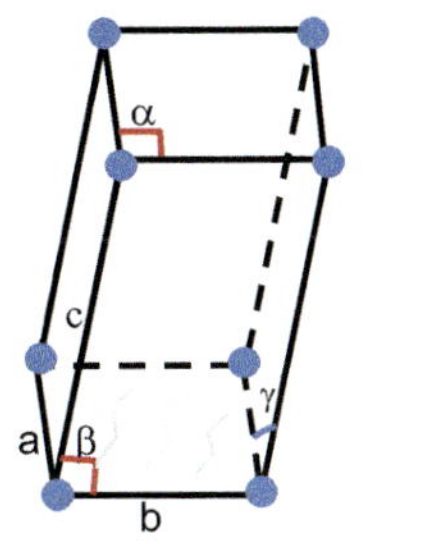
Simple Monoclinic

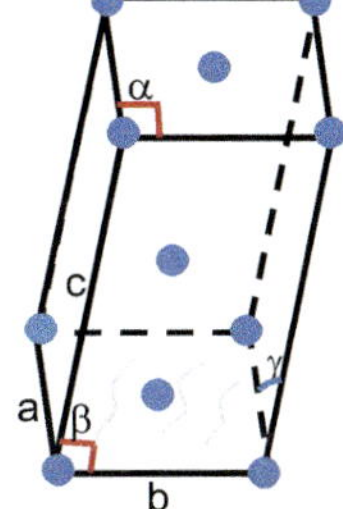
Base Centered Monoclinic

5. **Triclinic Crystal System**: $a \neq b \neq c$, $\alpha \neq \beta \neq \gamma \neq 90°$.

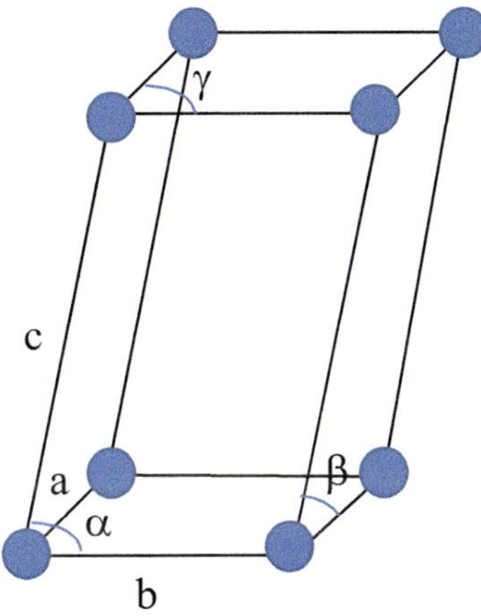

6. **Orthorhombic Crystal System:** $a \neq b \neq c$, $\alpha = \beta = \gamma = 90°$.

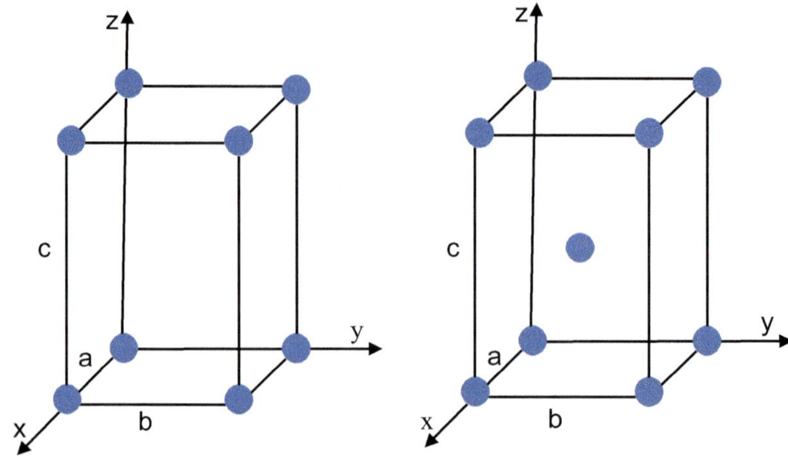

Simple Orthorhombic   Body-centered Orthorhombic

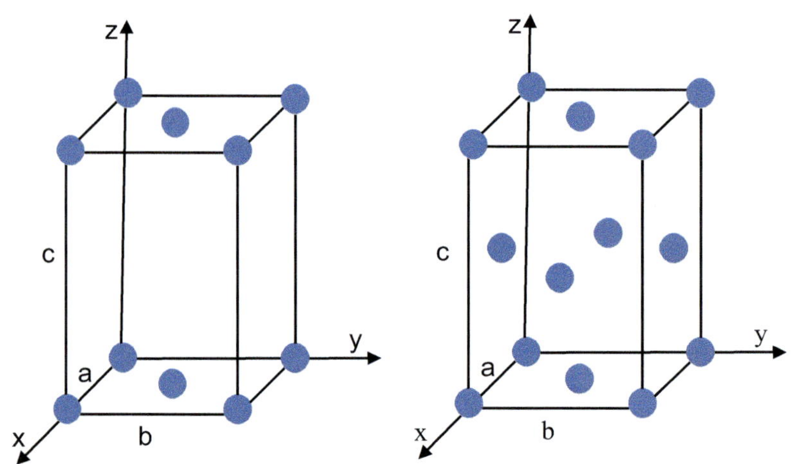

Base-centered Orthorhombic   Face-centered Orthorhombic

7. **Rhombohedral Crystal System** $a = b = c,\ \alpha = \beta = \gamma \neq 90°$.

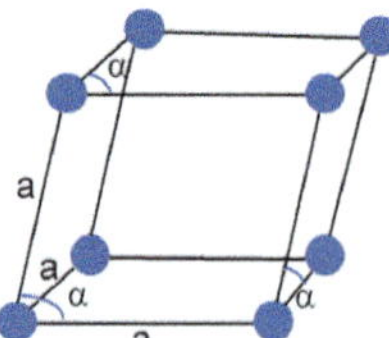

Table 4.1 Crystal structures and atomic (ionic) radii of some elements at room temperature and atmospheric pressure

Element	Crystal structure	Atomic (ionic) radius (A°)	Element	Crystal structure	Atomic (ionic) radius (A°)
Silver (Ag)	FCC	1.445 (1.26)	Alpha-Iron (α-Fe)	BCC	1.241 (0.74)
Gold (Au)	FCC	1.442 (1.37)	Tungsten (W)	BCC	1.371 (0.70)
Platinum (Pt)	FCC	1.387 (0.80)	Chromium (Cr)	BCC	1.249 (0.63)
Nickel (Ni)	FCC	1.243 (0.69)	Molybdenum (Mo)	BCC	1.363 (0.70)
Aluminum (Al)	FCC	1.432 (0.51)	Niobium (Nb)	BCC	1.426 (0.74)
Copper (Cu)	FCC	1.278 (0.96)	Tantalum (Ta)	BCC	1.43 (0.68)
Lead (Pb)	FCC	1.75 (0.84)	Cadmium (Cd)	HCP	1.49 (0.97)
Calcium (Ca)	FCC	1.976 (0.99)	Cobalt (Co)	HCP	1.253 (0.72)
Mercury (Hg)	Rhombohedral	1.55 (1.10)	Titanium (α-Ti)	HCP	1.475 (0.68)
Selenium (Se)	Monoclinic	1.15 (1.91)	Magnesium (Mg)	HCP	1.604 (0.66)
Gallium (Ga)	Orthorhombic	1.218 (0.62)	Zinc (Zn)	HCP	1.332 (0.74)
Potassium (K)	BCC	2.314 (1.33)	Boron (B)	Rhombohedral	0.46 (0.23)
Lithium (Li)	BCC	1.519 (0.68)	*Beta* tin (β-Sn)	BCT	1.405 (0.71)
Vanadium (V)	BCC	1.311 (0.74)	Arsenic (As)	Hexagonal	1.15 (2.22)

As can be seen, there can be different variations within each crystal system. For example, we have simple cubic, BCC, and FCC within the cubic system. Within the tetragonal system, we have the simple tetragonal and body-centered tetragonal, and so on for the other systems. When we add up all the different variations of crystals, we end up with **14**. These are referred to as the **Bravais lattices.**

Table 4.1 shows examples of materials that occupy different crystal structures at room temperature and atmospheric pressure. Temperature and pressure are essential to mention since a material's crystal structure may change at different pressures or temperatures (more about this later). The ionic radii of the elements are also listed. Note that an atomic radius differs from an ionic radius for the same element. This is because an ion is produced by giving away electrons (positively charged cation) or receiving electrons (negatively charged anion). The latter causes the ion to be larger than the atom, while the former causes the ion to be smaller. The size of the ion also depends on the number of ions surrounding it.

It is clear from Table 4.1 that certain elements or materials prefer to occupy specific crystal structures while others prefer different ones. For example, elements with a face-centered cubic crystal structure include aluminum, nickel, gold, silver, and platinum. Ones that have a BCC crystal structure would be chromium, vanadium, and iron. Ones that prefer HCP (hexagonal close-packed) include titanium and cobalt. Zirconia (ZrO_2) is monoclinic (not listed in Table 4.1), and so on. Materials choose these crystal structures for a reason. These materials are generally left with no choice but to choose the most energy-efficient configuration or structure if given the opportunity and means. In other words, materials do not like to have free energy lying about. Imagine if we humans (with our freedom of choice) behaved all the time like materials; we would be in a world with no energy crisis or waste. Just a thought .

Alex, next time you see a metal preferring to occupy the FCC crystal structure, understand that it is the most energy-efficient structure it can occupy at that temperature and pressure. Yes, I just said "at that temperature and pressure"

Does this really mean that the crystal structure of some materials can change if we change temperature or pressure?

Absolutely! Did you know that iron has a BCC structure at room temperature and atmospheric pressure but changes to FCC at atmospheric pressure above a temperature of 912 °C? Did you also know that graphite (the soft grey material your pencils are made of) becomes diamond when you subject it to extremely high pressures and temperatures?

Why do materials change their crystal structures?

*At the new temperature and pressure, the old structure is no longer the most energy-efficient, so it changes to the most energy-efficient crystal structure at the new conditions, per what we know as **Gibbs free energy**. So, next time someone asks you what the crystal structure of iron is, act smart and respond by asking, "At what pressure and temperature?"*

4.3 Benefits of the Unit Cell

The unit cell serves several important purposes, numbered below from 1 to 6:

1. It helps us visualize the positions of the atoms in a less crowded way compared with the much larger crystal lattice.

2. It helps us visualize the shape of the unit cell (e.g., cubic, tetragonal, etc.) and calculate its dimensions.

It is possible to determine the dimensions of a unit cell with knowledge of their relationship with the radius of the atom/ion from which the unit cell is made.

Consider the **FCC unit cell** in Fig. 4.4. Note that the dashed square intersects (centrally cuts through) the five atoms such that its corners are positioned on the centers of the four corner atoms. This square plane is then taken out and shown in Fig. 4.4b. If we observe the superimposed red triangle in Fig. 4.4b, we can see that its hypotenuse (**B**) equals **4.r** (where r is the atom's radius). The *adjacent* and *opposite* sides of the triangle are equal to **a** (the side of the cube). All we now need to do to determine a relationship between the radius of the atom (or ion) and the side of the cubic unit cell is to use the Pythagorean theorem.

$$B^2 = (4r)^2 = \mathbf{a}^2 + \mathbf{a}^2 = 2\mathbf{a}^2$$

Taking the square root on both sides gives.

$$B = 4r = \sqrt{2}\mathbf{a}$$

Hence, the side of the cube "a" is

$$= \frac{4}{\sqrt{2}} \cdot \frac{\sqrt{2}}{\sqrt{2}} \cdot r = 2\sqrt{2} \cdot r$$

Hence, **<u>FCC unit cells</u>**

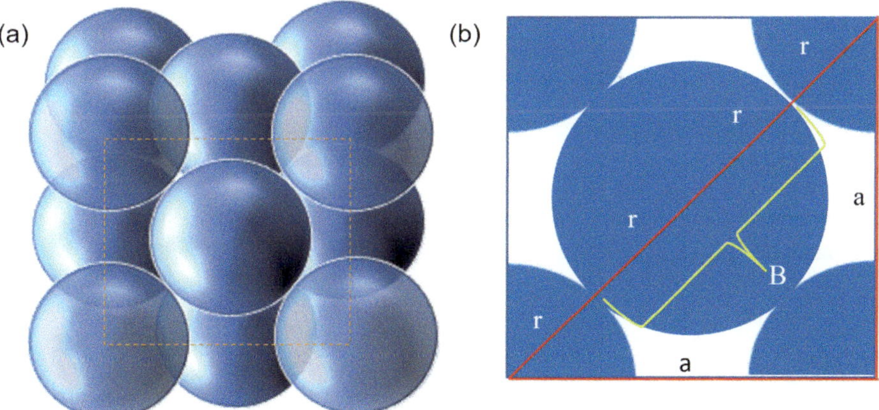

Fig. 4.4 a Hard sphere model of FCC unit cell with orange frame representing one side of the cubic unit cell (note the corners of the frame lie on the center of each of the atoms), **b** orange frame removed showing parts of the atoms that lie on that frame

$$\boxed{\mathbf{a} = 2\sqrt{2}r} \tag{4.1}$$

The relationship between the side of the cubic unit cell and the atom's radius will differ for **simple cubic** and **BCC** since they will have different atomic arrangements than the FCC.

Now, for a **BCC unit cell**, find out the relationship between the side of the cube and the radius of the atom. *Hint: The Pythagorean theorem can be of help here, too.* Our answer should be that shown in Eq. 4.2.

$$\boxed{\mathbf{a} = \frac{4r}{\sqrt{3}}} \tag{4.2}$$

Also, if we consider Fig. 4.1, we see that the side of a **simple cubic unit cell** is **2r**.

This means we can use trigonometry principles to arrive at relationships between the side of the unit cell and the atom's radius. Can we then find out the volume of an FCC unit cell? Of course, the volume of a cube is a^3, and we know how r is related to "a". As mentioned earlier, the radii of elements are all already known, so if we have, for example, gold with an atomic radius of 1.442 angstroms, we use Eq. 4.1 to find out the value of "a" and then take the cube of that value to arrive at the volume of the unit cell.

Although aluminum and nickel are FCC metals, the Al atom is smaller than the nickel atom, i.e., r will be smaller for Al. Therefore, although Al and Ni have the same relationship with r, the unit cell volumes will be numerically different once we input the actual radius values. In other words, the volume of the nickel unit cell will be numerically greater than that of aluminum.

3. The unit cell helps us visualize how many other atoms are touching any given atom. This is called the **coordination number** (i.e., the number of nearest-neighbor touching atoms). This value says a lot about how well packed the atoms of a material are.

We do this by picking "any" atom in the unit cell and finding out how many other atoms surround and touch it. Looking at the FCC unit cell in Fig. 4.4, let's pick the top center atom; it touches four corner atoms and the four face center atoms below it. That makes eight atoms. But when we place another unit cell on top of it, we will also find another set of four central atoms above. This gives a total of 12 touching atoms. Therefore, the coordination number for the FCC crystal unit cell is **12**.

If we look at the BCC unit cell and pick the body-centered atom, we see it is surrounded by 8 touching atoms. So, the coordination number here is 8. The HCP also has a coordination number of 12, like the **FCC**.

Example Problem 4.1

What is the coordination number of a simple cubic unit cell? Hint: Pick any atom and look around it. Try *using the hard-sphere model.*

(a) 8
(b) 12
(c) 10
(d) 6

Solution

By stacking eight simple cubic unit cells together, four at the bottom and four at the top, we can pick a central atom and see that the coordination number is 6. So, the answer is (d).

4. It helps determine how many "total" atoms are inside the boundaries of the unit cell.

Here, we need to understand that although an atom may be at the corner of a cubic arrangement of atoms, it does not mean that the whole atom's volume is inside the actual unit cell. Most of it will be sitting outside of it, and only a small part of it will be inside the unit cell boundaries.

To visualize this, let us consider an FCC crystal structure. Figure 4.5 clearly shows that a face-centered atom is shared by two unit cells. This means each unit cell claims one-half of each face-centered atom to be within its boundaries. Since we have six faces, ½ an atom from each face gives a total of three atoms from the faces that reside within the boundaries of the unit cell.

As a rule, a cubic unit cell can only claim 1/8 of each corner atom to be within its boundaries, i.e., each corner atom is shared by eight unit cells in total. This is seen in Fig. 4.6. So, from the eight corners, the total number of atoms inside the unit cell will be $1/8 \times 8 = 1$ atom. Hence, for an FCC unit cell, the total number of atoms that can be claimed to be within its boundaries is three from the faces and one from the corners, giving a total of four atoms.

Fig. 4.5 Illustration showing how a face-centered atom in an FCC unit cell is shared by two unit cells

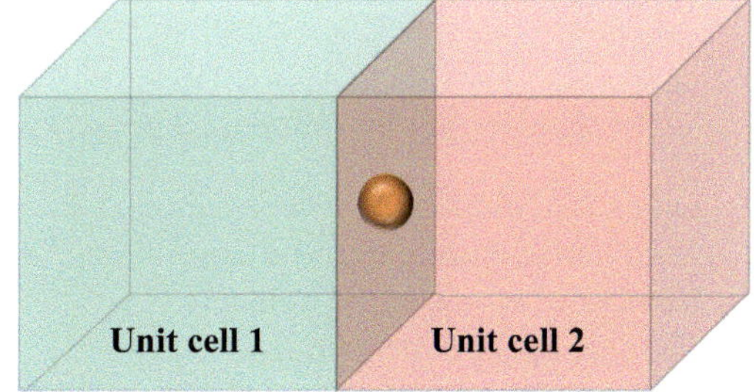

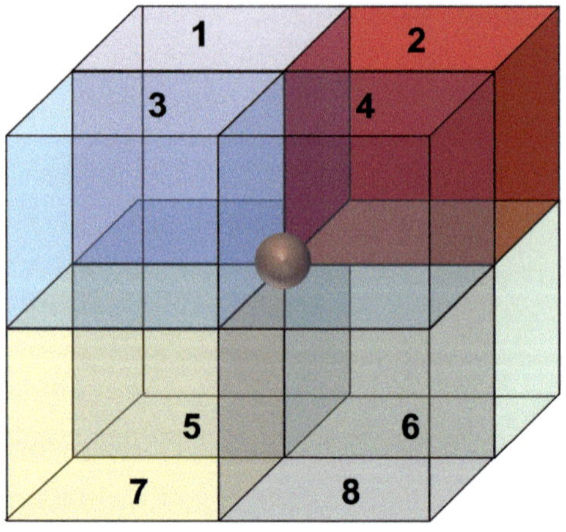

Fig. 4.6 Illustration showing how a corner atom in an FCC unit cell is shared by a total of eight unit cells

Likewise, BCC unit cells contain a total of two atoms, one atom from the corners and one body-centered atom.

Example Problem 4.2

How many atoms will be inside a simple cubic unit cell?

(a)   1
(b)   2
(c)   3
(d)   4

Solution

Each corner atom contributes 1/8 of its volume to being within the boundaries of the unit cell. Since we have eight corner atoms, then $1/8 \times 8$ corner atoms $= 1$. Hence, the answer is (a).

5. It also enables calculating the **theoretical density** of the material to which the unit cell belongs. Suppose we found the unit cell's volume (from its dimensions) and how many atoms are within its boundaries. Since atoms are the only entities that weigh, we divide the total mass of the atoms inside the unit cell by the unit cell volume to get the theoretical density of your material.

Let us take aluminum as an example, which has an FCC crystal structure.
The relative atomic mass of aluminum is 26.982 g/mol.
Number of atoms in the unit cell $= 4$.
The side of the cube $= = a = 2\sqrt{2} \cdot r$,
Volume of the cube $= a^3 = (2\sqrt{2} \cdot r)^3$.
The radius of the aluminum atom (from Table 4.1) $= 0.143$ nm.
Volume of the unit cell $= 6.61672512 \times 10^{-23}$ cm^3.
Total mass $= 4{*}26.982/6.023 \times 10^{23} = 1.791931 \times 10^{-22}$ g.

$$\rho_{Al} = \frac{mass}{volume} = 2.708\,g/cm^3 \text{ or } 2708\,kg/m^3$$

Alex, why don't you carry out the same calculation for nickel and iron and see what you get? I promise it will give you a lot of confidence if you manage to get it right! If not, try again and again

6. The unit cell can also tell us much about how atoms are packed. We can consider atomic packing on a 1D, 2D, or 3D level. The 3D version of atomic packing is called the **atomic packing factor (APF).** It is defined as the volume of atoms within the unit cell's boundaries divided by the unit cell's volume (calculated from its dimensions). For **FCC** and **HCP,** APF is **0.74**, while for **BCC**, it is **0.68**, and for **simple cubic**, it is **0.52**. You can see that the simple cubic unit cell is not well packed. However, the FCC and HCP are special. Did you know if I left you with ping pong balls and told you to take them and pack them most efficiently, no matter how long you take, the best you can do without squishing the balls is **0.74**. You may ask, how about the 1D and 2D versions of atomic packing? The 1D is related to packing along a line, and the 2D is atomic packing within a plane.

Consider Fig. 4.7 for a simple cubic unit cell. The yellow line represents a line drawn from one atom's center to the other's center.

If we consider the line as a pin that pierces centrally through each atom, it can be seen that the pin reaches halfway through each atom (by reaching its center). With this idea in mind, **Linear density (LD)** can be defined as the *number of atoms centered (or sitting) on a line divided by the actual length of that line*. Hence, for the yellow line drawn in Fig. 4.7:

$$LD = \frac{0.5 + 0.5}{a} = \frac{1}{a} = \frac{1}{2r}$$

Fig. 4.7 Illustration used in determining the linear density for the yellow line (drawn between the centers of two atoms) in a simple cubic unit cell

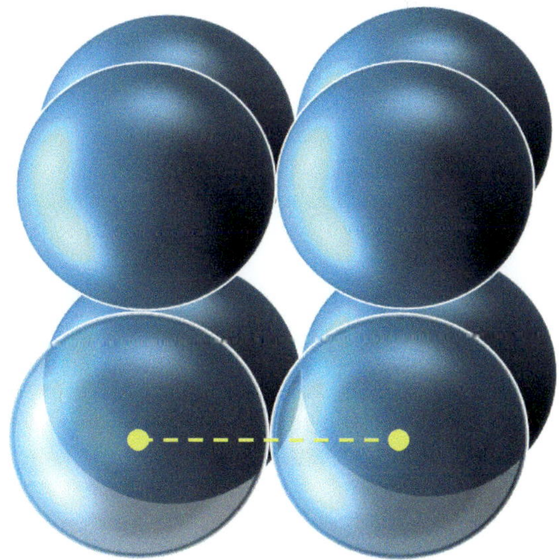

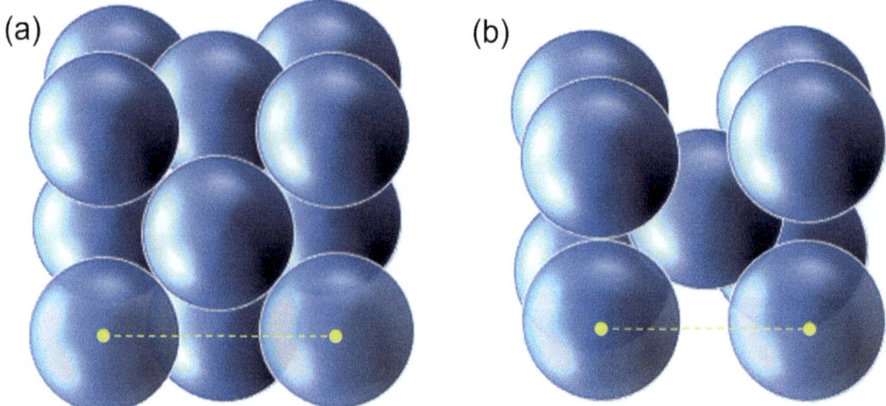

Fig. 4.8 Illustration to aid determining the linear density for the yellow line in (**a**) FCC and (**b**) BCC unit cells

Now, try this for the BCC and FCC (Fig. 4.8). Find out the LD of the yellow line in terms of the atomic radius, r

Answer: $\sqrt{3}/(4r)$ for BCC and $\sqrt{2}/(4.r)$ for FCC.

So, what about **2D atomic packing**? Imagine the dashed front frame of the FCC unit cell shown in Fig. 4.9a. Note, the corners of the frame lie on the 'center' of each of the atoms. When we remove that frame from the unit cell and observe it separately, as shown in Fig. 4.9b, the square frame (with side **a**) penetrates all the atoms centrally, but to different degrees. We can imagine the frame as a square, thin, and sharp sheet of glass that centrally cuts five ping pong balls (representing our atoms). We can see that the frame cuts each of the four corner balls a quarter of the way through, and it goes through or slices through the entire central ball.

In this case, if we were asked how many balls (or atoms) this face slices, we would reply ¼ of each corner ball and the whole of the central ball, giving us a total of two balls (or atoms) being sliced by (or sitting on) the face. Now, we come to the business of packing.

The **planar density (PD)** represents the 2D version of atomic packing. It is defined as *the number of atoms sitting on the face divided by the actual area of that face.*

So, for the example we discussed in Fig. 4.9, we have

$$\text{PD} = \frac{2\,\text{atoms}}{a.a} = \frac{2}{a^2}$$

We recall that "a" is the side of the FCC unit cell, and that there is a relationship between it and the radius of the atom. This means we can determine the planar

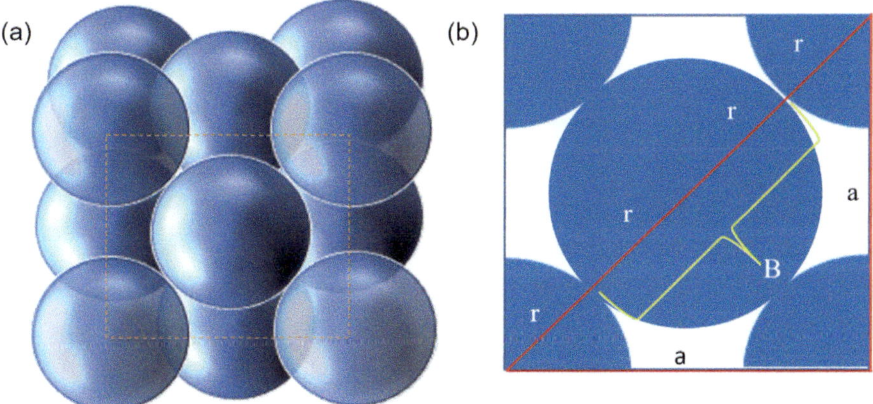

Fig. 4.9 a Hard-sphere model of FCC unit cell with orange frame representing the front face of the unit cell (*note the corners of the frame lie on the center of each of the atoms*), **b** orange frame removed from (**a**) showing parts of the atoms that lie on that frame

density in terms of the atomic radius. We will elaborate on this later when we discuss crystallographic planes in more detail. But for now, remember what **PD** refers to.

4.4 Close-Packed Atomic Arrangements

How closely atoms are packed together can play a critical role in the final properties of materials; for example, it can affect a metals density, strength, and ductility. In the games of pool and snooker, we typically stack the balls within a wooden or plastic triangle to set them up before we start the game. This ball arrangement demonstrates the closest and best packing we could have achieved with equal-sized balls and is undoubtedly the same for atoms. Additional atomic layers can then be stacked on this first layer, leading to different crystal structures. Figure 4.10a shows a typical snooker triangular stacking of balls analogous to the close-packed arrangement of atoms. Figure 4.10b shows how the second atomic layer, B, can be stacked on top of the first layer, A. It also shows two possible positions for atoms of the third layer, positions **1** and **2**. Different stacking arrangements can be produced depending on which position the atoms of the third layer choose to occupy, representing different crystal structures. For example, Fig. 4.10c shows that when the atoms of the third layer choose to occupy position 1, it leads to a third layer C. When the fourth layer atoms occupy position 3, it's as if layer A is being repeated. This leads to the atomic stacking sequence ABCABCABC, etc., in other words, the stacking sequence found in the FCC crystal structure. Suppose the atoms of the third layer prefer to occupy position 2. In that case, it's as if the first layer A is being repeated (see Fig. 4.10d), resulting in a stacking sequence of ABABAB, etc. found in HCP materials. In both cases, we are talking about close-packed arrangements. Since the atoms are in such close

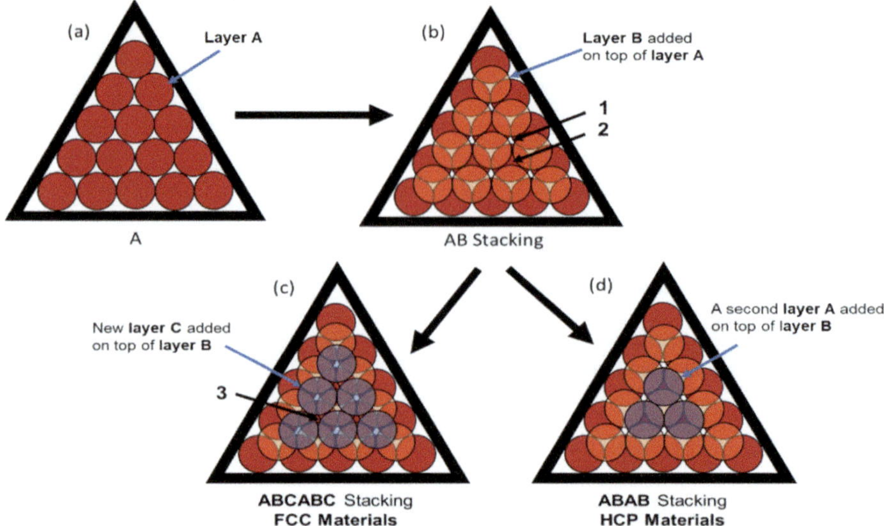

Fig. 4.10 Illustration of closed pack atomic stacking arrangements, **a** first atomic layer A, **b** second atomic layer B, **c** new layer C added on top of layer B, giving the ABCABCABC atomic stacking sequence found in FCC materials, and **d** third atomic (second layer A) which align with the initial layer A giving the ABABAB… atomic stacking sequence found in HCP materials

packing arrangements, it is no surprise that the coordination number for both FCC and HCP materials is 12, and the atomic packing factor is 0.74, the absolute theoretical maximum. However, these kinds of close-packed arrangements are not found in the BCC crystal structure, leading to the different observed properties between FCC and BCC materials, even though they belong to the same overall cubic system.

4.5   Point (or Fractional) Coordinates

Imagine if we were talking to someone over the phone and wanted to discuss with that person the location of a specific atom or point, a particular direction, or a specific plane of atoms in the unit cell. It would not be an easy task to do without an indexing system. To discuss this further, let's look at the simple cubic unit cell in Fig. 4.10. This approach can easily be applied to other crystal structures and systems. In the figure, we see two things:

1. The position or location of atoms.
2. Three axes drawn with $+ve$ and $-ve$ directions.

If we wanted to provide indices for the **positions of atoms** or a **point** within the unit cell, we could do so as follows. First, the position of the red atom is assigned as an arbitrary origin. The origin does not have to be located there, but we place it there

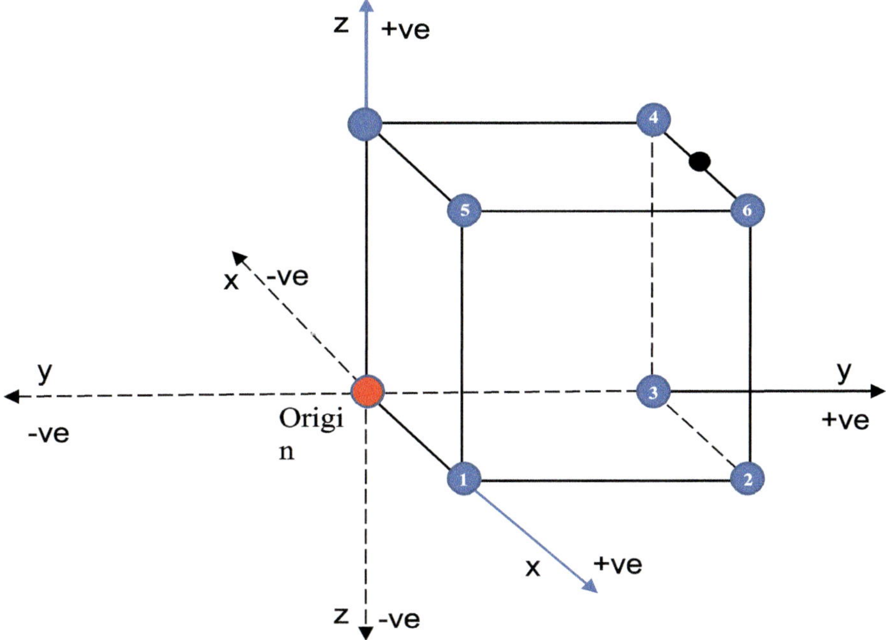

Fig. 4.11 Illustration showing a unit cell (containing one impurity back colored atom). The red atom is arbitrarily chosen as the origin, with coordinates 0, 0, 0

as a default unless it is more advantageous for us to position it elsewhere. Since the red atom is located at the origin, it is positioned at point zero along the x-axis, point zero along the y-axis, and point zero along the z-axis. In this case, its x, y, z fractional coordinates are 0, 0, 0. Note that the total length of a particular side of the unit cell along each axis is given the value 1.

As such, the following are the coordinates of the atoms numbered 1–6, in Fig. 4.11.

1. 1, 0, 0
2. 1, 1, 0
3. 0, 1, 0
4. 0, 1, 1
5. 1, 0, 1
6. 1, 1, 1

Example Problem 4.3

What are the black impurity atom's fractional coordinates in the unit cell in Fig. 4.11?

Solution

½,1, 1

The coordinates represent the fact that we had to move ½ of the side of the unit cell along the x-axis and one whole side along the y- and z-axes.

We do not enclose the final coordinates in brackets; just place commas between the numbers. This is the convention, and it is important to stick to it.

The unit cell in Fig. 4.12 is that of NaCl (sodium chloride), also known as common table salt. Interestingly, many materials occupy this crystal structure, such as AgCl, MgO, CaO, TiN, TiC, and ZrC in addition to ZrB. We can view the unit cell in Fig. 4.12 as having two interpenetrating FCC unit cells, one for Na cations (+ ve charged) and the other for Cl anions (− ve charged). The origin for the Na cation is found at $0, 0, 0$, while that of the Cl anion is at ½, $0, 0$. If we can imagine the Cl$^-$ unit cell being extended further along the x-axis, other blue (Cl) anions will be present in the locations where they are supposed to be for a Cl$^-$ FCC unit cell. Hence, the blue ions 1, 2, 3, and 4 are positioned in the center of the faces of the blue (Cl$^-$) FCC unit cell. We can see this by drawing three imaginary x-, y-, and z-axes from the origin at ½, $0,0$. Nevertheless, let us consider the whole of Fig. 4.12 as only one unit cell, with only one origin at $0,0,0$. In this case, the following would be the coordinates of ions 1–4.

1.   1, 1/2, 1
2.   1, 0, 1/2
3.   1, 1, 1/2
4.   1, 1/2, 0

Using the *fractional coordinate system,* we can identify the position of any atom within a unit cell and any location within the unit cell in general, even if an atom is not there.

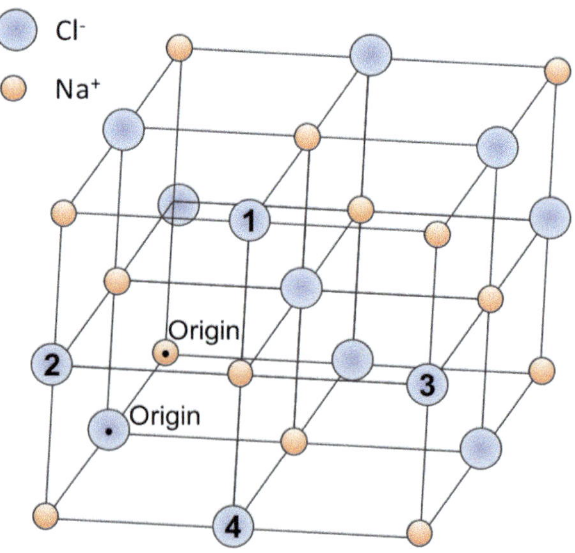

Fig. 4.12 Illustration showing unit cell of NaCl (table salt)

Example Problem 4.4

How many "total" ions of Na and Cl can be claimed to be inside the boundaries of this unit cell?

Solution

For a cube, corner atoms are shared by eight unit cells, and each face-centered ion is shared by two unit cells. So, let's count for Na, then Cl ions.

$\underline{Na^+}$:

Corner atoms: $8 \times 1/8 = 1$.

Face-centered atoms: $6 \times 1/2 = 3$.

So, the total number of Na ions inside the unit cell is 4.

$\underline{Cl^-}$:

The central Cl ion is totally inside this unit cell. All other atoms are shared in total by four NaCl unit cells; hence, for each of these atoms, only 1/4 can be claimed to be inside the NaCl unit cell. $12 \times 1/4 = 3$.

Total Cl ions inside the NaCl unit cell $= 4$.

Hence, the total number of Na and Cl ions inside the unit cell is 8 and 4 for Cl^- and 4 for Na^+.

No wonder the charge is balanced with these oppositely charged ions; we have four of each, four negative and four positive.

4.6 Crystallographic Directions

Now, let us see how we can deduce the indices of **crystallographic directions** (i.e., the vectors we see in Fig. 4.13). Let's start with the **black arrow** (vector) in Fig. 4.13. We do it by seeing how far the arrow traveled along the x-direction, y-direction, and z-direction, respectively. In this case, we can see that the arrow traveled the whole length of the unit cell in the X-direction. So, we will say 1. We do not see the arrow moving along the Y-axis or the Z-axis. In this case, the arrow traveled one unit along the X-axis and zero along the Y- and Z-axes. Hence, we can write.

$$\underline{X} \quad \underline{Y} \quad \underline{Z}$$
$$1 \quad 0 \quad 0$$

Since there are no fractions, we can place these indices between **square brackets,** i.e., [ ]. These square brackets clarify that we are talking about crystallographic directions, not coordinates or planes.

Hence, the **final direction indices** will be **[1 0 0].**

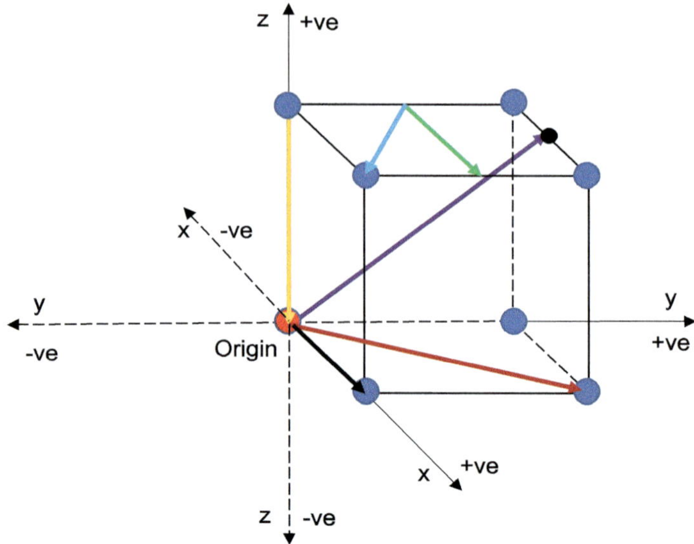

Fig. 4.13 Illustration showing a unit cell (containing one impurity back colored atom) and several vectors within it. The red atom is regarded as an *arbitrary* origin, with coordinates 0, 0, 0

Sometimes fractions can result during the process. Let's look at our solution for the **purple arrow**. We can see that it travels only ½ the unit cell side along the *X*-axis, the whole length of the side along the *Y*-axis, and the whole length along the *Z*-axis. Hence, we can write.

$$\underline{x}\ \ \underline{y}\ \ \underline{z}$$
$$\tfrac{1}{2}\ \ 1\ \ 1$$

Convention dictates that fractions should not be present in crystallographic direction indices. Hence, before placing the square bracket, we need to conduct a reduction operation by multiplying all the numbers by the common denominator, in this case, "2". By doing this, we get.

$$\underline{x}\ \ \underline{y}\ \ \underline{z} \qquad \xrightarrow{\ \times\,2\ } \qquad \underline{x}\ \ \underline{y}\ \ \underline{z} \qquad \xrightarrow{\text{place brackets}} \qquad \textit{Miller Indices}$$
$$\tfrac{1}{2}\ \ 1\ \ 1 \qquad\qquad\qquad 1\ \ 2\ \ 2 \qquad\qquad\qquad [1\ \ 2\ \ 2]$$

Ok, that's the fractions sorted out, but what about if a direction traveled backward along an axis? I.e., in the opposite direction than what we have considered so far. For that, let's take the yellow arrow as an example. The base of the arrow/direction can now be the new origin at 0,0,1 (remember the origin is always an arbitrary one, we can move it if it does not help us) and from that origin we can imagine three new x,y, and z axes from which we observe the direction. Hence the yellow direction is considered to be traveling along the negative z-axis from the new origin, and it does not travel along the new x or y axes. Hence, for the yellow crystallographic direction, we have.

X Y Z   No reduction needed   **X Y Z**   place brackets   *Miller indices*
0 0 -1   →   0 0 -1   → $[0\,0\,\bar{1}]$

By convention, the negative sign is not allowed to be placed to the left of any of the numbers making up the indices. Instead, it is placed on top of them. The **yellow arrow** represents this direction.

Alex, remember these things: you can lose points if you do not use these conventions (i.e., for putting the correct brackets and the negative sign). Always remember, that the origin is an arbitrary origin. This means, if an arrow emerges from some point then that point will be your new origin with brand new x,y and z axes. Then you observe how the direction travels along these axes.

Example Problem 4.5

Now, determine on your own the indices of the remaining green, blue, and red directions in Fig. 4.13.

Solution

Green: [1 0 0]
Blue: $[2\,\bar{1}\,0]$
Red: [1 1 0]

So far, we have shown how to deduce the Miller indices of a crystallographic direction drawn in a unit cell. Now, we try to do the opposite: provide Miller indices for a direction and draw the direction it. Let's draw the following indices:

$[112]$, $[101]$, $[0\bar{1}2]$, $[\bar{1}21]$. We will draw these in Fig. 4.14.

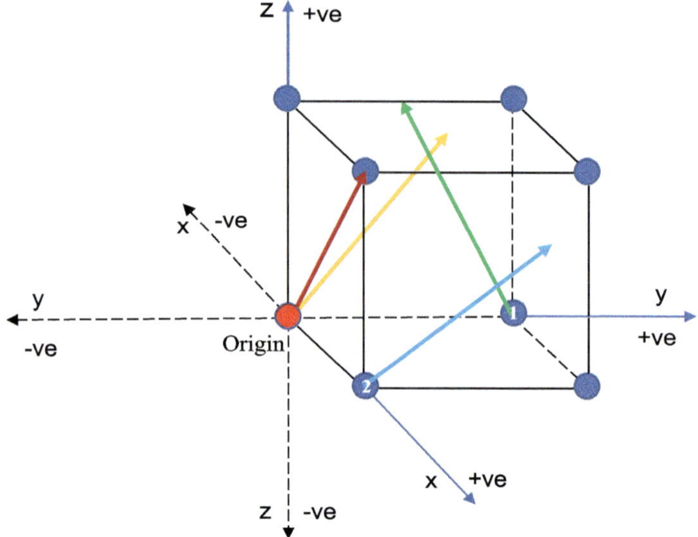

Fig. 4.14 Illustration showing a reduced-sphere model of a unit cell and several vectors within it. The red atom is regarded as an arbitrary origin, with coordinates 0, 0, 0

The [112] Direction

Here, we see that there is a number in the indices that is greater than 1. We first take that number (i.e., 2) out of the bracket, leaving us with **2 [½ ½ 1]**. So now we focus on drawing the direction [½ ½ 1], which will end up inside the unit cell and not outside it if we had drawn [1 1 2].

Hence, place the base or tail of the arrow at the origin and move the arrow ½ along the x-axis, ½ along the Y-axis, and 1 along the z-axis. This results in the yellow arrow.

The [101] Direction

The arrow travels 1 along the x-axis, 0 along the y-axis, and 1 along the z-axis, resulting in the red arrow.

The [0$\bar{1}$2] Direction

Here, we see two things in the bracket. A number greater than 1 (i.e., 2) and a negative number (− 1). For the number greater than 1, we take the 2 outside the bracket, leaving us with 2 [0$\frac{\bar{1}}{2}$1]. If we try to draw the [0$\frac{\bar{1}}{2}$1] direction, we will notice the direction being outside the unit cell along the negative y-axis. However, to keep the arrow inside the unit cell, we can move the origin to **point 1** and then draw the direction, resulting in the **green arrow** in the unit cell.

The [$\bar{1}$21] **Direction**:

Here we again take the 2 out of the bracket, leaving us with 2 [$\frac{\bar{1}}{2}1\frac{1}{2}$]. We then choose **point 2** as the **new origin** and draw the **blue arrow**.

It is sometimes relevant to know how well packed with atoms a particular crystallographic direction is. Here, we refer back to the linear density concept described in Fig. 4.7. It is possible to determine the linear density along a particular crystallographic direction. Let us consider Example 4.6.

Example Problem 4.6

Determine the linear density of the [111] direction in a BCC unit cell in terms of the atom's radius (r).

Solution

Figure E4.1 shows the [111] direction in a BCC unit cell. If we isolate and take out the direction along with the three atoms it intersects in the unit cell, we obtain the image on the right. In that image, the atoms are drawn as a hard sphere model. We see that the [111] direction pierces half of the edge atoms (in each case traveling a distance of r) and all of the central atom (which means it traveled a distance of $2r$). This means the total length of the direction is $2x\,(r) + 2r = 4r$. **Linear density** is defined as the *number of atoms centered (or sitting) on a line/crystallographic direction divided by the actual length of that line*. We have already determined that the length of that direction or line is $4r$. The green line has pierced halfway through the edge atoms, which means half of each of these atoms sits on that direction. Also, the whole of the middle atom sits on that direction as well. Hence, the total number of atoms that sit on the direction is $1/2 \times 2 + 1 = 2$. Hence, the linear density of the [111] direction in an FCC unit cell is given by

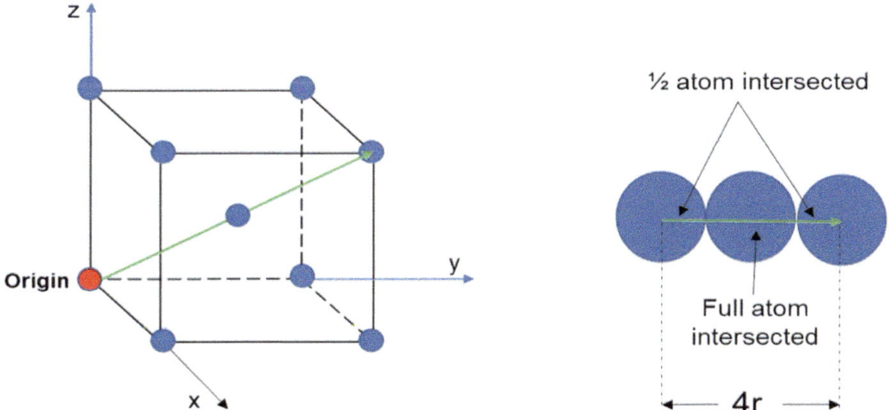

Fig. E4.1 The [111] crystallographic direction in a BCC unit cell, and an image showing atoms intersected by the direction

$$L_{[111]} = \frac{2}{4r} = \frac{1}{2r}$$

4.7   Crystallographic Planes

The steps for identifying plane indices are different than those for directions, and they are as follows:

1. For *each axis*, find out if the plane intercepts the axis <u>at a point</u> **or** <u>is parallel</u> to it. If it intercepts at a point on the axis, write it down as a fraction of that axis. If it is parallel, then we will state the intercept as infinity. This procedure is carried out for the *X*-, *Y*-, and *Z*-axes.
2. Then, take the *reciprocals* of the intercepts.
3. If step 2 results in any fractions, multiply everything by the common denominator (i.e., carry out a **reduction process**).
4. Finally, write the indices in between **parentheses**.

So, let us start with **plane A** in Fig. 4.15 to demonstrate how to do this.

As usual, we start with the origin at the **red** atom, and as seen, the three axes extend from it. From the perspective of the *X*-axis, plane A is parallel to the axis. It is also parallel to the *Y*-axis but intercepts the *Z*-axis at 1.

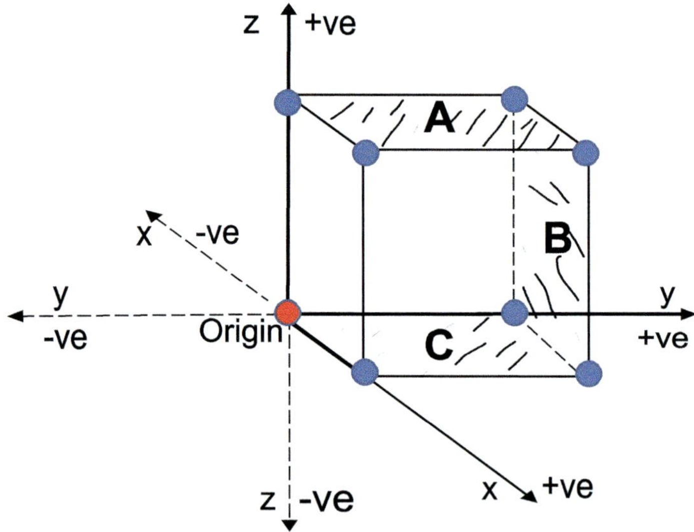

Fig. 4.15 Illustration showing planes A ,B and C in a unit cell

$$\underline{x}\ \underline{y}\ \underline{z} \xrightarrow{\text{Take Reciprocals}} \underline{x}\ \underline{y}\ \underline{z} \xrightarrow{\text{place brackets}} \textit{Miller indices}$$
$$\infty\ \infty\ 1 \qquad\qquad\qquad 0\ 0\ 1 \qquad\qquad\qquad (0\ 0\ 1)$$

Hence, we have

We note here that the intercepts for x and y were infinity. Then, we took the reciprocals, but there were no fractions. Hence, we **did not** need to carry out a reduction process. The final step was to insert the curved brackets or parentheses.

Similarly, for **plane B,** we have

$$\underline{x}\ \underline{y}\ \underline{z} \xrightarrow{\text{Take Reciprocals}} \underline{x}\ \underline{y}\ \underline{z} \xrightarrow{\text{place brackets}} \textit{Miller indices}$$
$$\infty\ 1\ \infty \qquad\qquad\qquad 0\ 1\ 0 \qquad\qquad\qquad (0\ 1\ 0)$$

Some issues immediately emerge when we come to **plane C** and start thinking about the intercepts. The plane lies everywhere on the x-axis and the y-axis. It is not intercepting at a point; it is also not parallel to any of the axes in the way we are accustomed to. This tells us that this origin will not help us, and we need to change it. When selecting our new origin, we need to realize that with the new origin comes new axes, and we should completely ignore the old ones. When we pick a new origin, we need to ensure that for the new axes belonging to the new origin, the plane needs to either intercept or be parallel to each of the axes; only then will that origin be valid. Let us consider the atom with coordinates 0,0,1 in Fig. 4.15 as our **new origin,** with its new axes. We can see that the plane will be parallel to the new X- and Y-axes and intercepts the Z-axis at -1, hence:

$$\underline{x}\ \underline{y}\ \underline{z} \xrightarrow{\text{Take Reciprocals}} \underline{x}\ \underline{y}\ \underline{z} \xrightarrow{\hspace{3cm}} \textit{Miller indices}$$
$$\infty\ \infty\ \bar{1} \qquad\qquad\qquad 0\ 0\ \bar{1} \qquad\qquad\qquad (0\ 0\ \bar{1})$$

This makes the indices of this plane $(00\bar{1})$.
What about **plane D** in Fig. 4.16?

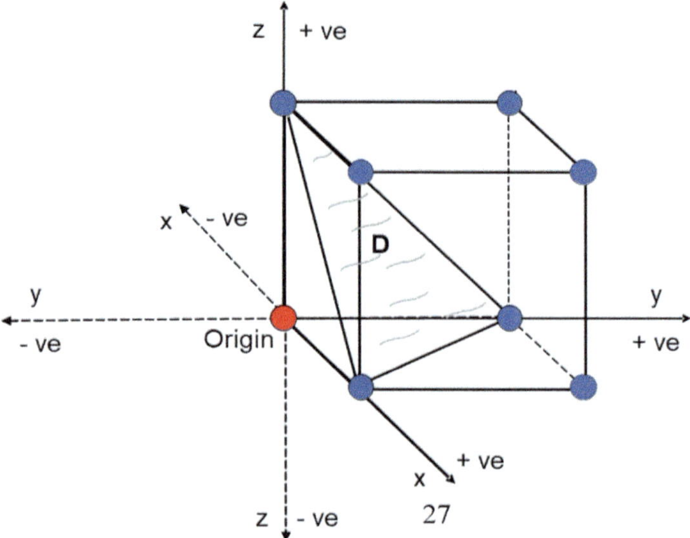

Fig. 4.16 Illustration showing a unit cell with plane *D* drawn in

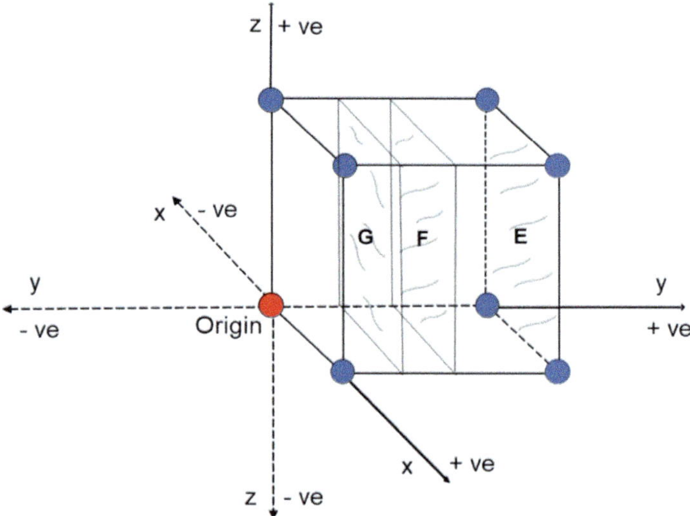

Fig. 4.17 Illustration showing a unit cell with planes E, F, and G drawn in

The next three **planes (E, F,** and **G)** in Fig. 4.17 are interesting.

$$\begin{array}{ccc} \underline{x} & \underline{y} & \underline{z} \\ 1 & 1 & 1 \end{array} \quad \xrightarrow{\text{Take Reciprocals}} \quad \begin{array}{ccc} \underline{x} & \underline{y} & \underline{z} \\ 1 & 1 & 1 \end{array} \quad \xrightarrow{\text{place brackets}} \quad \begin{array}{c} \textit{Miller indices} \\ (1\ 1\ 1) \end{array}$$

Plane E

$$\begin{array}{ccc} \underline{x} & \underline{y} & \underline{z} \\ \infty & 1 & \infty \end{array} \quad \xrightarrow{\text{Take Reciprocals}} \quad \begin{array}{ccc} \underline{x} & \underline{y} & \underline{z} \\ 0 & 1 & 0 \end{array} \quad \xrightarrow{\text{place brackets}} \quad \begin{array}{c} \textit{Miller indices} \\ (0\ 1\ 0) \end{array}$$

Plane F

$$\begin{array}{ccc} \underline{x} & \underline{y} & \underline{z} \\ \infty & \tfrac{1}{2} & \infty \end{array} \quad \xrightarrow{\text{Take Reciprocals}} \quad \begin{array}{ccc} \underline{x} & \underline{y} & \underline{z} \\ 0 & 2 & 0 \end{array} \quad \xrightarrow{\text{place brackets}} \quad \begin{array}{c} \textit{Miller indices} \\ (0\ 2\ 0) \end{array}$$

Plane G

$$\begin{array}{ccc} \underline{x} & \underline{y} & \underline{z} \\ \infty & \tfrac{1}{4} & \infty \end{array} \quad \xrightarrow{\text{Take Reciprocals}} \quad \begin{array}{ccc} \underline{x} & \underline{y} & \underline{z} \\ 0 & 4 & 0 \end{array} \quad \xrightarrow{\text{place brackets}} \quad \begin{array}{c} \textit{Miller indices} \\ (0\ 4\ 0) \end{array}$$

Planes (0 1 0), (0 2 0), and (0 4 0) are parallel. Let us name the numbers inside the brackets as h, k, and l, such that any plane we consider has indices $(h\ k\ l)$. In the above case, in (0 2 0), k is double that of (0 1 0). This is typically what occurs in the cubic system. A plane with $(nh\ nk\ nl)$ is always parallel to the plane $(h\ k\ l)$, where n is an integer. Hence plane (040) highlighted in yellow is parallel to the (010) and (020).

If we imagine another unit cell to the right of the unit cell in Fig. 4.16, it will have its own (0 1 0) plane. So, what would be the distance between these (010) planes? This is given by the following equation, which is **<u>only valid</u>** for cubic systems, where a is the side of the cube.

$$\frac{1}{d^2} = \frac{h^2 + k^2 + l^2}{a^2} \tag{4.3}$$

where "d" is referred to as the d-spacing, and "a" is the lattice parameter or the side of the cube. The lattice parameter is "a" $= 0.4046$ nm for aluminum with its FCC structure.

So, the **d-spacing** for the (010) planes is

$$\frac{1}{d_{(010)}^2} = \frac{(0)^2 + (1)^2 + (0)^2}{(0.4046)^2}$$

This leads to a d-spacing of 0.4046 nm or a for the (010) planes. If the same exercise is repeated for the (020) planes, we would see a d-spacing of 0.2023 or a/2. If it were repeated for the (040) planes, it would be equal to a/4.

We can also calculate the planar density of crystallography planes, as with linear density. Let's elaborate on this using Example Problem 4.7.

Example Problem 4.7
Determine the planar densities of the (100), (010), and (001) planes in a BCC unit cell in terms of the side of the cube "a".

Solution

Figure E4.2 represents the (100), (010), and (001) planes. As can be seen, the plane has four corners, each intercepting one-quarter of an atom. This leaves the total atoms being intercepted by the plane to be 1, and the area of the plane is $a \times a = a^2$. It follows that the planar densities, defined as *the number of atoms sitting on the plane divided by the actual area of that plane*, are

$$\mathrm{PD}_{(100)} = \frac{\frac{1}{4} \times 4}{a^2} = \frac{1}{a^2}$$

Similarly,

$$\mathrm{PD}_{(010)} = \frac{\frac{1}{4} \times 4}{a^2} = \frac{1}{a^2}$$

Fig. E4.2 Plane representing the (100), (010), and (001) planes

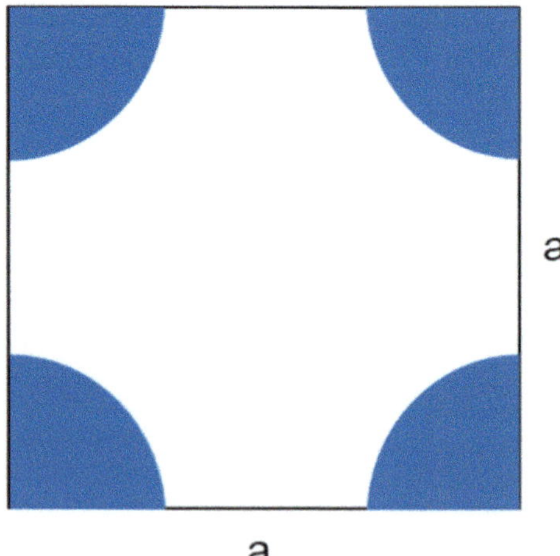

And

$$\mathrm{PD}_{(001)} = \frac{\frac{1}{4} \times 4}{a^2} = \frac{1}{a^2}$$

Here we see that the (100), (010), and (001) planes all have the same planar density.

So far, we have considered individual planes and directions. However, it is sometimes convenient to refer to a set of planes with the same planar density or directions with the same linear density. In this case, these sets of planes or directions are referred to as a **family of planes** and a **family of directions**, respectively.

We have seen in **Example Problem** 4.7 that for a BCC unit cell, the (100), (010), and (001) planes have the same planar density. Let's call this family the (100) family of planes. We will use a different bracket to differentiate the designation of a single plane from that of a family of planes. Recall that for an individual plane, we used parentheses to enclose the indices; however, for a family of planes, we use { }. Hence, the (100) family of planes mentioned above would now be referred to as {100}. Once we see these brackets, we will know that we are not talking about one plane but rather a set of planes that have the same planar density as the (100) plane. Similarly, directions with the same linear density will belong to the same family. Here, we use again different brackets, which are < > , and not the [] used for individual directions. All planes belonging to the same family are said to be **crystallographically equivalent**. Likewise, all directions belonging to the same family of directions are also said to be **crystallographically equivalent**.

It turns out that for a cubic system (simple cubic, BCC, or FCC) to find the planes belonging to the same family of a particular plane, all we need to do is change the order of the indices of that plane whichever way we want and even place negative signs on the indices again whichever way we want. We actually saw this in **Example Problem** 4.7. {100} included (100), (010), and (001). There are even more planes to be added to these three, for example, $(\bar{1}00)$, $(0\bar{1}0)$, and $(00\bar{1})$.

Likewise, for directions, a cubic system will provide the same simplification. The <100> family of directions includes [100], [010], [001], $[\bar{1}00]$, $[0\bar{1}0]$, and $[00\bar{1}]$.

Example Problem 4.8
Determine the planes belonging to the {123} family of planes for an FCC unit cell.

Solution

The answer is simple: basically, order as we please!

$(123), (132), (213), (231), (312), (321), (\bar{1}23), (\bar{1}32), (2\bar{1}3), (23\bar{1}), (3\bar{1}2), (32\bar{1}), (1\bar{2}3),$
$(1\bar{3}2), (21\bar{3}), (23\bar{1}), (31\bar{2}), (32\bar{1}), (\bar{1}\bar{2}3)$ …etc.

Professor, imagine being asked that on an exam. If the question involved any cubic crystal structure, it would be easy to solve

*I could not agree more with you, Alex. Let me give you another advantage of the cubic system. Imagine someone asking you to tell them the **crystallographic direction perpendicular** to a **particular plane** within a <u>cubic</u> crystal structure. Let that plane be the (123) plane, for example. Then, the answer will be the [123] direction. We use the same indices and replace the brackets of a plane with those of a direction. That direction will be perpendicular to that plane. Cool? Of course, you cannot do this for a crystal structure other than a cubic one. So, be careful*

**The Sphere
Layer 3**

Prof., how can I find out what layer 3 in the sphere is made of?

X-ray diffraction (XRD), my friend, do you remember the Braggs? It could be a start. We will need to analyze sample 3 (which represents layer 3). XRD is discussed next. Continue reading, my friend.

The incidence of monochromatic X-rays on a material's surface causes diffraction. We can consider the separation between crystallographic planes as gratings, confirming X-rays' wave nature. Figure 4.18 shows X-rays diffracted by a crystal with an ordered atomic arrangement. In this case, we only consider three atomic planes, *A*, *B*, and *C*, each separated by a distance d. The X-rays incident on a surface are all initially in phase and are of a single wavelength (called monochromatic X-rays). When an X-ray meets an atom, it can get diffracted, and the atom gives off X-rays. If all these emerging diffracted waves align (i.e., are in phase), this will be called constructive interference. We typically say the incoming beam is diffracted. **W. Bragg**, in 1913, found a fundamental relationship that lives with us to this day. With reference to Fig. 4.20, $2x = 2d \sin\theta$ is a simple equation derived from basic trigonometry. The distance $2 \times$ is the extra distance traveled by ray 2 compared to

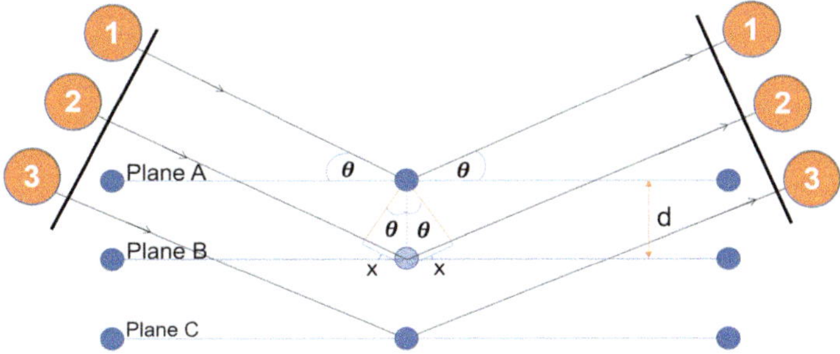

Fig. 4.18 Principles of X-ray diffraction

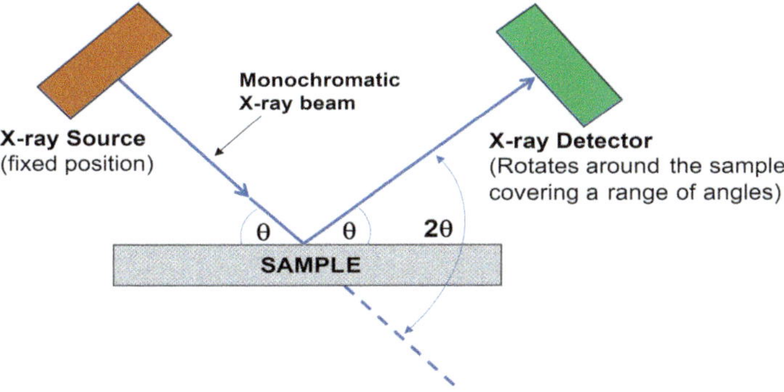

Fig. 4.19 Basic schematic of X-ray diffractometer operation

ray 1. So, if both rays were initially in-phase, for them to rejoin again after diffraction and still be "in phase", the distance "$2x$" needs to be equal to the wavelength (λ) or a multiple (n) of that wavelength, i.e., $n\lambda$.

This leads to **Bragg's Law** (Eq. 4.4):

$$n\lambda = 2d \, \sin \theta \tag{4.4}$$

The equation shows the conditions under which incoming X-rays can be diffracted constructively. The equation is also valid for different crystallographic planes, and it merely tells us the d-spacings of different planes that are diffracted at certain angles for an incoming X-ray with a known wavelength, λ. So, in a practical case, we would place a sample (typically in powder form) on a glass slide/sample holder inside an X-ray diffractometer.

Figure 4.19 shows a simplistic schematic of the main components of an X-ray diffractometer and its operation. The diffractometer consists of an X-ray source typically kept stationary, and its job is to produce a monochromatic X-ray beam (i.e., having a single wavelength), for example.

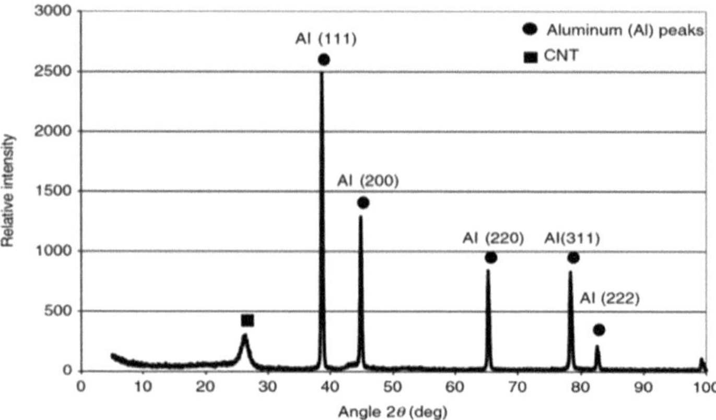

Fig. 4.20 X-ray diffraction scan of aluminum and carbon nanotube powder mixtures. (Fig. 6 in K. Morsi, A. M. K. Esawi, P. Borah, S. Lanka, A. Sayed, Characterization and Spark Plasma Sintering of Mechanically Milled Aluminum-Carbon Nanotube (CNT) Composite Powders, Journal of Composite Materials, Vol. 44, No. 16/2010, pp. 1991–2003, reproduced with Permission from Sage)

A typical wavelength is 0.154056 nm (however, other wavelengths can also be used), and the X-rays hit the sample surface at an angle of incidence, also called the **Bragg angle**. As can be seen from the figure, the incident beam is then diffracted and picked up by a detector. Recalling that a diffracted beam will only arise when Bragg's law is satisfied, which is picked up by the detector. Since the sample consists of numerous randomly oriented powder particles, planes will always be present to satisfy Bragg's law at the incident angle. Unlike the X-ray source, the detector does rotate around the sample, and by doing so, it can pick up diffracted X-rays at other incident angles. These diffracted x-rays represent different planes having different d-spacings. The intensity of the diffracted X-rays will be proportional to the amount of diffraction from planes that contribute to that reflection or intensity peak. The result will be a plot of 2θ versus relative intensity of the diffracted x-rays, where at specific 2θ values, peaks are generated. The scan thus will be like a fingerprint of a particular crystalline material that helps to identify it. Typically, the X-ray detector is made to rotate from a 2θ value of 30° to 140°, but this is not always the case, as it may be relevant to scan at lower angles to reveal specific peaks.

Figure 4.20 shows the results of an XRD scan on aluminum and carbon nanotube mixed powders. It clearly shows a peak at around $2\theta = 26°$ representing carbon nanotubes, while other peaks are identified as peaks (appearing at different angles) belonging to aluminum. Each peak represents diffraction from a different crystallographic plane within aluminum. In the figure, we see (111), (200), (220), (311), and

(222). There are also other peaks for aluminum that are not shown in the figure, but they would have appeared had the scan continued to above 100°.

So far, we have discussed crystalline materials and assumed absolute perfection in terms of the periodicity of the atomic positions within the lattice and throughout the material, and that "all" the atoms in the material belong to that same material. However, we typically find a small amount of impurity atoms (of elements other than the main element making up the material) present in addition to geometric irregularities within the material and crystal lattice. These are discussed in the next chapter.

Problems

4.1. Draw a hard-sphere and reduced-sphere model of an FCC unit cell.
4.2. List the advantages and disadvantages of the hard-sphere model.
4.3. List the advantages and disadvantages of the reduced-sphere model.
4.4. How many crystal systems are present on Earth?
4.5. List the 14 Bravais lattices.
4.6. What is the difference between the lattice parameters of a body-centered cubic unit cell and a body-centered orthorhombic unit cell?
4.7. How many atoms can be claimed to be within the boundaries of an FCC unit cell?
4.8. How many atoms can be claimed to be within the boundaries of a BCC unit cell?
4.9. Determine (from first principles) the densities (in g/cm^3) of nickel and lithium (you may use external sources).
4.10. Determine (from first principles) the relation between the side of a BCC unit cell and the atom's radius.
4.11. Inside a BCC unit cell, draw the [123], [$\bar{1}$01], [012], and the [121] directions.
4.12. Draw the [202], [431], and the [110] directions in separate simple cubic unit cells.
4.13. Determine the linear density of the [111] direction in an FCC unit cell in terms of the atomic radius, r.
4.14. Determine the linear density of the [100] direction in a BCC unit cell in terms of the atomic radius, r.
4.15. Determine three directions belonging to the <214> family of directions in a BCC unit cell.
4.16. Determine the color-coded directions in Fig. C4.1. Note that if a vector direction appears to end halfway along a side of the cube or in the middle of a plane, it is so.
4.17. Draw the (123), ($\bar{1}$01), (012), (002), (003), (112), and the (121) planes in a BCC unit cell. If that helps, you can use separate unit cells.
4.18. Determine the Miller indices of planes A, B, and C in Fig. C4.2
4.19. Determine the planar density of the (111) plane in a simple cubic unit cell.

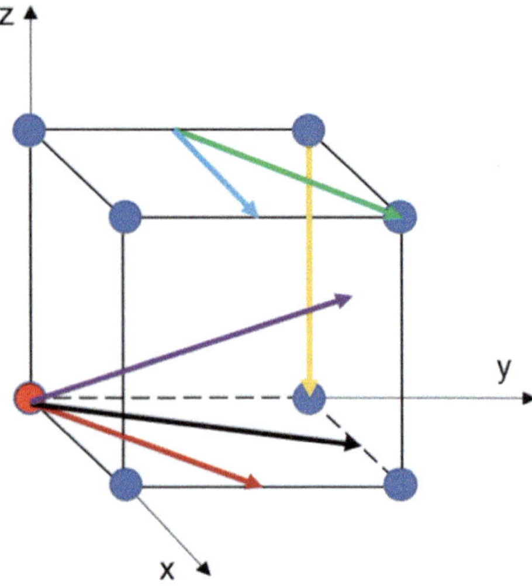

Fig. C4.1 Color-coded crystallographic directions in a simple cubic unit cell

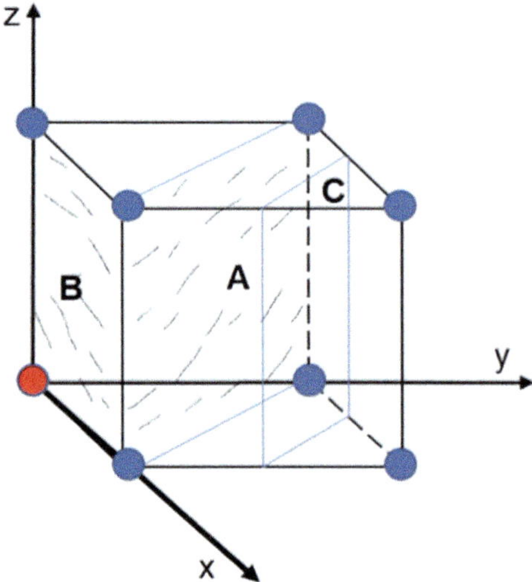

Fig. C4.2 Three planes, *A*, *B*, and *C*, in a simple cubic unit cell

4.20. Determine the planar densities of the (100) plane in an FCC, a BCC, and a simple cubic unit cell in terms of the atom's radius, *r*.

4.21. Determine planes belonging to the {100} family of planes in a tetragonal crystal structure.

4.22. What is the crystallographic direction perpendicular to the (100) plane in nickel?

4.23. What is the d-spacing between (111) planes in an FCC unit cell, in terms of the side of the cube "a"?

4.24. Determine the numeric value of the *d*-spacing between the (111) planes for copper.

4.25. X-ray diffraction (XRD) was conducted on aluminum powder. It was determined that the d-spacing of the (111) plane was 0.2338 nm. If the X-ray had a monochromatic wavelength of 0.154056 nm, determine the 2θ value on the XRD scan at which the peak for the (111) plane should appear.

Chapter 5
Alloys and the Perfection of Crystal Imperfections

5.1 Introduction

This chapter discusses irregularities that appear in the crystal lattice. Although typically referred to as crystallographic defects, we argue that such defects are necessary features of materials. The chapter also discusses alloying to produce solid solutions and several lightweight metals.

5.2 What Are Alloys?

We have all heard the words metal and alloy. Nowadays, car salespersons try to sell cars by saying, "They come with alloy wheels!" So, our society now views the word "alloy" as appealing.

The word alloy typically refers to metal alloys, where one or more impurity atoms are added to a metal and dissolve atomically in the metal. It is important to note that there are also ceramic alloys. An old example of alloys is iron meteorites, which are alloys of iron and nickel. Plain carbon steel is another notable example of an alloy, this time of iron and carbon. Carbon atoms here are the impurity atoms. Metal alloys are not limited to mixing only one impurity atom with a metal. The number of different impurity atoms added to a metal can be one, two, three, four, or more. If one atomically mixes **two** elements (the metal and the impurity atoms), the product will be a **binary alloy**. However, if one mixes **three elements**, it would be called a **ternary alloy**, and four elements produce a **quaternary alloy**. In contrast, an alloy containing five different types of atoms is a **quinary alloy**. This makes a pure metal a **unary "system"**, so we should not use the word alloy here. These definitions are summarized in Table 5.1.

Our current scientific understanding of unary, binary, and ternary systems is advanced; however, typically, we only add small amounts of alloying elements to the

K. Morsi, *An Engaging Approach to the Science and Engineering of Materials*,
https://doi.org/10.1007/978-3-032-06231-4_5

Table 5.1 Types of metals and alloys

Type	Number of elements
Unary system	1
Binary alloy	2
Ternary alloy	3
Quaternary alloy	4
Quinary alloy	5

main metal. Only recently, a new set of alloys has emerged in the scientific arena, called **High Entropy Alloys (HEA)**, or what are known as **Cantor alloys**, after **Prof. Brian Cantor**. These have been defined simply as alloys consisting of five or more elements added in approximately equal amounts (i.e., not in small amounts as we have been accustomed to). With the arrival of these new alloys, our previously established rules regarding phase predictions broke down, and suddenly new findings emerged; for example, instead of complex phases/crystal structures being produced, a simple crystal structure is formed, for example, a BCC crystal structure. The interest in these alloys is that they provide us with superior properties than conventional ones. One example of HEA is the CrMnFeCoNi alloy, where its constituents are added in equal atomic ratios.

5.3   The Perfection of Crystallographic Imperfections

We expect a crystal structure with all its atoms arranged perfectly in 3D, with no irregularities (Fig. 5.1a). However, we soon discover the presence of some irregularities, and we are quick to label them defects, impurities, or imperfections. Let us, however, tackle some of these "so-called" imperfections and delve deeper to see if they are redundant or serve a purpose.

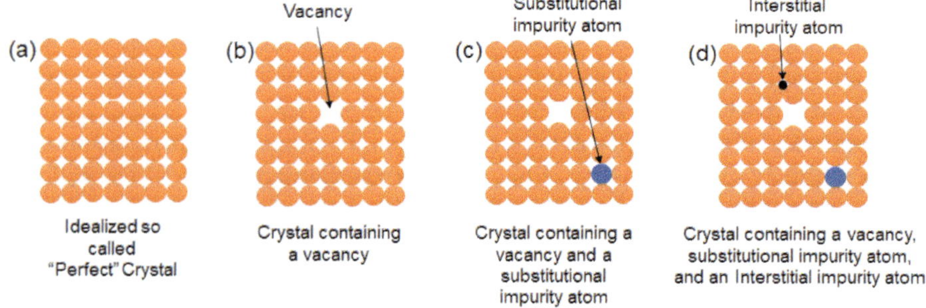

Fig. 5.1 Illustration of part of a crystal lattice showing **a** an idealized perfect crystal, **b** a vacancy; **c** a vacancy and a substitutional impurity atom; **d** a vacancy, a substitutional impurity atom, and an interstitial impurity atom

We will start our investigation with the so-called **point defects**. People refer to them as point defects since, geometrically, a point can represent them. For example, a missing atom (or vacancy), a foreign impurity atom (whether occupying the position of a host atom or in the interstitial space in between the host atoms), or a host atom that is not in its proper place (for example, a self-interstitial atom, where a host atom is lodged in the interstitial space between the host atoms within the crystal lattice). Figure 5.1(b-d) shows some of these point defects.

5.3.1 Point Defects

(a) **Vacancies**

As a recap, if we were presented with a block of a metal like aluminum, for example, we could imagine it as being made up of an enormous number of crystals (also called grains); each grain will be a crystal lattice where atoms are ordered in a specific long-range order, e.g., for aluminum, it will be a repeated FCC unit cell. While all the grains will contain such a crystal lattice, the grains are randomly oriented to one another. However, depending on the grain size, it can contain millions of billions of atoms. Let's do a quick and rough calculation. For example, if we imagine stacking three aluminum atoms in a row, with each atom having a diameter of 0.2864 nm, the length of this atomic arrangement from end to end will be 0.85 nm. For simplicity, let us assume that three atoms will be approximately 1 nm long. If we have a cube of side 50 μm, i.e., 50,000 nm, we will have $3 \times 50,000 = 150,000$ atoms just along one direction of the grain. To get the approximate number of atoms within the 3D grain, we have $(150,000)^3 = 3 \times 10^{15}$ atoms, i.e., 3×10^6 (i.e., million) $\times 10^9$ (i.e., billion) atoms. This is an enormous number of atoms, and the assumption of a perfect crystal would mean that all the atoms making up the crystal lattice would be where they are supposed to be. However, not all the positions of atoms are filled with atoms. A relatively small number of atoms will always be missing from their positions, leaving empty spaces behind (i.e., vacant lattice sites). An empty atomic site is referred to as a **vacancy,** as shown in Fig. 5.1b. Vacancies can be observed using modern instruments such as the **Scanning Probe Microscope** or the **High-resolution Transmission Electron Microscope**, and their concentration can be measured experimentally using **Positron Annihilation Spectroscopy**. The equilibrium number of vacancies present in a material can also be determined using the thermodynamic property Gibbs free energy, which is found to increase exponentially with an increase in temperature.

Equation (5.1a), derived from thermodynamic principles, shows how to calculate the equilibrium number of vacancies (n_v), typically per m^3, knowing the total number of atoms (n_a) per m^3, the activation energy (Q), which is the amount of energy needed to form one mole of vacancies, the temperature (T) in Kelvins, and the gas constant (R).

$$\frac{n_v}{n_a + n_v} = e^{-Q/RT} \tag{5.1a}$$

Since the number of vacancies is extremely small compared to the number of atoms, Eq. 5.1a can be approximated to

$$\frac{n_v}{n_a} = e^{-Q/RT} \tag{5.1b}$$

To get a rough idea about the relative number of vacancies we expect to find in a metal, let us consider aluminum (Al), with a Q value of 0.67 eV (~ 65 kJ/mol). Close to its melting point, there is approximately a vacancy for every 5000 atoms. At first glance, this looks pretty insignificant, but since we are speaking about 3D crystals, 4913 atoms would be present in a cubic arrangement of atoms with a side length comprising 17 atoms. Now, imagine that a vacancy would be present within such an infinitesimally small arrangement of atoms. Suddenly, a vacancy every 5000 atoms is not at all that insignificant.

Figure 5.2 plots the fractional vacancies versus temperature in Kelvin for aluminum up to a temperature just below its melting point. The exponential rise in fractional vacancies is easily seen.

Vacancies have essential roles. For example, one of the most fundamental mechanisms utilized in manufacturing is atomic diffusion, which can influence alloy design or even the sintering of powders to produce engineering products. Generally, atoms (e.g., impurity atoms of another element) may need to diffuse or move in the crystal lattice of the host material from one location to another. If that impurity atom is similar

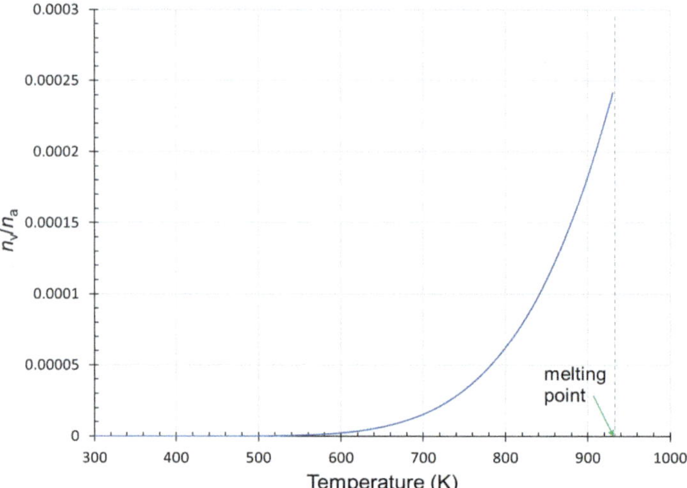

Fig. 5.2 A plot showing the exponential increase in fractional vacancies with temperature for aluminum

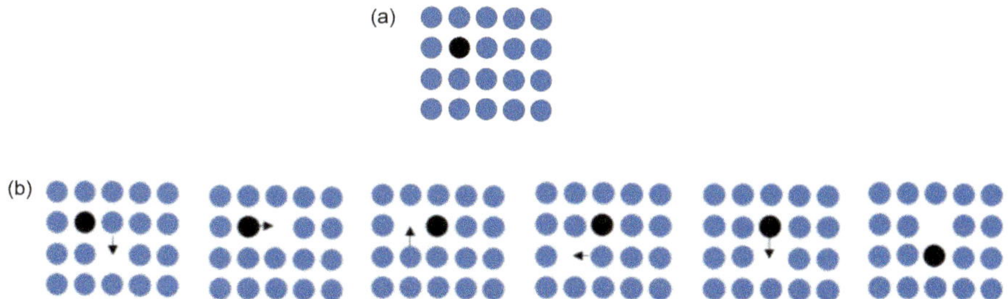

Fig. 5.3 Illustration of **a** an impurity atom (black) not able to diffuse due to the absence of a vacancy, and **b** an impurity atom able to diffuse due to the presence of a vacancy (Fig. 5.3b after R.A. Higgins *Properties of Engineering Materials, Hodder and Stoughton, 1986)*

to the host atoms, then atomic diffusion would be impossible since there is simply no space for the atom to diffuse to (Fig. 5.3a). However, looking at Fig. 5.3b, one can quickly realize how the presence of a vacancy could facilitate the diffusion of an impurity atom. This is a big deal and is a type of diffusion called **vacancy diffusion**. These vacancies can even strengthen materials at relatively low temperatures.

(b) Impurity Atoms

Another form of irregularity within the crystal lattice are impurity atoms from a different element than those of the host material.

For example, if we look at a metal such as silver, the first thing that comes to our minds is that all the atoms inside this metal are silver. However, it turns out that due to manufacturing and extraction processes, not all the atoms are silver, and a small number of atoms belong to other elements. These have been referred to as **impurity atoms** or **point defects** (since an atom represents a point geometrically) and are regarded as crystal imperfections. Incredibly, if we observe closely the influence of these "so-called" imperfections, we can easily claim that they are, in fact, at the **heart of perfection,** depending on the application. Let me explain. It turns out that having a small amount of **unintentionally** present impurity atoms within a metal can make it stronger than if it were 100% pure (i.e., 100% silver atoms). This fact has led us to intentionally add different impurity atoms to metals and produce what we know today as alloys, stronger than the pure metal form. The word purity here also needs some clarification. It is virtually impossible to produce a material that is 100% pure. Hence, people have arrived at terms such as **commercially pure (CP)** to establish some expected purity for metals, typically with a minimum of 99% by weight of the intended metal. For example, we have commercially pure grades such as CP–Ti grades 1–4 for titanium. Where grade 1 contains impurities 0.2 wt% Fe and 0.18 wt% O, grade 2 (0.3 wt% Fe and 0.25 wt% O), grade 3 (0.3 wt% Fe and 0.35 wt% O) and grade 4 contains (0.5 wt% Fe and 0.4 wt% O).

Outside of the CP metals, there are very important alloys where alloying elements have been intentionally added in significant amounts. For example, one of the most popular Ti alloys is the Ti–6Al–4 V (i.e., 6 wt% Al and 4 wt% V). This alloy produces a unique microstructure and provides properties suitable for biomedical and aerospace applications. Other examples of alloys are steel (made up of at least iron and carbon), aluminum alloys, copper alloys, and magnesium alloys. Each will have its advantages and attributes.

A fundamental question then arises. How exactly does the addition of impurity atoms make a material stronger? Before answering this question, we must discuss the alloying concept further. If we have a commercially pure metal, apart from a few impurity atoms, all the atoms in the crystal lattice(s) are made up of that metal, which we call **host** atoms. If we want to add atoms from other elements to the lattice, these foreign/impurity atoms can mix with host atoms in two ways. Either the incoming atoms take up a lattice position of one of the host atoms (i.e., become a **substitutional impurity atom**), as seen in Fig. 5.1c, or the impurity atom could sit in the always available vacant **interstitial** space between the host atoms within the lattice and become an **interstitial impurity atom** (Fig. 5.1d). In both cases, we have dissolved the impurity atoms (also called the solute atoms) inside the crystal lattice (also called the solvent), and just like dissolving sugar or salt in water to produce syrup or saline solutions, we now have formed a **solid solution**. Solutions don't have to be liquids; solids also have solutions. If the solute atoms substitute the position of host (solvent) atoms, then the solution formed would be called a **substitutional solid solution**. On the other hand, if the impurity atoms only occupy the interstitial space between the host atoms, the solution will be called an **interstitial solid solution**. These are shown schematically in Fig. 5.4a, b. Whether impurity atoms enter the lattice and produce a substitutional solid solution or not one will depend on several factors. These are the **Hume-Rothery Rules**, after University of Oxford Professor **William Hume-Rothery**.

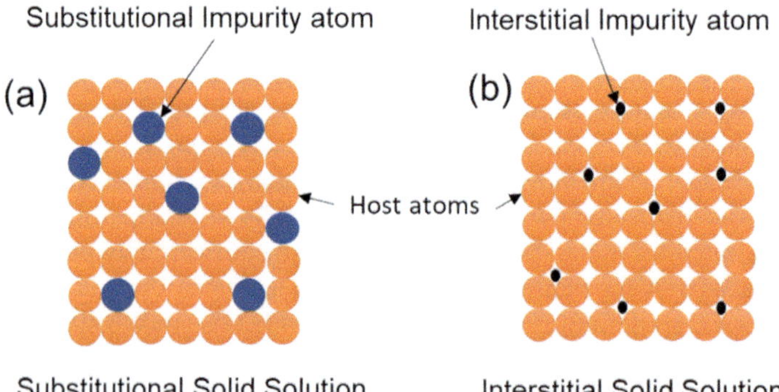

Fig. 5.4 **a** Substitutional solid solution, **b** Interstitial solid solution. In both cases, most atoms (orange) belong to the solvent. Whereas the added atoms (black) are the solute

Alex, I am struggling to meet a research grant proposal deadline, so I must go. Tomorrow, please learn about the Hume-Rothery Rules, one of the most fundamental rules in alloy design

The Hume-Rothery Rules

As can be imagined, the number of alloys that can be produced is considerable if we consider all the different elements that can be mixed. One fundamental question arises. If we mix any two elements, is it necessary that they dissolve in one another and that their atoms mix on the atomic level in one crystal lattice? The answer to that question came from the pioneering work of Oxford Professor Hume-Rothery; he laid down the rules for producing substitutional solid solutions. Let's have a look at what he said.

Rules for Substitutional Solid Solutions

The following will favor the formation of substitutional solid solutions and extended solubility of substitutional impurity atoms in the host atom crystal lattice:

1. If the impurity atoms being added have radii within $\pm 15\%$ of the radius of the host metal atoms. The closer their size is, the less lattice distortion when substituting the host atoms. This leads to the conclusion that the more the solute atom size differs from the host atom size, the less its maximum solubility.
2. If the impurity atoms naturally crystallize in the same crystal structure as the host atoms (e.g., both are FCC or BCC, for example).
3. The impurity atoms have a similar electronegativity to the host atoms.
4. The impurity atoms have the same or higher valency than the host atoms. However, if the impurity atoms have lower valency than the host atoms, it would not be favorable for them to go into a solution. This rule applies mainly to alloys of silver, copper, and gold (i.e., monovalent) metals when other higher valency atoms are added to them.

The Hume-Rothery rules, they make sense, don't they? If a lattice wants to welcome impurity atoms, don't you think it will want to favor atoms like its own? Now, to the point about the valency, since metals predominantly bond using metallic bonding, each host atom in the crystal lattice supplies its share of its valence electrons to the electron cloud, for example, two electrons. If an impurity atom wants to join the lattice and only offers 1 electron to the electron cloud, that wouldn't be good business, would it? However, if an impurity atom comes in and offers three electrons instead of two, I would say the lattice made a good deal. If it could talk, it would say to the impurity atom, welcome home

When these rules are applied, we see some interesting things. Some alloys that highly conform to these rules have complete solubility in one another (or are said to be completely miscible). Such alloys include Cu–Ni, Au–Pt, Ag–Au, As–Sb, and Ti–Zr. Table 5.2 shows how gold (Au) and silver (Ag) adhere very well to the Hume-Rothery rules.

What about alloys that do not conform so highly to the Hume-Rothery rules? With reference to Table 5.3, let's see how the Hume-Rothery rules apply when divalent (valency 2) (*solute*) atoms that crystallize as hcp (i.e., beryllium (Be), cadmium (Cd), and Zinc (Zn)) try to dissolve in FCC lattices of copper (Cu) or silver (Ag), respectively. When Be is added to Cu (valency 1 or 2), where the Be atom is 12.9% smaller than the Cu atom (i.e., below the $\pm$ 15% size limit), its maximum solubility in Cu is 16.6 at.%. In contrast, when Be is added to Ag (valency 1), where the Be atom is 22.9% smaller than silver atoms, its maximum solubility is only a mere 3.5 at. %. Let's move on to Cadmium (Cd); here, Cd again is divalent and is 16.5% larger than Cu atoms (i.e., exceeds the size limit), and hence we see only a solubility of 1.7 at.% Cd in Cu. The Cd atom is only 3.1% larger than the Ag atom, resulting in a much higher maximum solubility of 42.5 at.%. Finally, the **zinc (Zn)** atom happens to be 4.2% larger than the Cu atom and 8% smaller than the Ag atom, i.e., the size factor is satisfied for both Cu and Ag, resulting in significant and similar maximum

Table 5.2 Application of Hume-Rothery rules: Au and Ag have *complete* solid solubility in one another

Element	Crystal structure	Valency	Electronegativity	Atomic radius (nm)
Au	FCC	1	2.54	0.1445
Ag	FCC	1	1.93	0.1442

Table 5.3 Demonstration of the Hume-Rothery rules (part of the data extracted from Table 14.1 in *A. Cottrell, An Introduction to Metallurgy, Second edition (reprint), The Institute of Materials (1995)*)

Solute atom (valency, electronegativity)	Solvent atom (valency, electronegativity)	Atomic size factor (%)	Maximum solubility of solute in solvent (at.%)
Beryllium, Be (2, 1.57)	Copper, Cu (1 or 2, 1.9)	− 12.9	16.6
Beryllium, Be (2, 1.57)	Silver, Ag (1, 1.93)	− 22.9	3.5
Cadmium, Cd (2, 1.69)	Copper, Cu (1 or 2, 1.9)	+ 16.5	1.7
Cadmium, Cd (2, 1.69)	Silver, Ag (1, 1.9)	+ 3.1	42.5
Zinc, Zn (2,1.65)	*Copper, Cu (1 or 2, 1.9)*	*+ 4.2*	*38.4*
Zinc, Zn (2,1.65)	*Silver, Ag (1, 1.9)*	*− 8.0*	*40.2*

solubilities of 38.4 at.% and 40.2 at.% in Cu and Ag, respectively. This exercise demonstrates the strong influence of the size factor on the maximum solubility of substitutional solid solutions. Now, we don't have complete solubility because other Hume-Rothery rules have not been satisfied; for example, for a substitutional solid solution, solute and solvent atoms should normally crystallize in the same crystal structure, which they don't in this case.

Interstitial Solid Solutions

Regarding interstitial solid solutions, we need to consider the size of the available interstitial space between the host atoms. This will depend on the type of interstitial site and the crystal structure. For example, within the FCC crystal structure are two types of interstitial sites: an **octahedral** site and a **tetrahedral** site. These are presented in Fig. 5.5 using reduced-sphere models of an FCC crystal structure.

As it happens, the relative radius of the interstitial sites in relation to the radius of the host atoms can be easily found if we employ the hard-sphere model. These are **0.414 R_{host}** and **0.225 R_{host}** for the **octahedral** and **tetrahedral sites**, respectively. Hence, the tetrahedral sites are about 46% smaller than the octahedral ones in the FCC crystal structure. Moreover, as the figure shows, the coordination number for the octahedral sites is 6, while that of the tetrahedral is 4. The names tetrahedral and octahedral refer to the number of faces these shapes have. This can also be seen in Fig. 5.5, the octahedral shape has eight faces, and the tetrahedral has four faces. An FCC crystal structure has four octahedral sites and eight tetrahedral sites. Regarding the BCC crystal structure, there are **6 octahedral sites** and **12 tetrahedral** ones. The octahedral sites in the BCC crystal structure are unlike those in the FCC counterpart. Two of the six atoms are closer to the interstitial site than the other four. This means the coordination number of the octahedral interstitial sites in the BCC crystal structure is two rather than 6. Additionally, the relative radii of the

Fig. 5.5 Interstitial sites in FCC crystal structure **a** octahedral site and **b** tetrahedral site

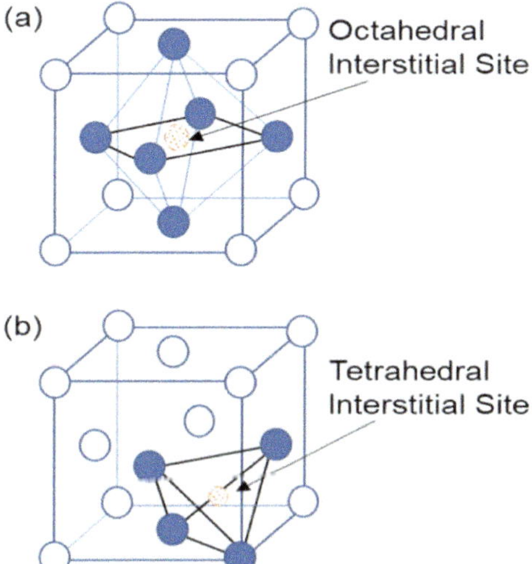

Table 5.4 Summary of octahedral and tetrahedral interstitial site information in BCC and FCC crystal structures (multiple sources)

	FCC	BCC
Number of **tetrahedral sites**	8	12
Relative radius of interstitial site	$0.225\,R_{host}$	$0.291\,R_{host}$
Coordination number	4	4
Number of **octahedral sites**	4	6
Relative radius of interstitial site	$0.414\,R_{host}$	$0.154\,R_{host}$
Coordination number	6	2

interstitial sites in relation to the radius of the host atoms for both the **octahedral** and **tetrahedral** sites in BCC crystal structures are **0.155 R_{host}** and **0.291 R_{host}**, respectively. As opposed to the FCC structure, here we have the tetrahedral sites that are not smaller than the octahedral ones but rather ~ 88% larger. Table 5.4 provides a summary of the above facts. So, knowledge of the size of the available interstitial space is essential; however, it turns out that interstitial atoms that occupy some of these spaces have a radius that is a little larger than the interstitial space, meaning that the atoms around it will be pushed, resulting in some lattice strain. The above information can have implications for solubility limits in interstitial solid solutions and the strengthening effectiveness of interstitial impurity atoms by solid solution strengthening.

5.3.2 *Line Defects*

When theoreticians calculated the theoretical shear strength of metals, which, if exceeded, would lead to the plastic deformation of the metal, their calculations were found to overestimate the experimentally observed strength values, i.e., metals would plastically deform (permanently change shape) at a stress value ~ 10,000 × less than what their calculations predicted. Incredibly, in 1934, three independent papers by **G.I. Taylor, E. Orowan,** and **M. Polanyi** were published, suggesting that the discrepancy can be explained by the presence of imperfections (what we know today as dislocations) that require a low stress to move and generate permanent deformation that can change the shape of a material. One type of these dislocations is called an **edge dislocation,** as seen in Fig. 5.6, which can be viewed as an extra half plane of atoms.

The remarkable vision of these scientists was confirmed years later (in the early 1950s) through the observation of dislocations in silver-halide crystals by **Hedges** and **Mitchell** using a decorative method. This was quickly followed by the first-ever observation of dislocations using electron microscopy (following the invention of the electron microscope) and proof that they indeed move under the application of shear stress. This remarkable work was carried out in 1956 (more than 20 years after the proposition of Orowan, Taylor, and Polanyi) by the University of Oxford Professor

Fig 5.6 Example of an edge
dislocation

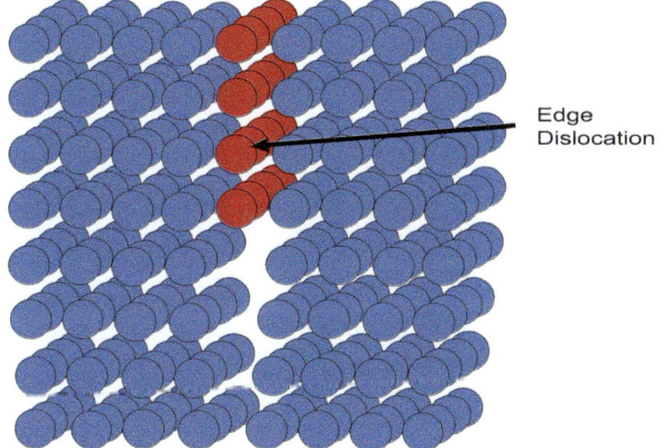

Sir Peter Hirsch and co-workers. These dislocations are ultimately responsible for the observed permanent change in the shape of metals and alloys when subjected to a significant stress. As one can imagine, our technological world would have been greatly affected without these dislocations, particularly manufacturing. For example, manufacturing processes such as forging, extrusion, drawing, and rolling would not have been possible without these "so-called" imperfections. Not only that, but metals are also **forgiving materials**, which means a metal component in service would bend or dent rather than fracture, unlike a ceramic, if an overload occurred for some reason. Again, this forgiveness stems from the presence of dislocations. This is what I mean by saying they are at the heart of perfection. They serve a purpose, don't they? Another type of dislocation is the **screw dislocation** which, although it travels in a different way than an edge dislocation, nevertheless produces the same effect as an edge dislocation. Both dislocations exist within metals and alloys and contribute to the observed permanent deformation of metals and alloys under significant stress levels. Figure 5.7 shows an ultrahigh-resolution image of an edge dislocation in a thin copper film.

As mentioned, dislocations give rise to permanent deformation and the malleability of metals. At the same time, if we engineer the material to hinder the motion of dislocations, we will make the material less malleable and stronger. More about that later. Since these entities are essential, we must know their abundance in a given material. The way dislocations are quantified is through what is called the **dislocation density**. The dislocation density is defined as either **the total length of dislocations** (in m, cm, or mm) **per volume of material** (in m^3, cm^3, or mm^3) or **the number of dislocations intersecting a unit area** (in m^2 or mm^2). This leaves the overall units of dislocation density for both definitions as **m^{-2} or mm^{-2}**. Typically, the lowest dislocation density is found in metals annealed at high temperatures and cooled very slowly. In this case, we can expect a dislocation density of $\sim 10^6$ cm/cm^3, which is 10 km/cm^3. Whereas much more significant amounts are found in a material that has been mechanically worked at relatively low temperatures, say by forging or rolling, here, dislocation densities on the order of $\sim 10^{10}$ cm/cm^3 or 100,000 km/

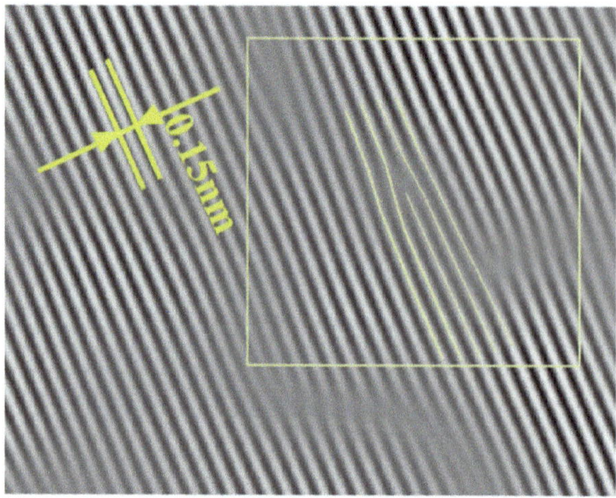

Fig. 5.7 Edge dislocation in electrodeposited thin film of copper (the dislocation is shown in the middle of the yellow box). Image from: Ni, K.; Wang, H.; Guo, Q.; Wang, Z.; Liu, W.; Huang, Y. The Construction of a Lattice Image and Dislocation Analysis in High-Resolution Characterizations Based on Diffraction Extinctions. Materials 2024, 17, 555. https://doi.org/10.3390/ma17030555. *Image reproduced with modification under the Creative Commons Attribution (CC BY) license* (https://creativecommons.org/licenses/by/4.0/)

cm^3 are possible. Now, we could look at these numbers, move on, or ponder and contemplate instead. We are saying that we can have a total dislocation length of 100,000 km within a 1 cm^3 worked metal (the size of a sugar cube). To provide a better perspective, the circumference of our planet is ~ 40,000 km. This means that a worked metal (i.e., shaped by force) the size of a sugar cube could contain total lengths of dislocations that are 2.5 times the circumference of the Earth. Nevertheless, this exercise strongly suggests that dislocations are abundant and should never be ignored.

5.3.3 *Planar/Interfacial Defects*

Other types of irregularities include the following:

(a) **Grain Boundaries**

We previously discussed grain boundaries between randomly oriented grains. Within these grain boundaries, atoms lacked any order and displayed poor packing compared with atoms within the grain. We also defined the formation of a low-angle grain boundary between grains as being misaligned by < 15°, and a high-angle grain boundary with $a > 15°$ misalignment. We also concluded that high-angle grain boundaries have more energy than low-angle ones. The unique structure of grain boundaries allows them to influence atomic diffusion, strength, corrosion, and even optical and

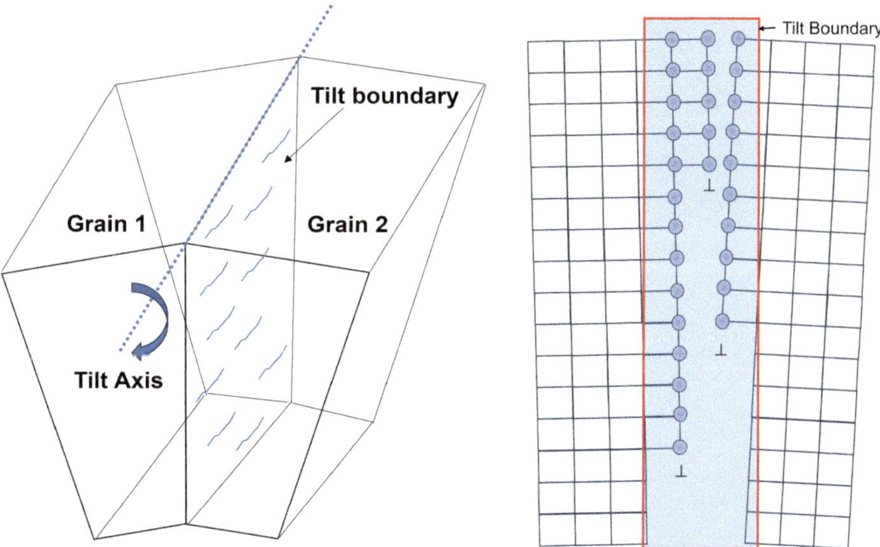

Fig. 5.8 Simple schematics of tilt boundary (schematic on the right is redrawn and modified from W.T. Read, Dislocations in Crystals, Legare Street Press, (2022))

electrical properties. They play an even more significant role in nanostructured materials (i.e., with grain sizes ≤ 100 nm), for example, render a typically brittle material ductile.

Other boundaries besides these traditional grain boundaries between randomly or otherwise oriented grains of the same material are also important. These include tilt boundaries, twist boundaries, twin boundaries, and phase boundaries, as shown below.

(b)   **Tilt Boundaries**

These low-angle grain boundaries can be produced through the alignment of edge dislocations. This kind of boundary is shown in Fig. 5.8. As can be seen, it is as if two separate crystals (grains) have been tilted slightly and then joined. The simplified schematic on the right exemplifies how a series of edge dislocations could be seen to form the boundary eventually.

(c)   **Twin Boundary**

An example of twin boundaries that can occur inside a grain is shown in Fig. 5.9. The figure shows two twin boundaries, which represent mirror planes. The lattice regions above and below each of these boundaries are seen as mirror images of one another. Between the two twin boundaries is a region where atoms have been displaced by shear compared with the rest of the lattice. This region is called a twin or twin band. Note that the plane of atoms closer to the twin boundary is shifted a certain distance,

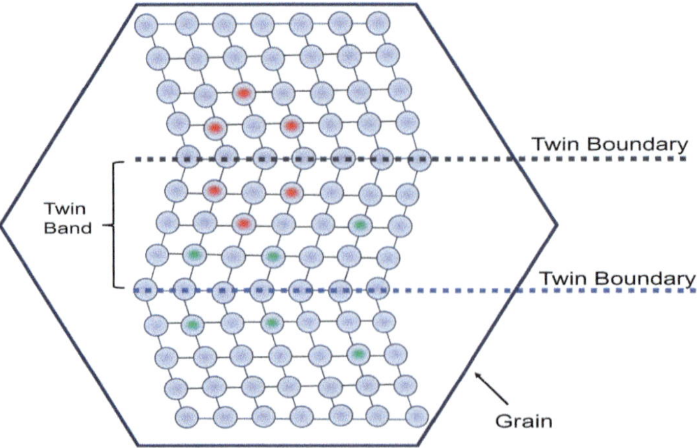

Fig. 5.9 Schematic of a twin and accompanied twin boundaries. Red and green marked atoms are mirror images of other atoms of the same color on the other side of their respective twin boundaries (based on multiple sources)

but as we move away from the twin boundary, the consecutive planes are shifted to larger and larger distances. As one can imagine, twin boundaries like dislocations can also contribute to plastic deformation and shape change of materials. However, the stress required for twins to form is higher than that needed to move dislocations. This means that twinning typically takes a secondary role for ductile materials where dislocations are free to move. However, twinning can be prominent in materials where dislocation motion is hindered. Figure 5.10 is a schematic showing how permanent deformation is possible with twinning.

There are two primary types: **Annealing twins**, which can appear after a metal is deformed at low temperatures and then annealed at high temperatures. These twins occur as part of the grain growth process; more about this later. For metals, typically, FCC materials can exhibit annealing twins. On the other hand, **mechanical twins** appear, for example, in BCC and HCP metals; these typically are favored by rapid deformation (for example, rapid bending) at low temperatures. One notable

Fig. 5.10 The deformation of materials via twinning (*after R.A. Higgins Properties of Engineering Materials, Hodder and Stoughton, 1986*)

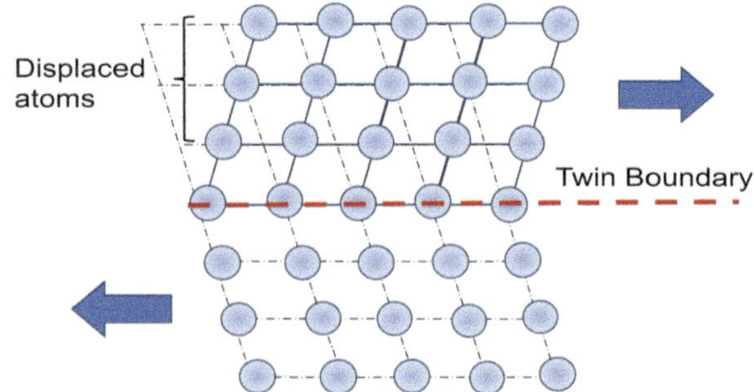

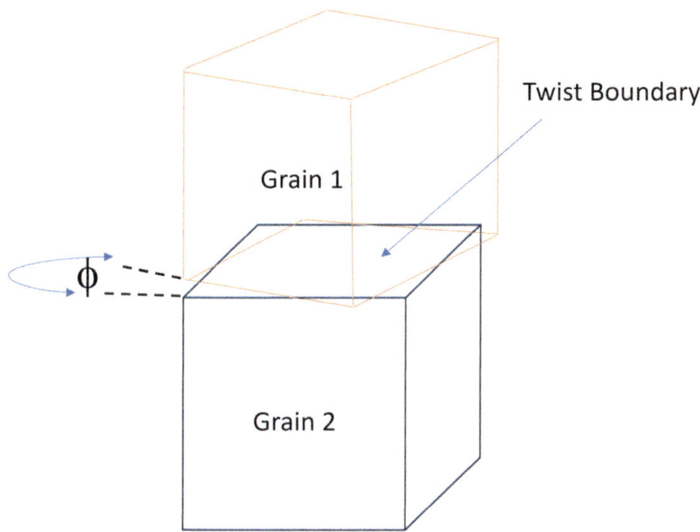

Fig. 5.11   Schematic illustrating a twist boundary of angle ϕ (multiple sources)

example of twinning is the deformation of β-tin, which has a body-centered tetragonal crystal structure. Here, the deformation of tin generates considerable twinning with an accompanying crackling sound. Our ancestors have referred to this sound as "the cry of tin".

(d)   **Twist Boundary**

Twist boundaries form if the two grains separated by the boundary are twisted relative to each other by an angle, as shown in Fig. 5.11.

(e)   **Phase Boundaries**

While grain boundaries are boundaries between grains of the same material and crystal structure, **phase boundaries** exist between grains of different phases. These can be different materials or the same material but with a different crystal structure. Figure 5.12 shows this schematically. Depending on the interfacial energies and whether the energies are isotropic around any given phase, as can be seen, the gray-colored phase (phase 1) in the microstructure can appear as a full-blown grain, as a precipitate inside a white phase 2 grain, or at the grain boundary between two grains of phase 2.

Several situations could arise when these two phases meet at a boundary. Either the two lattices of the phases have the same orientation and crystal structure and match perfectly at the boundary, or they may not have the same crystal structure but still join continuously at the boundary. Such a phase boundary would be considered **coherent** (Fig. 5.13a). These boundaries have been referred to as strain-free and typically have very low phase boundary energies, for example, 0.05 J/m^2. An example of coherent phase boundaries can be found between the hcp silicon-rich phase and

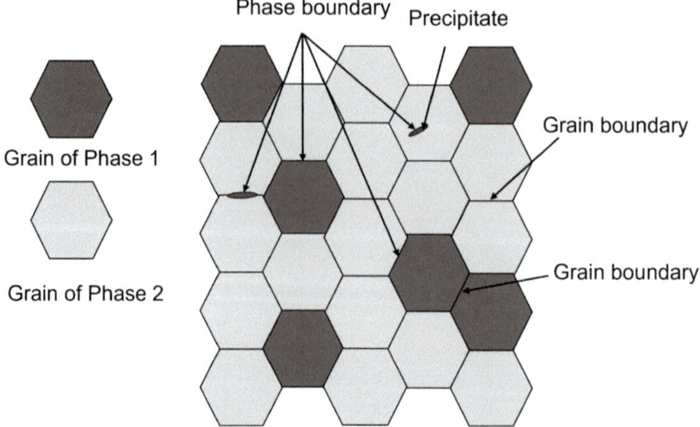

Fig. 5.12 Difference between grain boundaries and phase boundaries

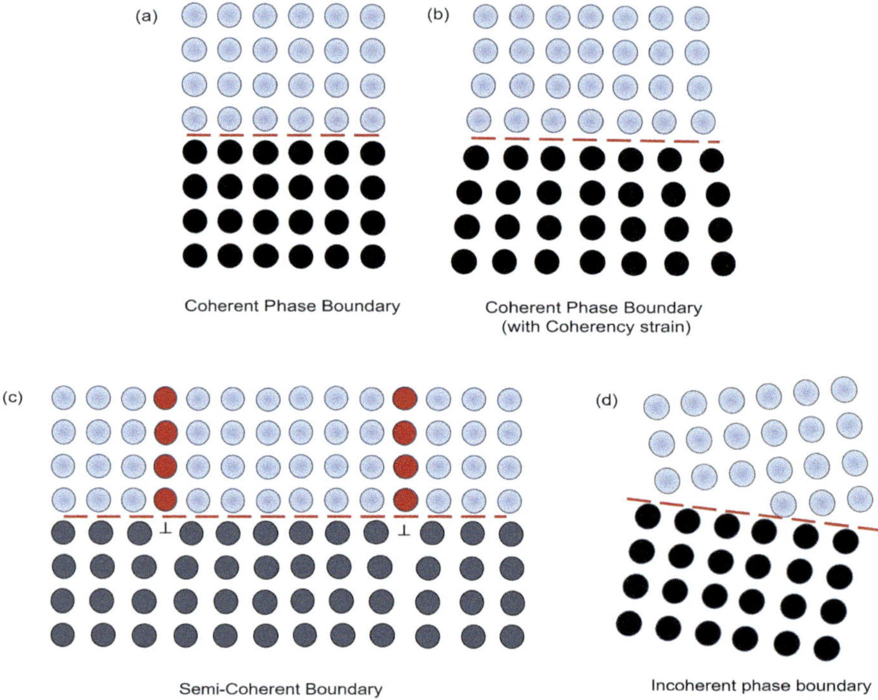

Fig. 5.13 Types of phase boundaries (dashed red) between two phases **a** coherent phase boundary, **b** coherent phase boundary with coherency strain, **c** semi-coherent phase boundary, and **d** incoherent phase boundary (*after M. F. Ashby, D. R. H. Jones, Engineering Materials 2, second edition (Butterworth Heinemann (2001))*)

the FCC copper-rich phase in Cu-Si alloys. If the atomic spacing is different in each grain, it will lead to *coherency strain* at the boundary (Fig. 5.13b). If the strain becomes too large, the creation of dislocations helps to accommodate the situation, resulting in semi-coherent boundaries (Fig. 5.13c). When the boundaries are between two phases that are chemically and structurally different, they are typically **incoherent phase boundaries** (Fig. 5.13d). These incoherent phase boundaries tend to have boundary energies similar to those of regular grain boundaries, i.e., 0.5 J/m^2. Therefore, depending on the coherency at the phase boundary, the energy of these boundaries will be different.

(f)   Stacking Faults

We previously saw close-packed arrangements, where FCC and HCP had the arrangements ABCABC... and ABAB..., respectively. We may find stacking faults in some local crystal regions, as shown in Fig. 5.14. Here, either part of a plane is missing, referred to as an intrinsic stacking fault, or an extra part of a plane is added, which is an *extrinsic stacking fault*. Stacking faults increase the energy of the material.

(g)   Surfaces

With reference to Fig. 5.15, surfaces can be present in a material either externally, i.e., at the outer extremities of the material, or internally, for example, the inside of a pore or the internal surface of a crack. It is important to note that surfaces play an important role in materials.

   A surface can be viewed as a planar irregularity from the point of view of atomic bonding. With reference to Fig. 5.16, we can categorize two types of atoms for a hypothetical crystalline material: interior atoms (color-coded blue) and surface atoms (color-coded red). If we analyze the bonding experienced by a given interior atom, we will find that the bonding is satisfied; for example, for an FCC or HCP material, each atom will be *surrounded* and bonded to 12 other atoms (hence their

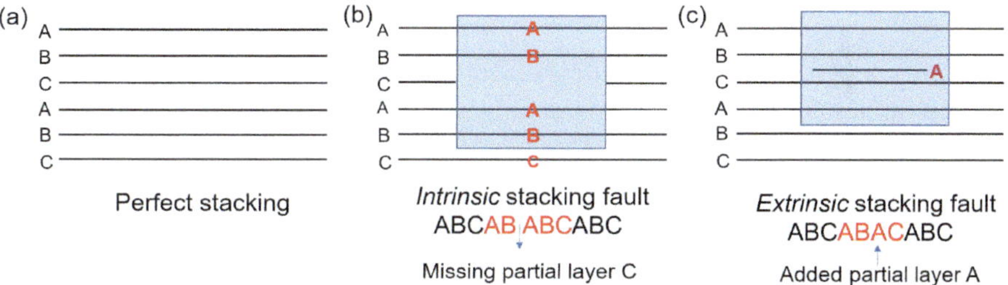

Fig. 5.14   a Crystal with perfect stacking of close-packed planes ABCABC..., **b** a region with a missing part of the C plane resulting in an *intrinsic* stacking fault, **c** a region where an extra A plane is inserted causing an *extrinsic* stacking fault. Shaded areas are regions of significant strain (based on multiple sources)

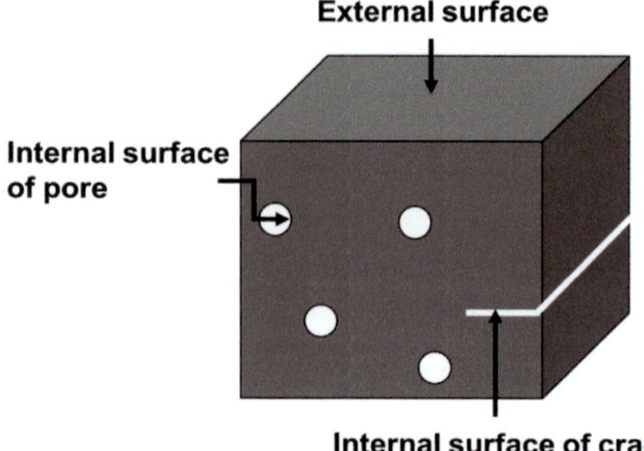

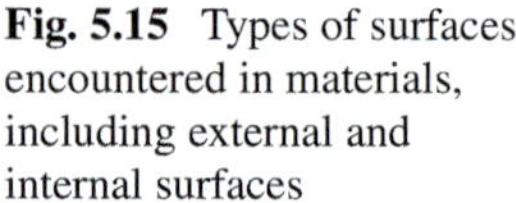

Fig. 5.15 Types of surfaces encountered in materials, including external and internal surfaces

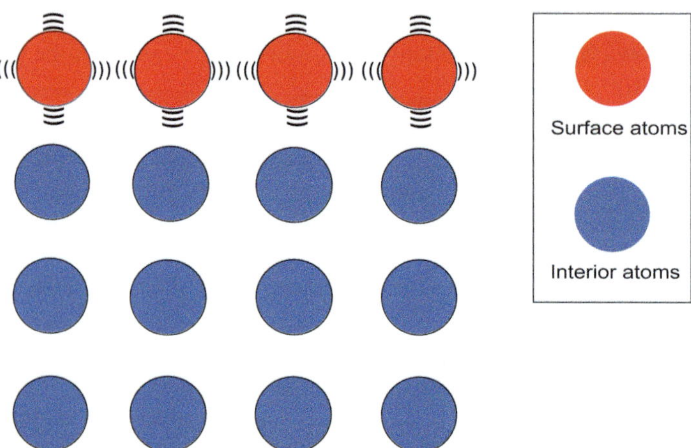

Fig. 5.16 Difference between the energy state of surface and interior atoms ones

coordination number is 12). The situation for a surface atom (red) is different since, due to its location, this kind of atom can only bond with atoms below it and to its sides since there are no atoms above it. This means the interior atoms are more restricted or bound than the surface atoms, and hence surface atoms are found to be more energetic and vibrate more than the interior atoms. In other words, the coordination number of surface atoms is less than that of interior ones. For example, if the surface plane in an FCC metal happens to be (111) plane, the coordination number of these surface atoms is 9. This leaves 3 atoms without bonding. These broken bonds give rise to what we know as **surface energy** and can be computed predominantly using this approach. However, nanoscale materials are a little more involved. All surfaces (external or internal) possess surface energy, typically in units of $\mathbf{J/m^2}$. It can be difficult to experimentally obtain accurate values for surface energies partly because surfaces are very sensitive to impurities and temperature. This led theoreticians to adopt computational approaches to determine surface energies from first principles. If we assume that surface energy arises from the broken bonds at the surface, one could imagine that depending on which crystallographic plane terminates at the surface, the number of broken bonds (and hence surface energy) will change. Surface energy values of materials are typically ~ 0.5–2 J/m^2.

Prof., you mentioned that imperfections can be at the heart of perfection, but I fail to figure out how a surface, pore, or crack can be at the heart of perfection. How can they ever be helpful?

Tell that to chemists. Surfaces are crucial for processes such as catalysis; moreover, if not for the large surfaces of powders, the sintering process would not have been possible; there goes most of the engineering ceramic components and a great deal of powder-based engineering components. You have a surface covering your face and body, right? What about heterogeneous nucleation in the solidification of metals and alloys that is enabled by the presence of surfaces? As for pores, did you know there are materials today based on porosity? Looking at the natural materials around us, one can easily see that bones and wood are porous

We also have synthetic foams, including polymer, ceramic, and metallic ones. These materials exhibit unique properties regarding sound and vibration absorption, ultra-lightweight, buoyancy, and crashworthiness. We can now produce foams with porosity levels of around 99%, which means only 1% is actual material. Imagine a material like aluminum, which is already considered a lightweight metal (with a density of 2.7 g/cm^3); on the other hand, an aluminum foam containing 99% porosity can have a density of 0.027 g/cm^3. This means it would be 100 times lighter than aluminum. Crazy.

Now, are you really saying that cracks may not be helpful? Well, I beg to differ, my friend. Can you cut a rough diamond down to a controllable size and make the most precious diamond rings without cracks? Just research the cleaving of diamonds. Do you think using explosives to make way for mountain roads does not depend on the formation of cracks in the mountains? What about bullet-proof vests that form cracks upon the impact of a high kinetic energy bullet? These cracks require energy to form; hence, the bullet's energy is dissipated that way. I can continue, but I'll leave you to learn more about alloys. Have fun!

5.4   Lightweight Alloys

Alloys are important materials; they can improve the properties of the base metal and expand its usefulness. Here, we focus on lightweight metals and alloys (having densities of 4.5 g/cm^3 and lower). If we examined the number of lightweight metals, we could count about 14. Aluminum with a 2.7 g/cm^3 density would certainly qualify, as would magnesium (1.7 g/cm^3) and titanium (4.5 g/cm^3). These alloys are competitive for **structural applications**. Structural applications are those that require the component to carry a load. The remaining lightweight metals are either highly reactive, toxic, or tend to have a low melting point, thus limiting their engineering applications. An example is lithium; despite its incredibly low density (0.53 g/cm^3), it is typically used in lithium-ion batteries due to its high energy density, and as an alloying element for aluminum rather than a standalone material. As explained below, beryllium is another light metal used only to a limited extent.

5.4.1   Aluminum and Its Alloys

Aluminum in the periodic table has an atomic number of 13 (i.e., 13 protons and 13 electrons), 14 neutrons, and an atomic mass of 26.982 g/mol. We recall that, only the mass of the protons and neutrons counts toward the atomic mass.

Where would we be today without aluminum and its alloys? This metal is uniquely positioned as a lightweight metal that can be used in applications requiring lightweight materials. Its ~ 2.7 g/cm^3 density is lower than many metals, including nickel (8.9 g/cm^3) and titanium (4.5 g/cm^3). Could we imagine an airplane or a racing car would want to be made out of a heavy material so that it would still need to carry it or drag it along? Of course not. Are there materials out there that are stronger than aluminum? Of course. But are they all lightweight? Certainly not! Imagine a scenario where a component is needed to carry a load or force of 50kN within a machine. Assuming the dimensions of the component allow it, steel comes along and says, "I can do that, no problem", and at the same time, an aluminum alloy also throws its hat in and says, "I can also do that job". I know you must be wondering: Do materials talk? Just humor me for now. Of course, cost can play a part in our decision, but if our objective is to use the lightest material possible to do the job needed, then the choice is clear! Aluminum alloy wins.

Classification of Aluminum Alloys

Today, aluminum and aluminum alloys are important materials in many industries, including aerospace, automotive, food packaging, electronics, power, and sports. Space exploration relied on aluminum, including the International Space Station and the Mars Curiosity Rover mission. According to the Aluminum Association, aluminum cans contain 12 times as much recycled content as plastics and 3 times as much recycled content as glass. Aluminum "beverage" cans or aluminum window

Table 5.5 Wrought aluminum designations and applications (multiple sources, including *K.Budinski, M.Budinski, Engineering Materials, Properties and Selection, 8th edition, Pearson, Prentice-Hall (2005)*)

Designation	Major alloying element(s)	Applications
1000	None	Cooking foil, power transmission lines, fin stock, welded tanks, chemical equipment
2000	Copper	Aircraft skins, rivets
3000	Manganese	Cooking pans and beverage cans, pressure vessels
4000	Silicon	Filler material in welding and brazing
5000	Magnesium	Pressure vessels, marine applications, appliances, sheet metal parts, boats
6000	Magnesium and Silicon	Frames of windows, truck and marine structures, pipelines, extended architectural shapes
7000	Zinc	Aircraft forgings, railway carriage shells
8000	Other elements	Al–Li used in aerospace

frames can be, in most cases, recycled back to themselves, hence giving us an endless supply of these products. Another interesting fact is that the energy needed to recycle aluminum is about 5% of what is needed to produce new aluminum. The metal is lightweight, recyclable, corrosion-resistant, and, in some cases, can be as strong as some steels. The low electrical resistivity of unalloyed aluminum prompted its use as electric power transmission lines.

Aluminum alloys can be produced in **wrought** or **cast** form. Wrought means the alloy has been subjected to mechanical work, for example, by extrusion, forging, rolling, etc. A cast alloy typically has a composition that allows fluidity and good casting characteristics. Typically, Al-Si alloys are a good choice, finding applications such as engine blocks and pistons. Other casting alloys also exist. Wrought aluminum alloys have been classified into eight groups, which we call series. They start from the 1000 series to the 8000 series. Table 5.5 summarizes these classifications for wrought alloys with examples of some applications. The 1000 series represents unalloyed aluminum with a minimum purity of 99% by weight; these are typically quite ductile and find applications as cooking foils and electric transmission lines, among others. The 2000, 3000, 4000, 5000, 6000, and 7000 series are aluminum alloys with major alloying elements: copper, manganese, silicon, magnesium, magnesium + silicon (Mg_2Si), and zinc, respectively. The 7000 series alloys are known for their very high strengths. The 8000 series can contain other elements, such as lithium, and Al-Li alloys are examples of the 8000 series. These alloys take advantage of the low density of lithium (0.5 g/cm^3) compared with aluminum (2.7 g/cm^3) in producing low-density aluminum alloys. It becomes evident that aerospace is their natural target application.

5.4.2   Titanium and Its Alloys

Titanium is the fourth most abundant metal in the Earth's crust; however, due to its high reactivity, it is mainly found as an element within a compound, for example, in TiO_2 (rutile) or $FeTiO_3$ (ilmenite). The presence of titanium as a metal in these compounds was suspected as early as 1791. Still, it was the German Chemist **Klaproth** in 1795 who gave titanium its name after studying rutile and identifying it as an oxide of an unknown metal. This metal is now known as titanium. To obtain it as a standalone metal, we needed to extract it. This was an extremely difficult task until the **Kroll process** was introduced 1937–1940, where titanium tetrachloride ($TiCl_4$) compound was reduced with magnesium under an inert atmosphere to produce sponge (i.e., porous) titanium. The extremely high reactivity of titanium with oxygen requires care in extraction and component manufacture, which adds to its cost and is currently a significant obstacle to its widespread use. Recently different approaches have emerged that reduce the negative environmental impact and cost of the Kroll process.

As a material, titanium brings several important properties, including strength, lightweight, and excellent corrosion resistance. An immediate advantage titanium has over aluminum as a lightweight material becomes apparent if we consider its melting point (1670 °C), compared with aluminum (660 °C). Whereas aluminum is typically limited in its applications to a maximum temperature of 150 °C, titanium/titanium alloys, on the other hand, can be used up to 600 °C. Excessive oxidation at any temperature greater than this will, however, result in a decline in its properties. Another advantage of titanium (depending on whether alloying elements are included) is its biocompatibility. This means the human body would not reject it if placed inside it as an implant.

Titanium is used in a commercially pure form (CP–Ti) and as an alloy. The types of alloying elements introduced to Ti or their lack of can affect how the Ti atoms arrange themselves in the material. α-titanium alloys are HCP, while β-titanium alloys are BCC. In α/β titanium alloys, both crystal structures can be found. The point is that alloying changes titanium's properties and expands its range of applications. An example of α-titanium is commercially pure titanium, which is used mainly in applications that require corrosion resistance, such as a bioimplant. Very recently, in Japan, it has also been used as a roofing material for all types of weather. An example of a β-titanium alloy is Ti 15–3 (which is Ti–15 V–3 Cr–3Al–3Sn, all in weight percent), which finds use in the consumer product industry as handlebars and seat posts for mountain bikes, as well as aerospace structures, among others. A main advantage is that it can be shaped cost-effectively at room temperature into sheets. The α/β titanium alloy Ti64 (Ti–6Al–4V) is the most popular of all titanium alloys; it is known as an aerospace alloy and, at the same time, a biomaterial. Alternative titanium alloys (containing nontoxic alloying elements, e.g., Nb, Mo, Zr) as biomaterials are gaining interest after possible toxic effects of V to the body have been suspected in Ti64. Ti64 also finds application as golf club heads; its low density is a factor.

Additionally, one of the main advantages of titanium alloys is their use as a high-temperature structural material.

5.4.3 Magnesium and Its Alloys

Unlike many elements extracted from ores found on land, magnesium is extracted from magnesium chloride found in our oceans, but it can also be extracted from salt mines or rocks. This silvery-white metal is considerably lighter than aluminum, the lightest structural (load-bearing) metal on earth, with a 1.74 g/cm^3 density. It was commercially produced around 1920, and today, it and its alloys have many applications, including those in the aerospace, automotive, and electronics industries. It is also used as an alloying element for aluminum. Pure metal is generally relatively weak; hence, typically it is its alloys that find structural applications. Potentially, the most successful alloying element with Mg is Al. Other alloying elements, such as manganese and zinc, are also used. Magnesium is less stiff than aluminum, has a slightly lower melting point (650 vs. 660 °C), has excellent machinability, and is biocompatible. Magnesium is more expensive than aluminum and experiences burning when exposed to high temperatures due to a reaction with oxygen. Hence, manufacturing processes such as casting and machining require a protective environment, which adds to the cost. However, Mg's atmospheric corrosion properties are superior to steel's but not as good as aluminum's.

5.4.4 Beryllium and Its Alloys

The metal has a density of 1.85 g/cm^3, slightly higher than magnesium but lower than aluminum. The metal is, however, highly toxic, brittle, reactive, and, on top of this, also very expensive. However, it finds limited applications within the aerospace and nuclear industries. The latter application is due to its transparency to electromagnetic radiation.

Problems

5.1. If a Ni–Cu system is called a binary system, what would we call a pure metal and a Ni–Cu–Ti system?

5.2. State three examples of point defects and explain how each one could be useful.

5.3. If the energy required to form a vacancy is 0.9 eV per atom for copper, what would be the fraction of vacancies at 900 °C?

5.4. What is the difference between a substitutional and an interstitial solid solution?

5.5. What are the four criteria put forward by Hume-Rothery to determine the extent to which certain alloying elements will form a substitutional solid solution?

5.6. Using electronegativity, vacancy, crystal structure, and ionic radius data for nickel and copper, determine the extent of their solubility in one another.

5.7. What is the radius of the octahedral interstitial site in nickel and iron?

5.8. What is the radius of the tetrahedral interstitial site in nickel and iron?

5.9. Do dislocations have a function or functions? If so, explain?

5.10. It was determined that the total length of dislocations in a 5 mm^3 piece of metal is 5×10^9 mm. What is the dislocation density? What does the dislocation density tell you about this piece of copper?

5.11. Give three examples of interfacial defects, using sketches as appropriate.

5.12. What type of twins do FCC materials form?

5.13. What is the difference between a grain boundary and a phase boundary?

5.14. Explain the source of surface energy.

5.15. Which phase boundary will have a higher energy, coherent or semi-coherent? And why?

5.16. What is a stacking fault?

5.17. Explain the difference between an intrinsic and an extrinsic stacking fault.

5.18. What is the difference between a wrought and a cast alloy?

5.19. Name three lightweight metals or alloys.

5.20. Which aluminum alloy is suitable for use as power transmission lines, and why?

5.21. What is the advantage of an Al–Li alloy?

5.22. What is meant by the cry of tin?

Chapter 6
Diffusion and Phase Diagrams

6.1 Introduction

Atomic diffusion is crucial in alloy design, manufacturing, and material performance during service. This chapter introduces atomic diffusion and phase diagrams.

6.2 Introduction to Atomic Diffusion

I am sure you were standing in a room at some point in your life, and suddenly, someone came in wearing strong aftershave. If that person was standing at position x_1 in the room, and even if you were standing on the other side of the room (at position, x_2), you would have eventually smelled the aftershave. You would have smelled it without anyone pushing air toward you using a fan, but because of the atomic/molecular diffusion of the aftershave molecules from where the person was to where you were. Initially, the room was only filled with air molecules, and then new aftershave molecules were introduced where the person stood (x_1). The room suddenly had a region with more aftershave molecules than anywhere else. This inequality gave rise to a **concentration gradient** of aftershave molecules between the person and you. The word concentration refers to the amount of aftershave molecules (C). The word gradient refers to the difference in concentration of aftershave molecules (C_1) where the person was standing and the concentration of aftershave molecules (C_2) where you were standing in the room (x_2), divided by the distance between you both ($x_2 - x_1$). Of course, initially, the concentration of aftershave molecules where you stood would have been zero, but as time progressed, you would have smelled more and more of these molecules as they arrived. This was an example of diffusion in gases.

Another example of diffusion, this time in liquids, is that of an ink drop in a glass of water. Suppose we gently place a drop of blue ink on the water's surface. We

K. Morsi, *An Engaging Approach to the Science and Engineering of Materials*,
https://doi.org/10.1007/978-3-032-06231-4_6

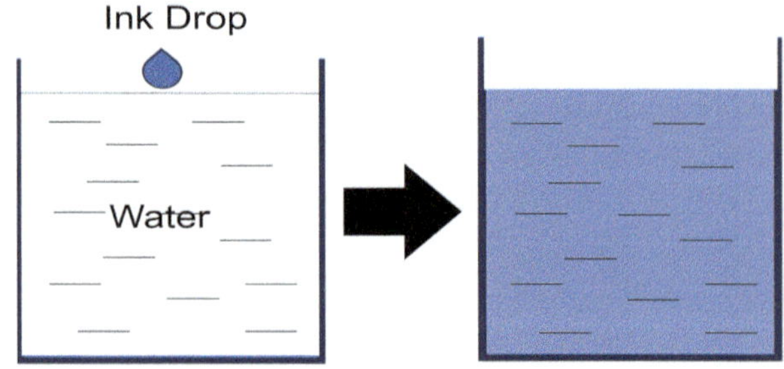

Fig. 6.1 Example of diffusion in liquids, molecules from an ink drop manages to travel throughout the glass of water, without stirring

intuitively know that the color of that glass of water would eventually turn light blue without any stirring action to mechanically mix the ink with the water, as shown in Fig. 6.1. Why is that? Well, as soon as the molecules of the ink drop joins the water at the surface, they become part of that system, which now has some inequality in terms of its molecular distribution. Over time, the ink molecules will diffuse (travel) from the region where they are in high concentrations (the top surface) to regions in where they are in low concentrations until the distribution of the ink molecules balances out, and the inequality disappears.

The **concentration gradient** is the **driving force** behind atomic diffusion, meaning that whenever a concentration gradient is established, diffusion of atoms or molecules occurs, where they travel from locations with a high concentration level to locations with lower concentrations. How fast the diffusing atoms and molecules travel will depend on several things. One of them will be the concentration gradient, i.e., the greater the concentration gradient, the greater the inequality, the greater the driving force for diffusion, and the faster the diffusion will be. Despite the well-established fact that atoms or molecules travel from areas where their concentration is high to ones where their concentration is low, there are notable examples where the opposite is true. For instance, atoms travel from regions where they are in low concentrations to regions where they are in high concentrations; this is the case in material systems with a miscibility gap. But let's not bother ourselves with these situations right now.

The concentration gradient (which plays a major role in atomic diffusion) is defined in Eq. 6.1.

$$\text{Concentration gradient} = \frac{dC}{dx} = \frac{C_2 - C_1}{x_2 - x_1}. \tag{6.1}$$

The most important thing to know is that the fundamental scientific reason why diffusion occurs is to reduce the system's energy. With their freedom of choice, humans can be very energy inefficient, but not atoms, molecules, or materials. Materials will always try to reduce their energy whenever possible. Previously, we introduced Gibbs free energy; it turns out that by the impurity atoms (ink molecules or aftershave molecules) mixing with the atoms or molecules of the host (water or air), the Gibbs free energy of the system is reduced.

Diffusion occurs in liquids and gases, as we saw in the previous examples, but it also occurs in solids. One example of this is in the process of **carbonization** (also called carburization), which is the enrichment of the surface of a component with carbon. An example of a component that would benefit from carbonization is a steel gear. Steel is an alloy of iron and carbon; if the surface of the steel gear is enriched with excess carbon, it makes the gear stronger and harder at that surface location. By only enriching the surface of the gear with carbon, we are avoiding the interior being enriched, which could have otherwise made the gear brittle.

Carbonization involves heating the component, for example, a steel gear, in an environment that can supply carbon to its surface at high temperatures. An example of such an environment would be carbon monoxide/carbon dioxide gas (Fig. 6.2). The advantage of using a gas as the surrounding carbon-enriching medium is that it can reach crevices and intricate features of the component and enrich its surface with carbon at difficult-to-reach areas. Carbonization can also be carried out using solid and liquid mediums. Since the component's surface now has more carbon than its interior, a concentration gradient is set up. This provides a driving force for carbon atoms to diffuse from the surface to the interior of the gear. Suppose we control the depth at which carbon diffuses by controlling the temperature and time. In that case, we can produce a gear with a surface concentration of carbon that is higher than its concentration anywhere else in the gear. This will make the gear very hard at the surface where it is needed to resist wear while retaining toughness in the rest of the gear.

We have so far discussed the driving force for diffusion, in which atoms travel from one region of the material another, but how does this physically happen? What is the **mechanism** by which atoms diffuse? If we consider a crystalline metal or an alloy possessing a crystal lattice. If foreign atoms are to diffuse inside this lattice, then we are faced with two possible situations. If the impurity atom is much smaller than the host lattice atoms, and can reside in the interstitial space within the crystal

Fig. 6.2 Gear surrounded by carbon-rich gas that deposits carbon at the surface of the gear at high temperatures (image of gear generated using ChatGPT 4.0 by K. Morsi)

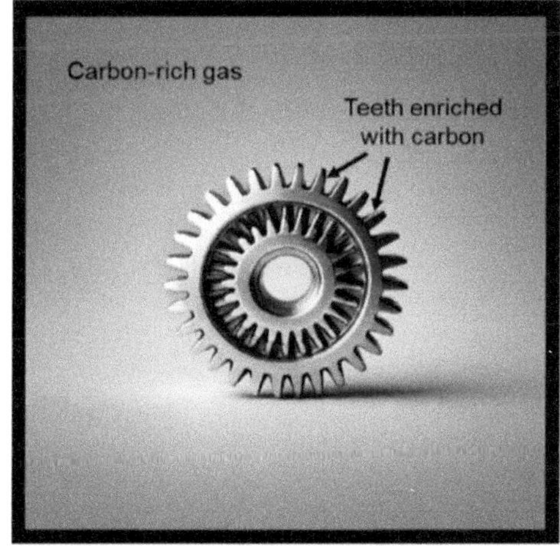

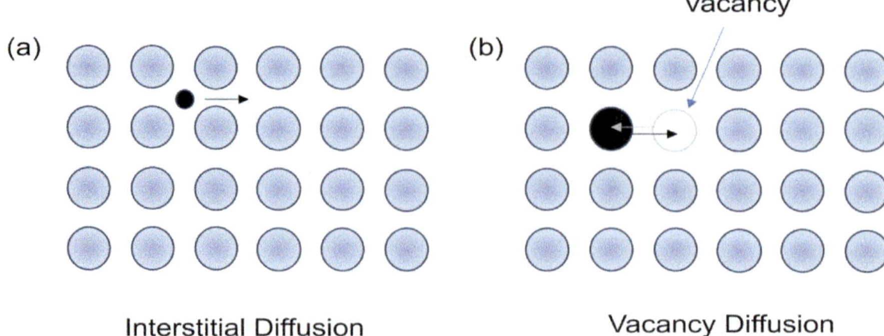

Fig. 6.3 Schematic showing **a** interstitial diffusion where interstitial impurity atom always has a place to jump to, and **b** a substitutional impurity atom where a vacancy is needed for the impurity atom to jump to before diffusion is even possible. Note that as the black substitutional impurity atom moves from left to right, it will leave behind a vacant lattice site. It's as if the vacancy moved in the opposite direction to the impurity atom

structure, it will become an interstitial impurity atom and will diffuse via **interstitial diffusion**. This is shown schematically in Fig. 6.3a, where an interstitial impurity atom can jump to another neighboring interstitial site; in other words, there is always a site to which it can jump.

On the other hand, if the impurity atom is a substitutional one, it has no choice but to wait for a vacancy to appear next to it, and then it jumps to that vacancy. This type of diffusion would be called **vacancy diffusion**. One can easily imagine that if the substitutional atom jumps into a vacancy, it will leave behind a vacant lattice site, in other words, a vacancy. It can, therefore, be concluded that vacancies travel in the opposite direction to substitutional atoms. This can be seen in Fig. 6.3b.

Does this mean that if we cover the surface of a steel gear with carbon at room temperature, it will diffuse into the gear?

Not quite, Alex. You see, what has not yet been discussed is how the atoms would physically be able to move. Don't you think that these atoms are already bonded to the surrounding atoms? Before they move, don't they have to break away from these bonds? Let's say they managed to break away; don't they need energy to jump? Come on, we would need energy if we wanted to go from point A to point B, right? Whether we take a bus, a plane, a car, a bicycle, or even walk. Why do you think it's any different for an atom? Continue reading, my friend, and you will soon know the answer to your question.

Having a concentration gradient is one thing; all it does is establish the motivation for atoms to diffuse. Whether they can diffuse or not is a different issue. Any atom that diffuses requires energy that enables it to travel or jump. Let us take the diffusion of an interstitial atom as an example, with reference to Fig. 6.4. The figure schematically shows the interstitial atom's equilibrium position and its initial Gibbs free energy at point 1. If the atom is to diffuse to point 3, it needs to pass through point 2. This means it needs to pass between the two host atoms, which can generate a significant amount of strain and distortion in the lattice. This lattice strain requires work, resulting in an increase in the Gibbs free energy. This increased Gibbs free energy is called the **activation energy**. In other words, if the interstitial atom is to go from position 1 to position 3, it needs to be supplied with this activation energy. For this to happen, the material must be at a temperature where thermal energy can supply this activation energy. As can also be seen in Fig. 6.4, the activation energy for vacancy diffusion is greater than that of interstitial diffusion, in other words, the diffusion of a substitutional atom will cause more lattice distortion than an interstitial one.

Several factors can increase the rate of atomic diffusion. Increasing temperature or the concentration gradient (the driving force) will cause atoms to diffuse faster.

Since vacancy diffusion requires a vacancy to be present next to the impurity atom, with all things equal (i.e., same temperature, concentration gradient, and host lattice), interstitial diffusion will be orders of magnitude faster than vacancy diffusion. This is because interstitial atoms always have a place to jump to and do not require the presence of a vacancy. Moreover, the activation barrier for the diffusion of interstitial impurity atoms is lower than that of substitutional impurity atoms.

In the case of the diffusion of impurity atoms within a given lattice at a fixed temperature, the concentration gradient will dictate how fast the atoms will diffuse. Two conditions can be considered. The first is if the concentration gradient remains constant during diffusion, while the second considers a concentration gradient that changes during diffusion. The former leads to what is known as **steady-state diffusion**, while the latter leads to non-steady-state **diffusion**. We will tackle each one separately.

6.3 Steady-State Diffusion

This is a diffusion condition in which the concentration gradient of the diffusing species (atoms or molecules) is kept constant with respect to time during diffusion. In this case, the driving force for atomic diffusion would remain constant, so we would not expect any change in the diffusion rate to occur during the diffusion process. Atoms will move as fast at the start as they would at the end of the diffusion process. Figure 6.5 shows an example of a block that experiences steady-state diffusion only in one direction (the x-direction). Here, the left side of the block (at $x = 0$) is enriched with a high concentration (C_H) of impurity atoms that is kept constant with time, while the right side (at $x = x_L$) is also maintained at a low concentration of C_L. The concentration gradient is thus kept constant with respect to time and is given by

$$\frac{\mathrm{d}C}{\mathrm{d}x} = \frac{C_2 - C_1}{x_2 - x_1} = \frac{C_L - C_H}{x_L - 0}$$

Fig. 6.4 Effect of position of an interstitial atom and a substitutional atom on the Gibbs free energy, showing a higher Gibbs free energy when the interstitial atom is in between the host atoms. This increased energy is called the activation energy (adapted and modified from P. G. Shewmon, in Physical Metallurgy, 2nd edition, R. W. Cahn (Ed.), North-Holland, Amsterdam, 1974)

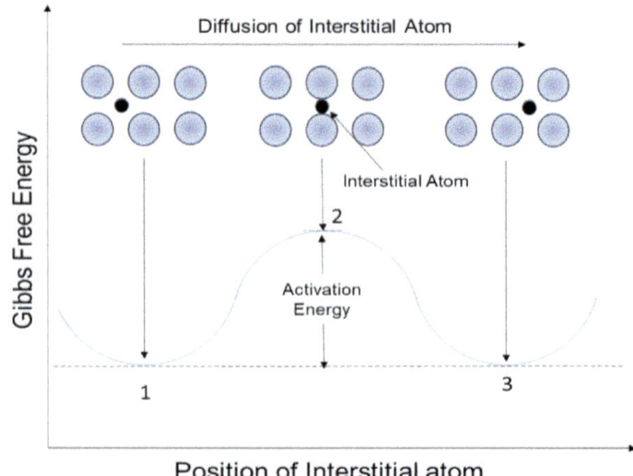

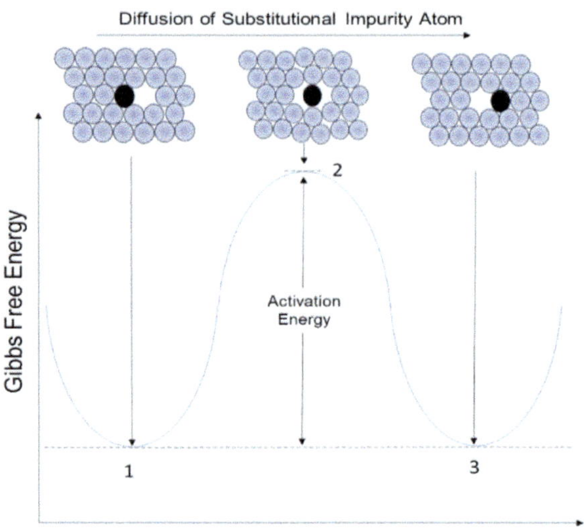

Figure 6.5 shows that concentration is only a function of distance x; it changes only with distance and not time. Once established, this concentration profile will remain the same, whether we leave it for 5 min or 5 months.

We now define a new term, the **diffusion flux, J**. This is simply the mass of atoms diffusing through a unit area per unit time and typically has the units $kg m^{-2} s^{-1}$. It becomes intuitive that a high magnitude of the concentration gradient would lead to a higher driving force for diffusion and, therefore, a higher J. In other words, J is directly proportional to dC/dx:

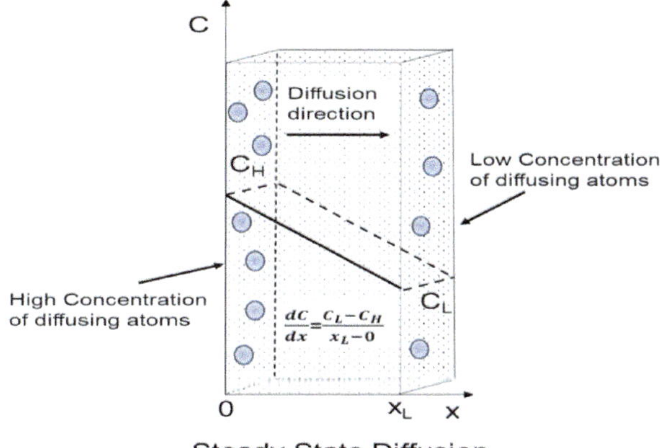

Fig. 6.5 Steady-state diffusion, showing left side of block with higher "constant" concentration of the diffusing impurity atoms and the right side with a lower "constant" concentration of diffusing impurity atoms. The concentration gradient (dC/dx) here does not change with time. Note the linear relation *between* concentration and distance x from left to right

$$J \propto \frac{dC}{dx}$$

We can now remove the proportionality sign and replace it with a constant of proportionality, D. This leads to **Fick's first law**, as shown in Eq. 6.2:

$$J = -D\frac{dC}{dx} \tag{6.2}$$

where D is the diffusion coefficient, sometimes called the **diffusivity**, typically in units of m^2/s. The negative sign is there to counteract the fact that diffusion goes from high concentration to low concentration, and hence C_2 will always be lower than C_1, making dC/dx always negative. Therefore, **Fick's first law** deals with **steady-state diffusion**.

D can be considered a constant if we carry out the diffusion process at a constant temperature. If, however, we decide to carry out diffusion at a higher or lower temperature, then the value of D will change. This makes D what we call a *temperature-dependent* constant. This temperature dependence of D is expressed in Eq. 6.3:

$$D = D_o e^{\left(\frac{-Q}{RT}\right)} \tag{6.3}$$

where D_o is the pre-exponential (a temperature-independent constant), Q is the activation energy for diffusion (in kJ/mol), R is the gas constant, and T is the temperature in Kelvins.

The **activation energy** is the *energy needed for one mol of atoms to diffuse*. This energy is treated as a constant in Eq. 6.3. However, we must understand that when atoms diffuse from one part of the material to another, they have options regarding their paths. Figure 6.6 is an example of this, which shows the different mechanisms by which impurity atoms can travel across a block of material. Assuming a higher

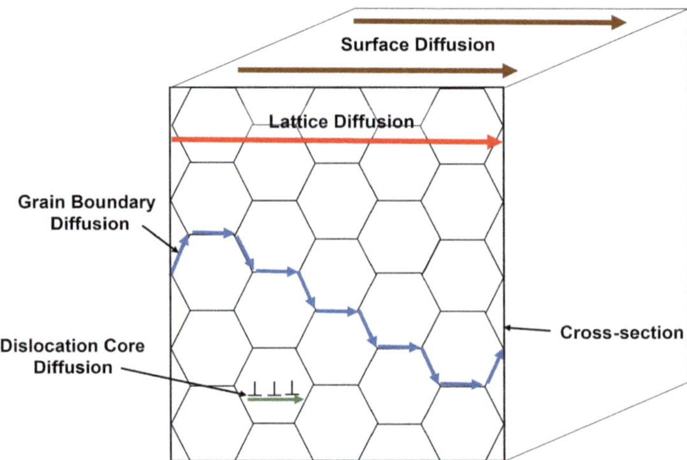

Fig. 6.6 Schematic of a cross section of a material, showing surface, grain boundary, lattice (bulk), and dislocation core diffusion mechanisms

concentration of impurity atoms on the left surface of the block than on the right, atoms will diffuse from left to right down the concentration gradient. Here, atoms have several choices: they can travel over the surface, which we call **surface diffusion**, through the block via the lattices of the grains, which we call **lattice diffusion**; along grain boundaries, which we call **grain boundary diffusion**; or through the cores of dislocations, which we call **dislocation core diffusion** (Fig. 6.7). We need to understand that the atoms will pursue all these options. Whenever there is an opportunity to diffuse, it will take it; remember they are trying to reduce the Gibbs free energy of the system, and beggars can't be choosers. This means all the above-mentioned diffusion mechanisms will be active simultaneously. It becomes obvious that a substitutional impurity atom traveling along a grain boundary (with all its available space due to the poor atomic packing there) will have an easier job than one traveling through a grain where a vacancy needs to be present, not to mention the energy barrier resulting from the distortion it would cause. Likewise, an atom diffusing on the free external surface will have an even easier job. Atoms traveling through the core of a dislocation or a grain boundary are said to undergo **short-circuit diffusion**. This is because a dislocation core, or a grain boundary, provides a much faster corridor for atoms to diffuse through.

There are a couple of points we need to understand. Different mechanisms would naturally have different activation energies; for example, the activation energy of a given atom diffusing in a material will be higher if diffusion occurs through lattice diffusion rather than grain boundary diffusion. Less energy is needed to travel through grain boundaries with their low density of atoms. The other important fact is that

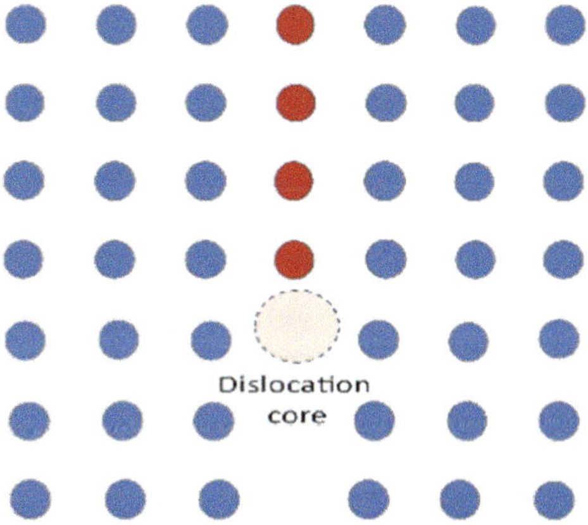

Fig. 6.7 Edge dislocation (red) beneath which lies a hollow dislocation core, extending into the plane of the image, in which atoms can diffuse through. In essence it acts as a diffusion corridor for atoms

although all mechanisms take place simultaneously, there will typically be one dominant diffusion mechanism, which most atoms will follow. Let me explain. Atoms can diffuse much faster through grain boundaries than through grains. At low temperatures, one would expect that the dominant mechanism would be grain boundary diffusion. However, lattice diffusion is still active; since the number of vacancies is low at low temperatures, the contribution from lattice diffusion will be limited. This, however, will change at higher temperatures, where the number of vacancies increases exponentially. Here, although grain boundary diffusion will be faster, we must consider how much of the material volume is occupied by grain boundaries; we will immediately see that most of the material volume is made up of grains instead. Hence, lattice diffusion becomes the dominant diffusion mechanism only when there is a high concentration of vacancies. Therefore, when we use a value for the activation energy, it is assumed that it is for the dominant diffusion mechanism.

Did you get the idea behind activation energy, Alex? If so, then answer the following.

If a substitutional impurity atom and an interstitial impurity atom were to diffuse via lattice diffusion in the same crystal lattice and at the same temperature, which one do you think would have the lower activation energy?

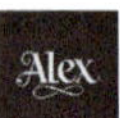

Although interstitial atoms must overcome an energy barrier because their migration causes distortion between the host atoms, one would expect more distortion caused by substitutional atoms trying to diffuse. Also, for atoms to move, they need to break away from their bonds; the bonds between substitutional impurity atoms and host atoms are typically stronger than those between interstitial impurity atoms and the same host atoms. Hence, strongly bonded substitutional impurity atoms would tend to have higher activation energies. Therefore, for these reasons, the diffusion of the interstitial atom will have a lower activation energy than that of the substitutional atom, and interstitial diffusion is faster than vacancy diffusion under the same conditions.

Well done! You've been reading up on the subject from multiple sources! Now let's test you with a problem from "As a Matter of Fact".

Example Problem 6.1

A 5-mm-thick iron sheet is exposed on one side to a carbonizing environment that maintains the concentration there at 0.5 kg/m^3 of C. On the other side, an oxidizing environment is set up to maintain the composition at 0 kg/m^3 C. What would the diffusion flux be if the temperature is maintained at 1000 °C? $D_o = 20 \times 10^{-6}$ m^2/s, $Q = 148$ kJ/mol.

Solution

This diffusion situation is typical of steady-state diffusion because the concentration gradient does not change with respect to time (since they used the word maintained to refer to the concentrations of the impurity atoms on both sides of the sheet). Therefore, it is solved using Fick's first law:

$$J = -D\frac{dC}{dx} = \frac{C_2 - C_1}{x_2 - x_1}$$

However, the question does not provide the value for the diffusivity, D, but instead, it provides D_o (the pre-exponential), the activation energy (Q), temperature ($1000 + 273 = 1273$ K), and $R = 8.314$ J/(K mol) from which D can be calculated using the following equation:

$$D = D_o e^{\left(\frac{-Q}{RT}\right)}$$

Hence

$$D = 20 \times 10^{-6}.e^{-148000/(8.314 \times 1273)},$$

$$D = 1.69 \times 10^{-11} \text{ m}^2/\text{s}.$$

Then using Fick's first law, given $C_2 = 0$, and $C_1 = 0.5$ kg/m^3, and $x_2 - x_1 = 5 \times 10^{-3}$ m, we have

$$J = -D\frac{dC}{dx}$$

$$J = -1.69 \times 10^{-11} \times (0 - 0.5)/5 \times 10^{-3}$$

$$\boxed{J = 1.69 \times 10^{-9} \text{ kg/m}^2 \text{ s}}$$

6.4 Non-steady-state Diffusion

In non-steady-state diffusion, the concentration gradient does change with time, and one would expect the diffusion flux to do the same. At any given diffusion temperature, unlike steady-state diffusion in which the concentration of the diffusing atoms depended only on the position x within the sample, the concentration of the diffusing atoms, C, in non-steady-state diffusion depends on both the position, x, and time, t. This is expressed by the partial differential equation, in Eq. 6.4, which assumes the diffusion coefficient does not depend on x:

$$\frac{\partial C}{\partial t} = D\frac{\partial^2 C}{\partial x^2} \tag{6.4}$$

We can solve this equation by considering the solution for a semi-infinite solid and using boundary conditions. The idea behind theoretically assuming a semi-infinite solid is that, unlike an infinite solid, a semi-infinite solid will have a surface with which we can work. This means we can assume some fixed composition of the diffusing impurity atoms at the surface (C_s) of a semi-infinite solid, as seen in Fig. 6.8.

At $t = 0$: $C_x = C_0$, at $0 \leq x \leq \infty$
At $t > 0$: C_x at $x = 0$ (i.e., at the surface) $= C_s$, and $C_x = C_0$ at $x = \infty$

The resulting solution of the differential equation is given by Eq. 6.5:

$$\frac{C_x - C_0}{C_s - C_0} = 1 - \text{erf}\left(\frac{x}{2\sqrt{Dt}}\right) \tag{6.5}$$

x is the position beneath the surface, x is 0 at the surface and increases in magnitude beneath the surface, C_x is the concentration of the diffusing atoms at position x beneath the surface, C_0 is the concentration at the surface, D is the diffusion coefficient, and t is time. The erf in the equation represents the **Gaussian error function**, which is solved numerically. In other words, if we assume whatever is inside the brackets of the error function to be z, then the equation now becomes Eq. 6.6:

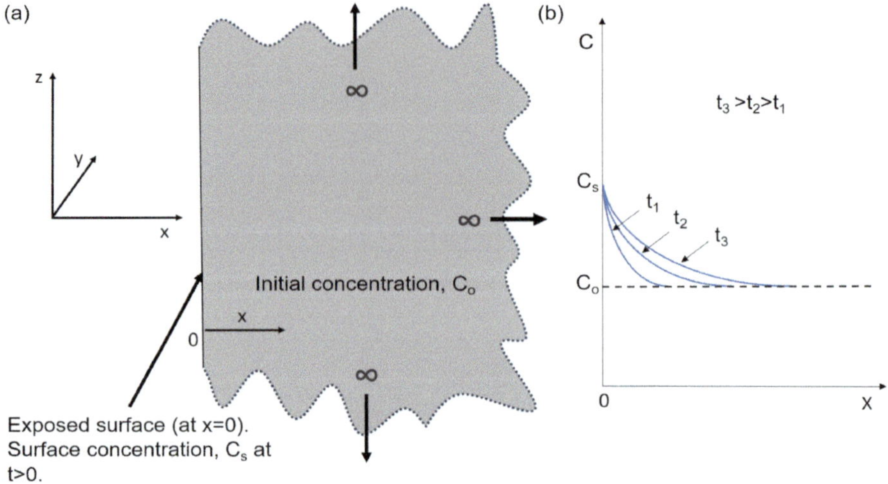

Fig. 6.8 Schematic of **a** Semi-infinite solid, note that the dimension perpendicular to the plane of the sketch of the solid also goes to infinity, **b** concentration profile at different times of heat treatment, note, only 1D diffusion is assumed in the x-direction

$$\frac{C_x - C_o}{C_s - C_o} = 1 - \mathrm{erf}(z) \tag{6.6}$$

where $z = \frac{x}{2\sqrt{Dt}}$.

Table 6.1 provides z values for the corresponding erf(z) values and vice versa. The equation can be very powerful in predicting how temperature and time affect the distance atoms can diffuse from the impurity-rich surface into the material. This could be helpful in situations where we want to enrich the surface with alloying elements, for corrosion-resistant or wear-resistant applications.

Equation 6.6 can be used in one of the following two ways:

1. To determine the **time** needed for the impurity concentration to reach a certain level at a particular distance (x) beneath the surface.
2. To determine the **impurity concentration at x** distance beneath the surface after a specific diffusion time and temperature.

It should be understood that the value of D already incorporates the temperature at which the diffusion occurs (see Eq. 6.3). The following examples demonstrate the usefulness of Eq. 6.6.

Example Problem 6.2

Consider a non-steady-state diffusion process in 1020 steel. How long will it take for carbon to diffuse from the surface (initially maintained at a carbon concentration of 0.5 wt% C) and raise the carbon concentration at 0.5 mm beneath the surface to 0.3 wt% C? Assume $D = 1.9 \times 10^{-11}$ m^2/s (at 1000 °C).

Table 6.1 Error function table

z	erf (z)	z	erf (z)	z	erf (z)
0	0	0.5	0.5205	1.3	0.93401
0.02	0.02256	0.55	0.5633	1.4	0.95229
0.04	0.04511	0.6	0.60386	1.5	0.96611
0.06	0.06762	0.65	0.6420	1.6	0.97635
0.08	0.09008	0.7	0.6778	1.7	0.98379
0.1	0.11246	0.75	0.7112	1.8	0.98909
0.15	0.1680	0.8	0.7421	1.9	0.99279
0.2	0.2227	0.85	0.7707	2.0	0.99532
0.25	0.2763	0.9	0.79691	2.1	0.99702
0.3	0.32863	0.95	0.8209	2.2	0.99814
0.35	0.3794	1.0	0.8427	2.3	0.99886
0.4	0.42839	1.1	0.88021	2.4	0.99931
0.45	0.4755	1.2	0.91031	2.5	0.99959

Solution

Since the steel is 1020. This means it contains 0.2 wt% C.

$C_x = 0.3$ wt% C
$x = 0.5 \times 10^{-3}$ m
$C_o = 0.2$ wt% C (since 1020 steel)
$D = 1.9 \times 10^{-11}$ m^2/s
$C_s = 0.5$ wt% C.
$t = ?$

Using Eq. 6.6,

$$\frac{C_x - C_o}{C_s - C_o} = \frac{0.3 - 0.2}{0.5 - 0.2} = \frac{0.1}{0.3} = 1 - \text{erf}(z)$$

$$\text{erf}(z) = 1 - \frac{0.1}{0.3} = 0.6667$$

So, we know the value of erf(z) but not z. Note that the time, t, is actually incorporated inside z.

Remember, $z = \frac{x}{2\sqrt{Dt}}$.

Within z, we know the values of x and D but not t, which is our objective. Hence, the idea is to determine first what value of z would give us an erf$(z) = 0.66667$. This can be found using Table 6.1 and mathematical interpretation.

As can be seen, the calculated erf (z) of 0.6667 lies between the erf values of 0.6420 and 0.6778 in the table. This means the corresponding z values must lie between 0.65 and 0.7. To obtain the exact z value, we need to employ mathematical interpolation as follows:

$$\frac{0.6778 - 0.6420}{0.7 - 0.65} = \frac{0.6667 - 0.6420}{z - 0.65}$$

which gives a z-value of 0.6845.

Since, $z = \frac{x}{2\sqrt{Dt}}$.

Therefore $t = \left(\frac{x}{2z}\right)^2 \Big/ D$.

Substituting for the values of x, z, and D gives

$$\boxed{t = 7021 \text{ s or } 117 \text{ min.}}$$

Example Problem 6.3

Considering the diffusion problem presented in Example Problem 6.2. What would be the composition at a distance x $= 0.5 \times 10^{-3}$ m, given a surface concentration of (C_s) of 0.5 wt% C, diffusion temperature of 1000 °C (i.e., D $= 1.9 \times 10^{-11}$ m²/s), and diffusion time $= 7021$ s?

Solution

This is a non-steady-state diffusion problem; hence, the governing equation will be

$$\frac{C_x - C_o}{C_s - C_o} = 1 - \text{erf}(z)$$

where $z = \frac{x}{2\sqrt{Dt}}$

$X = 0.5 \times 10^{-3}$ m
$C_s = 0.5$ wt% C
$C_o = 0.2$ wt% C
$D = 1.9 \times 10^{-11}$ m²/s
$T = 7021$ s
$C_x = ?$

It is clear that from the data provided, the parameters inside the effort function of z are all given. That means x, D, and t. This means we can work out what z is. Then all we need to do is work out, from the error function table, the erf(z). Let's go through this.

$$z = \frac{x}{2\sqrt{Dt}} = \frac{0.5 \times 10^{-3}}{2 \times \sqrt{1.9 \times 10^{-11} \times 7021}} = 0.6845.$$

From the error function table, z lies between the z values of 0.65 and 0.70.

z	erf (z)
0.65	0.6420
0.6845	erf (z)?
0.7	0.6778

$$\frac{0.6778 - 0.6420}{0.7 - 0.65} = \frac{\text{erf}(z) - 0.6420}{0.6845 - 0.65}$$

$$\text{Erf}(z) = 0.666702$$

Since,

$$\frac{C_x - C_o}{C_s - C_o} = 1 - \text{erf}(z).$$

Therefore,

$$\frac{C_x - 0.2}{0.5 - 0.2} = 1 - 0.666702 = 0.333298$$

$$\boxed{C_x = 0.29999, \ \text{i.e.,} \ 0.3 \ \text{wt\% } C.}$$

Professor, I understand that assuming a semi-infinite solid helps to simplify the solution of the differential equation, but how can that ever relate to diffusion in an actual material component, which is, in fact, finite?

Great question! The semi-infinite solid concept is used to help us understand many things. Let's take, as an example, atoms diffusing in a finite component, as you mention. As long as the atoms do not reach the other side of the component at the end of the diffusion process, the component can be treated as a semi-infinite solid. You can use a semi-infinite analysis on a bar if its length is $> 10\sqrt{(Dt)}$. Did you know that our Earth can be treated as a semi-infinite solid? You are practically standing on its surface.

6.5 Equilibrium Binary Phase Diagrams

Alloys are important engineering materials, generally stronger than pure metals. While a unary metal (only one element) will have a distinct melting point, a binary alloy containing two elements will exhibit different behavior. Before we start, let us first define a few terms. A binary phase diagram is one that has two components which can be atomically mixed to different degrees and produce an alloy. These components can be elements, for example, nickel and copper. Hence here, copper and nickel would be the two **components** of the binary phase diagram. As for the word "**phase**" in phase diagrams, it is defined as a *region of the material's microstructure that has uniform chemical and physical properties or characteristics.* For example, in Fig. 6.9a, we see ice and water existing together in a container. This container is said to have two phases (ice and water). While ice and water share the same chemical formula (H_2O), ice is a solid and water a liquid. Despite them having the same chemical characteristics, they do not share the same physical ones and, hence, are deemed to be two separate phases. Figure 6.9b shows another example of two separate phases. It is well known that **allotropy** exists in iron, which means iron can exist in more than one crystal structure. For example, iron exists in the BCC crystal structure (called ferrite, α) at room temperature and atmospheric pressure. In contrast, at higher temperatures, specifically above 912 °C, it changes to an FCC crystal structure (called austenite, γ).

Here, too, we can say that ferrite and austenite are two separate phases of iron; while both have the same chemistry (iron), their different atomic arrangements (crystal structures) result in different physical properties. For example, FCC materials are known to be more ductile than BCC ones; they also behave differently than BCC materials at low temperatures (more about that later). In the case of iron, the temperature can affect the crystal structure and result in a change in the phase from BCC ferrite to FCC austenite; not only that, but if the temperature is increased yet again, the FCC austenite (γ) will then switch back to a high-temperature BCC version of ferrite (δ), yet another phase. Iron will melt at an even higher temperature and generate a new phase, liquid iron, L.

The situation becomes more involved if, instead of having only iron atoms, we have an iron "alloy" (i.e., iron atoms atomically mixed with atoms of another element). Here, the type of phases that can result will depend on the temperature and composition, i.e., how much of the other element was added to iron. Throughout this discussion, we assume that pressure is kept constant at atmospheric pressure since it, too, can cause phase changes.

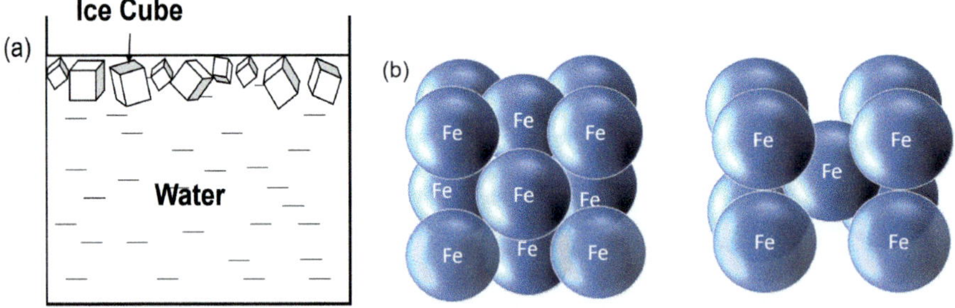

Fig. 6.9 **a** Water and ice are separate phases and **b** FCC iron (austenite) and BCC iron (ferrite) are separate phases

Therefore, alloy systems generally have more complicated dependencies on temperature and composition than pure metals. These dependencies are presented within equilibrium phase diagrams that identify which phase exists at any given temperature and composition. The word equilibrium in equilibrium phase diagrams refers to the fact that the phases mentioned are stable and would be present if enough time is provided for them to form.

6.5.1 Isomorphous Phase Diagrams

Phase diagrams can be classified into different types. One of the simplest of all is what we call the **isomorphous binary phase diagram**. This is an equilibrium phase diagram in which the two components being added are entirely soluble in one another. Unlike sugar and water, in which we can only dissolve a certain amount of sugar in water before sugar is no longer soluble, in isomorphous phase diagrams, both elements will dissolve in one another with no limit. Examples of elements that have this complete solubility in one another include silver and gold, nickel and copper, gold and copper, molybdenum and tungsten, chromium and iron, chromium and vanadium, and titanium and zirconium. We will now represent the two mutually soluble elements by hypothetical elements A and B. A schematic of a hypothetical equilibrium isomorphous phase diagram is shown in Fig. 6.10, where atoms of the two elements A and B are atomically mixed, producing alloys of different compositions. The y-axis is temperature, and the x-axis is the composition, measured in weight percent of element **B**. Since there are only two elements (A and B), the weight percent of A will be 100 minus the weight percent of B. For example, a composition of 40 wt% B on the x-axis is an alloy of 40 wt% of atoms from element B and 60 wt% of atoms from element A. It follows that at 0 wt% B, we have 100 wt% A, meaning that the material is nothing but pure element A, and likewise 100 wt% B is nothing other than the pure element B.

Phase diagrams consist of regions called **phase fields** in which a particular phase or phases exist. Any phase present within a phase field is identified by a letter. All solid phases are represented by a Greek letter, and liquid phases are represented by the letter L. Hence, if any phase in a phase diagram is identified by a Greek letter, it is always a solid (or solid solution); likewise, L is always a liquid (molten phase). The α in Fig. 6.10 is, therefore, a solid phase; in fact, it can be described as a solid solution at compositions > 0 wt% B. What we mean is that if element A is an FCC material, for example, then all the atoms in its crystal structure will belong to element A. Still, as we add atoms from element B (assuming that element B satisfies the Hume-Rothery rules for substitutional solid solutions), the B atoms will dissolve in the crystal structure of element A and simply substitute or replace A atoms in the lattice, forming a solid solution. Since the material experiences complete solid solubility, then as we keep adding more and more B atoms, the A atoms will keep getting substituted with B atoms until we substitute all of them at composition 100 wt.% B, resulting in pure element B.

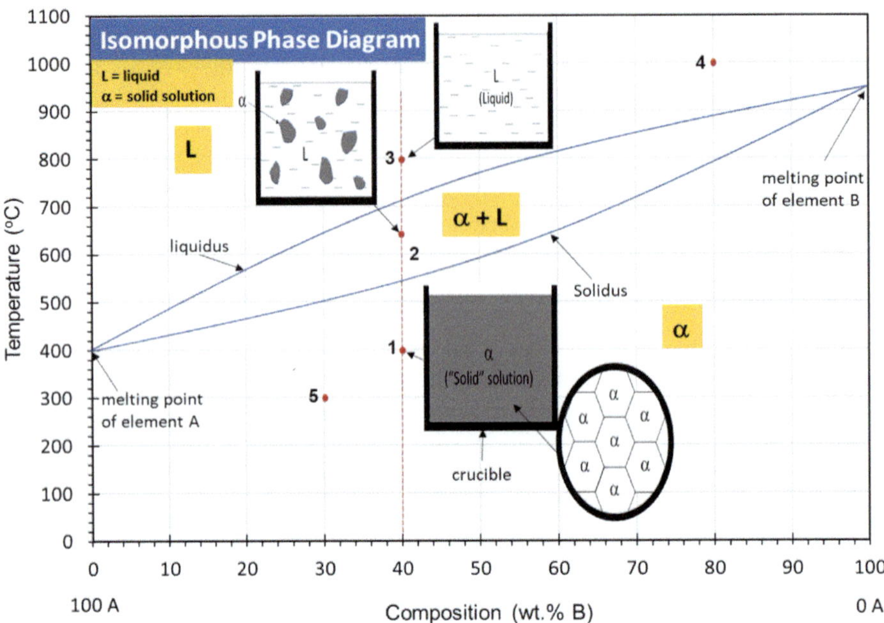

Fig. 6.10 A schematic of a hypothetical equilibrium isomorphous phase diagram

The phase diagram tells us that for pure element A (0 wt% B), above 400 °C, the phase that should form is L (a liquid); in other words, element A would have melted, making 400 °C the melting point of element A. Likewise the melting point of element B (100 wt.% B) will be ~ 950 °C. The phase diagram also shows three distinct phase fields (regions), identified in yellow boxes as, α, L and α + L separated by lines. As seen in the figure, the **solidus** is the lower line, *with everything below being a solid*, while the **liquidus** *is the upper line, with everything above being a liquid*. For a binary alloy of any composition, phase diagrams can tell us which equilibrium phase (s) to expect at any given temperature. This is quite powerful because, with knowledge of phase diagrams and how to use them, we would be in a position to design and control alloy microstructures and, hence, their properties and applications.

Phase diagrams can tell us many things about an alloy. If we identify a point on a phase diagram (for example, points 1, 2, and 3 in Fig. 6.10), we can deduce at least four important facts about that alloy. These are as follows:

1. The number of phases existing at that point.
2. What these phase(s) are.
3. What the composition(s) of the phase(s) is.
4. The mass fraction(s) of the phase or phases present.

Let's find out these facts about two of the points (1 and 3) identified in Fig. 6.10.

Point 1

1. How many phases exist at this point?

   Answer: 1

2. What is this phase?

   Answer: α, which is a solid phase.

3. What is the composition of that phase?

   Answer: The phase at this point has a composition of 40 wt% B–60 wt% A.

4. What is the mass fraction of the phase?

   Answer: Since there is only one phase (α), then its mass fraction is 1, i.e., only one phase is present in the crucible. We can also see from the figure that the material is polycrystalline with all the grains being of the α phase.

Point 3

1. How many phases exist at this point?

   Answer: 1

2. What is this phase?

   Answer: L, which is a liquid (molten) phase.

3. What is the composition of that phase?

   Answer: The phase at this point has a composition of 40 wt% B–60 wt% A.

4. What is the mass fraction of the phase?

   Answer: Since there is only one phase (L), then its mass fraction is 1, i.e., only one phase is present in the crucible.

Before we answer these same questions for point 2, let us explain first a few things. As far as questions 1 and 2 are concerned, they were easily solvable by simply looking at the phase diagram. Questions 3 and 4 were also very easy to answer since all we were considering was a single phase, whether it was a solid or a liquid. If we, however, consider a point within the two-phase field ($\alpha + L$), then we must be a little careful. Practically speaking, one would normally take the alloy and place it in a crucible; the crucible would then be placed in a furnace and heated to high temperatures. As we can see in Fig. 6.10, there is a red dashed line connecting **points 1, 2, and 3**, and the physical state of the alloy inside a crucible is also shown. The figure shows that at point 1, the material inside the crucible is still solid and totally composed of the α phase. As the temperature is raised, as soon as the alloy enters the $\alpha + L$ two-phase phase field, a very small amount of liquid will appear, and as the temperature is raised further inside the two-phase field, more and more liquid will be formed until the alloy becomes 100% liquid at the end of the two-phase field. Any increase in temperature beyond this point will simply just increase the temperature of the liquid (molten) phase. It can, therefore, be seen that at **point 2**, the material is in the two-phase region, and thus both α and liquid are present, while at **point 3**, we see the crucible only filled with liquid (molten alloy).

We should understand that whether we are talking about points 1, 2, or 3, the solid alloy that was initially placed in the crucible had a certain overall composition (in this case, 40 wt.%B-60wt.% A). This alloy composition will not change if we simply heat the crucible from point 1 through to point 3, since we are not adding any more atoms or removing any atoms during heating. This is true, unless, of course, the alloy experiences significant oxidation or evaporation processes, which we are not concerned with here. This means that the total number of atoms (atoms of A and atoms of B), even in the two-phase field, will remain the same as before, i.e., 40 wt% B–60 wt% A. However, the individual phases will both have different compositions which will not be the 40 wt% B–60 wt%. This means that the liquid and α that form will each take different percentages of A and B atoms from the initial solid α. The question then becomes, how do we determine the compositions of L and α within the two-phase region? This is where the phase diagram can again be very useful, as explained below.

- **Phase Composition Determination in a Two-Phase Field**

The process is to draw a line at **point 2**, parallel to the x-axis, called the **tie line**, see Fig. 6.11. The composition at which the tie line intercepts the solidus (C_α) will be the composition of the α phase and where it intercepts the liquidus (C_L) will be the composition of the liquid phase within the two-phase field.

Ok, so it looks like you just learned how to determine the composition of each phase in a two-phase field. Again, by composition, we mean the weight percent of A and B atoms inside each phase. Let's now consider point 2 in Fig. 6.11.

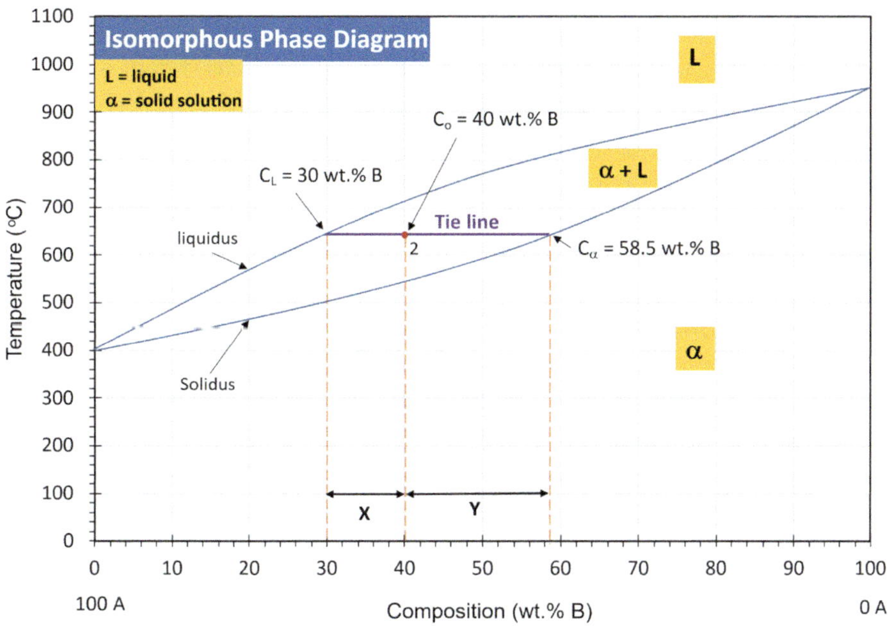

Fig. 6.11 A schematic of a hypothetical isomorphous phase diagram, showing a tie line

Point 2

1. How many phases exist at this point?

   Answer: 2

2. What are these phase(s)?

   Answer: L and α.

3. What is the overall composition of the **whole** "alloy"?

   Answer: That would be C_o, which is the composition 40 wt% B–60 wt% A.

4. What is the composition of the liquid phase?

   Answer: That would be C_L which is the composition 30 wt% B–70 wt% A.

5. What is the composition of the solid α phase?

   Answer: That would be $C_α$, which is the composition 58.5 wt% B–41.5 wt% A.

Professor, I'm a little confused. I can see that the composition of the α and L phases (i.e., how much A and B atoms are in both phases) in the two-phase field are completely different, and if I add the two compositions together (30 wt% B–70 wt% A, and 58.5 wt%B–41.5 wt% A) it gives me 88.5 wt% B–111.5 wt% A, and if we instead take their average it would be 44.25 wt%B–55.75 wt% A. Hence, it does not result in the base overall composition of the alloy (40 wt% B–60 wt% A) from which these two phases came in the first place. Am I missing anything?

This is a common misconception, Alex. Don't worry, I will explain. You made a mistaken assumption by adding up the compositions or averaging them. You assumed that both phases were present in equal amounts, i.e., equal mass fractions. Remember, when the alloy is first heated into the α + L two-phase field, only a small amount of liquid is formed first, and as the alloy is raised to higher temperatures, more and more liquid is formed (which means the solid is also decreasing in mass fraction). So, you cannot assume the two phases are present in equal amounts. If they were present in equal mass fractions, then if you would simply average the compositions of the two phases, to get the overall composition you started out with (which is the one if you only had a single phase). The following section in As a Matter of Fact explain mass fractions.

- **Mass Fraction Determination in a Two-Phase Field**

Phase diagrams can also help us determine the **mass fraction** of each phase within a two-phase field using the **lever rule** as follows:

Considering an alloy of base/overall composition, C_o, heated inside a two-phase field (α + L), in Fig. 6.11. The compositions of the α and L phases, as mentioned, will be C_α and C_L, respectively. The lever rule states that the mass fraction of the liquid phase (X_L) can be determined using Eq. 6.7a:

$$X_L = \frac{C_\alpha - C_o}{C_\alpha - C_L} \tag{6.7a}$$

In other words, if we wanted to determine the mass fraction of the liquid of composition C_L, we would subtract the base composition C_o from the composition of (i.e., the solid phase) and divide by the difference between the compositions of the liquid and phases.

Likewise, the mass fraction of the α phase (X_α) can be determined using Eq. 6.7b.

$$X_a = \frac{C_o - C_L}{C_\alpha - C_L} \tag{6.7b}$$

Now, with reference to Fig. 6.11 and point 2.

$$X_L = \frac{Y}{X + Y} = \frac{58.5 - 40}{58.5 - 30} = 0.649$$

and

$$X_a = \frac{X}{X + Y} = \frac{40 - 30}{58.5 - 30} = 0.351.$$

Notice that when we add up the mass fractions X_L and X_α, we end up with 1. The solidified 40 wt% B–60 wt% A alloy will be a polycrystalline material, with all the grains belonging to the α phase (as seen in Fig. 6.10 point 1).

Do you see this, Alex? This means that at the end of an exam, if the professor is telling everyone, "You have 30 s remaining," depending on the professor, it may be a good idea if you have already calculated X_L to simply subtract X_L from 1 to get X_α and save time.

6.5.2 The Eutectic Phase Diagram

Another type of phase diagram of importance is the eutectic phase diagram. Eutectic comes from the Greek "Eutektos" meaning "easy to melt". This can be seen in Fig. 6.12, which is part of the H_2O–NaCl phase diagram. As can be seen, ice has a melting point of 0 °C; however, if we add 23 wt% NaCl, the melting point decreases to − 21.3 °C, above which a liquid solution (Brine) is formed. This has practical implications; in fact, that is why whenever we see black ice on a road surface during subzero temperatures, the solution is always to place salt on it; the salt will allow traction for vehicles and the ice to melt. This means if the temperature is above − 21.3 °C, the ice would melt; if the temperature is below, however, another type of salt may be needed. An effective salt is calcium chloride, which can allow the melting of ice at temperatures down to around − 50 °C.

Similar phase diagrams to the H_2O-NaCl diagram can be found in metal alloy systems. For example, the lead–tin, aluminum–silicon, and silver-copper systems.

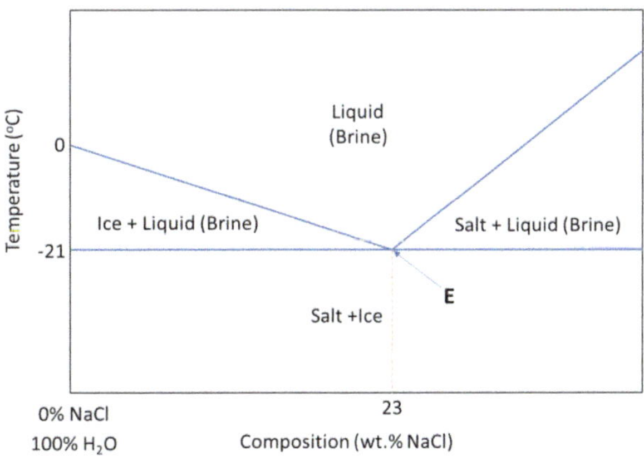

Fig. 6.12 A schematic of part of the H_2O-NaCl phase diagram

However, let us start with a *hypothetical* eutectic phase diagram with components A and B, as seen in Fig. 6.13. Let us go through this phase diagram to learn more about it. On the far left and far right of the phase diagram, we see α and β phases, respectively, which are, in fact, solid solutions. As seen, α will have mainly atoms from element A and very few atoms from element B as part of its crystal structure. Likewise, β would consist mainly of B atoms, with some A atoms also dissolved in the B crystal lattice. Since these phases exist at the two extreme ends of the phase diagram, they are called **terminal solid solutions**. The melting point of element A would be ~ 820 °C, and that of element B would be ~ 1020 °C, since, above these temperatures, only liquid (L) exists. On the far left, we see the **solvus**, which shows a solubility limit of B in A that increases with an increase in temperature. This feature is utilized to strengthen some materials, as we will explain later. Moreover, both the **liquidus** and the **solidus** lines are also identified in the figure.

What is unique about eutectic phase diagrams is that if we start at either end of the phase diagram, the melting point will come down as we add more of the other elements. We see in the figure that as we add element B to A (moving from left to right on the diagram), the temperature above which we enter the liquid phase field (i.e., the melting point) decreases until a composition of 60 wt% B is reached, where the temperature is 620 °C. Likewise, as we add more of element A to element B, we also observe a decline in the melting point until the composition of 60 wt% B (i.e., 40 wt% A) is reached, also at 620 °C. Hence, they both decrease down to a single point; this point is called the eutectic point. The composition there (60 wt%

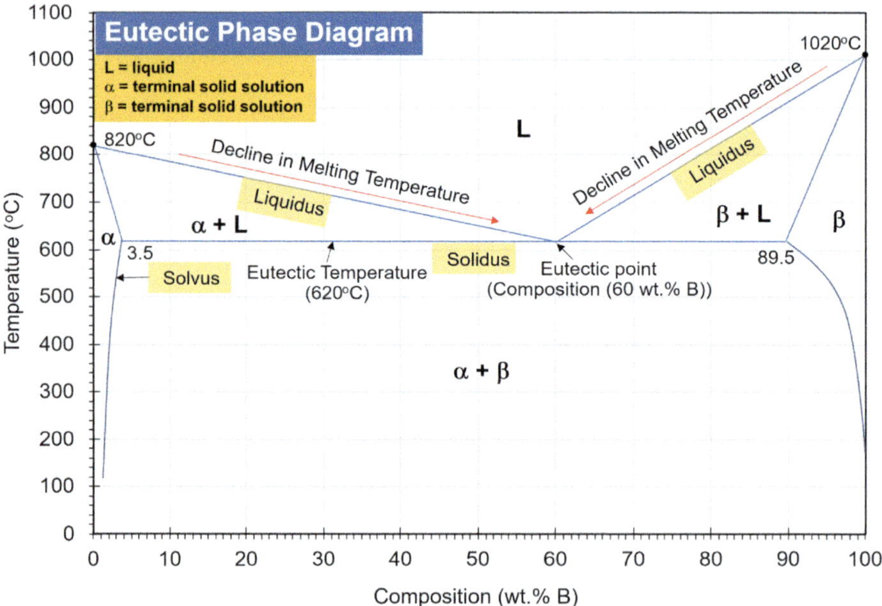

Fig. 6.13 A hypothetical eutectic phase diagram between components A and B

B) is called the **eutectic composition**, and the temperature (620 °C) is called the **eutectic temperature**. Suppose an alloy has a composition lower than the eutectic composition. In that case, it is called a **hypoeutectic alloy**, and an alloy with a composition higher than the eutectic composition is called a **hypereutectic** alloy. The eutectic temperature represents the lowest melting point in the phase diagram and occurs at the eutectic composition. This phase diagram consists of six-phase field, α, β, L, α + β, L + α, and L + β. As expected, all that we have learned in determining phase compositions and mass fractions can also be applied to the eutectic phase diagram and any binary phase diagram, for that matter. What is truly unique about the eutectic phase diagram is the eutectic reaction. Imagine an alloy of composition 60 wt% B, at a temperature of 700 °C, the alloy will be liquid. However, as the alloy is cooled below the eutectic temperature, suddenly the liquid, which is of a composition 60 wt% B, will need to disappear, as it is not welcome (stable) in the α + β phase field. Hence, it transforms into two phases, which is called the **eutectic transformation** or **the eutectic reaction**. The transformation occurs just below the eutectic transformation temperature of 620 °C. As can be seen from the figure, the α and β phases have compositions of 3.5 and 89.5 wt% B, respectively.

This means the A and B atoms in the liquid will need to be divided unequally between α and β. Since diffusion requires time, the liquid finds it much easier to make that transaction locally in many spots along the liquid/solid interface as solidification proceeds. This is seen in Fig. 6.14, where more atoms of element B are sent to the β layers and more atoms of element A are sent to the α layers from the solidifying liquid.

Now we will consider a slightly different eutectic phase diagram (Fig. 6.15), with a slightly different eutectic temperature and melting points of the elements or components, and then see how microstructures are developed if we start with point 1 in Fig. 6.15, for an alloy with composition 3 wt% B, we can see that at this temperature and composition, the alloy will be inside a single-phase field of α, this

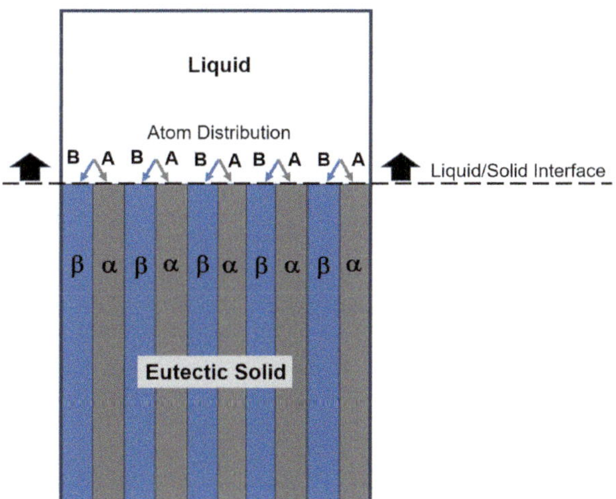

Fig. 6.14 Solidification of a liquid of eutectic composition, into a lamellar (layered) eutectic solid microstructure

means 3 wt% of atoms of element B have dissolved in the crystal lattice of element A, resulting in a single solid solution, α. If we decrease the temperature to point 2, the alloy crosses the solvus line (i.e., the solubility limit) and enters a two-phase field of α and β. This means that the total 3 wt% of B atoms can no longer dissolve in α, the way they did at the higher temperature, and some of the B atoms will have to leave the solid solution. What happens is that the extra B atoms will pick up some A atoms and form a B-rich phase identified in the microstructure as β precipitates.

Figure 6.15 clearly shows that if we use the dashed tie line we see that the composition on the α phase in the α + β phase field is 2 wt% B and the composition of the β phase is 98 wt% B.

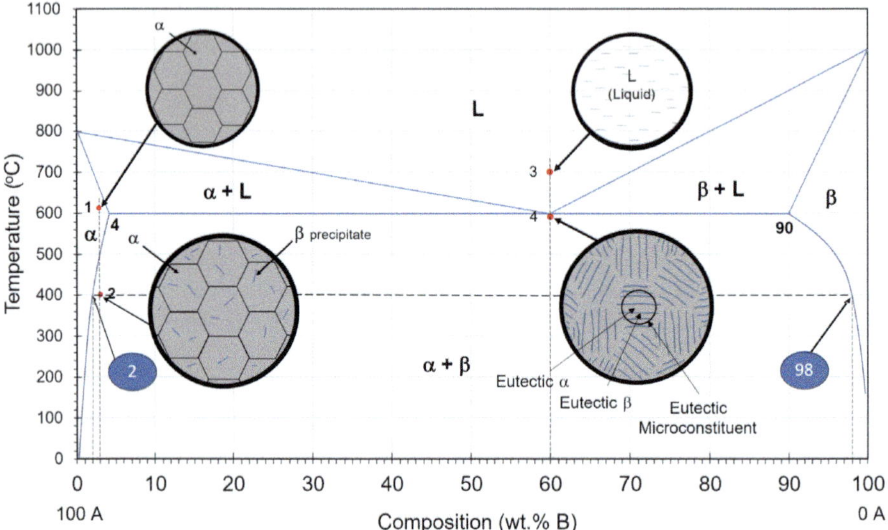

Fig. 6.15 Hypothetical eutectic phase diagram, showing β precipitates formed upon cooling from point 1 to point 2, and a eutectic microstructure (a lamellar structure) formed as the liquid of eutectic composition is cooled below the eutectic temperature, i.e., from point 3 to point 4

Alex, what are the mass fractions of the α and β phases in the 3 wt% B alloy at point 2?

For this we would have to use the lever rule as follows:

$$X_\alpha = \frac{98 - 3}{98 - 2} = 0.99$$

$$X_\beta = \frac{3 - 2}{98 - 2} = 0.01$$

Great job, now in the same figure, composition 3, is none other than the eutectic composition, and at 700 °C it is clear that the whole alloy is completely molten, i.e., L. If you however, cool to point 4, which is just underneath the eutectic temperature, you get the α and β phases forming in a layered or a lamellar structure, which is called the **eutectic microconstituent***. This eutectic microconstituent always, <u>and I mean always</u>, has the eutectic composition. Whether you find it present in small or large quantities in the microstructure. In his particular case 100% of the alloy microstructure is made up of this eutectic microconstituent. We can call it a eutectic microstructure. If I asked you what the mass fraction of this eutectic microconstituent is, you will tell me it is 1. Which means all of the alloy is made up of it.*
However, if I asked you what about the α and β phases that make up this microconstituent, what are their individual mass fractions?

Well, here, too, we will use the lever rule. Since **point 4** *is very close to the horizontal solidus line, I will use the line as my tie line to determine the compositions of α and β. The line intersects the β solvus line at 90 wt.% B, and intersects the α solvus line at 4 wt.% B. Which means the compositions of α and β are 4 wt.%B and 90 wt.% B respectively. Now we can apply the lever rule as follows to find out the mass fractions of α and β:*

$$X_\alpha = \frac{90 - 60}{90 - 4} = 0.349$$

$$X_\beta = \frac{60 - 4}{90 - 4} = 0.651$$

Very well done!! Now let us consider Fig. 6.16, *in As a Matter of Fact. The figure can teach us several additional things about eutectic phase diagrams. Let us first consider* **points 1–4**, *at composition 20 wt.% B. At* **point 1**, *the alloy is molten liquid. Its composition is still 20 wt.% B and its mass fraction is 1 since all the material is made up of it. As the temperature is lowered to point 2, the alloy enters the α + L phase field, which means the α phase needs to precipitate out of the liquid. The composition of the α phase will be different from that of the liquid phase; this can easily be seen if we draw a tie line. However, the overall composition is still 20 wt.% B, the composition of the liquid phase at* **point 2**, *is 29 wt.% B and that of the α phase is 2 wt.%B. The α phase at this point is referred to as* **primary α**, *to differentiate it from another α that will form later. The mass fraction of the liquid phase will be (20 − 2) / (29 − 2) = 0.667, and that of the α phase will be 0.333. Interestingly, as we cool further to point 3, the composition of the liquid becomes 55 wt.% B which is approaching the eutectic composition of 60 wt.%B, and its mass fraction is (20 − 4) / (55 − 4) = 0.314. At a temperature lower than that of* **point 3** *but still within the α + L phase field, a simple application of the tie line will show that the composition of the liquid is the eutectic composition of 60 wt% B. This is important since it shows that at that point, although we have two phases, primary α and liquid, the liquid will always be of a eutectic composition, and it does not matter which composition the main alloy is. This is the point before the alloy enters into the α + β phase field. Suppose we conduct a simple exercise by moving the main alloy composition along the solidus line within the α + L phase field. In that case, we will easily deduce that although the liquid mass fraction may differ, just above the solidus line, the liquid composition will always be the eutectic composition of 60 wt.%B. Upon cooling the alloy below the solidus line into the α + β phase field, it becomes clear that the liquid is not stable there, but the primary α phase is. Therefore, upon cooling below the solidus to* **point 4**, *the primary α originally formed does not need to transform since it is stable in the new phase field and thus remains in the microstructure. This is not the case, though, for the liquid phase. Since the liquid, however much there is, is always of the eutectic composition, therefore when it does transform below the solidus, it transforms to the eutectic microstructure or micro-constituent (α + β). This is clearly seen in the microstructure of point 4 in the figure. Now since the liquid transformed to the layered or lamellar structure of alternating layers of α and β, the α layers that are formed are now called* **eutectic α**, *to differentiate them from the* **primary α** *formed earlier. They both have the same composition, though, as the tie-line would confirm.*

Example Problem 6.4

Consider an alloy of composition 20 wt% B–80 wt% A, as shown in Fig. 6.16, at a temperature just below the solidus line. What does the microstructure consist of? What are the phases or microconstituents present and their relative mass fractions?

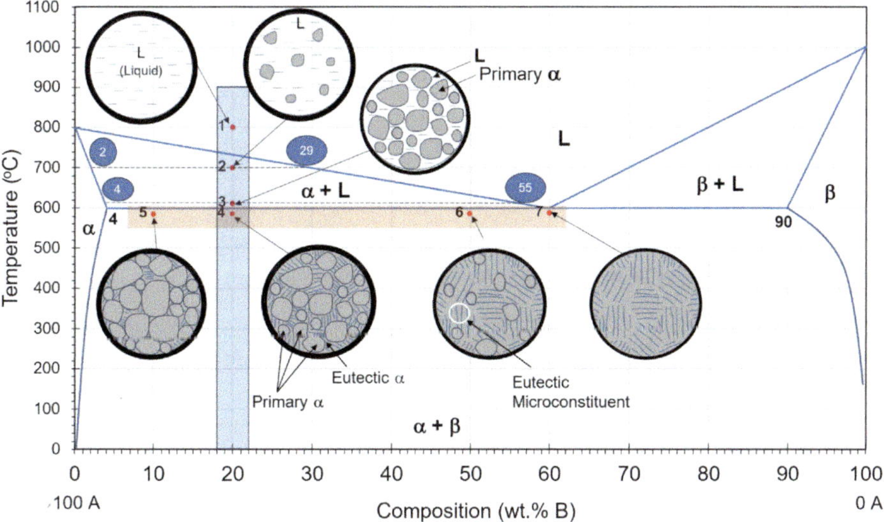

Fig. 6.16 Hypothetical eutectic phase diagram, showing formation of primary α upon entering the α + L phase field at point 2, which increase in amount as the temperature is reduced to point 3. At point 4, all the liquid present at point 3 transforms to the eutectic microconstituent as seen in point 4. As we go from point 5, to 4, to 6 to 7, the amount of eutectic microconstituent increases and the primary α decreases until at the eutectic composition the whole microstructure is made up of the eutectic microconstituent

Solution

From the figure, since the temperature of interest is just below the solidus, we can use the solidus line as our tie line. In this case, we can approach the solution in two ways, depending on how we want to look at the microstructure. As it stands, the microstructure consists of primary α particles and eutectic microconstituent, consisting of alternating layers of α and β. Hence, we can approach the problem by saying we have a microstructure consisting of primary α of composition 4 wt% B and a eutectic microconstituent of composition 60 wt% B. Alternatively, we can say the microstructure consists of two phases, α (i.e., primary and eutectic α) and β. The α has a composition of 4 wt% B while the β has a composition of 90 wt% B. Let's first solve the problem assuming the first approach.

Microstructure Consisting of Primary α and Eutectic Microconstituent

The composition of primary α ($C_{P\alpha}$) is 4 wt% B, and that of the eutectic microconstituent (CE) is 60 wt% B; we recall that the composition of the alloy is 20 wt% B (C_o). To obtain the mass fraction of primary α phase (X_α) and that of the eutectic microconstituent (X_E) we use the lever rule again:

$$X_{P\alpha} = \frac{C_E - C_o}{C_E - C_\alpha} = \frac{60 - 20}{60 - 4} = 0.714$$

$$X_E = \frac{C_o - C_\alpha}{C_E - C_\alpha} = \frac{20 - 4}{60 - 4} = 0.286$$

So, we see that the mass fraction of primary α is 0.714 while that of the eutectic microconstituent is 0.286. Remember, these are the mass fractions and not the volume fractions, which would require knowledge of the densities of the phases.

Microstructure Consisting of Total α (Primary α + eutectic α) and β

Whether we talk about primary or eutectic, they are still the same. They both have the same composition; it's just that one exists as particles and the other as layers within the lamellar eutectic structure.

The composition of total α ($C_{T\alpha}$) is 4 wt% B, and the composition of β (C_β) is 90 wt% B; we recall that the composition of the alloy is 20 wt% B (C_o). To obtain the mass fraction of total α phase ($X_{T\alpha}$) and β (X_β) we use the lever rule again:

$$X_{T\alpha} = \frac{C_\beta - C_o}{C_\beta - C_{T\alpha}} = \frac{90 - 20}{90 - 4} = 0.814$$

$$X_\beta = \frac{C_o - C_{T\alpha}}{C_\beta - C_{T\alpha}} = \frac{20 - 4}{90 - 4} = 0.186$$

So, we see that the mass fraction of total α is 0.814 while that of the β phase is 0.186.

Okay, Alex. You found the mass fraction of the total α and the mass fraction of the primary α. What is the mass fraction of the eutectic α ($X_{E\alpha}$)?

That's easy! Since total α is primary α plus eutectic α, then if we have the mass fractions of total α and primary α, we simply subtract primary α mass fraction from the mass fraction of total α to get the mass fraction of eutectic α.

$X_{E\alpha} = X_{T\alpha} - X_{P\alpha} = 0.814 - 0.186 = 0.628$

Hence, the mass fraction of eutectic α is 0.628.

With reference to Fig. 6.16, as we move from **points 5, 4, 6, and 7**, we see that more and more of the eutectic microconstituent is formed. This is easily understood if we determine the mass fraction of the liquid before solidifying. Using the lever rule reveals an increase in the liquid phase from left to right on the composition axis. The liquid phase is of a eutectic composition in all cases, and hence when it transforms, it transforms to the eutectic microstructure.

Suppose we consider an alloy having a composition greater than the eutectic composition of 60 wt% B and follow the solidification of a liquid with that composition, we will cross the L + β phase field. In this case, as with the hypoeutectic alloy, the β phase will be called **primary β**, and upon reducing the temperature to just below the solidus temperature, the liquid phase (which will also be of eutectic composition) will transform to the eutectic microconstituent, consisting of the same layers β and α, which would be called eutectic α and eutectic β. All the procedures conducted for the hypoeutectic alloy are equally applicable to the hypereutectic alloy.

*Professor, I conducted the compositional analysis for **sample 3** from **layer 3,** and it showed the following elemental composition:*
Aluminum: 85.5%
Zinc: 8.2 wt%
Magnesium: 2.9 wt%
Copper 2.3%
Iron, zirconium, silicon, manganese, and titanium are at 0.1–0.15 wt range each.
Other trace elements were also found.
Looking at Table 5.5, *it seems this is a 7000 series aluminum alloy, but I'm not sure exactly which one.*

Great job, Alex, you are almost there. The composition is, as you said, a 7000 series aluminum alloy.

Time lapses

⋮

*I know I've been away for a bit, but I went ahead and carried out my own tests. The strength of this material is high, in fact, the highest of any commercial aluminum alloy today. Specifically, its tensile strength is ~ 709 MPa, and its yield strength is 681 MPa. Your next chapter in As a Matter of Fact will teach you about strength. Given that its density is also 2.85 g/cm^3, it is now clear what this material is. I'm sure you are anxious to find out. The material is **Tennalum**® **7068** (also known as 7068-T6511). Indeed, this is the strongest aluminum around, with excellent corrosion resistance. Imagine a material with a strength similar to steel but almost 2/3 lighter. It does have some disadvantages, such as being very expensive and challenging to make currently, resulting in limited applications. If we combine its advantages of strength and low density, we find that this material's strength-to-weight ratio exceeds almost all engineering alloys, including titanium alloy, Ti6Al4V. However, I believe its inclusion in the Sphere was based more on accomplishments rather than widespread use. This alloy was developed in the mid-1990s by Tennalum Inc., a Division of Kaiser Aluminum Corp. Materials. Its addition to the sphere may have been to indicate its potential future impact and use. We find examples in history where a metal was extremely expensive to use at first, but at some point, a new manufacturing method became available and made it cost-effective. You don't have to look far to find such an example; let's look at aluminum and how it was first treated as a precious metal but then became widely used.*

Now, let us keep thinking about **layer 4**.

Problems

6.1. With all things being equal, which is faster, interstitial, or vacancy diffusion, and why?

6.2. What is meant by carbonization?

6.3. Consider two steel plates: plate A is 5 mm thick and plate B is 10 mm thick. Both have a concentration of the same impurity atoms on their left side of 0.5 wt% C, and on the right side 0.1 wt% C. Calculate the concentration gradient of the two iron plates. Which one is greater in magnitude and why?

6.4. Is the activation energy for atomic diffusion always a constant value, or can it change under certain conditions? Explain your answer.

6.5. A 10-mm-thick plate of 1030 steel (0.3 wt% C) was enriched with carbon on one side to raise the composition there to 0.5 wt% C, and on the other side a carefully controlled oxidizing environment is established that maintains the concentration there at 0.2 wt% C. If the diffusion coefficient is 1.4×10^{-12} m^2/s, calculate the diffusion flux. The density of steel is 7870 kg/m^3.

6.6. Calculate the activation energy for atomic diffusion for impurity atoms diffusing in a crystal lattice, if the diffusion coefficient is 1.1×10^{-11} m^2/s, the D_o pre-exponential is 2×10^{-5} m^2/s, and the temperature at which diffusion is occurring is 1200 K.

6.7. Discuss the different types of diffusion mechanisms. Which one of them do you think will have the lowest activation energy and why?

6.8. A 1020 (0.2 wt% C) steel gear is enriched at its surface with carbon to raise its surface concentration to 0.9 wt% C. If the gear is heated to 1273 K, how much time (in minutes) would it take for the carbon concentration to become 0.5 wt% C at 0.9 mm under the surface? *Note:* $D = 1.9 \times 10^{-11}$ m^2/s.

6.9. The surface of a nickel component is enriched with copper in a way to maintain its concentration there to be 5 wt% copper. If the component is left at 723 K for 2 h, what would be the concentration of copper at 1 mm beneath the surface? *Note:* $D = 1.3 \times 10^{-22}$ m^2/s.

6.10. Give two examples of short-circuit diffusion.

6.11. Explain the significance of the word equilibrium in an equilibrium phase diagram. What does it say about the phase diagram itself?

6.12. What is the difference between an isomorphous phase diagram and a eutectic phase diagram?

6.13. Is it correct to call a eutectic microstructure a eutectic phase? If not, then why? And what would you call it?

6.14. α-titanium has a hexagonal close-packed (hcp) crystal structure and β-titanium has a body-centered cubic crystal structure. Are they considered different phases of titanium?

6.15. What is the difference between composition and mass fraction?

6.16. Describe the room temperature microstructure of an alloy quenched (rapidly cooled) to room temperature from the $\alpha + L$ two-phase field.

6.17. With reference to the A-B isomorphous phase diagram in Fig. 6.10. Consider an alloy of composition 70 wt% B at 800 °C, and answer the following:

   (a) Identify the phase(s) present.
   (b) What is the composition of each phase?
   (c) What is the mass fraction of each phase?

6.18. With reference to Fig. 6.13, consider an alloy of composition 80 wt% B at a temperature of 620.5 °C. Answer the following:

   (a) Identify the phase(s) present.
   (b) What is the composition of each phase?
   (c) What is the mass fraction of each phase?

6.19. With reference to Fig. 6.13, consider an alloy of composition 80 wt% B at a temperature of 619.5 °C. Answer the following:

(a)   Sketch the microstructure and identify its phases on the sketch.
(b)   What is the composition of eutectic microconstituent?
(c)   What is the composition of primary β?
(d)   What is the mass fraction of the eutectic microconstituent and primary β?
(e)   What is the mass fraction of total β?

6.20.   With reference to Fig. 6.13, consider an alloy of composition 60 wt% B at a temperature just below the eutectic point. Answer the following:

(a)   Sketch the microstructure and identify its phases on the sketch.
(b)   What is the composition of each phase?
(c)   What is the mass fraction of each phase?

Part III
Layer 4: A Notorious Material Indeed

The Sphere

Layer 4

"A notorious material indeed."

Layer 4 presents an interesting proposition. Initial investigations by the professor's graduate research assistant (GRA) using EDX reveal that it primarily consists of iron, which is also magnetic. At least, this is the only information that the GRA was allowed to say at this point.

It seems like some iron alloy, but to be honest, I'm not sure where to start, Prof.

It looks like you may need to conduct further investigations to arrive at a conclusive finding as to the identity of this material. I suggest reading Chaps. 7 and 8 in As a Matter of Fact. I think it could be quite useful.

Chapter 7
Mechanical Behavior of Metals and Alloys

7.1 Introduction

If a material component is required to carry a load during service, we must be fully aware of its mechanical properties. These properties may include hardness, tensile/compressive properties (strength, ductility, and modulus), fracture toughness, wear, creep, and fatigue resistance. This chapter discusses some of these properties, starting with hardness.

7.2 Indentation Methods for Determining Mechanical Behavior (Hardness Tests)

Different materials behave differently when subjected to a force; some may appear stiff (and resist the force), and others may appear compliant (compliance is the opposite of stiffness). Some may be stronger than others; some may be easily formed into a new shape by applying force, while others may fracture rather than change shape. All these may be quantified by using specialized tests. First, we will discuss the **Hardness Test**. This is one of the simplest tests used to determine a material's mechanical response to a localized applied load. Figure 7.1 shows a simple schematic of such a test. Here, a hardened steel spherical indenter is used to contact the surface, and under the influence of the applied force, the spherical ball sinks into the material surface; after it is removed, it leaves behind an indent (i.e., an impression). The deeper the indenter is allowed to sink into the sample, the softer the sample is, and vice versa; a hard material will resist the penetration of the indenter, thus leaving behind a smaller indent or indentation impression.

What is so special about this test is that it is cheaper and easier to perform than many of the other tests, and yet it can still provide us with helpful information about the "hardness" or strength of a material. We only need a sample with an exposed

K. Morsi, *An Engaging Approach to the Science and Engineering of Materials*,
https://doi.org/10.1007/978-3-032-06231-4_7

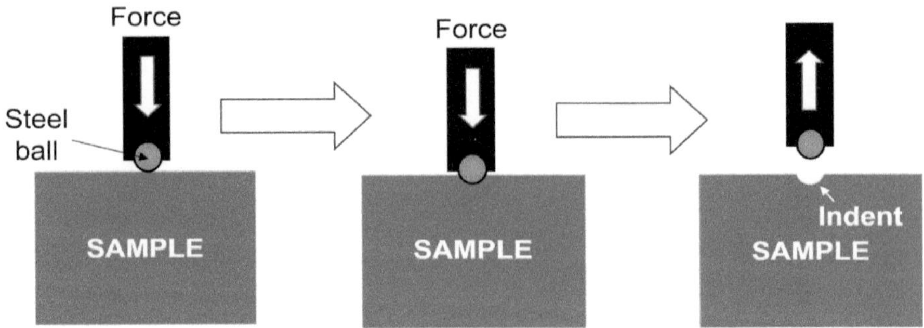

Fig. 7.1 Illustration of a typical hardness test conducted on a sample surface, note the impression or indent left behind on the surface of the sample

surface that we can test. Also, if the test was carried out on a component, we can still use the component after testing. We call this type of test a ***non-destructive test***, as opposed to other tests, which tend to destroy a material in the process of testing it, i.e., they are *destructive tests*. One of the most significant advantages of hardness testing is that we only need a small amount of material to conduct the test. This became very useful when the field of nanomaterials first started. At that time, nanomaterials were quite expensive to produce, and scalability was a major issue, so only small amounts of very expensive materials were available. One would hardly think of destroying such a precious commodity while trying to first learn about it, right? Hence, hardness testing became the mechanical test of choice for nanomaterials. So, what is the deal with this so-called *hardness test*? It involves contacting the sample's surface with a small indenter and applying force.

Several hardness tests are available today, which differ in the type and shape of the indenter and the magnitudes of loads (forces) being applied. Some tests measure different things to arrive at a hardness value. For example, some tests measure the size of the indent left behind by the indenter (Fig. 7.2). In contrast, others do not need to do that and instead inherently measure how deeply an indenter manages to sink into the material using a given force and indenter type.

Before we discuss these in greater detail, we need to remind ourselves that things were not always as sophisticated as we see today. In fact, as far back as around 300

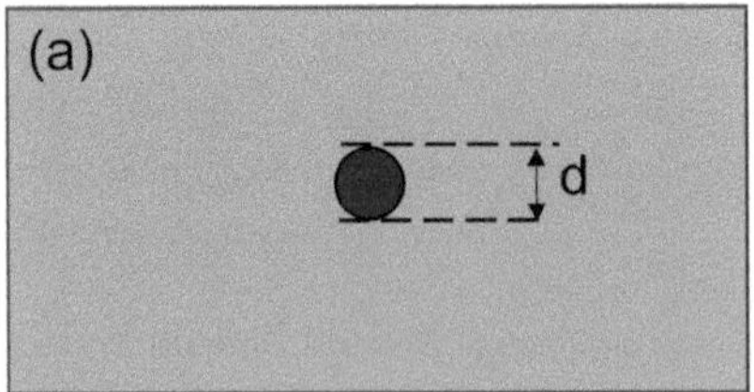

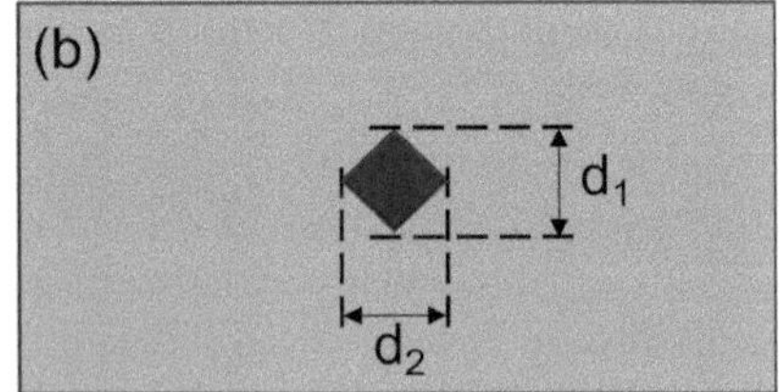

Fig. 7.2 Illustration of a typical impressions on sample surface left after indentation by **a** a spherical indenter, **b** a pyramidal indenter (multiple resources)

BC, people used to compare the hardness of one material to another by seeing which one scratches the other. In 1812, **Friedrich Mohs**, a German Geologist, devised a hardness scale that could classify different minerals not by their chemical characteristics but rather by their physical ones. This is known as the **Mohs scale**. Specifically, he introduced a Hardness scale ranging from 1 to 10 (10 for the hardest material on earth). Here, a harder material was defined as one that would scratch a softer one. In this scheme, diamond was sitting proudly at the top of the scale with a value of 10 (being able to scratch all materials), while Talc was confined to the absolute bottom. Between these two lies the rest of the materials, each with a given Mohs hardness indicating its ability to scratch all materials with a lower Mohs hardness number. Hence, Gypsum had a Mohs hardness of 2, while quartz (silicon dioxide, SiO_2) had a value of 7. The Mohs hardness of fingernails that we sometimes use to scratch things is around 2–2.5. However, most of the engineering industry resorts to other, more quantitative hardness tests, as described below.

7.2.1 The Rockwell Hardness Test

This is perhaps the most beloved test in the industry due to its ease of operation and the least labor required to arrive at the hardness value. In this test, an indenter first indents the sample surface using a minor force (10 kg) followed by a major force. The hardness tester then measures the difference in depth penetration of the indenter into the material. Deeper penetration signifies a soft material instead of a material that only allows a small penetration of the indenter under the same applied force or load. Hence, all the operator needs to do is place the sample on the testing stage underneath the indenter and press a button. The indenter will come down and press into the sample, then go up, and a hardness value is displayed. There is no need to measure the size of the indentation impression left behind, which is a great appeal of the test. Hence, the hardness value is related to the penetration depth.

However, the materials available to us are vast, and understandably, one load value or even one type of indenter may not be enough to cover all the ranges of penetrations needed to test all materials. As such, several indenters are used for the Rockwell Hardness test. For example, a diamond cone or a hardened steel ball of 1/8th inch or 1/16th inch in diameter. More specifically, while the magnitude of the initially applied minor load is fixed at 10 kg, the magnitude of the major load varies from 60 to 150 kg. Table 7.1 shows the different loads (minor and major) and indenters used together with typical materials suitable for testing within each scale, per ASTM E18 standards. Hence, a particular combination of force and indenter type will constitute one of the Rockwell hardness scales.

A particular scale is selected depending on a material's hardness. If the hardness value appears close to the scale's upper limit, one should move to a higher hardness scale to obtain a more accurate value. Similarly, a lower scale is used if the hardness value lies close to the bottom of a scale.

Table 7.1 Rockwell Hardness Scales and their definitions (from ASTM E18)

Rockwell scale (abbreviation)	Minor load (kg)	Major load (kg)	Shape and type of indenter	Materials suited for scale
A (HRA)	10	60	120° spheroconical diamond (i.e., Brale)	Cemented carbides
B (HRB-S) or (HRB-W)	10	100	1/16th inch diameter steel ball (s) or tungsten carbide ball (W)	Copper alloys, malleable iron
C (HRC)	10	150	120° spheroconical diamond (i.e., Brale)	Titanium
D (HRD)	10	100	120° spheroconical diamond (i.e., Brale)	Medium case-hardened steel, thin steel
E (HRE)	10	100	1/8th inch diameter ball	Magnesium and aluminum alloys, Cast iron
F (HRF)	10	60	1/16th inch diameter ball	Copper alloys (annealed)
G (HRF)	10	150	1/16th inch diameter ball	Malleable irons and cupro-nickel alloys

So, if someone tests the hardness of a material using the Rockwell hardness test and reports the hardness as **HRA 60**, according to Table 7.1, this will immediately tell us that the A-scale was used, and the Rockwell hardness number was 60. Now we can imagine that, according to Table 7.1, not all scales are the same. For example, scale C requires an application of 150 kg as opposed to the 60 kg used in scale A; this tells us that the material to be tested must be hard for it to need 150 kg to penetrate its surface, enough to register output. If one conducts the same hardness test on the same material but using the C-scale, the value registered on the display would be much lower than the 60 registered using the A-scale. This is because the A-scale is typically used to measure the hardness of relatively softer materials. Hence, we have several scales to ensure we cover the wide range of hardness values expected for all metals and alloys. The balls are typically made of hard steel (hence, S is written after the test abbreviation, i.e., HRB-S). Still, tungsten carbide (WC) balls have also been used and thus indicated by a W after the hardness number abbreviation, i.e., HRB-W. A **superficial Rockwell hardness test** is used for smaller samples, where the minor and major loads are reduced compared with the regular Rockwell hardness test. This helps to reduce the extent of damage to the sample.

7.2.2 The Brinell Hardness Test

This is a test that *does* require us to measure the size of the indentation produced. In this case, the indenter is a hardened steel ball (or tungsten carbide) of diameter 1 cm, and the loads applied vary from 500 to 3000 kg. To gain a deeper appreciation of this level of force, 3000 kg could be the weight of two small cars. 3000 kg is typically used to measure the hardness of materials such as cast irons and steels, while 500 kg is used to measure the hardness of non-ferrous alloys. Equation 7.1 shows how the Brinell hardness number is calculated; note that D is the diameter of the indenting ball, d is the diameter of the circular indentation impression produced on the surface, and P is the applied load in kg.

$$\mathrm{HB} = \frac{2P}{\pi D\left(D - \sqrt{D^2 - d^2}\right)} \tag{7.1}$$

As can be seen from Eq. 7.1, we must measure the size of the indent, which is an extra step compared to the Rockwell hardness test. The output will be a Brinell hardness number (**HB**) in units of kg/mm^2. The advantage is that we only need one scale that can measure various materials with varying hardness values (unlike the Rockwell scale). For example, using the same scale, we can measure the hardness of a soft lead alloy and a much harder titanium alloy.

7.2.3 The Vickers (Macro) Hardness Test

Whereas for Rockwell hardness, we need to use different scales to capture the hardness of different materials with varying hardness values, the Vickers test can be used for all materials in one go! This hardness test usually uses loads up to 100 kgf and a diamond pyramid indenter.

Equation 7.2 can be used to obtain the Vickers hardness (HV), where d is the average of the two diagonals d_1 and d_2 in Fig. 7.3a.

$$\mathrm{HV} \sim \frac{1.854P}{d^2} \tag{7.2}$$

Equation 7.2 was derived by considering the applied load divided by the surface area of the indentation left behind. Note that it is not the projected area but rather the surface area.

The result is usually reported as the Vickers hardness number, such as 200 HV. The 200 is 200 kgf/mm^2. As such, the HV number can also be converted to units of Pascal (N/m^2). However, although Pascal is also the unit for pressure, hardness should not be confused with pressure, as it is inherently different. Sometimes, the

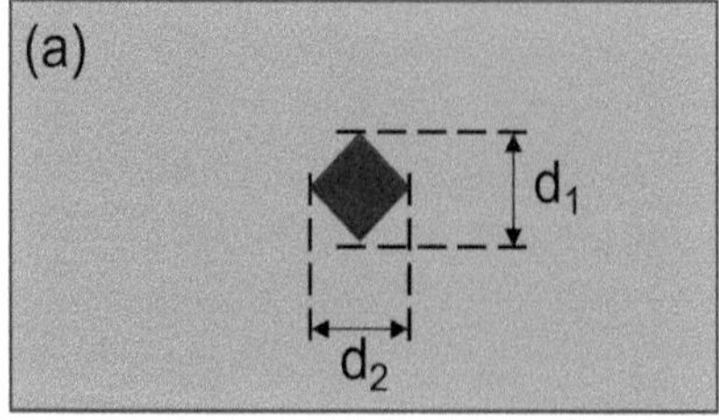

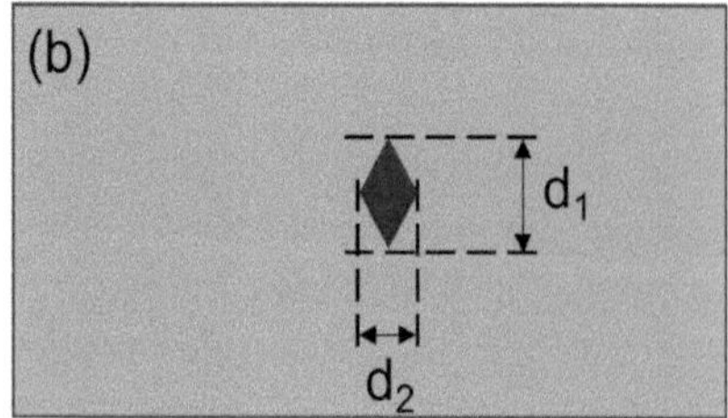

Fig. 7.3 Illustration of a typical impression on sample surface left by indentation using **a** Vickers indenter and **b** using a Knoop indenter

number is written as 200 HV50, where 200 is the Vickers hardness number, and 50 represents the 50 kg load used.

7.2.4 The Microhardness Test

There comes a time when you only have a small sample, which is too small to use on any other tests. Since Rockwell and Brinell use loads of more than 60 kg, the sample may be damaged easily or even destroyed. As such, people resorted to using a test that requires lower loads, in fact, less than 1000 g.

These low loads result in extremely small indentations on the sample's surface (schematically presented in Fig. 7.3a), requiring a microscope to measure their sizes. Since the indentation size is so small, the surface must be very well polished to avoid scratches and see the indent properly. Not only that, but the surface also needs to be prepared specially so as not to contain any residual stress that may affect the reading. For example, during grinding operations, the harsh contact and removal of material from the surface may leave behind subsurface damage (and residual compressive stresses) that may introduce unintended effects and may skew the results, as the material in this vicinity would not properly represent the material condition elsewhere on the sample. One must carefully remove material during grinding and polishing to remove the subsurface damaged region introduced by the previous grinding operation before proceeding to the next grinding/polishing stage.

Microhardness testing encompasses two types of tests:

a. The **Vickers hardness test** and
b. The **Knoop hardness test**.

These tests use loads equal to or less than 1000 g (1 kg). They also both use diamond indenters. The difference comes in the shapes of the indenters. Both have the shape of a pyramid, but the pyramid for Vickers has a square base with equal diagonals, while that of the **Knoop** has a base with a diamond shape, i.e., one diagonal is larger than the other. As such, when a Vickers indentation and a Knoop indentation are produced, they leave impressions on the sample's surface, as shown in Fig. 7.3a

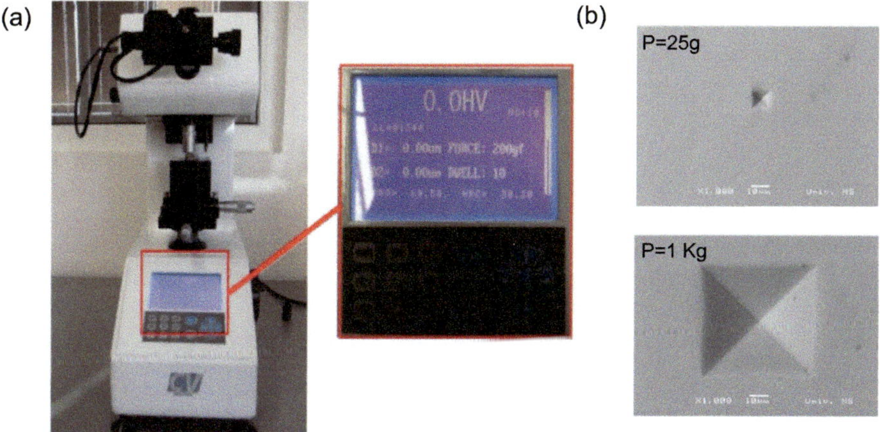

Fig. 7.4 **a** Vickers microhardness tester (image from Monka, P. P.; Monkova, K.; Vasina, M.; Kubisova, M.; Korol, M.; Sekerakova, A. Effect of Machining Conditions on Temperature and Vickers Microhardness of Chips during Planing. Metals 2022, 12, 1605. https://doi.org/10.3390/met12101605, **b** Vickers indentation impressions on iron aluminide, using 25 g and 1 kg indentation loads. Image from: Balos, S.; Pecanac, M.; Trivkovic, M.; Bojic, S.; Hanus, P. Load-Independent Hardness and Indentation Size Effect in Iron Aluminides. Materials 2024, 17, 2107. https://doi.org/10.3390/ma17092107. All images reproduced and modified under the terms and conditions of the Creative Commons Attribution (CC BY) license (https://creativecommons.org/licenses/by/4.0/)

and b. Figure 7.4 shows a Vickers microhardness tester and impressions left behind on an iron aluminide sample. Note that the impression size increases significantly when the indentation load increases from 25 g to 1 kg (Fig. 7.4b).

Example Problem 7.1

A Vickers (macro) hardness test was conducted on a material. A load of 50 kg was used, resulting in an indentation, with an average diagonal length of 0.5 mm. Determine the Vickers hardness for that material.

Solution

$$\mathrm{HV} = \frac{1.854 \times 50}{(0.5)^2} = 370.8 \mathrm{HV}50$$

Some Practical Aspects

If someone gives us a sample to test and asks us how hard it is? Would we simply test the sample one time? Of course not. We could indent it, for example, 5–10 times and calculate an average and standard deviation. That is because we want to ensure the hardness is consistent across the material's microstructure. So, where do we place our second indentation after we have indented for the first time? We must be cautious not to place it too close to the first indentation. Note, the first indentation materialized only because the material beneath it flowed away, allowing the impression to be placed there. This material flow will change the local properties of the materials, and as such, we do not want to make another indentation close to that region; otherwise, we would be measuring the properties of a region with characteristics different from the remaining materials. So, when carrying out multiple indentations, we need to separate indentations by a distance ~ 3 times that of the indentation diagonal, as shown in Fig. 7.5 for Vickers hardness indentations and Knoop ones. The figure shows the advantage that Knoop testing could bring if we wanted to see how hardness varies across a small region of a material microstructure. We can typically fit in more Knoop indentation in the same length as a Vickers indentation, obtaining more data points to work out the hardness profile.

Equation 7.2 can also be used to obtain the Vickers microhardness number. On the other hand, Eq. 7.3 can be used to obtain the Knoop hardness (HK), where d_2 is the major diagonal in Fig. 7.3b.

$$\text{HK} = \frac{14.2P}{d_2^2} \tag{7.3}$$

The type of hardness test we use may depend on factors such as which machine we have at our disposal. However, it is possible to convert one hardness number to another, for example, HRA to HB or HV. Figure 7.6 provides a summary of hardness abbreviations.

Fig. 7.5 Illustration of a typical multiple impression on sample surface left by indentation using **a** Vickers indenter and **b** using a Knoop indenter, while keeping a separation of 3 indentation diagonals. More indents can be placed using the Knoop tester, which would be very useful in determining hardness profiles over short distances

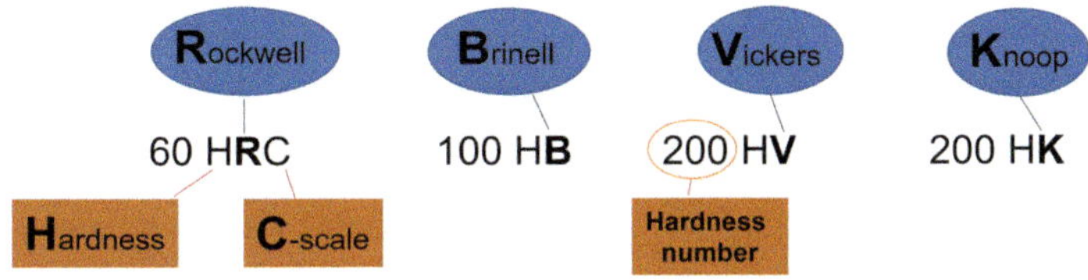

Fig. 7.6 Hardness abbreviations explained

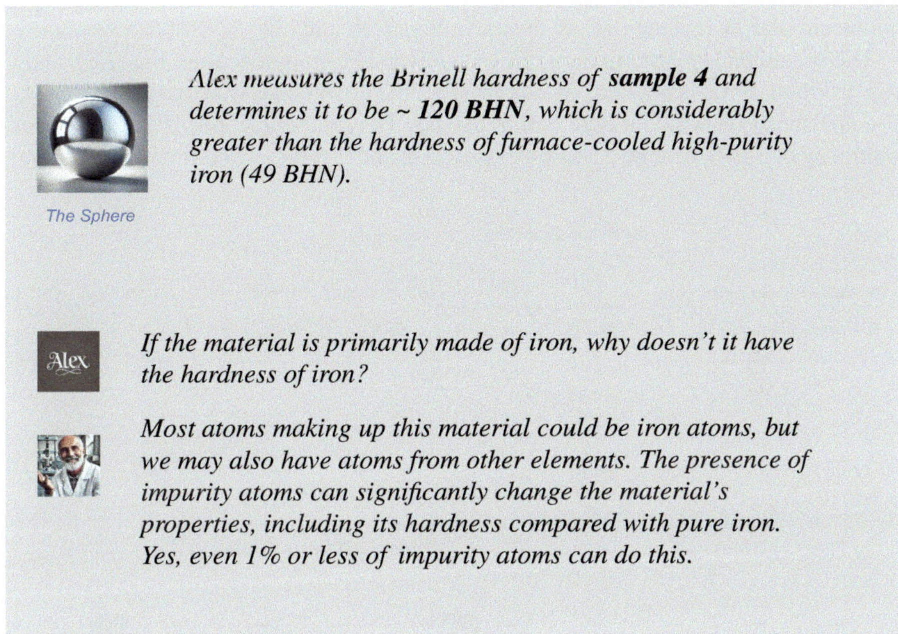

7.3 Tensile Testing

Apart from the non-destructive hardness test, other tests are destructive but provide a great deal of information about the material. One such test is the **Tensile test**.

Let's imagine that someone brings a steel block and asks, how stiff, strong, or ductile is this material in tension. We would initially be speechless, as we wouldn't know where to start. Luckily, some testing standards are available that describe how to test a metal in tension. One example of such a standard is the ASTM E8 "Tension Testing of Metallic Materials". In this standard, we are told what geometry we should machine our sample to and even what speed we should test it at. It provides several choices for geometries and samples of different sizes that one can use, depending on the situation. An example of a typical tensile testing specimen geometry is presented in the schematic in Fig. 7.7a. As can be seen, the specimen consists of two gripping sections, where the specimen is firmly gripped by a machine that would ultimately pull it apart.

As the specimen cross-sectional changes from the gripping section to the mid-section (reduced section), it does so gradually through a radius. Had it been an abrupt change in geometry, it would have left a sharp corner at the transition, which would have been a potential failure site during testing. The reduced section contains the **gage length** (because we gage the material within this length, i.e., how the gage length and diameter change during testing). Finally, the cross section of the sample can be circular or rectangular, as shown in Fig. 7.7b and c.

The specimen is placed in the grips of a universal testing machine (Fig. 7.8), and a high gripping force is applied. As can be seen, the universal testing machine consists of a crosshead on which a load cell is placed. The machine controls the crosshead motion up or down through a computer program and is typically commanded to move

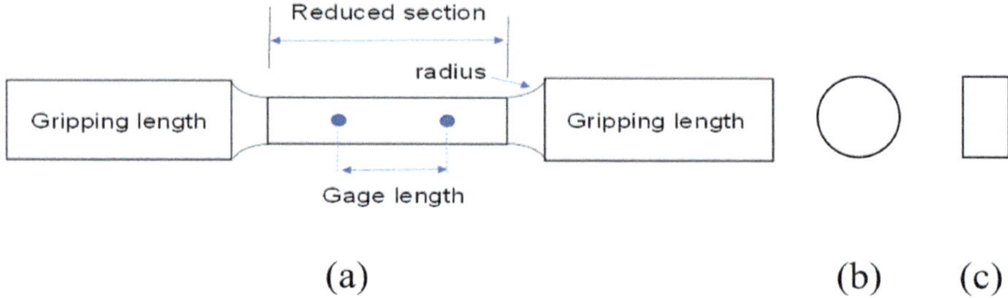

Fig. 7.7 **a** Schematic of tensile testing sample showing gage length and other important features of the sample geometry, **b** circular cross-section, **c** rectangular cross section

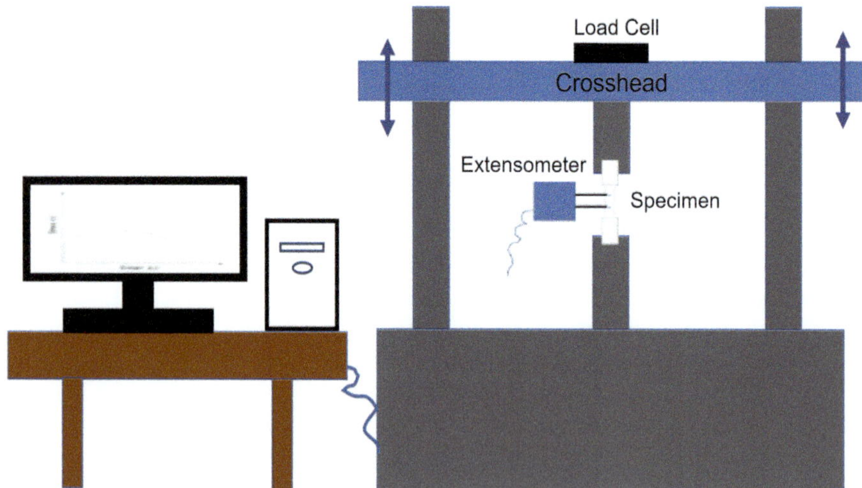

Fig. 7.8 Schematic of universal testing machine

at a set speed (per the ASTM standard). In the case of tensile testing, the crosshead is commanded to move vertically up at the prescribed speed. As the crosshead moves, the gripped specimen must comply with the movement, i.e., stretch, since the machine is much stronger than the specimen. The universal testing machine then continuously outputs two important data sets during the test: the **instantaneous load** applied at any given instant and the corresponding **instantaneous extension** of the original gage length.

Alex, how would you know the extension the specimen is experiencing during the test?

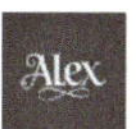

I know how to find the extension at any given time during the test since I know the speed at which the crosshead moves. Since speed is distance over time, if the testing speed, for example, is 5 mm/min, then after half a minute (30 s), the gage length would have extended by 2.5 mm, and so on. Am I right?

*Unfortunately, you are not correct. Remember that other parts/ components exist between the crosshead and the specimen. What if once the crosshead starts to move upwards, the grips align better, i.e., the slack is taken up by the crosshead motion? Isn't the displacement of the crosshead now being taken up by the grips aligning? The specimen would not feel anything yet, right? You must also consider the compliances (opposite of stiffnesses) of all the components between the specimen and the crosshead. So, it's not as easy as it looks. You need to measure the extension of the specimen exactly where you need it, i.e., at the gage length. For that, you will need another instrument called an **extensometer**. This instrument can measure the extension to less than a micrometer accuracy. This is very important, as you will see later when measuring certain properties. If you used the crosshead speed to calculate the specimen extension, the extension you measured could be entirely in error. Many have fallen into this pitfall, but now you know to avoid it and maybe advise others.*

The fundamental output of the test is *instantaneous* force data (from the load cell) versus *instantaneous extension* data (from the extensometer). A typical plot would look something like that in Fig. 7.9. As we can see, it consists of an initial linear part (showing a linear relationship between force and extension), and then the curve becomes nonlinear.

Imagine if we had two aluminum specimens, one had double the cross-sectional area of the other (i.e., thicker), and then we tested both samples. Would the magnitude

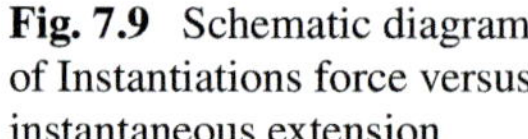

Fig. 7.9 Schematic diagram of Instantiations force versus instantaneous extension

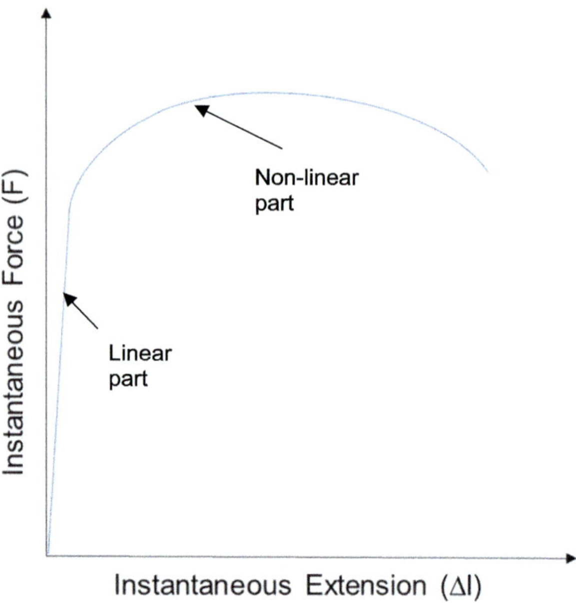

of the force needed to stretch both samples to the same extension be the same? Intuitively, we know deep in our hearts that the thicker specimen will require more load. That is because it is not only the force that is important, but also the force per unit area. If instead, we applied the same force per unit area to both samples, we should expect a similar response. Force per unit area is what we call **Stress**. Stress is unique because we have now considered the specimen's cross-sectional area. That is why instantaneous force is typically converted to stress and not used independently.

Typically, the **instantaneous force (F)** is divided by the **original cross-sectional area (A_o)** to define what we call the **Engineering stress (σ)**, Eq. (7.4).

$$\text{Engineering stress} = \frac{F_i}{A_o} \tag{7.4}$$

That can't be right. Why should I keep dividing the instantaneous force output during the test by the "original" cross-sectional area to find out the force per unit area at that point in the test? Shouldn't the cross-sectional area of the sample be getting smaller as the test proceeds? Shouldn't I divide the instantaneous force by the "actual" or instantaneous cross-sectional area of the specimen at that point during the test? I can see that the force being applied to the specimen changes during the test from Fig. 7.9, *and so would the diameter of the sample (and hence the cross-sectional area). Since the specimen is being stretched, the diameter should get smaller. Am I missing anything, Professor?*

"Indeed, you are correct; I'm proud of you; you should always be critical of information that comes your way and test it with your logic. If you want true stress at any given point during the test, you should divide the **instantaneous force** *(F_i) by the* **instantaneous area** *(A_i) to get the* **true stress** *(σ_T), as you imply. However, these guys are not delusional; you see, if we simplify matters and divide by the original cross-sectional area (which we already know) and assume it does not change within the linear part of the curve, our error for the value of the stress would be around ~ 1%. So, it saves us the hassle of measuring the specimen diameter during the test (and hence the instantaneous area). Now, I'll leave to continue reading "As a Matter of Fact".*

As stress considers the cross-sectional area of the specimen, likewise, imagine if we had two specimens with the same cross-sectional area, one with a 1-inch gage length and the other with a 2-inch gage length. If we applied the same tensile stress to both specimens, the larger gage length would extend more than the shorter gage length. If we try this with two elastic bands, one smaller than the other, and if we hang the same weight on both, we will see that the actual "extension" of the longer elastic band in terms of mm will be larger than the shorter one. Hence, like in the case of force, we need to convert the extension to something else that considers this gage length effect. This leads us to define a new term, which is **Strain**. As with engineering stress, we now define **engineering strain (ε) as the instantaneous extension (l) divided by the original gage length (L_o)**. In this case, we divide an extension (e.g., in mm or inches) by another length (also measured in mm or inches). Hence, the **engineering strain** as defined in Eq. 7.5 **has no units**,

$$\text{Engineering strain} = \frac{l_i - l_o}{l_o} = \frac{\Delta l}{l_o} \tag{7.5}$$

This means that we can replot the *instantaneous force-extension* curve in Fig. 7.9 by dividing the instantaneous force value by the original cross-sectional area to obtain

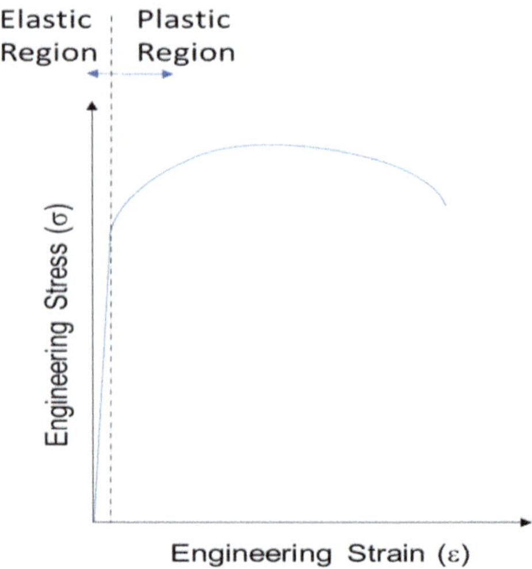

Fig. 7.10 Schematic diagram of engineering stress versus engineering strain, showing two regions, an elastic region, and a plastic region

the engineering stress (σ) and dividing the instantaneous extension by the original gage length to obtain the engineering strain (ε), as seen in Fig. 7.10. Note that since we are dividing the instantaneous force and extension values by constants, i.e., original cross-sectional area and original gage length, respectively, the shape of the curve does not change.

As shown in Fig. 7.10, the curve can be divided into two regions: a linear region, called the **elastic region**, and a nonlinear region, called the **plastic region**. Incredibly, such a short test (which could be over in less than a few minutes) can give us many important properties used in engineering design today. Let us see how many properties we can measure from this test. We will start by investigating the elastic region and then move to the plastic one.

7.3.1 *The Elastic Region*

The word *elastic* means precisely that. If we pull an elastic rubber band, it will get longer, but if we decide to let go of it, it will go back to its original length. Likewise, suppose materials such as metals are subjected to an applied stress that still keeps them in the elastic region of the stress–strain curve; these materials will behave very similarly to the elastic band, i.e., stretch (but by much smaller amounts), and the gage length will extend. However, if we remove the stress, the material will rebound to its original length (i.e., the original gage length). We should not be too surprised by this behavior. Aren't all materials made up of atoms? Aren't these atoms bonded together, with spring-like bonds?

Figure 7.11a is a schematic of the elastic region of an engineering stress-engineering strain curve. For metals, this region is linear and occurs over very small strains. It is generally very steep but enlarged and expanded here for clarity. The material behavior within the elastic region is often called **linear elastic** when the elastic region is linear. At very small strains (< 0.001), virtually all materials will behave linearly elastic. Many polymers and other materials deviate from this behavior at larger strains (discussed later). Taking the slope of the elastic region will give us our first property, **Young's Modulus (E)**; sometimes, it is referred to as the **Modulus of Elasticity**.

Note that within the elastic region, the atoms are also being pulled away from each other (see Fig. 7.11b), and the bonds act like springs that can stretch to varying amounts depending on the applied load, following **Hooke's law** (Eq. 7.6):

Where

$$\text{Force (F)} = \text{Spring Constant (k)} \cdot \text{distance } (x) \tag{7.6}$$

In a similar fashion,

$$\text{Stress (s)} = \text{Young's Modulus (E)} \cdot \text{Strain } (\varepsilon) \tag{7.7}$$

In other words,

$$\boxed{E = \frac{\sigma}{\varepsilon}} \tag{7.8}$$

So here we have it: the definition of **Young's modulus (E)** or the **modulus of elasticity**, which is similar to the stiffness of a material. We can expect a material with a high Young's modulus to be stiffer than one with a lower modulus. This modulus is a material property, meaning a given material will have a Young's modulus value. The modulus of a material can change if the temperature changes or porosity is present within the material. Increasing either temperature or pore content leads to a decline

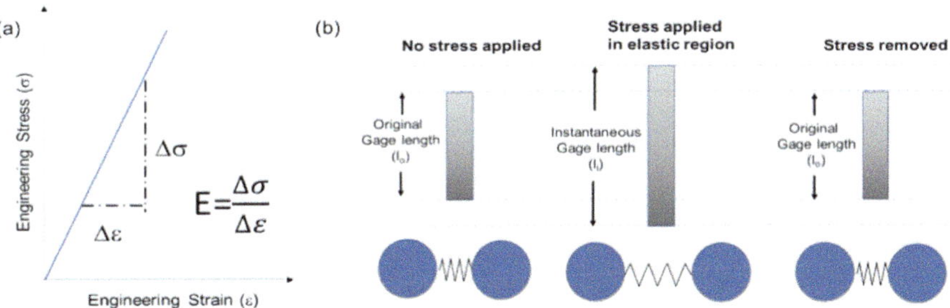

Fig. 7.11 **a** Schematic diagram of an enlarged elastic region (for clarity) of an engineering stress-engineering strain curve, **b** Schematic showing that if the gage length extends within the elastic region during tensile testing, the atoms will be pulled apart, and the imaginary spring will stretch, and when the load is released, the atoms will return to their proper positions (Note that extensions are exaggerated in the figure for clarity, and in metals and ceramics extensions are much smaller)

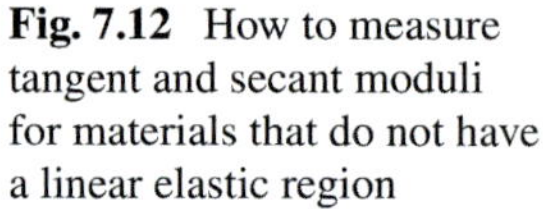

Fig. 7.12 How to measure tangent and secant moduli for materials that do not have a linear elastic region

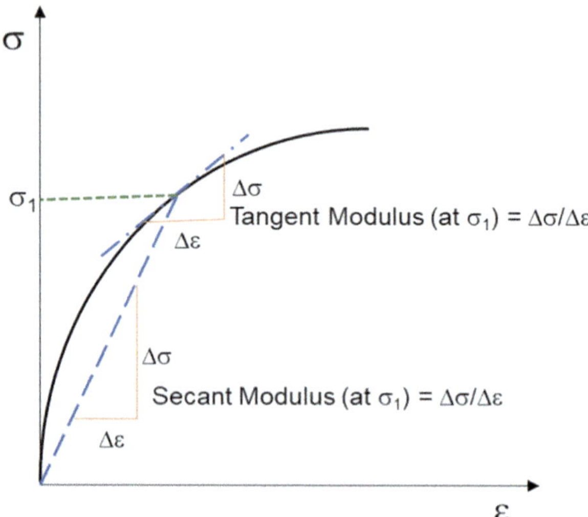

in the modulus. Equation 7.15 can be viewed as a relationship between **engineering stress** and **engineering "elastic" strain**; in other words, tell me what stress you want to apply to a material, and I will tell you how much it will strain elastically, if I knew its Young's modulus. For example, aluminum has a Young's modulus of ~ 70 GPa. If we wanted to apply 50 MPa of stress to an aluminum component, then since $E = \sigma/\varepsilon$, then $\varepsilon = \sigma/E = 50 \times 10^6/70 \times 10^9$ giving an elastic strain of 0.71×10^{-3}. This information could be very helpful when designing components expected to carry a load during service. Their new dimensions will now be known if the load is applied. Table 7.2 lists Young's moduli of several materials. Note that, for metals, the modulus values obtained under compressive loading would usually be the same as if the load were a tensile one.

Table 7.2 Approximate Young's moduli values of some common materials (multiple resources)

Material	Young's modulus (GPa)
Wood	12
Human bone	25
Magnesium (Mg)	45
Aluminum (Al)	70
Glass	70
Titanium (Ti)	110
Copper (Cu)	110
Iron (Fe)	207
Aluminum oxide (Al_2O_3)	380
Tungsten (W)	408
Carbon nanotubes	1000

It should be said that the measurement of Young's modulus is not as simple as taking the slope of the linear elastic part of the curve for all materials. Some materials do not have a linear elastic region, so we cannot simply measure a slope. For example, concrete and many polymers. They would exhibit a nonlinear elastic region similar to that shown in Fig. 7.12. To overcome this problem, the **Secant modulus** and the **Tangent Modulus** were introduced. In both cases, a stress is first identified. From that stress value on the curve, we either draw a line to the origin and take its slope to obtain the *Secant Modulus*, or at the stress value on the curve, we draw a tangent and take the slope of that tangent and obtain the *Tangent modulus*. Not only will both the secant and tangent moduli depend on the stress at which we take the measurement, but if we measure both moduli at the same stress, we will end up with different values for the slope and therefore the modulus (as seen in Fig. 7.12). Hence, we need to indicate the stress at which we took the measurement when we report the modulus and which modulus it is (secant or tangent). This way, we can properly compare one material to another that displays similar nonlinear elastic regions. On another note, although normally the purely elastic region in metals and alloys is confined to very small strain values (< 0.01), when it comes to materials such as rubber, the material can remain purely elastic for very large strain values upwards of 4. Thus, when a rubber band is pulled to that level and released, it can act as an effective catapult.

Another way of measuring Young's modulus is by measuring a material's fundamental resonance frequency. As outlined in ASTM C1259-15, the sample (in this case, a ceramic) can be a rectangular bar, a cylindrical rod, or a disc struck once with an impulse tool. For example, with the rest of the setup, the impulse tool could be a 5 mm steel ball fixed to a 10 cm long polymer rod, as shown in Fig. 7.13. A transducer (which could either be a non-contacting microphone or a contacting accelerometer) senses the vibrations and transforms them into electrical signals. For example, the relation used to calculate the dynamic Young's Modulus (E) for a rectangular bar is provided in Eq. 7.9 from ASTM C1259-15.

$$E = 0.9465 \times \left(\frac{m f_f^2}{b} \right) \times \left(\frac{L^3}{t^3} \right) \times T_1 \tag{7.9}$$

m mass of sample bar
b width of bar
L length of bar
t thickness of bar
f_f fundamental resonant frequency of bar
T_1 correction factor

In addition to determining Poisson's ratio (as explained below), the dynamic shear modulus can also be measured. As with ceramics, certain procedures are also used to obtain Young's modulus of metals using resonance frequencies.

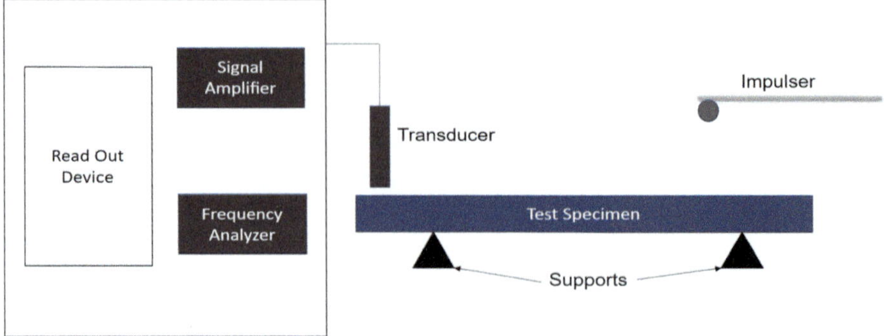

Fig. 7.13 Setup for measuring Young's modulus, shear modulus and Poison's ratio for ceramics (after ASTM C1259-15)

Is Young's modulus an important property, Prof.?

Are you kidding me? This is probably one of the most important properties in engineering. Without it, we would be in no man's land with regard to building engineering structures or products. Incredibly, though, this property was recognized and used in engineering design only after the death of Thomas Young (1773–1829), who introduced it. Could you believe he resigned from his lectureship at the Royal Institution possibly because his work was not appreciated? Ironically, he published his paper in 1807, and only years later, after his death, did people use this very important property in design. Without it, we could not design bridges, airplanes, or countless engineering products. For example, a plane's wings would flex upwards if Young's modulus were not high enough. Imagine that.

Other properties can also be deduced from the elastic region. For example, **Poisson's ratio**. This is another essential property in design. With reference to Fig. 7.14, if we apply a tensile stress (σ_z) to a material along the z-axis, we expect the material to elongate or stretch along that axis. Hence, the instantaneous length (l_{zi}) would be greater than the original length (l_{zo}) along that axis. What about the dimensions along the other axes (x- and y-axes)? Our intuition tells us that the material should shrink in these other directions (i.e., $l_{xo} > l_{xi}$ and $l_{yo} > l_{yi}$).

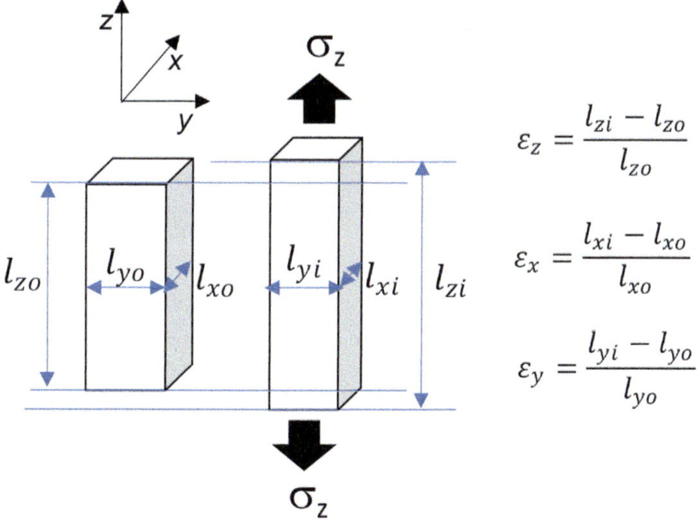

Fig. 7.14 Elastic deformation along the 3 axes due to application of stress along the z-axis (based on W. D. Callister and D. G. Rethwisch, Materials Science and Engineering, 10th edition, Wiley (2018))

The corresponding strains along the three axes are given in Eqs. 7.10 (a–c). Note that the strains along the x- and y-axes will be negative, since the material is expected to contract in these directions.

$$\varepsilon_z = \frac{l_{zi} - l_{zo}}{l_{zo}} \tag{7.10a}$$

$$\varepsilon_x = \frac{l_{xi} - l_{xo}}{l_{xo}} \tag{7.10b}$$

$$\varepsilon_y = \frac{l_{yi} - l_{yo}}{l_{yo}} \tag{7.10c}$$

Poisson's ratio (v) is thus defined as:

$$v = -\frac{\varepsilon_x}{\varepsilon_z} = -\frac{\varepsilon_y}{\varepsilon_z} \tag{7.11}$$

There is an inherent assumption in Eq. 7.11, which is that as we apply the stress along the z-axis, the material will shrink by the *same amount* along the y- and x-axes, hence, $\frac{\varepsilon_x}{\varepsilon_z} = \frac{\varepsilon_y}{\varepsilon_z}$. Materials that behave in this way are called **isotropic materials**. These are materials that have the same properties in different directions. These properties could be stiffness, strength, electrical conductivity, thermal conductivity, optical properties, magnetic properties, etc. Since no load is applied along the x- and

y-axes, the material along these directions is subjected to the same condition, and if it is isotropic, then it is expected to have the same strain along both axes. Hence, Eq. 7.11 fully applies.

Materials that have different properties in different directions are called **anisotropic materials**. In that case, we can't expect the material to contract to the same amount along the x- and y-axes, i.e., $\frac{\varepsilon_x}{\varepsilon_z} \neq \frac{\varepsilon_y}{\varepsilon_z}$. Note that both isotropic and anisotropic materials find applications in engineering. Table 7.3 lists Poisson's ratios for some materials.

It is important to remember that most materials have *positive* Poisson's ratios, meaning they will shrink laterally if they are elastically stretched, and expand laterally when they are elastically compressed. However, some materials display negative Poisson's ratios, such as single-crystal iron pyrites (anisotropic), 2D honeycomb, other structures, and some foams. In other words, if compressed, they tend to shrink laterally and expand laterally when stretched. Materials that display negative Poisson's ratios are called **auxetic** materials.

Effect of Temperature and Porosity on Young's Modulus

The values of Young's moduli for the metals and glass presented in Table 7.2 are room temperature values for fully dense (i.e., containing no porosity) materials. However, an increase in temperature results in a decline in a material's modulus, and so does

Table 7.3 Room temperature Poisson's ratios for different materials

Material type	Material	Poisson's ratio (v)
Metallic crystalline solid	Ag	0.38
	Al	0.34
	Au	0.42
	Cu	0.34
	α-Fe	0.29
	Ni	0.31
	W	0.29
Covalent crystalline solids	Ge	0.28
	Si	0.27
	Al_2O_3	0.23
	TiC	0.19
Covalent ionic solid	MgO	0.19
Covalent glass	SiO_2	0.20
Network polymer	Bakelite	0.20
Elastomer	Natural rubber	0.49
Chain polymer	Polyethylene	0.40
	Polystyrene	0.33

After: T. H. Courtney, Mechanical Behavior of Materials, 2nd edition, McGraw-Hill, 2000

an increase in porosity. The **Mackenzie Equation 7.12** was introduced to predict the effect of porosity on Young's modulus for porous materials.

$$E = E_o\left(1 - 1.91P + 0.91P^2\right) \tag{7.12}$$

E_o is Young's modulus for the pore-free material, and P is the fractional porosity. Note, a fractional porosity of 0.2 means the material contains 20% porosity. The equation is valid for materials having Poisson's ratio of 0.3 and up to a fractional porosity of 0.5. Recently, **Pabst and Gregorova** proposed a more refined equation for brittle materials, considering a much wider range of relative porosity, and it is probably our current best equation. The equation also indicates that the pore fraction does not have to be 1 for Young's modulus to be zero for a ceramic material. The material will not be mechanically stable when the relative porosity reaches a critically high value (P_c). Hence, Young's modulus will still have a zero value for relative porosity equal to or greater than P_c.

The **Pabst-Gregorová** equation (Eq. 7.13), which is built on previous equations, is as follows:

$$E = E_o \times \left(1 - M \times P + (M - 1) \times P^2\right) \times \left(\frac{1 - P/P_c}{1 - P}\right) \tag{7.13}$$

where

$$M = \frac{3(1 - v)(9 + 5v)}{2(7 - 5v)}$$

E_o Young's modulus of a fully dense material with no porosity.
P pore fraction. (0 for zero porosity and 1 for all porosity (i.e., no material).

Alex, why don't you test this equation for the ceramic, alumina (Al$_2$O$_3$)? You can use porosity values of P = 0, and 1, and a P$_c$ of 0.9.

From Tables 7.2 and 7.3, I can see that the pore-free alumina (E$_o$) modulus is 380 GPa, and Poisson's ratio is 0.23. A 0.23 Poisson's ratio gives an M value of 2.004
Using Eq. 7.13, when P = 0, E = E$_o$, and when P = 1, E = 0, When P = P$_c$ = 0.9, E = 0. You know what? I can also see that you plotted the Mackenzie and the Babst-Gregorova models for relative porosity values from 0 to 1, Fig. 7.15.

Figure 7.15 plots both the **Mackenzie** and the **Pabst-Gregorová** equations. Researchers have also suggested equations that predict the influence of temperature on Young's modulus and even the combined influence of temperature and porosity on Young's modulus (E).

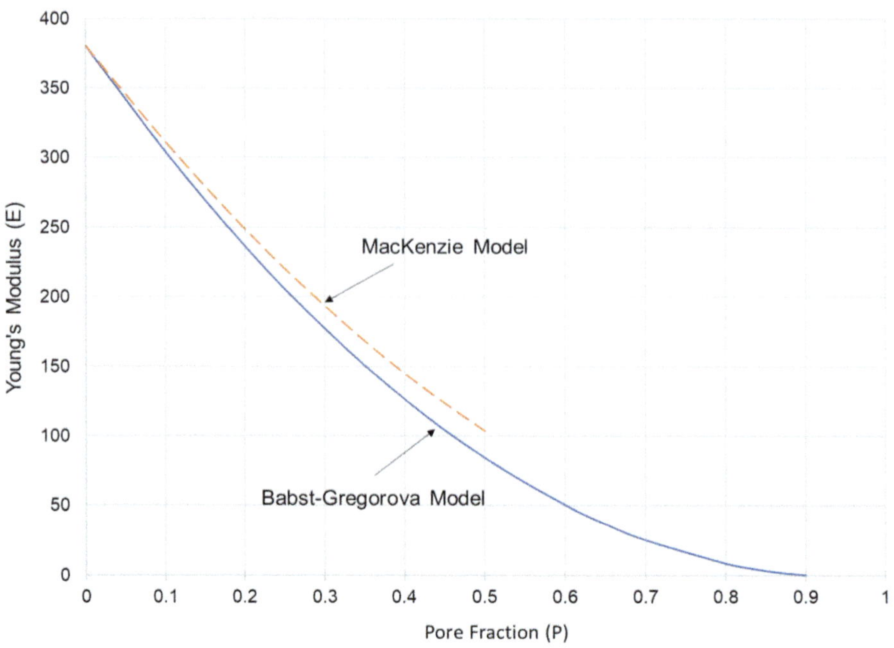

Fig. 7.15 Effect of porosity on the Young's modulus of alumina (Al_2O_3) ceramic, using the Mackenzie model (only valid up to 0.5 pore fraction), and the Babst-Gregorova model valid over larger pore fractions and includes the critical pore fraction at which the material loses its mechanical integrity (assumed here to be 0.9)

Anelastic Behavior

When loading a specimen into the elastic region, as the stress is applied, we assume that the material will strain instantaneously. Typically, there is a lag of strain, i.e., it takes a finite time for the strain to reach its full equilibrium value. This behavior is called **Anelasticity**, and it happens at least to a small extent in all materials; however, it is more exaggerated in some materials.

Figure 7.16 shows how such a strain lag (**elastic hysteresis**) results in energy dissipation within the elastic region's loading and unloading cycle. As can be seen, the area under the loading curve (and hence energy) is larger than that under the unloading curve; the difference in energy is shown by the shaded gray region, which is the amount of energy dissipated in the cycle. This behavior opens the door to applications where damping is needed, such as noise or mechanical vibration reduction. Here, energy is dissipated due to internal friction. In metals, the motion of dislocations allows this energy dissipation to occur. Of course, we are not talking about the macroscopic yielding phenomenon, where gross plastic deformation results, but rather the local movement of dislocations within grains. Hence, only soft metals that allow easy movement of dislocations are suited for damping applications, such as magnesium, lead, and very pure aluminum. By no means are dislocations the only way

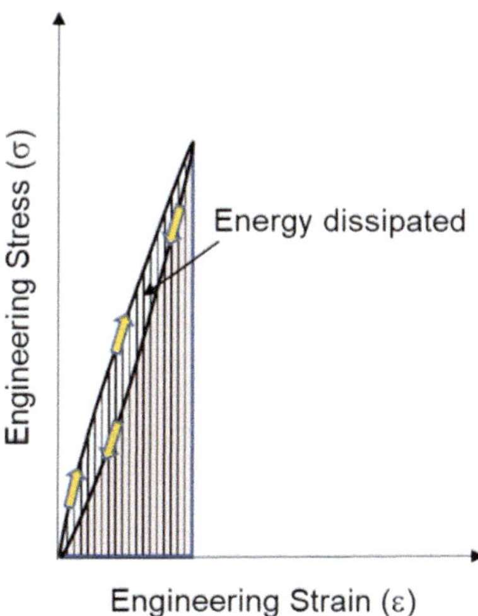

Fig. 7.16 Energy dissipation (hysteresis) due to anelasticity, providing a basis for damping applications

damping can be facilitated. In ceramics and cast iron, where microcracks are present, damping can be facilitated by the rubbing action of crack faces, which provides an internal friction mechanism for energy dissipation. Recently, grain boundary friction has also been suggested as a mechanism for energy dissipation, for example, in magnesium. Here, a reduction in grain size increases the grain boundary density, increasing damping.

On the flip side, if a metal or alloy has mechanisms that hinder dislocation motion, say by pinning them, it will have very low damping capacities and will be more suited for applications that should not damp sound or vibrations, for example, bells. Here, the objective is to make a ringing sound when the bell is struck. Bronze (an alloy of Cu and Zn) is well suited for this application since the Zn solute atoms will impede dislocation motion. Crack-free brittle materials would also provide a ringing sound when struck.

The Modulus of Resilience (U_r)

The elastic region transitions to the plastic region after reaching a certain stress level, called the **Yield Strength**, σ_y (more about that later). The corresponding strain value at the yield strength is called the yield strain (ε_y). We can calculate one more property if we examine the elastic region again. The shaded area underneath the linear part of the engineering stress-engineering strain curve up to the yield strength in Fig. 7.17 represents the amount of energy per unit volume the material can absorb before it reaches the plastic regime. Since we do not typically want a material to deform plastically during service, this energy indicates how much the material can resist this transition and is represented by the property **Modulus of Resilience (U_r)**, also sometimes called the **resilience**. U_r is defined in Eq. 7.14.

$$U_r = \frac{1}{2}\frac{\sigma_y}{\varepsilon_y}$$

$$(7.14)$$

Fig. 7.17 The shaded area under the elastic region represents the modulus of resilience

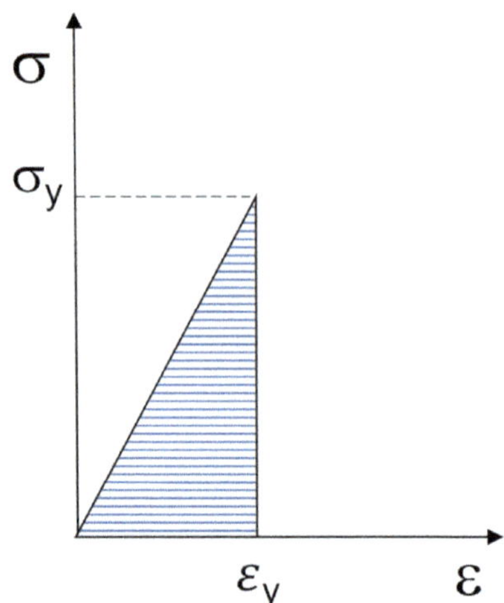

But since $= \frac{\sigma}{\varepsilon} = \frac{\sigma_y}{\varepsilon_y}$. Then Eq. 7.14 can be rewritten as:

$$U_r = \frac{1}{2}\frac{\sigma_y^2}{E} \tag{7.15}$$

Lastly, we have been discussing situations where a material is uniaxially deformed. That means the stress was applied along only one axis, and specifically, the stress was applied in tension. Imagine if a shear stress (τ) instead was applied. That would result in a shear strain (γ) as represented in Fig. 7.18. Hence, we can similarly define a shear modulus (G) as the shear stress divided by the shear strain.

$$\text{Shear Modulus} = G = \frac{\tau}{\gamma} \tag{7.16}$$

7.3.2   The Plastic Region

In the previous section, we discussed the *elastic region* of the engineering stress-engineering strain curve. Note that numerous metals and alloys gradually transition from elastic to plastic behavior. A question then arises. At which point does the material transition to the plastic region, i.e., when does the elastic region end? Here, we define the **elastic limit** as the stress above which plastic deformation starts. The term **proportionality limit** is also mentioned in the literature, which means it is the stress level beyond which we cannot say the stress and strain are linearly dependent. The elastic limit and the proportionality limit have very close values, but measuring

Fig. 7.18 2D view of shear deformation, showing shear stress and shear strain

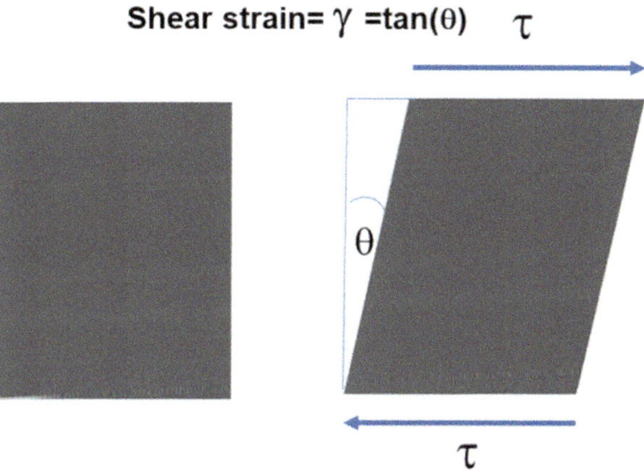

them accurately is not an easy task. For this reason, we have developed another concept called the *0.002 offset yield strength* (more about that later).

Now, let us look at the situation more fundamentally by imagining what a crystal lattice of a single-crystal specimen experiences during elastic and plastic deformation. Previously, we defined an edge dislocation (i.e., a line defect) as an extra half-plane of atoms squeezed between two consecutive atomic planes. We also mentioned that without these so-called line defects, alloys and metals as we know today would generally be as brittle as ceramics. We didn't discuss how these dislocations enable the change of shape of metals and alloys under applied stress. It turns out that only when shear stress is applied to dislocations will they move and allow the collective transfer of matter. However, during a tensile test, we are externally applying a tensile force and not a shear force, so why do we get a permanent change in dimensions (plastic deformation) through the motion of dislocations? **Schmid's law** answers this question. As can be seen from Fig. 7.19, dislocations glide on a glide (or slip) plane. These planes of atoms are typically atomically highly packed planes with high planar densities.

We can see that although the tensile testing specimen is pulled uniaxially in a vertical direction, the applied force (F) can be resolved into a normal and parallel component to the glide (slip) plane. This parallel component of the force is, in fact, a shear force component (F_s), which generates the shear stress that acts on the dislocations, eventually forcing them to move. The shear force is thus given by:

$$F_s = F.\mathrm{Cos}\theta$$

where F_s is the shear force, F is the applied uniaxial force and θ is the angle between the slip direction and the applied tensile force. To find the shear stress (τ), we must divide F_s by the cross-sectional area (A_s) on the slip plane on which it is acting. We can relate the cross-sectional area on which the tensile force is acting (A_o) with the projection of area A_s, as follows:

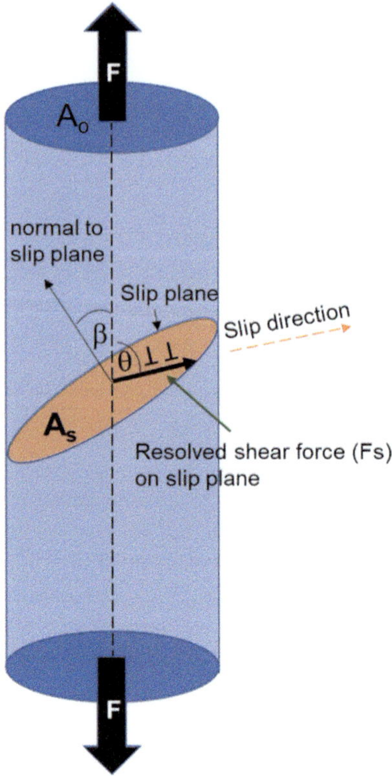

Fig. 7.19 Slip plane in single crystal and resolved shear stress (multiple sources)

$$A_o = A_s . \mathrm{Cos}\beta$$

$$\text{i.e.,} \ \ A_s = A_o / \mathrm{Cos}\beta$$

If we divide F_s in Eq. 4.14 by A_s we obtain the following:

$$\frac{F_s}{A_s} = \tau_r = \frac{F.\mathrm{Cos}\theta}{A_o/\mathrm{Cos}\beta} = \sigma.\mathrm{Cos}\theta.\mathrm{Cos}\beta$$

$\tau_r =$ resolved shear stress
$\sigma =$ applied tensile stress
Schmid factor $= Cos\theta.Cos\beta$.

Hence,

$$\boxed{\tau_r = \sigma.\mathrm{Cos}\theta.\mathrm{Cos}\beta} \tag{7.17}$$

This means that during the tensile test, even within the elastic region, two things are simultaneously happening: a slight increase in the separation between atoms (i.e., elastic strain) and a resolved shear stress pushing on the dislocations. The higher the applied stress, the higher the magnitudes of the elastic strain and the resolved shear stress. When the applied stress reaches the **yield strength (σ_y)** value, the resolved shear stress reaches a critical value, called the **critical resolved shear stress (τ_{crss})**, enough for bonds ahead of the dislocation to start breaking, as shown in Fig. 7.24, reconnecting with the dislocation and restoring it to a complete plane of atoms. However, this leaves the plane of atoms to the right of the initial dislocation with only half a plane (i.e., an edge dislocation). This process's continuation leads to the motion of an edge dislocation from left to right, as seen in Fig. 7.20. Now, humor me a little. If I were a dislocation within the elastic region, I would feel the push of the initial low-resolved shear stress at the start of loading. As the applied stress increases within the elastic region, I would feel a stronger and stronger push, but it's only until the critical resolved shear stress is reached that I would start to move. Similarly, during elastic deformation within the elastic region, the springs (bonds) between the atoms are pulled, and the dislocations also feel the resolved shear stress. Still, when the applied stress (and hence the resolved shear stress) reaches a particular (critical) value, only then do the dislocations start to move. This signifies the start of the plastic region (plastic deformation), which would result in permanent residual strain in the sample upon unloading. An important point needs to be made. If the material is loaded to a stress level within the plastic region, it will experience a corresponding strain. Part of that strain will be plastic, and part will be elastic (since we are still pulling on the atoms at that stress and therefore also the springs/bonds between them). The elastic component of the strain will disappear if we remove the applied stress, but the plastic component will remain as residual strain in the sample.

Hence, at the point of yielding, Eq. 7.17 can be rewritten as:

$$\tau_{crss} = \sigma_y . \mathrm{Cos}\theta . \mathrm{Cos}\beta \tag{7.18}$$

Let's look at the **transition** between the **elastic** and **plastic regions**. We can encounter two types of transitions (shown in Fig. 7.21). In Fig. 7.21a, the transition

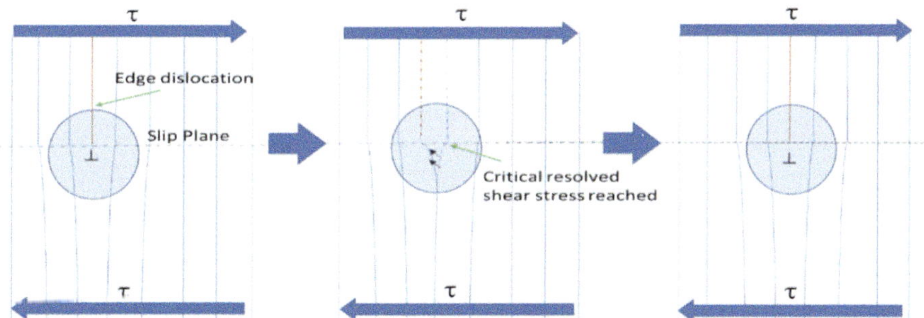

Fig. 7.20 Motion of an edge dislocation under a critical resolved shear stress

appears smooth; a smooth transition is typically observed for ductile metals and alloys and most polymers. The transition appears abrupt in Fig. 7.21b, showing an **upper** and **lower yield point** phenomenon. For instance, iron, low-carbon steel, and some polymers can display this behavior.

We must always be aware of the stress value at which plastic deformation (i.e., yielding) occurs (the **Yield Strength**), so we do not subject a component during service to a higher stress value. Otherwise, it could change shape or dimensions permanently during service, which could result in disastrous consequences. Identifying the yield strength for materials that exhibit an upper and lower yield point is easy. Typically, the stress level at which the lower yield point appears is the yield strength. Of course, while still in the elastic region, we can reach stress levels higher than the lower yield point stress levels (but lower than the upper yield point), and the material should still be within the elastic region. So why are we quoting the lower yield point, not the upper one? The answer is that we are being conservative.

Let's start by discussing Fig. 7.21a. As can be seen, the transition from elastic to plastic is gradual, which may cause difficulty in locating the yield strength. A 0.002 (i.e., 0.2%) offset yield strength is usually reported to avoid this issue. Before we discuss the 0.2% offset yield strength, let us first cover an essential related topic. Figure 7.22a shows a typical stress–strain curve; the application of the stress (σ_1) produces a corresponding strain of ε_1. Since the stress took the specimen into the nonlinear plastic region, dislocations have certainly moved. So, the observed strain is not purely elastic anymore, nor is it purely plastic, but as explained earlier, it is a combination of both (i.e., elastic + plastic strain). Figure 7.22b demonstrates that if we initially stress the specimen to σ_1 (which generates both elastic and plastic strain components) and then decide to unload, only the elastic strain component of strain ($\varepsilon_{1)}$ will be reduced.

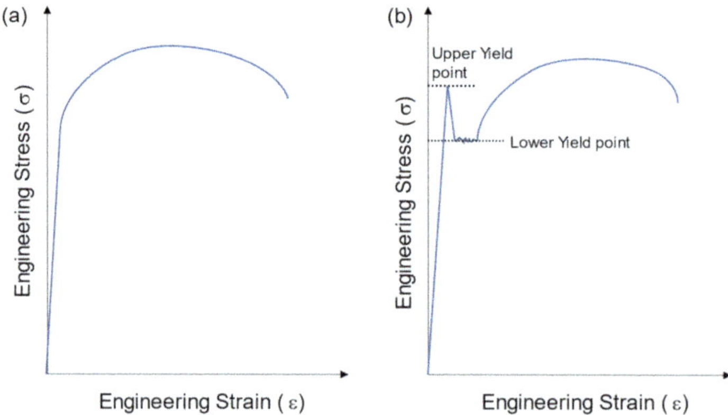

Fig. 7.21 Tensile engineering stress-engineering strain curve: **a** with a gradual transition from the elastic to the plastic region, and **b** sharp ending to elastic region, showing an upper and lower yield point

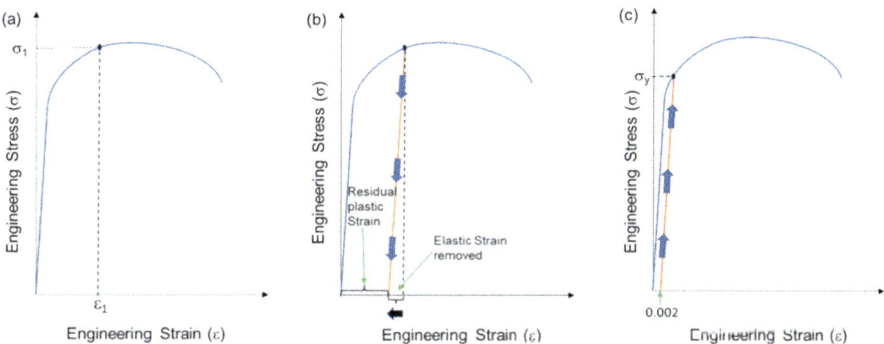

Fig. 7.22 Stress-Strain curve: **a** where the sample is loaded to stress σ_1 taking it inside the plastic regime, the strain (ε_1) at this stress level is partly elastic and partly plastic, **b** by removing the stress (unloading the specimen gradually until the stress is zero, the elastic part of the strain is removed and the only strain remaining is purely plastic and permanent, **c** the idea is extended to find the 0.002 offset yield strength, by going to the 0.002 strain location on the engineering strain axis and drawing a line parallel to the linear elastic region. The stress value where the line intersects the curve is the **0.002 offset yield strength**

Since we already have an established relationship between engineering stress and engineering (elastic) strain (Eq. 7.8), we can calculate the amount of elastic strain recovered if we reduce the stress by a certain amount. This means if we wanted to know if we reduce the stress by a certain amount and how much elastic strain will be removed, we only need to draw an unloading line parallel to the elastic part of the curve. This is shown in Fig. 7.22b. As can be seen, when the stress is removed completely, we see that at a stress value of zero, the unloading line does not go back to zero strain; rather, it intersects the strain axis at a particular strain value. This strain is the **permanent** or **residual plastic strain** in the specimen. If we load our specimen solely within the elastic regime and decide to unload, the unloading line will return to zero strain, and the specimen's instantaneous gage length will return to the original gage length. However, if we load the specimen into the plastic regime and decide to unload, only the elastic strain will be removed. Still, we will be stuck with the residual plastic strain, and our specimen will have a longer gage length compared to the original one. In Fig. 7.22c, we use the knowledge gained from Fig. 7.26b to solve the Yield strength identification problem for stress–strain curves that gradually transition from elastic to plastic. If all that is needed is to know when we enter the plastic region, then we can produce a stress level that will guarantee that we have entered the plastic region, but by a very tiny amount. The way to do this is to identify a point on the strain axis where the strain reads 0.002, then draw a line parallel to the linear part of the stress–strain curve. The stress value where this line intersects the curve will be the **0.002 offset yield strength (σ_y)**.

 *Tell me, Alex, which properties can you find out from the elastic region of the engineering stress–strain curve?*

 *The modulus of elasticity (also known as Young's modulus), the modulus of resilience, Poisson's ratio, the elastic limit, the limit of proportionality, and, of course, the 0.002 offset yield strength*

 *Well done!!! Did you know that if you were in England, instead of answering the 0.002 offset Yield strength, you would be saying the 0.1% Proof Stress? Yes, that's what they call it over there. They follow the same procedure but for a 0.001 (i.e., 0.1%) strain*

With reference to Fig. 7.21b, a question might arise: why is there an upper yield point and a lower yield point? Why is there a plateau after the lower yield point? As soon as we get out of the elastic region, the stress needed to continue plastically deforming the specimen is less. Let's investigate this a little further. As mentioned earlier, a good example of a material that exhibits this type of engineering stress-engineering strain curve is plain low-carbon steel (plain carbon steel is an alloy of only iron and carbon). We observe a sudden drop in stress after the upper yield point due to the effect of what we call **solute atmospheres** or **Cottrell atmospheres,** after **Sir Alan Cottrell**. To explain solute atmospheres, let us start by considering an edge dislocation. It could be viewed as imposing itself on the crystal lattice. I mean an extra half-plane of atoms is inserted between two planes of atoms and will tend to crowd the atoms in that region (resulting in a region of compression). It also effectively wedges open the same atomic planes below the slip plane (resulting in a region of tension), as seen in Fig. 7.23a. If atoms above the slip plane around the dislocation are under a state of compression, they would be pushed closer together, resulting in an elastic strain and, hence, elastic strain energy. Likewise, an elastic strain energy is also established in the tensile region below the dislocation. Hence, lattice distortion results from the presence of a dislocation. Since materials don't like lattice distortions since they increase their internal energy, any opportunity to reduce such energy will be welcomed. As such, during the solidification of alloys (where impurity atoms are present), impurity atoms will diffuse to the location of the dislocation where the lattice is in distress. Suppose the impurity atoms are larger than the host atoms (while not exceeding the 15% size difference established by Hume-Rothery). In that case, the impurity atoms will preferentially take up host atoms' positions in the tensile region to reduce the tension and strain energy in that location (Fig. 7.23b). Likewise, suppose the impurity atoms are slightly smaller than the host atoms. In that case, they will substitute the host atoms in the dislocation region above the slip plane where compression is present, thereby reducing the strain energy there (Fig. 7.23b).

So now we get to uncover the mystery behind the upper and lower Yield points that appear in the engineering stress-engineering strain curves of low-carbon steels. Solute atmospheres can be present in low-carbon steels since the carbon atom is

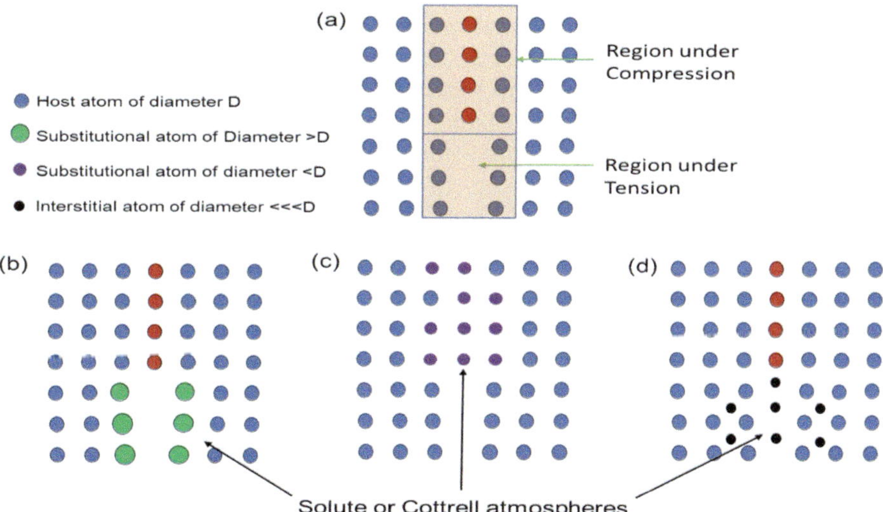

Fig. 7.23 **a** The presence of an edge dislocation causing distortion in the lattice, **b** Impurity atoms of a size slightly greater than the size of the host atoms substitute host atoms below the dislocation to reduce tension in that region, **c** impurity atoms slightly smaller than the host atoms substitute the host atoms in and around the dislocation to reduce compression in that region, **d** interstitial impurity atoms, fill interstitial sites below the dislocation to reduce tension in that region (based on multiple sources)

much smaller than the iron atoms; within the BCC iron lattice, carbon atoms can occupy interstitial sites between iron atoms. Beneath the dislocation, the interstitial carbon atoms diffuse to interstitial sites below the slip plane to reduce tension in that region, as seen in Fig. 7.23d.

The crystal lattice greatly welcomes the reduction in strain energy around a dislocation, enabled by the presence of solute atmospheres. If a critical resolved shear stress is applied and the dislocation is allowed to escape this region (i.e., slip), it would leave behind a chaotic situation, basically a mess. This is because the solute atoms would remain in their positions and not move with the dislocation. Therefore, if a dislocation is allowed to escape the solute atmosphere, it would regain its strain energy (which we had previously reduced through the solute atmosphere) and leave behind the solute atmosphere, resulting in significant lattice distortions in that region. We only have to imagine a collection of solute atoms of a different size than the host atoms left behind; there would only be one outcome: excess strain energy. Hence, if a dislocation is allowed to escape, the lattice will increase in energy. Now, let's put ourselves in the position of the lattice, would we want to go from a low-energy configuration to a higher energy one? Of course, not; hence, the solute atmosphere will try to pin or keep the dislocation from slipping and escaping. The only way for the dislocation to pry itself free from the solute atmosphere is to overcome that pinning effect. This leads to the requirement of a high critical resolved shear stress and, consequently, a high applied tensile stress (i.e., the upper Yield point). Once the dislocation is free, less stress (lower yield point) is needed for it to glide. But why does the stress become almost constant for a short period (i.e., the curve exhibits a small plateau) at the lower yield point? This is because the presence of solute atmospheres causes a phenomenon called **Lüder bands** (which are localized slip bands). Here, small diagonal bands, typically at ~ 45° to the applied stress, travel through the tensile testing specimen from a region of high-stress concentration. In other words, slip is localized in these traveling bands instead of simultaneously occurring throughout the whole volume of the gage length. After this period of slip band motion, the stress starts to increase similarly to that seen in Fig. 7.21a.

For either of the two stress–strain curves discussed, once the material fully enters the plastic region, we need to continuously apply higher and higher forces to continue the deformation. This means the material is getting stronger and stronger the more we deform it. Here, the metal is said to be experiencing **work hardening** or **strain hardening**. In the engineering stress-engineering strain curve, this translates to an increase in engineering stress until the **tensile strength** (the maximum stress in the engineering stress-engineering strain curve) is reached. The stress *appears* to go down after that until the specimen fractures. During this period a neck starts to appear in the gage length, which continues to reduce in size until the point of fracture (see Fig. 7.24)

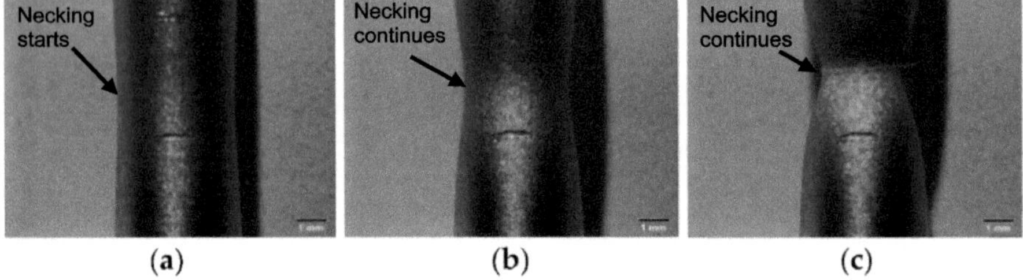

Fig. 7.24 Ferritic steel tensile testing specimen showing start of necking and continued development of necking, as the specimen is deformed further into the plastic regime (from Dzioba, I.; Lipiec, S.; Pala, R.; Furmanczyk, P. On Characteristics of Ferritic Steel Determined during the Uniaxial Tensile Test. Materials 2021, 14, 3117. https://doi.org/10.3390/ma14113117. Reproduced with modification under the terms and conditions of the Creative Commons Attribution (CC BY) license (https://creativecommons.org/licenses/by/4.0/)

Prof., why do metals get stronger when we plastically deform them?

Do you want to hear the textbook-style answer or my philosophical take? You know what, let me tell you both. Let's start with the textbook explanation (which you must write down in an exam to earn credit). As you know, during plastic deformation, dislocations move, but you didn't know that every time you extend the material further into the plastic region, more and more dislocations are generated. If you initially needed a particular stress level (yield strength) to move a certain number of dislocations, now you will need a higher stress level to move the now larger number of dislocations. Let me give you an analogy. Imagine if you need to evacuate a room quickly, let's say, due to a planned test fire drill. If there was only one person in front of you at the door, you could easily nudge or gently push that person out so you could leave safely on time. However, if there were 10 people ahead of you? You will not be strong enough to push them out; you will need someone like the Incredible Hulk who can push with a larger force (stress) than you. Likewise, more dislocations will need a higher force to get them moving and continue the plastic deformation, so the material will get stronger and stronger as plastic deformation continues. I hope that brings the point across

Yes, it does, but what about your philosophical take on it?

You know me, I don't like to get bogged down with only conventional explanations. I want to remain as a free spirit. So don't tell anyone about my philosophical ideas; otherwise, they will think I'm crazy. I don't see how you can be surprised that a material will get stronger when it is plastically deformed. Haven't you heard of the expression "When the going gets tough, the tough get going"? I promise you that metals are known to be tough even by engineering standards (more about that later). Also, haven't you heard the expression, "What doesn't kill you makes you stronger"?

If I want to know how a metal will behave when subjected to force or stress, I place myself in its shoes. If someone tries to push me around, I will resist and shout, "HOW DARE YOU!" I believe that is how the metal responds. When we try to stretch it within the elastic region, it resists through its bonds, but when we further take it into the plastic region, we've overstepped our boundary. We are trying to permanently change its existence as a specimen with its unique dimensions, and we have, in fact, just waged war on the metal and are in the process of inflicting severe damage to it. As such, the metal will fight back, generating new soldiers (dislocations) to resist further deformation. However, since the universal testing machine is much more powerful than the specimen, there can only be one eventual winner. Yet the battle continues, and the deformation continues, and when it does so, the metal generates more and more dislocations. This fight continues until the specimen dies (i.e., fractures). In other words, the metal work hardens up to the point of fracture, or the metal resists and fights to the death (how honorable). It's funny; it's as if the dislocations are like our white blood cells that increase in number to fight infections or injuries.

Prof., this all sounds great, but when I look at Fig. 7.21a, I see the stress initially increases with an increase in strain in the plastic region, as you say, but only to a point. Then, it starts dropping with further deformation. So, how can you say that the metal work hardens and continues to get stronger and stronger to the point of fracture when the stress is going down after necking?

The answer to that is that looks can be deceiving. Remember that when initially calculating the engineering stress, we divided the instantaneous force by the original cross-sectional area; this has returned to haunt us now. You and I know that the cross-sectional area doesn't remain constant during the test, and as the gage length continues to extend permanently, the cross-sectional area decreases. Also, after the maximum stress observed in the engineering stress-engineering strain curve is reached (i.e., the tensile strength), an unusual thing happens. A type of instability occurs in the specimen inside the gage length region, namely, a neck forms (as seen in Fig. 7.24*). Once this neck is formed, with its smaller cross-sectional area compared with the remaining cross-sectional area inside the gage length, all the deformation becomes confined there, and it simply keeps reducing in area as necking continues. Although not visible from the outside, inside the neck, we see the appearance of many pores or micro-voids, particularly around impurities. These pores will grow as necking continues until the pores start to link up together and form a crack. At that point, the specimen fractures into two pieces.* Figure 7.25 *shows this clearly within the fracture surface of an aluminum-carbon nanotube (CNT) composite. Here, the CNTs act as impurities, resulting in microvoid formation around each of them*
Indeed, the fracture surface can reveal much about the type of failure the material experienced. Had the surface of a material appeared shiny, for example, without any voids, that material would have failed in a brittle fashion. Let me back up my claim that a metal resists and work hardness until it breaks. If we instead divided the instantaneous force by the instantaneous (actual) cross-sectional area to obtain the true stress, that true stress would continue to increase with strain until the point of fracture

Figure 7.25 shows pore formation during necking for aluminum-carbon nanotube composites. The fact that we used the original cross-sectional area and original gage length to arrive at an engineering stress and engineering strain, respectively, leaves something to be desired. Logic tells us we must divide the instantaneous force by the instantaneous cross-sectional area within the gage length. This would be called the **true stress**, which is defined in Eq. 7.19:

$$\sigma_T = \frac{F_i}{A_i} \tag{7.19}$$

where A_i is the instantaneous area.

Likewise, the **true strain** is defined as:

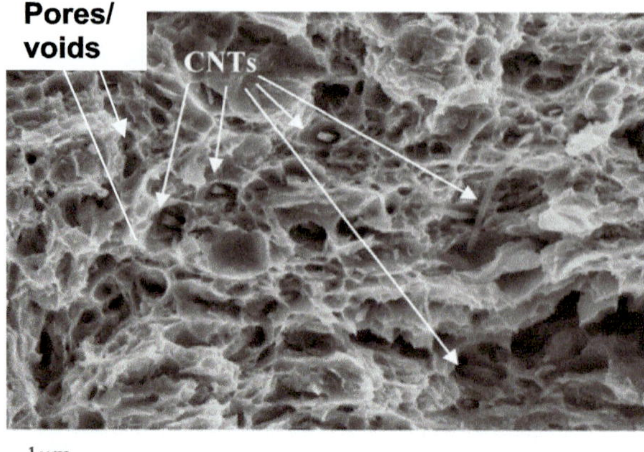

Fig. 7.25 Fracture surface of aluminum-CNT tensile testing specimen that experienced necking before fracture, showing microvoid formation around CNTs, since they are perceived as impurities. (Fig. 10 (c) in A. M. K. Esawi, K. Morsi, A. Sayed, A. Abdel Gawad, P. Borah Fabrication and properties of dispersed carbon nanotube–aluminum composites, Materials Science and Engineering A 508 (2009) 167–173 (reproduced with modification with permission from Elsevier)

Fig. 7.26 Engineering stress versus engineering strain and true stress versus true strain

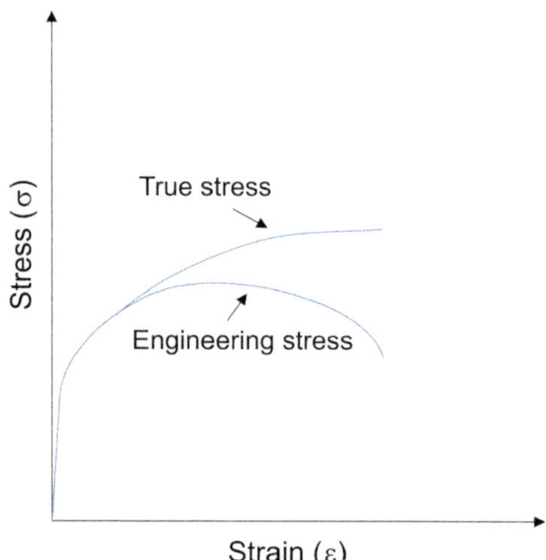

$$\varepsilon_T = \ln\left(\frac{l_i}{l_o}\right) \tag{7.20}$$

Figure 7.26 shows the difference between an engineering stress–strain curve and a true stress-true strain curve. The true stress is higher than the engineering stress, but this time the true stress increases with strain until the point of fracture, i.e., the specimen work hardens to the point of death or failure. So, can we claim that true stress is always greater than engineering stress? This is the case under tensile loading, where the tensile stress causes the original cross-sectional area to decrease as the material is extended. However, if we were to compress the material with compressive stress instead, the original cross-sectional area would increase (not decrease); in that situation, the true stress would be lower than the engineering stress. So, it depends on the mode of loading being applied, namely, tensile versus compressive.

It is possible to obtain true stress and true strain data from engineering stress and engineering strain data through the following derivations:

Since

$$\varepsilon = \frac{\Delta l}{l_o} = \frac{l_i - l_o}{l_o} = \frac{l_i}{l_o} - 1$$

Hence, $\varepsilon + 1 = \frac{l_i}{l_o}$.

Since,

$$\varepsilon_T = \ln\left(\frac{l_i}{l_o}\right)$$

Taking the natural log of both sides gives:

$$\boxed{\varepsilon_T = \ln(1 + \varepsilon)} \tag{7.21}$$

Noting that the original gage length is l_o and the original cross section is A_o, as the tensile test begins, the gage length is extended to a new length called the instantaneous gage length (l_i), and the original cross-sectional area is decreased to A_i, the instantaneous cross-sectional area.

During the test, the value of l_i increases, and that of A_i decreases. If we apply constant volume considerations within the gage length, we find:

$$A_o.l_o = A_i.l_i$$

Therefore, $A_i = \frac{A_o l_o}{l_i}$.

Substituting the above expression for A_i in the following equation

$$\sigma_T = \frac{F_i}{A_i}$$

Gives,

$$\sigma_T = \frac{F_i l_i}{A_o l_o} = \sigma\left(\frac{l_o + \Delta l}{l_o}\right) = \sigma(1 + \varepsilon)$$

Hence,

$$\boxed{\sigma_T = \sigma(1 + \varepsilon)} \tag{7.22}$$

Since we used a constant volume assumption, Eq. 7.22 is only valid until necking. After necking, voids appear within the neck region, nullifying the constant volume assumption.

Interestingly, the true stress-true strain curve can yield important insights into the degree of work hardening in metals and alloys. This is depicted in the **Hollomon equation** (Eq. 7.23), where n is the work-hardening exponent. The higher the n value, the more a material can work harden during plastic deformation.

$$\sigma_T = K\varepsilon_T^n \tag{7.23}$$

The n value is typically lowest for hcp materials (~ 0.05) and highest for FCC materials (~ 0.5), with BCC materials lying in between (~ 0.2). Also, the constant K equals the true stress when $\varepsilon_T = 1$.

When Does Necking Occur?

Interestingly, many materials, such as metals and plasticine, experience necking; however, other materials, like chewing gum, resist necking. Once the necking process starts, it is the beginning of the end for the specimen, i.e., the process of necking continues until fracture.

The condition when necking occurs is:

$\frac{d\sigma_T}{d\varepsilon_T} = \sigma_T$ (on the true stress-true strain curve)

or

$\frac{d\sigma}{d\varepsilon} = 0$ (on the engineering stress-engineering strain curve)

The necking (plastic instability) phenomenon has practical implications. For example, it is vital to suppress plastic instability in certain manufacturing processes, such as deep sheet metal drawing. The manufacturing of car bodies is an example. Here, mild steel is well suited since, in its stress–strain curve, necking appears after a significant amount of plastic strain. However, this is not the case for aluminum alloys, which neck sooner than mild steels.

After the specimen fractures, we have an opportunity to place the two parts of the specimen back together again and measure the **final gage length (l_f)** and the **final diameter (D_f)** within the necked region at the fracture point and compare with the **original diameter (D_o)** and **original gage length (l_o)**, as seen in Fig. 7.27. In doing so, we can now define the material's **ductility**. This property typically can be defined in two ways: **the percent elongation** to fracture or **the percent reduction in area** (calculated from the measured diameters). These two definitions are presented in Eqs. 7.24a and b.

$$\text{Ductility} = \frac{l_f - l_o}{l_o} \times 100 \tag{7.24a}$$

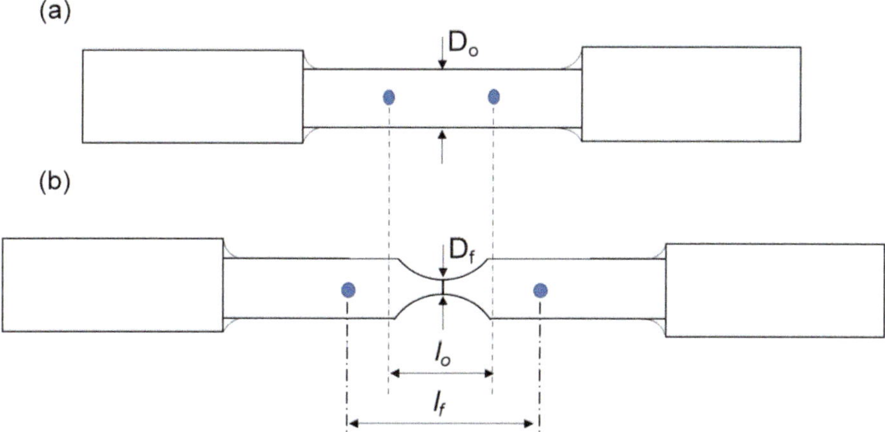

Fig. 7.27 Schematic of tensile testing specimen: **a** before testing, showing initial gage length and diameter and **b** after testing showing two halves of broken sample brought back together, showing final gage length and final diameter

$$\text{Ductility} = \frac{A_o - A_f}{A_o} \times 100 \tag{7.24b}$$

The ductility value calculated using these two equations will differ even for the same specimen broken in the same test. Therefore, if we wanted to compare the ductility of two different materials, we need to compare their respective percent elongation to fracture or their percent reduction in area. Sometimes, we hear the word **malleability**, which refers to how much a material can plastically deform in compression before failure. It should not be confused with **ductility**, which is in tension. As it happens, every ductile material is malleable, but not every malleable material can necessarily be ductile; an example is silly putty.

When discussing resilience, we calculated the area under the engineering stress-engineering strain curve within the elastic region. We even acknowledged that the area represents energy per unit volume. Similarly, we can calculate the area under the engineering stress-engineering strain curve till the point of specimen fracture. As can be imagined, this area will be considerably greater than that under the elastic region. Again, the area represents the energy per unit volume that the material absorbed until it fractured. This is called the **toughness** of a material, which, in simple terms, is the area under the stress–strain curve. In other words, a tough material endures or absorbs considerable energy before it fractures. A high strength and high elongation to fracture favors a large area, which translates to a high toughness. Isn't the word tough? How would we characterize a person who is hard to break down, even mentally? He or she would be called a tough cookie. Hence, materials are not too different from us in that respect.

Now that the discussion on tensile testing has ended, let's recap some important properties. These are presented in Table 7.4.

Table 7.4 Definition of terms, properties, and units

Engineering terms and properties	Definitions	Units
Engineering stress (σ)	F_i/A_o	N/m^2 or Pa
True stress (σ_T)	F_i/A_i	N/m^2 or Pa
Shear stress (τ)	F_s/A	N/m^2 or Pa
Engineering strain (ε)	$\Delta l/l_o$	No units
True strain (ε_T)	$\ln(l_i/l_o)$	No units
Shear strain (γ)	$\tan(\theta)$	No units
Young's modulus (E)	$\Delta\sigma/\Delta\varepsilon$	N/m^2 or Pa
Shear modulus (G)	τ/γ	N/m^2 or Pa
Poisson's ratio (**v**)	$-\dfrac{\varepsilon_x}{\varepsilon_z} = -\dfrac{\varepsilon_y}{\varepsilon_z}$	No units
Modulus of resilience (U_r)	$\dfrac{1}{2}\dfrac{\sigma_y{}^2}{E}$	N/m^2 or Pa
Yield strength (σ_y)	0.002 offset yield strength or stress at lower yield point	N/m^2 or Pa
Tensile strength	Max. stress in σ–ε curve	N/m^2 or Pa
Ductility	$\dfrac{l_f-l_o}{l_o} \times 100$ or $\dfrac{A_o-A_f}{A_o} \times 100$	No units
Toughness	Area under σ–ε curve	N/m^2 or Pa

Table 7.5 lists selected materials' yield strength, tensile strength, and ductility. The values include those for pure metals and steel. It also shows how heat treatment, such as normalizing, can increase the strength of steel, e.g., 4150 steel. The low strengths of lead and polymers are clearly seen. Ceramics such as alumina and titanium carbide are brittle and have a range of strengths rather than a single value. We will discuss the reason for this in a later chapter.

7.3.3   Impact Testing

The mechanical tests we have discussed (hardness and tensile tests) involve subjecting the material to a force applied at relatively slow speeds or strain rates. Some of the properties obtained from the tensile test, such as yield strength and ductility, will change if the strain rate is increased significantly, for example, when the material is subjected to rapid impact loading conditions. Our understanding of how materials fracture and fail under different conditions was acquired after several major historical disasters. One example was that of the **Liberty ships**, which were cargo ships employed during World War II. These ships were being mass-produced to serve the needs of the allied forces quickly and to achieve high rates of production; steel sheets were welded together rather than being riveted (which would have been the proper way of assembly, but time-consuming). The steel had very good room temperature properties; however, when the temperature declined, so did the ductility

Table 7.5 Selected mechanical properties for selected materials

Material	Yield strength (MPa)	Tensile strength (MPa)	Ductility (%)
Aluminum	15–20	40–50	50–70
Iron	130	265	43–48
1015 steel (annealed)	284	386	37
1060 steel (annealed)	372	626	22.5
4150 steel (annealed)	379	730	20
4150 steel(normalized)	734	1155	12
Lead	9	15	48
Copper	33	209	33
Magnesium	21	90	2–6
Titanium	170–480	240–550	–
Molybdenum	200	600	60
Nickel	59	317	30
Nylon 6	–	86	5–50
PTFE	–	7–28	250–350
Polystyrene (PS)	–	6–7.3	1–2.3
Rigid PVC	–	35–55	1–10
Flexible PVC	–	10	200–450
Polypropylene	–	31–41	30–200+
Alumina (Al_2O_3)	–	200–310	–
Titanium carbide (TiC)	–	240–275	–
Tungsten carbide (WC)	–	895	–
Single wall carbon nanotube (SWCNT)	–	5000	–

Multiple sources including J.R. Davis, Structure and Properties of Metals, Metals Handbook Desk Edition, Second Edition, ASM International, pp. 118, 219, 577, (1998), Michelle M. Gauthier, Engineered Materials Handbook, Desk Edition, ASM International, pp. 92, 104 (1995)

of the steel and its fracture resistance, unknown to the designers at the time. Indeed, after production, cracks started to initiate at discontinuities within the ship structure, such as at the corners of doorways. The hull also suffered such damage that some ships broke in two even before departing. We needed to follow a testing standard in the tensile test, for example, ASTM E8. As part of the standard, the testing strain rate was typically specified between ~ 8×10^{-4} and ~ 8×10^{-3} s^{-1}, in addition to specifying appropriate specimen shapes and dimensions. Likewise, ASTM E23 is used for impact testing. The standard specifies specimen dimensions; an example is shown in Fig. 7.28a, where a square cross-sectional bar is presented, in which a 45° V-notch is machined 2 mm deep. The bar is placed in a holder, and a heavy pendulum with a hammer/striker attached at the end can swing and fracture the bar Fig. 7.28b.

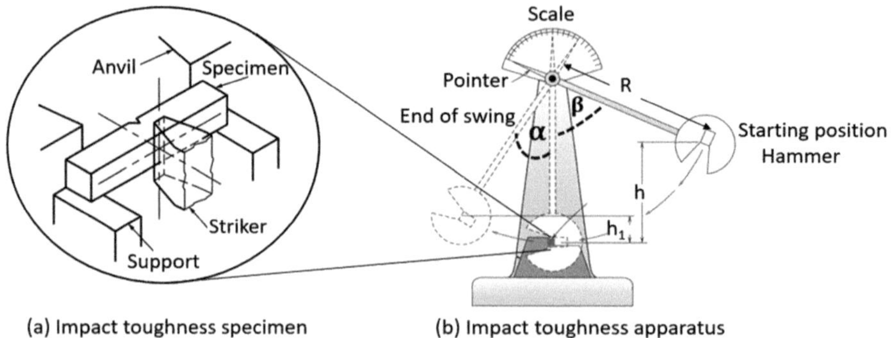

(a) Impact toughness specimen   (b) Impact toughness apparatus

Fig. 7.28 Charpy Impact toughness test: **a** Striker from incoming pendulum arm hits sample causing fracture at the notch site. **b** Charpy impact test apparatus. Image from Monaheng, L. F.; du Preez, W. B.; Polese, C. Towards Qualification in the Aviation Industry: Impact Toughness of Ti6Al4V(ELI) Specimens Produced through Laser Powder Bed Fusion Followed by Two-Stage Heat Treatment. Metals **2021**, 11, 1736. https://doi.org/10.3390/met11111736. Reproduced under the terms and conditions of the Creative Commons Attribution (CC BY) license (https://creativec ommons.org/licenses/by/4.0/)

The pendulum arm is initially raised to a specified height, with corresponding potential energy. When released, the arm converts potential energy to kinetic energy, eventually impacting and fracturing the bar. Since fracturing the bar requires energy, some of the pendulum's potential energy will be expended in fracturing the specimen, and hence, the arm does not reach the initial height (and hence the potential energy) it started with.

This difference in potential energy is taken as the energy absorbed in the fracturing of the specimen and is referred to as the **Impact Energy**. The speed of the hammer/ striker as it hits the specimen is generally ~ 5.5 m/s, which generates strain rates in the specimen that can reach 10^3 s^{-1} (depending on the material being tested. These are orders of magnitude higher than the strain rates commonly observed in tensile testing. Two tests are usually conducted for impact testing: the **Charpy V-Notch** and **IZOD** tests. The only difference between the two is how the specimen is positioned inside the specimen holder with respect to the incoming impacting hammer/striker. In the IZOD test, the specimen is positioned upright with the notch facing the hammer. When the hammer strikes the upper part of the specimen, it generates a tensile stress on the surface facing it, and failure is initiated there at the notch. As for the Charpy test, the specimen is laid down with the notch facing away from the incoming hammer/ striker; likewise, within this configuration, the surface where the notch is placed is also subjected to tensile stress, resulting in failure initiation at the notch.

Generally, the energy measured in this test has no intrinsic meaning in material design. However, what the test can reveal is a transition between ductile and brittle behavior. Ductile materials require more energy to fracture than brittle ones. It turns out that some metals and alloys can transition from being ductile to becoming brittle as the temperature is reduced (hence requiring less energy to fracture), which would be picked up by the impact test. The steel from which the liberty ships were made experienced such a transition when they were subjected to cold environments.

In impact testing, samples are heated to different temperatures and maintained there for enough time to ensure a homogenous temperature throughout each sample's volume. Each sample is then removed and immediately impact tested. Typically, the ASTM standards dictate that the impact test should be carried out within 5 s of removing the sample from its temperature-controlled environment (excluding room temperature). The temperatures are then plotted on the x-axis, and the impact energies on the y-axis. A material that does not exhibit a DBTT will show an impact energy that does not significantly change with a decline in temperature. However, a material that does experience a DBTT may show a significant drop in impact energy with a decrease in temperature. Here, determining such a temperature becomes important, as it reveals the temperature below which the material will approach brittle behavior.

Not all materials undergo a ductile-to-brittle transition. Figure 7.29 shows that FCC (and some HCP) metals and alloys do not experience this transition even at extremely low temperatures. Hence, materials such as aluminum are generally suited for low-temperature applications, such as cryogenic applications. Another interesting viewpoint is that all brittle materials, including ceramics and high-strength alloys, also do not experience such a transition. This is because they are already brittle and require little energy to fracture.

For materials that exhibit a DBTT, such as BCC metals, the question becomes, how can we define a **ductile-to-brittle transition temperature (DBTT)** below which the material behaves in a brittle manner? Here, the approaches differed. Below are possible options for its definition:

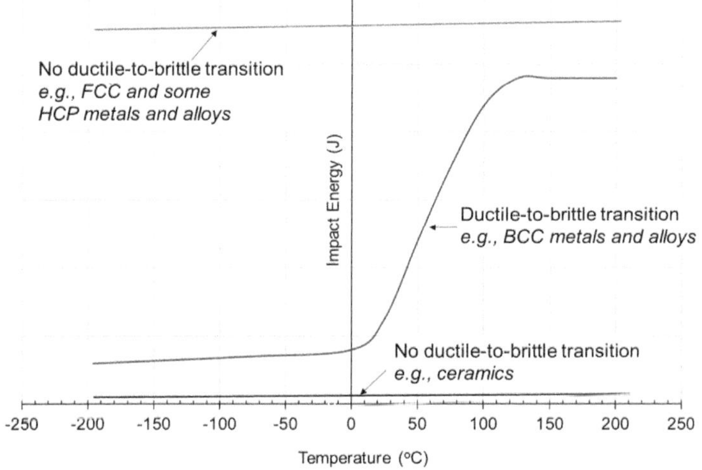

Fig. 7.29 Impact energy versus temperature showing ductile-to-brittle transition behavior for BCC metals and alloys, and not for FCC and some hcp metals and alloys. Ceramics also do not have a DBTT (*Note* high-strength alloys will have a similar behavior to ceramics)

1. The temperature at which the impact energy starts to dip to lower values. Some conservative-minded folks (like me) would agree with this view.
2. The temperature corresponding to the average impact energy between the highest and lowest observed energies.
3. The temperature at which an average impact energy value occurs, for example, 20 J.
4. Examine the fracture surface of the broken specimen to determine what percentage of the fracture area showed features of a ductile fracture and what percentage appeared as a brittle fracture. It could be the temperature at which 50% of the fracture surface showed brittle fracture, as will be explained.

Each of these definitions would result in a different value for the DBTT, as shown in Fig. 7.30. If we adopted definitions 1, 2, or 3, the DBTT would be 133 °C, 53 °C, and 17 °C respectively.

The last option listed above for determining the ductile-to-brittle transition temperature involved analyzing the fracture surface of the broken impact testing specimens. There is a whole field called **fractography** that deals with this science. A lot can be learned about how a component fails by analyzing its fracture surface through fractography. It is well known that a **ductile fracture** is sometimes referred to as a fibrous or shear fracture. It typically has a dull appearance. Fractured surfaces of brittle materials are generally shiny and bright. So, if a material that displays a DBTT, when impact tested at high temperatures (i.e., before it transitions to becoming more and more brittle), requires a high impact energy to break, the fracture surface will be 100% dull or fibrous (meaning ductile fracture). On the other hand, as the temperature is lowered and the material starts to transition to gradually more brittle behavior, more and more of the surface becomes shiny. A simple image analysis of the fracture surface can reveal the percentage of ductile (dark) and brittle (bright) fractures. This area analysis can also identify the DBTT; for example, it can be defined as the temperature at which 50% of the fracture surface appears ductile. The DBTT

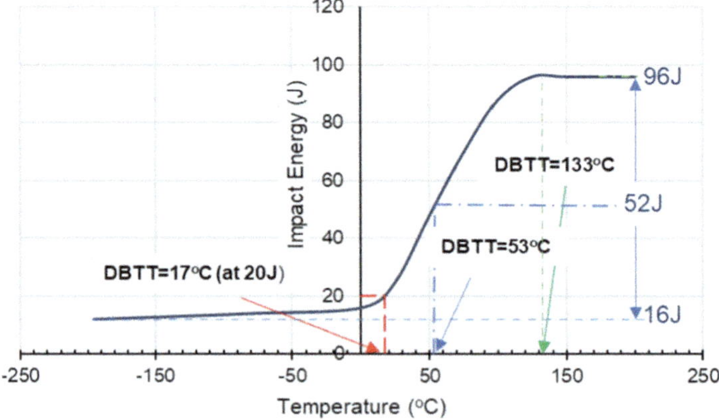

Fig. 7.30 DBTT values according to three different definitions

deduced from this approach could also differ from those determined using the other three approaches shown in Fig. 7.30.

We are still identifying a ductile-to-brittle transition temperature (DBTT) in all the above definitions. A fundamental question, however, arises. Would we rather use a material at a temperature below or above its DBTT? The answer is easy. Why would we want to use a material below its DBTT when it is on its way to becoming increasingly brittle? Hence, we would prefer to use materials at temperatures above their DBTT; hence, the lower the DBTT of a material, the better.

Problems

7.1. The hardness of a material was quoted as HR 70. Is this information sufficient? If not, what is missing?

7.2. Which hardness test does not require the measurement of an indentation size?

7.3. What is the difference between Vickers macrohardness and Vickers microhardness?

7.4. What is the difference between Knoop microhardness and Vickers microhardness?

7.5. What are the advantages and disadvantages of hardness testing?

7.6. A Brinell hardness tester was used to measure the hardness of a material. A load of 500 kgf was applied using a 10 mm steel ball. The impression left behind had a diameter of 0.5 mm. What is the Brinell hardness?

7.7. A 0.07 MPa tensile stress is applied to an aluminum sample; how much will the sample strain be (assume elastic deformation)?

7.8. If the sample in problem 7.7 had a gage length of 1 inch, how much would it extend under that stress?

7.9. A material is said to have negative Poisson's ratio. What does this mean?

7.10. What would be the expected Young's modulus of an alumina (Al_2O_3) sample containing a 0.05 pore fraction?

7.11. A polymer is said to have a secant modulus of 50 MPa. Is this information complete? If not, then why?

7.12. The secant and tangent modulus of a polymer are reported. Do you expect them to have the same value? If not, then why?

7.13. Sketch the shapes of two engineering stress-engineering strain curves you could encounter in ductile metals/alloys. Make sure you label the axes correctly and illustrate how you would calculate Young's modulus on one of the curves. Also, identify on the curves as appropriate the *0.002 offset Yield strength, the upper and lower yield points*, and the *tensile strength.*

7.14. Someone reported to you that the yield strength of a metal is 200 MPa, and its tensile strength is 155 MPa. Is this information reliable? If not, then why?

7.15. A 1 mm copper wire is subjected to an engineering stress of 50 MPa. What would be the applied force?

7.16. Why does stress increase beyond the yield strength during tensile testing?

7.17. An engineering stress of 50 MPa is applied to a magnesium sample having Young's modulus of 45 GPa and a yield strength of 21 MPa. The resulting engineering strain is 0.02. Determine the elastic and plastic strains.

7.18. A tensile testing specimen with an original gage length of 25 mm is loaded in tension using a 90 kN tensile force, which causes it to extend by 0.01 mm. The applied force generates an engineering stress of 100 MPa on the specimen, keeping it in the elastic region. *Determine the following:*

   (a) The original cross-sectional area of the sample.
   (b) The engineering strain at the applied stress.
   (c) The true strain
   (d) The true stress
   (e) How much strain will remain in the specimen if the applied force is removed?

7.19. What is the difference between resilience and toughness?

7.20. A metal was tensile tested, and after the specimen was fractured, the two pieces of the broken specimen were brought back together again. It was determined that the gage length increased from 1 inch to 1.2 inches. Determine the metal's ductility.

7.21. Would you expect the true stress to be lower or higher than the engineering stress during tensile loading? Is it the same in compression loading? Explain.

7.22. You are offered two materials, one with a ductile-to-brittle transition temperature of $-50\ °C$ and the other with a ductile-to-brittle transition temperature of $100\ °C$. You eventually want to use the material at room temperature. Which one would you initially consider using, and why?

7.23. A metal is loaded to an engineering stress of 200 MPa, resulting in an engineering strain of 0.08. What are the corresponding true strain and true stress values, respectively?

7.24. An aluminum sample is loaded in tension within the elastic region. If the elastic engineering strain is 0.001, what will the engineering and true stress be? Note that Young's modulus is 70 GPa. Calculate the percentage error by using the engineering stress instead of the true stress.

Chapter 8
Ferrous Alloys

8.1 Introduction

This chapter discusses ferrous alloys, including cast iron and steel, and the effect of heat treatments on their microstructure and properties.

8.2 Ferrous Alloys

Iron-based alloys are called ferrous alloys, encompassing steels and cast irons. Interestingly, most metals used today in engineering are ferrous alloys. The most basic type of steel is an alloy of iron and carbon. If we trace back our earliest encounters with iron, it was indeed with iron meteorites. The material was not pure iron but iron with nickel and other elements. In other words, it was an alloy from which mankind used to make valuable artifacts. Later, iron smelting from iron ores took place in the Iron Age to produce **Bloomery iron**, where iron was reduced from its ores but not melted. With the introduction of blast furnaces capable of generating exceedingly high temperatures, it became possible to reduce and melt iron from its ores and produce what we refer to as **pig iron**, with typical carbon contents of ~ 4 wt%. This carbon content is, however, double the maximum carbon content that defines steels.

It can easily be claimed that steel is one of the most successful materials of all time. It is abundant and economically produced but can also exhibit properties extending over a wide range through various alloying, processing, and heat treatment approaches. This makes the material quite versatile and suitable for various applications. Where would our world as we find it today be without ferrous alloys, which are used to produce machinery, tall buildings, bridges, cutting tools, etc.? The following sections discuss these important alloys (steels and cast irons).

K. Morsi, *An Engaging Approach to the Science and Engineering of Materials*,
https://doi.org/10.1007/978-3-032-06231-4_8

8.3 Steels

As mentioned, steel in its simplest form is an alloy of iron and carbon (typically also containing less than 1 wt% Mn), referred to as **plain carbon steel**. However, many other steels can emerge when this plain carbon steel is alloyed with other alloying elements and heat-treated in various ways. Figure 8.1 shows one way of classifying the steels available to us today.

Let us break down these different classifications. The alloy content of a steel classified as **low alloy** or **high alloy** is sometimes subjective. However, alloying elements (other than carbon) typically added to plain carbon steel to reach a maximum combined total of 10 wt% content is considered low alloy, and high-alloy steel is above this value. Under low-alloy steels, high-strength low-alloy steel stands out as steel with greater strength than plain low-carbon steels. Additional alloying elements in HSLA steels are copper, vanadium, nickel, and molybdenum. These steels are also ductile and formable; they are heat treatable, and can have strengths (following heat treatment), greater than 480 MPa (Table 8.1).

Iron is one of a number of metals that are allotropic; it has a BCC crystal structure (ferrite) at atmospheric pressure and ambient temperature but switches to an FCC structure (austenite, γ) above 912 °C. If this allotropy phenomenon had not been present in iron, we would not have been able to conduct all these ingenious heat treatments we conduct for steel and obtain all this incredibly wide range of properties we utilize today. Something to think about. Interestingly, we now have thousands of different types of steel if we consider their different compositions and the heat treatments they have been subjected to. The iron atom is much larger than the carbon atom, with a radius of 0.1241 nm versus 0.071 nm for carbon; this means that carbon dissolves in the iron lattice interstitially to produce an interstitial solid solution with a relatively small solubility limit. Figure 8.2 is a schematic of a small part of the iron-carbon phase diagram up to a carbon content of 6.7 wt% carbon. In it, we can see several phases: **ferrite (α)**, a high-temperature **ferrite (δ)**, **austenite (γ)** (which is nonmagnetic), and a line compound called **cementite (Fe$_3$C)**. It is called

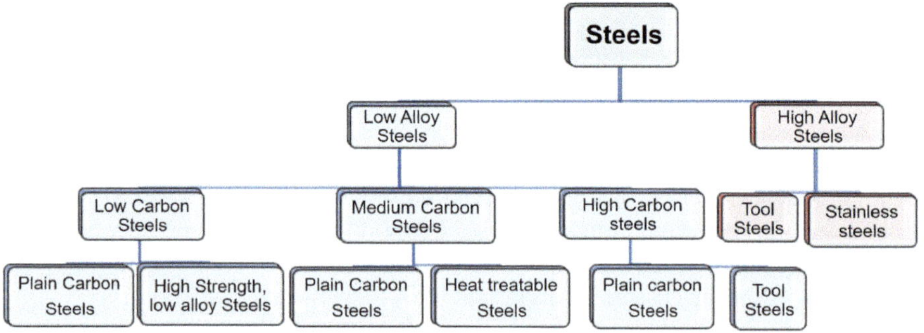

Fig. 8.1 Classification of steels (after W. D. Callister, D. Rethwisch, Materials Science and Engineering-An Introduction, 8th edition, Wiley, 2010)

Table 8.1 Summary of the compositions of certain steels (multiple sources)

Steel	Composition
Low-carbon steel	Less than 0.25 wt% C
Medium-carbon steel	0.25–0.6 wt% C
High-carbon steel	0.6–1.4 wt% C
Low-alloy steel	$\leq$ 10 wt% total alloying elements other than carbon
High-alloy steel	> 10 wt% total alloying elements other than carbon
Stainless steel	Contains a high content of chromium, typically between 11 and 21 wt% Cr, promoting corrosion resistance due to chromium oxide thin dense protective layer formation
Tool steel	It can be low alloy, high carbon steel, or high-alloy steel. Contains hard carbide-forming alloying additions, e.g., Cr, Mo, W, and V
High-speed steel (HSS)	A subset of tool steel, used as cutting tools, e.g., 0.75 wt% C, 4.25 wt% Cr, 18 wt% W, 1.2 wt% V
High-strength low-alloy (HSLA) steel	For example, A440 0.28 wt% C, 1.35 wt% Mn, 0.3 wt% Si (max), 0.2 wt% Cu (max), 0.04 wt% P (max), 0.05 wt% S (max)

a line compound because 100% Fe_3C can only be obtained along the vertical line in the phase diagram instead of a phase field. The phase diagram shows that the solubility limit of carbon in BCC ferrite iron is 0.02 wt% C at 727 °C, whereas it is 2.14 wt% C in FCC (austenite) at 1147 °C. Despite the BCC structure having a lower atomic packing factor than FCC (0.68 vs. 0.74) and having more vacant space inside its unit cell to accommodate more interstitial atoms, presumably, this does not translate to better solubility of carbon. This is because the interstitial sites in the BCC structure, although more than those in the FCC structure, are not all large enough to accommodate the carbon atom freely; on the contrary, the FCC austenite has larger interstitial sites that can accommodate more carbon than ferrite. The phase diagram is restricted to a carbon content of 6.7 wt% and not greater than that, since beyond that point, materials produced will have limited mechanical properties and would be of limited use.

Another point worth mentioning is that the Fe_3C compound is metastable. This means it is not the equilibrium phase that is supposed to form; hence, it should eventually transform to iron and graphite. However, this would take a long time, far beyond our lifetimes. Even if the transformation were accelerated by heating steel containing Fe_3C to a temperature just below the eutectoid temperature (A_1), it would take years to revert to iron and graphite. So, for all intents and purposes, Fe_3C is the phase we consider for our engineering applications.

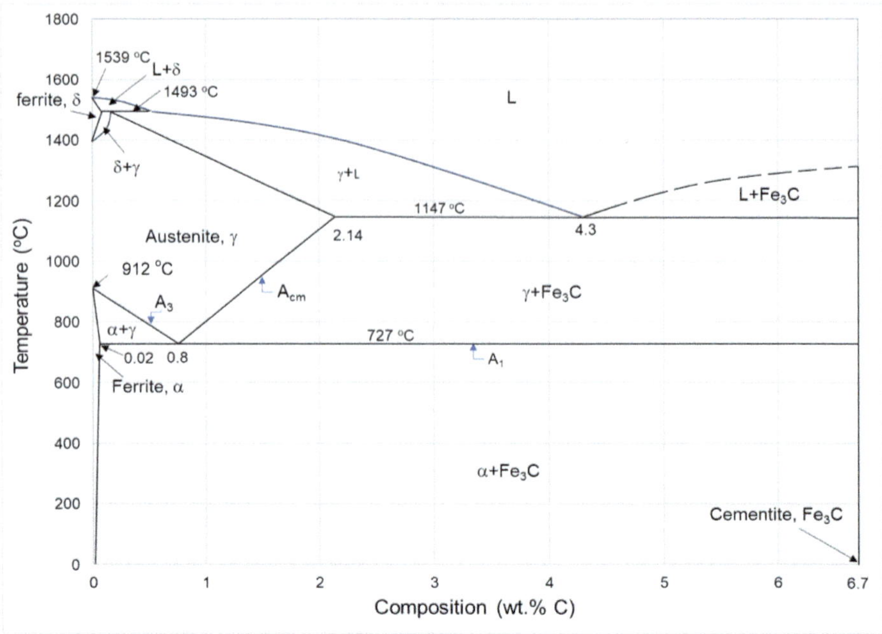

Fig. 8.2 A schematic of the iron-carbon diagram (from multiple sources including ASM Handbook, Volume 4, Heat Treating of Irons and Steels, J. Dossett and G. E. Totten, editors, https://doi.org/10.31399/asm.hhb.v04d.a005995, ASM International (2014))

Alex, how many eutectic points do you see in Fig. 8.2?

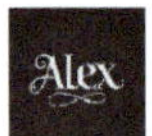

I'm a little confused, Prof. I can see a clear eutectic point, which shows a liquid going into two solids at a composition of 4.3 wt% C and a temperature of 1147 °C, at which I see L ⇔ γ + Fe₃C. So, my answer is that there is one eutectic reaction. However, two other points caused me confusion. For example, at 0.8 wt% C and 727 °C, I see a reaction where a solid solution (γ) goes into two other solids and Fe₃C. It almost looks identical to the eutectic reaction, but this time, it is a solid (not a liquid) going into two other solids of different compositions, i.e., γ ⇔ α + Fe₃C. There is another reaction at a higher temperature (1493 °C) and low-carbon content, with an L + δ ⇔ γ. I can't say that they are eutectics, but what are they?

*You are right, Alex. These two reactions are called **eutectoid** and **peritectic** reactions, respectively. In phase diagrams, we can encounter several types of phase transformations, some of which are summarized in Fig. 8.3.*

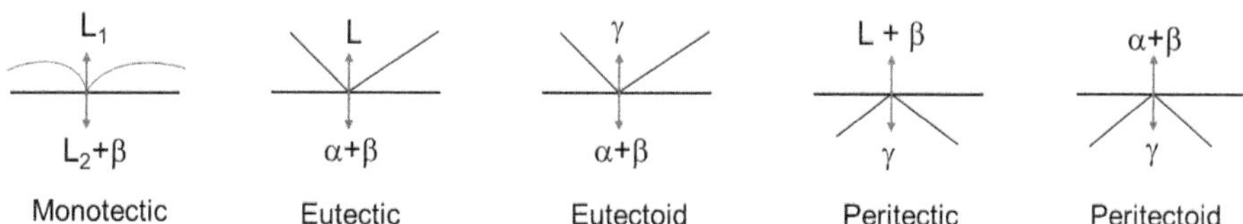

Fig. 8.3 Different transformations that can occur in binary phase diagrams

The eutectoid point lies at a composition of 0.8 wt% C and a temperature of 727 °C. During cooling from a temperature above 727 °C, the austenite (γ) transforms to ferrite (α) and cementite (Fe_3C), which is a very important phase transformation in steels.

Figure 8.3 shows three lines of importance. A_3 is the boundary between austenite and the two-phase field, austenite (γ) and **proeutectoid ferrite (α)**. The proeutectoid phase α is the phase that forms first above the eutectoid temperature, which is a similar concept to the primary α phase, in eutectic phase diagrams. A_{cm} is the boundary between austenite (γ) and $\gamma + Fe_3C$. It should be noted that steel covers up to 2 wt% of carbon content; above that, cast irons are produced.

The most important point in the iron-carbon phase diagram is the **eutectoid** point, which occurs at a composition of approximately 0.8 wt% C at a temperature of 727 °C. This composition is called the **eutectoid composition,** and the temperature is called the **eutectoid temperature.** Any composition less than 0.8 wt% C is called a **hypoeutectoid** composition, and a composition above 0.8 wt% is called a **hypereutectoid** composition. Like the eutectic reaction, the eutectoid reaction also produces a lamellar microstructure, in this case, of alternating layers of ferrite and Fe_3C. The layers here are typically finer than those usually produced in eutectic reactions. This is because the diffusion of atoms occurs in the solid state, which is slow compared with atomic diffusion in a liquid, i.e., as in the case of the molten phase (L) in an eutectic reaction.

This lamellar (layered) structure is called **pearlite (P)**, resulting from the eutectoid reaction. The name is given because its color resembles that of the mother pearl. Although pearlite always has a composition of 0.8 wt% C it is not considered a phase; instead, it is referred to as a **microconstituent**, consisting of alternating layers of ferrite and cementite phases. A simple exercise to determine the relative mass fractions of ferrite and cementite in pearlite in a plain carbon steel of eutectoid composition (0.8 wt% C) reveals the following:

$$X_{Fe_3C} = \frac{0.8 - 0.02}{6.7 - 0.02} = 0.119$$

$$X_\alpha = \frac{6.7 - 0.8}{6.7 - 0.02} = 0.88$$

This means the mass fraction of ferrite is 7.4 times that of cementite. Since the densities of cementite and ferrite are close, the layers of ferrite in the lamellar structure are about seven times thicker than those of cementite. The ferrite appears to be a somewhat continuous phase with layers of cementite embedded in it. Figure 8.4 shows a schematic of the layered pearlitic microstructure.

Like the eutectic phase diagrams discussed earlier, the microstructure highly depends on the composition. Figure 8.5 shows how the microstructure develops for a 0.4 wt% C and 1.0 wt% C plain carbon steel alloy, respectively.

For the 0.4 wt% C alloy, the microstructure starts in the Liquid phase field (**point 1**), then cools to **point 2**, which coincides with the formation of austenite (γ) solid

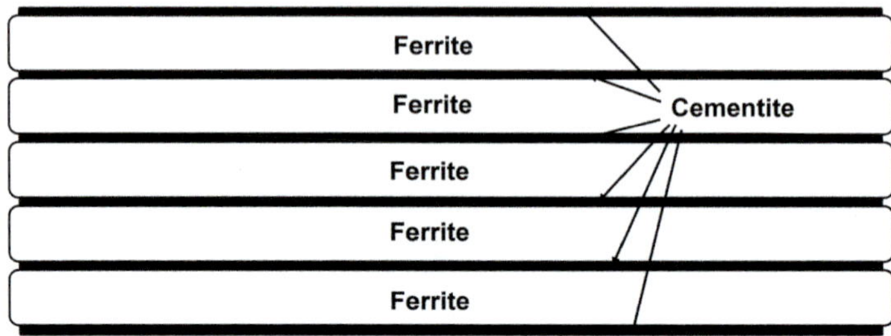

Fig. 8.4 Schematic of pearlite, showing alternating layers of ferrite and cementite. Note the ferrite layers are approximately 7 times thicker than the cementite layers

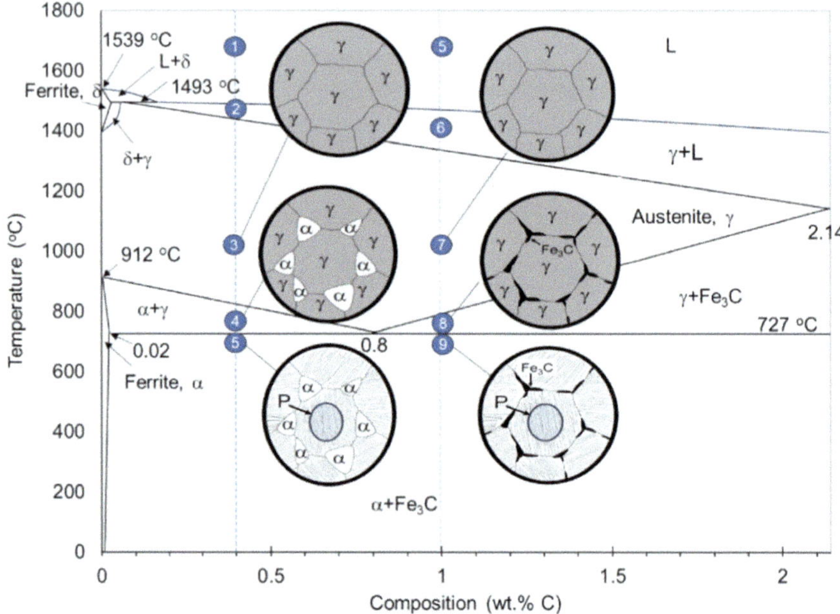

Fig. 8.5 Schematic of part of the Fe–C phase diagram, showing alloys with compositions of 0.4 wt% C and 1.0 wt% C, and their microstructures at different temperatures. P is pearlite (diagram based on multiple sources including ASM Handbook, Volume 4, Heat Treating of Irons and Steels, J. Dossett and G. E. Totten, editors, (2014))

phase; at **point 3**, all the grains in the material are austenite. The alloy is cooled to the austenite + proeutectoid ferrite phase field (point 4), and proeutectoid ferrite (α) is formed. Below the eutectoid temperature of 727 °C, the only phase to transform is the austenite, which transforms to pearlite (P). This results in a microstructure consisting of proeutectoid ferrite and pearlite at point 5.

In the case of the 1.0 wt% C alloy, the **proeutectoid phase** is cementite (Fe_3C), forming in the austenite + Fe_3C phase field (point 8). After cooling below the eutectoid temperature (point 9), the austenite again transformed into pearlite. In this case, the final microstructure is **proeutectoid** cementite and pearlite, as seen in the figure. We can see cementite growing at the grain boundaries.

Example Problem 8.1

A 1040 hypoeutectoid alloy is cooled to 726 °C, i.e., just below the eutectoid temperature. The resulting microstructure consists of pearlite and proeutectoid ferrite. What would be the mass fraction of each?

Solution

A 1040 steel is a plain carbon steel of composition 0.4 wt% C. It is hypoeutectoid since its composition is lower than the eutectoid composition of 0.8 wt% C. Since the composition of proeutectoid ferrite is 0.02 wt% C, and that of pearlite is 0.8 wt% C, then the mass fraction of pearlite, X_p is:

$$X_p = \frac{0.4 - 0.02}{0.8 - 0.02} = 0.487$$

And that of proeutectoid, ferrite, $X_{\alpha'}$

$$X_{\alpha'} = \frac{0.8 - 0.4}{0.8 - 0.02} = 0.513$$

Example Problem 8.2

In Problem 8.1, what would be the mass fraction of cementite and the total ferrite (i.e., proeutectoid ferrite + eutectoid ferrite) in the alloy?

Solution

Since the composition of cementite is 6.7 wt% C (CFe_3C), and that of ferrite (whether proeutectoid or eutectoid) is 0.02 wt% C, the mass fraction of cementite in an alloy of composition 0.4 wt% C (C_o) is:

$$X_{Fe_3C} = \frac{C_o - C_\alpha}{C_{Fe_3C} - C_\alpha} = \frac{0.4 - 0.02}{6.7 - 0.02} = 0.057$$

The mass fraction of the remaining ferrite, which we can refer to as total ferrite (proeutectoid + eutectoid) ferrite, α_T

$$X_{T\alpha} = \frac{C_{Fe_3C} - C_o}{C_{Fe_3C} - C_\alpha} = \frac{6.7 - 0.4}{6.7 - 0.02} = 0.943$$

If we wanted to find out the mass fraction of the eutectoid ferrite, it would be easy since we already know from Example Problem 8.1 that the mass fraction of proeutectoid ferrite is 0.513 and the total ferrite from Example Problem 8.2 is 0.943. The eutectoid ferrite will be the difference between them.

Mass fraction of Eutectoid ferrite = Mass fraction of total ferrite-mass fraction of proeutectoid ferrite = $0.943 - 0.513 = 0.430$.

It should be mentioned that although austenite is unstable below the eutectoid temperature, there is a period where the microstructure is still 100% austenite and is referred to as **unstable austenite (A_u)**. Over time, pearlite starts to nucleate at grain boundaries and grows at the expense of the unstable austenite phase; then, eventually, all the unstable austenite would transform to pearlite. The nucleation of pearlite at grain boundaries is logical since the atoms at grain boundaries are at a higher energy state than those in the interior of a grain and, therefore, easier to break away and participate in the growth of a new phase.

There are two things that we need to consider. The first is the driving force for the austenite to pearlite transformation, which increases as the temperature is reduced further below the **eutectoid temperature (T_E)**, and the austenite becomes more and more unstable. The second is the thermal energy available for atomic diffusion, which is needed to make this transformation happen. The thermal energy decreases as the temperature is reduced further below the eutectoid temperature. Hence, we need to consider both effects when considering such transformations. This leads to what we call a **Temperature-Time-Transformation (TTT)** phase diagram. In fact, by exploring and utilizing this diagram, we can produce new phases that are not even present in the Fe–C equilibrium phase diagram.

As mentioned earlier, imagine a scenario for a eutectoid alloy of 0.8 wt% C composition, in which austenite is cooled from an initial temperature of T_o to an **isothermal transformation temperature (T_{trans})**, below the eutectoid temperature (T_E). As mentioned, there is a period where no phase transformation occurs, and the microstructure still consists of grains of unstable austenite; at some point, though, pearlite starts to nucleate, and over time, it grows, consuming the unstable austenite until the microstructure is 100% pearlite grains. We can hence identify times at T_{trans}, when the transformation to pearlite **begins**, when **50%** of the microstructure has transformed to pearlite, and when the transformation to pearlite has just been **completed** (i.e., reaching 100% pearlite). This provides the basis for the TTT diagram. Hence, if we plot temperature on the Y-axis and time on the X-axis, we can plot three points at the isothermal transformation temperature (T_{trans}): one point at which the transformation starts, another when it is ~ 50% complete, and finally, a point when the transformation to pearlite is ~ 100% complete. If this process is repeated for a range of transformation temperatures, by then connecting all the start points, all the 50% points, and all the 100% points, it would result in the green, blue, and red lines in Fig. 8.6, which represent the start of, 50% and 100% pearlite formation for a range of isothermal transformation temperatures.

Fig. 8.6 Sketch of TTT diagram (isothermal) for a plain carbon steel alloy of eutectoid composition (For the real TTT diagram, refer to Atlas of Isothermal Transformation and Cooling Transformation Diagrams, American Society for Metals, Metal Park, OH, 1971)

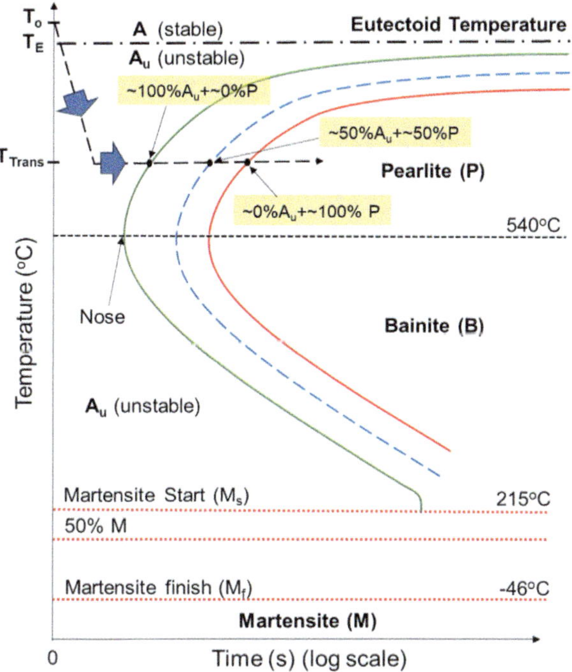

If we examine the TTT diagram further, we can see that as T_{trans} is reduced further below 727 °C, the time to start forming pearlite also progressively decreases. We must remember that below 727 °C, Austenite is unstable, and the stable phase, according to the Fe–C equilibrium phase diagram, is pearlite. The lower the transformation temperature, the higher the driving force for the unstable austenite to transform to the pearlite. This leads to a decline in the time needed for the transformation to start. This happens until a transformation temperature of 540 °C.

Below 540 °C, the transformation starts after longer and longer times. The reason for that is as follows. Although the driving force for the unstable austenite to pearlite transformation increases further below 540 °C, we are faced with a kinetic problem, since atomic diffusion is significantly reduced due to the very low temperatures involved. This leads to an overall increase in the time needed for the unstable austenite to pearlite transformation to start, as seen in Fig. 8.6.

This results in a TTT diagram, also called a C-curve, because of its shape and the unique isothermal transformation temperature, where the time to start the transformation is shortest. Essentially, the diagram has some sort of nose at that transformation temperature. For steel of eutectoid composition, this nose occurs after only ~ 1 s at 540 °C; hence, the microstructure is still unstable austenite at holding times less than one second.

With reference to Fig. 8.6. Suppose a component (presumably a very thin one) is rapidly cooled from T_o to an isothermal transformation temperature below 540 °C and held there for a long period. According to Fig. 8.6, pearlite is no longer the outcome of the transformation but rather a new phase called **bainite.** Unlike pearlite, which consists of alternating layers of cementite (Fe_3C) and ferrite (α), bainite appears acicular/needle-like or plate-like in the microstructures. It consists of a ferrite matrix that contains *elongated* particles of Fe_3C. These particles are extremely small and typically difficult to see using an optical microscope; hence, an electron microscope with a much higher magnification and resolution is required. As seen from the TTT diagram, we produce bainite if we quench and hold for a long time at a temperature between 215 °C and 540 °C. At the higher temperatures within this temperature range, the cementite formed will be coarser than that formed closer to 215 °C. This results from higher diffusion rates at higher temperatures within this temperature range. As such, bainite formed at the higher end of the temperature range is called **upper Bainite**, while those produced at the lower end are called **lower bainite**. Since the diffusion becomes very slow at the lower end of the temperature range, we expect the cementite formed to be finer in scale, which has significant implications for the mechanical properties. The TTT diagram also shows another feature: a **Martensite start temperature (M_s)**. Assuming that the cooling rate was so high that the cooling curve does not cross the transformation start curve, even at the nose, then at a temperature just above the martensite start temperature (M_s), the microstructure would be entirely made up of unstable austenite grains. As the temperature is reduced below M_s, **martensite** starts to form in the microstructure at the expense of unstable austenite. Below the **Martensite finish temperature (M_F)**, the microstructure becomes totally made up of martensite. For eutectoid steels, this is somewhere around -46 °C. This means that if we quench eutectoid steel to room temperature, the microstructure will largely consist of martensite and some residual unstable austenite.

So, what exactly is **martensite**? Unlike the formation of pearlite or bainite, which both involve atomic diffusion. The transformation of unstable austenite to martensite has been described as a **diffusionless transformation** or a **displacive transformation**. But how exactly can such a transformation change the crystal structure, from unstable austenite with an FCC crystal structure to martensite with a body-centered tetragonal one (BCT)? We must simply understand that since the steel is cooled rapidly to a very low temperature, the opportunity for atoms to diffuse and cause structural change has now been taken away. However, the low temperature that the "unstable" austenite finds itself in results in a very high driving force for the austenite FCC crystal structure to transform to the BCC structure of ferrite. We recall that the maximum solubility limit of carbon in BCC iron is much smaller than that in FCC austenite. This means that the rapid decline in temperature also prevents the significant carbon present in FCC austenite from leaving the FCC unit cell. The excess carbon remains as the FCC structure tries to transform into the expected BCC ferritic structure. This leads to a highly supersaturated solid solution with a distorted BCC crystal structure extending in one direction, in other words, becoming the BCT crystal structure of martensite. The carbon supersaturation is said to be more than 1000 times, which makes it very difficult for dislocations to glide. This is the primary reason why martensite is so hard and brittle. In this phase transformation, the crystal structure changes from FCC to BCT through a displacive reaction where very slight coordinated movements of atoms enable the shift to the new BCT crystal structure of martensite. Martensite appears needle-like or plate-like in the microstructure. It is a metastable (non-equilibrium) phase. Hence, if there is any opportunity for it to convert back to a more stable phase or phases, it would do so, specifically if heated to higher temperatures, where the atoms acquire thermal energy. However, it is treated as being stable in many steels at room temperature.

Note that if the cooling rate is not high enough to avoid crossing over the transformation start boundary in the TTT diagram, other phases could be formed, such as pearlite. It is important to mention that only the unstable austenite transforms into all these phases. When unstable austenite partially transforms to a new phase(s) during cooling, we are stuck with that newly formed phase, and only the remaining unstable austenite in the microstructure can transform to martensite. This means we cannot obtain a microstructure with 100% martensite.

Professor, you keep talking about rapidly cooling the material to a very low temperature from the austenitic phase field, keeping the material at this isothermal temperature, and observing which phases form over time. But how would you practically do this, and how would even people be able to physically obtain the data that makes up the TTT diagram in the first place?

Alex, imagine we produce the eutectoid alloy in the dimensions and shape of a thin and small one-cent coin. We can then prepare different salt baths maintained at various temperatures. We then dip the coin in these constant-temperature baths; since the coin is thin, it reaches the bath temperature quickly. Then, after the prescribed times, the coin is removed and quickly dipped in, for example, water, which essentially rapidly cools the coin to room temperature, preventing any further transformations. The microstructure is then examined. So, the trick is to use very thin and small samples.

It should be mentioned that quenching a component to a particular temperature and keeping the temperature constant to control the microstructure can be highly impractical. Continuous cooling is used instead for the heat treatment of steels. This has implications for even the shape and details of the TTT diagram produced under isothermal conditions, as follows:

1. Transformation start and transformation end times can shift to higher times.
2. The transformation temperatures typically shift to lower values, except for martensite formation temperatures, which typically remain approximately the same.
3. Due to the nature of these cooling curves, it becomes impossible to produce Bainite, and the typical microstructures developed are either Martensite, Martensite and Pearlite, Coarse Pearlite, or fine Pearlite.

Figure 8.7 is a sketch showing the difference in curves for isothermally produced diagrams and those that depend on continuous cooling. The transformation start and

Fig. 8.7 Sketch of TTT diagrams (isothermal and continuous cooling) showing microstructures resulting from continuous cooling curves. After Atlas of Isothermal Transformation and Cooling Transformation Diagrams, American Society for Metals, Metal Park, O. H., 1971, which contains the real TTT diagram

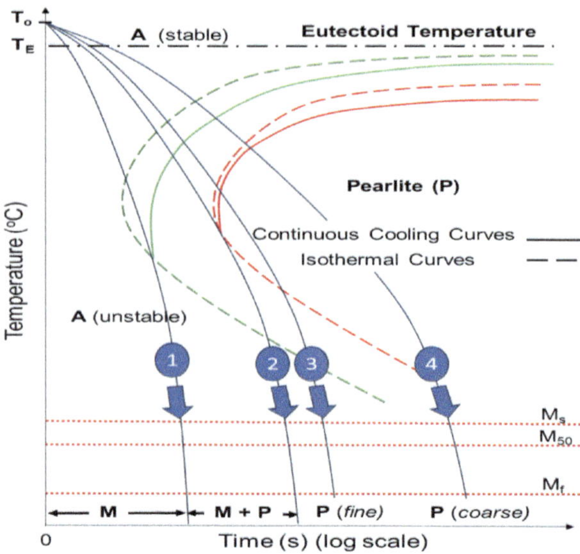

finish boundaries shift to lower temperatures and longer times than the isothermally produced boundaries. Also, for a continuous cooling diagram, if the cooling rate is higher than that shown by cooling curve 1, then martensite will be the resultant phase; a cooling rate between curves 1 and 2 will result in a mixture of martensite and pearlite, and cooling curve 3 will result in fine pearlite. In contrast, curve 4 will be coarse pearlite.

Figure 8.8 is a sketch of a TTT diagram showing how different microstructures can be produced by changing the cooling schedule. In schedule 1, the component is rapidly quenched from the austenitic phase field to a temperature below the M_f, forming 100% martensite. In schedule 2, we quench and remain within the unstable austenite field without crossing the transformation start line for a period of time, then we cool to below the M_f temperature. Here, the outcome will be the same as in schedule 1, which is 100% martensite. In schedule 3, we see an initial quench to a temperature below the nose but within the unstable austenite region until we reach the 50% transformation dashed line, followed by a quench to a temperature below M_f. The first hold to the 50% line generates a microstructure that consists of 50% unstable austenite and 50% bainite. However, once that microstructure is further quenched below the M_f temperature, only the unstable austenite transforms into martensite. This results in a microstructure that consists of 50% bainite and 50% martensite. We recall, only the unstable austenite transforms in this process, and whatever phase we partially form during cooling, we are stuck with. Schedule 4 is similar to schedule 3 but differs in the initial quench, which takes the material to a temperature above the nose and still within the unstable austenite region, this time holding to the 50% dashed line, producing 50% pearlite and 50% unstable austenite, which transforms to 50% pearlite and 50% Martensite upon quenching to below M_f. Schedules 5 and

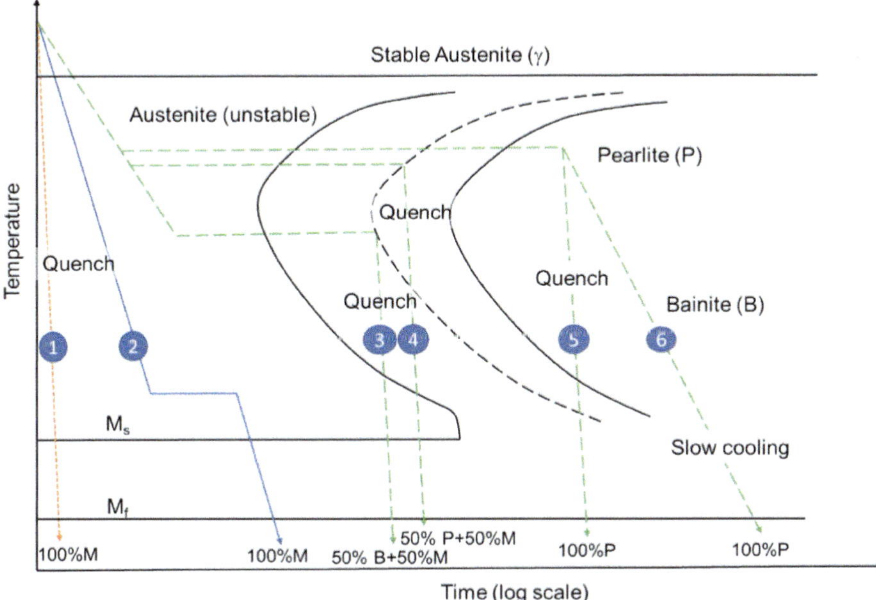

Fig. 8.8 Sketch of TTT diagram showing different microstructures produced due to different cooling schedules (multiple sources)

6 are similar until it is time to cool the sample finally. Schedule 5 cools faster than schedule 6. However, both were quenched to a temperature above the nose in the unstable austenite region and held at that temperature until the microstructure entered the 100% pearlite region. Upon cooling, both of these materials will result in 100% pearlite.

8.4 Heat Treatments in Steel

Normalizing

Sometimes, during steel manufacturing, the developed microstructure is coarse. This has negative consequences in terms of the mechanical properties at room temperature. It is possible to heat that component to a few tens of degrees (say 50 °C) above the A_3 upper critical temperature, thereby entering the austenitic phase field. When this happens, all the microstructure converts to austenite of a small grain size (which can be retained upon cooling). This procedure is used primarily for hypoeutectoid alloys.

Spheroidizing anneals

Machining components with high carbon content compositions can be quite challenging. When pearlite or bainite is formed, we can heat the component to a temperature just below the A_1 lower critical temperature (i.e., lower than the eutectoid temperature) for prolonged periods. For example, 650–700 °C for times greater than 12 h. When this happens, the layered and fine cementite structure is transformed to a more spherical geometry; by doing so, the interfacial area and, therefore, the interfacial energy between cementite and ferrite are significantly reduced. This kind of structure, which is called **spheroidite**, consists of a continuous ferrite matrix reinforced with sphere-like particles of cementite. Spheroidite is much softer than the previous pearlitic or bainitic microstructures and can be easily machined. After machining is conducted, the component can be re-strengthened by normalizing, producing a finer microstructure that is not spheroidite.

Tempered Martensite

Although martensite is extremely hard, it is also brittle. To produce a microstructure that has appreciable ductility while still being strong, the martensite can be further heat-treated to produce **tempered martensite**. As an example, for a eutectoid plain carbon steel alloy, it is possible after forming the martensite to heat the component to a temperature between 250 and 650 °C and leave it there for a given amount of time. Here, an equilibrium microstructure will develop, where cementite particles (not layers) are surrounded by a ferrite matrix. This is in marked contrast to the expected pearlitic microstructure we are used to seeing for that composition. To understand why this is the case, we need to remember that the formation of pearlite with its layered structure and its exceedingly large cementite (Fe_3C)/ferrite (α) interfacial area only formed because, during cooling, austenite did not have time to form cementite as distinctly separate particles within a ferrite matrix. It had to resort to forming a layered structure, where atoms were distributed at many locations at the austenite/pearlite interface. In the case of tempered martensite, all the time in the world is available to produce cementite as distinct particles within a ferrite matrix, with a much lower cementite (Fe_3C)/ferrite (α) interfacial area than pearlite. This, hence, represents a more energetically favorable microstructure. Nevertheless, there is still interfacial energy that exists in this tempered martensite microstructure, and naturally, if one tempers the material at a higher temperature and/or longer times, the cementite formed will become coarser, thus reducing the interfacial area and energy, and in the process reducing the yield strength and improving ductility. Hence, even within tempered martensite, a range of properties can be obtained depending on the heat treatment cycle. The microstructure of tempered martensite, although it may appear similar to the spheroidite microstructure, is, in fact, much finer in scale and hence results in significantly higher strength and hardness values than the soft spheroidite,

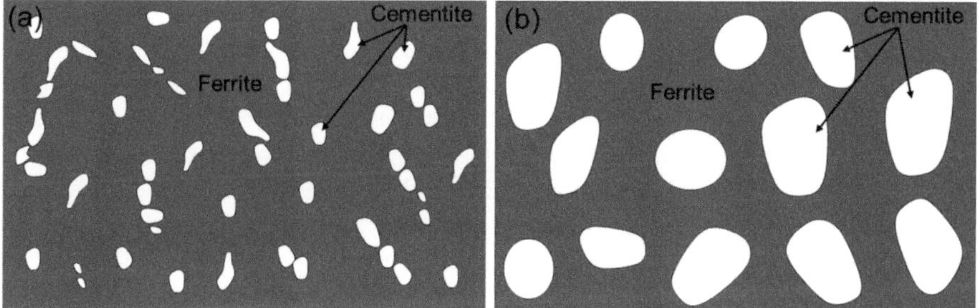

Fig. 8.9 Schematic of microstructure of **a** tempered martensite, **b** spheroidite

with its coarse, rounded cementite particles. Figure 8.9 shows a schematic of the microstructure of tempered martensite and spheroidite, to drive this point home.

Martempering

As previously discussed, producing martensite requires a rapid quench that avoids crossing the nose of the TTT diagram. Even if one avoids the nose, the component is subjected to significant strains and stresses, as the component's surface is cooled at a significantly higher rate than its interior. With its body-centered tetragonal crystal structure (as opposed to the FCC austenitic crystal structure), martensite would start forming at the surface before the component's interior, resulting in significant stresses and strains. To avoid this problem, it is possible to quench the component only to a temperature above the martensite start temperature, i.e., still within the unstable austenite region. Initially, the interior of the component is still at a higher temperature than the surface of the component; however, if left at that temperature for an appropriate amount of time, the temperature throughout the component will be homogenized. Following this, the component is cooled slowly to below the M_f temperature by cooling in air, for example. This provides a milder cooling process compared to rapid quenching. After producing this martensite, a tempering process is then conducted. This whole process is called **Martempering**. It is typically only applicable to thin components since we still have to initially bypass the nose of the TTT diagram to avoid the formation of pearlite.

Austempering

The conventional method of tempering, where martensite is first formed, followed by a tempering heat treatment, is a two-stage process. It is possible to avoid the challenging martensite formation stage and the tempering stage (with its additional heating step) altogether by simply producing bainite instead of tempered martensite. This process is called **austempering**, and involves again, quenching the component to a temperature above the martensite start temperature and leaving it there until the temperature homogenizes inside the component (while still in the unstable austenite region of the TTT diagram), then leaving the component at that temperature until it crosses over to the bainite region. Interestingly, the properties of bainite are similar to those of tempered martensite, without the need to produce martensite and the subsequent heating to temper it. Another interesting fact is that the toughness and ductility of austempered steels are higher than those obtained for traditional tempered martensite, which has similar hardness values. Again, the austempering process is limited to thin sections that can avoid crossing the nose of the TTT diagram.

Figure 8.10 summarizes these heat treatments that can be used to modify the microstructure and therefore the properties of steels.

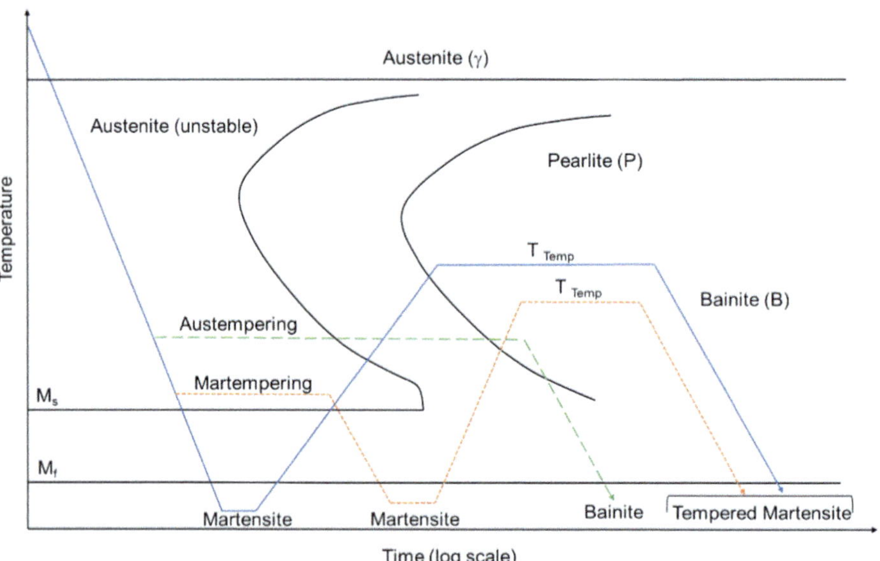

Fig. 8.10 Summary of heat treatments that manipulate the microstructure and hence properties of steels (multiple sources)

Alloying Additions to Steel

One of the main problems in forming 100% martensite in plain carbon steels is the very short time (1 s or less) before crossing the nose in the TTT diagram. This makes it more probable that during quenching, the cooling curve could cross over into the transformation start boundary and form some pearlite, for example. The cooling rate must be exceedingly high to avoid the nose and favor 100% martensite formation, which can require very aggressive cooling mediums such as agitated water. Agitated water breaks the vapor layer between the component and the cooling medium to establish direct component-cooling medium contact and increase the heat transfer coefficient, thereby increasing the cooling rate. This, however, could impose stresses and strains on the component that can lead to cracking (or warping for ductile materials). One of the significant advantages of adding an alloying element other than carbon is that it increases the amount of time during cooling until the nose is reached. Apart from cobalt, all alloying elements tend to favor this, which relaxes the otherwise stringently high cooling rate requirement to produce martensite. Martensite can then be produced simply by quenching in a milder medium, such as oil instead of water or "agitating" water. Table 8.2 shows the relative cooling effectiveness of different cooling mediums. The higher the H coefficient (equivalent to the heat transfer coefficient in convective heat transfer), the more effective it is at extracting heat from the hot steel component. The cooling rate is also shown for a 2.5 cm diameter bar at the central location. Note that the surface would experience a higher cooling rate than the values listed for the center of the bar.

As mentioned, plain carbon steels contain iron and carbon as the main constituents (with typically less than 1 wt% Mn). Adding alloying elements other than carbon can profoundly influence the microstructure and phases developed in steels. For example, adding certain elements can stabilize the high-temperature FCC austenite phase by hindering its transformation to ferrite upon cooling. Many of those elements, such as copper and nickel, also have an FCC structure at room temperature. Other elements would favor the stabilization of the BCC ferrite phase instead; these include

Table 8.2   Effect of different cooling mediums on the cooling rate of a 2.5 cm diameter bar

Cooling medium	H coefficient	Cooling rate at center of a 2.5 cm bar (°C/s)
Oil (no agitation)	0.25	18
Oil (agitation)	1	45
Water (no agitation)	1	45
Water (agitation)	4	190
Brine (no agitation)	2	90
Brine (agitation)	5	230

Cooling rate provided is that at the center point of the bar
After D. R. Askeland, W. J. Wright, Essentials of Materials Science and Engineering, 3rd edition, Cengage Learning, 2009

Table 8.3 Ferrite and austenite stabilizing elements in steels

Ferrite (α) stabilizers	Austenite (γ) stabilizers
Cr, Mo, W, V, Al, Si	Ni, Mn, Cu, Co

chromium, silicon, tungsten, vanadium, aluminum, and molybdenum, many of which crystallize as BCC at room temperature. Table 8.3 summarizes such elements.

Alloying elements can also affect the formation of cementite. Some elements can destabilize cementite, causing carbon to exist as graphite instead of being part of cementite—for example, silicon, aluminum, and nickel cause such an effect.

While some of the alloying elements will only dissolve in ferrite, for example, nickel, silicon, and aluminum, other elements will additionally form their carbides when added (many of these carbides are even harder than cementite, Fe_3C), including carbides of Mn, Cr, Mo, W, Ti, and Nb. Logically, such elements could be found in **tool steel**, where high hardness is needed for cutting tools and dies. Table 8.4 summarizes some elements' relative propensity to form carbides instead of dissolving in ferrite.

Adding alloying elements can shift the eutectoid point in the Fe–C phase diagram. For example, adding elements such as Ti would increase the eutectoid temperature and reduce the composition needed to obtain 100% pearlite, i.e., the eutectoid composition is now at less than 0.8 wt% C. Such is also the case for Cr and Mn. For Mn, a pearlitic microstructure is formed at a carbon content 0.65 wt% and a Mn content of 2.5 wt%. Likewise, cutlery steel, which contains 13 wt% Cr would have a eutectoid composition of 0.35 wt%. Since microstructure, i.e., whether pearlitic or ferritic and pearlitic, etc., affects properties, typically alloy steel can still have very good properties even with less carbon than plain carbon steel. Later, we will discuss strengthening metals through alloying and other means.

One of the primary outcomes of adding high levels of alloying elements (i.e., High alloy steel) is forming stainless steel. These have significant chromium contents, typically between 11 and 21 wt%, and even more. Here, at least 11 wt% Chromium

Table 8.4 Relative propensity of alloying elements to dissolve in ferrite as opposed to forming carbides

Relative proportion dissolved in Ferrite	Element	Relative proportion present as carbide
	Nickel	
	Silicon	
	Aluminum	
	Manganese	
	Chromium	
	Tungsten	
	Molybdenum	
	Vanadium	
	Titanium	
	Niobium	
Maximum 0.3 wt.%	Copper	

After R. A. Higgins, Properties of Engineering Materials, Hodder and Stoughton, London, 1986

Table 8.5 Summarizes some characteristics of each stainless steel

Stainless steel type	Magnetic	Heat treatable
Ferritic stainless steel	Yes	No
Austenitic stainless steel	No	No
Martensitic stainless steel	Yes	Yes

addition is needed to enable a continuous, adherent, thin, and dense chromium oxide film to form at the component surface, rendering the material oxidation resistant. If only chromium (BCC) is added, since Cr is a ferrite (α) stabilizer, the stainless steel produced will be called **ferritic stainless** steel. They typically have a carbon content of less than 0.12 wt% and are not heat-treatable because they do not form martensite upon quenching from the austenitic phase field. The presence of chromium can lead to grain growth and embrittlement if heated to high temperatures. Ferritic stainless steels have moderate ductility and good strength; these properties and compositions result in relatively low-cost ferritic stainless steels having very good resistance to corrosion while exhibiting moderate formability.

Adding elements like nickel and vanadium can hinder grain growth caused by adding Cr, which are added to counter such an effect. Adding nickel (an austenite stabilizer) to the chromium can result in **austenitic stainless** steel. Due to the absence of ferrite at room temperature, these steels cannot be quenched to produce martensite and tempered. They are, however, quite ductile, thus offering great formability in addition to their corrosion resistance. One such alloy is the 316 stainless steel (0.08 wt% C, 17 wt% Cr, 12 wt% Ni, 2.5 wt% Mn), which finds applications as cardiovascular stents due to its biocompatibility. The 18–8 stainless steel (also called 304 stainless steel) is another austenitic stainless steel, the most widely used stainless steel.

Another important stainless steel to mention is **Martensitic stainless steel**. Here, carbon contents between 0.1 and 1 wt% C can be present, with Cr contents less than 17 wt%. These steels can be quenched in oil to produce martensite, followed by tempering. The steel boasts high strength and corrosion resistance. Its use as ball bearings, valves, and knives is hence understandable (Table 8.5).

8.5 Surface Treatments

Sometimes, increasing a component's wear, corrosion, oxidation resistance, or hardness is important. If we increase the hardness of the whole component, we will reduce its ductility and toughness, which could be detrimental. If we, however, ask where in the component the improvement of all the above-mentioned properties is needed? The answer would be the surface and not the whole component. For a gear, only the teeth of the gear need to be hard and wear-resistant (not the component's interior); likewise, if we are worried about corrosion or oxidation, the first attack by the outside hostile environment will certainly be at the component's surface. Hence, it makes sense to concentrate on the surface and somehow engineer it in such a way as to limit wear resistance or oxidation/corrosion of the component. This is the foundation of the field of **surface engineering**. Here, we typically have two options: the first is to modify the surface of a component by alloying or subjecting it to a mechanical treatment, for example, shot peening. This is called **surface modification**. The second is **surface coating**, where a component is coated with another material that protects it. An example of this would be thermal barrier coatings.

Let us consider the following processes related to steel.

Case Hardening

Carburizing/Carbonizing

The objective is to develop a hard case at the surface of a steel component to impart hardness and wear resistance only at that location. The idea builds on the fact that if the surface of a tough steel component, say low-carbon steel, is somehow enriched with excess carbon, it will be hard and wear-resistant while the component still retains its toughness. This carbon enrichment of the surface can be achieved by carrying out a **carbonization process**.

In gas carburization, the material is placed in a furnace and heated to a high temperature in a carbon-rich gas environment. The temperature is high enough to take the material inside the austenitic phase field. At this temperature, the carbon-rich gas deposits carbon at the material's surface, and a concentration gradient is established, which drives the diffusion of carbon from the surface into the interior of the material. By controlling the temperature and time, it is possible to produce a surface layer of controlled thickness with a higher carbon content (and therefore harder) than the component's interior. Therefore, the process of **carburization or carbonization** involves the enrichment of the surface with carbon. The component can also be subjected to subsequent heat treatments. For example, quenching to produce martensite only at the surface, which could also be further tempered.

Generally, the carbon source that enriches the component's surface with carbon can be in solid form (charcoal), liquid form, or gaseous form. Gas and liquid carburization/carbonization have the advantage over solid carburization in that liquid and gas can get to intricate features (and enrich carbon there) while a solid carburizing medium cannot.

Nitriding

Nitriding is when we enrich the surface with nitrogen. Here, nitrogen enters the steel surface and, therefore, its crystal lattice as an interstitial impurity. However, suppose the steel already contains small amounts of alloying elements such as aluminum, chromium, molybdenum, or vanadium. In that case, nitrides of these elements may also be formed, which profoundly increases the hardness at that surface location. The nitrogen is introduced through ammonia; for example, at 500 °C, ammonia (NH_3) dissociates into hydrogen and nitrogen according to the following Eq. (8.1):

$$NH_3 \leftrightarrow 3H + N \tag{8.1}$$

Nitriding has several advantages. First, only low temperatures are needed compared with carburizing (~ 500 °C vs. 900–950 °C); furthermore, unlike carburizing, which could involve subsequent heat treatments such as quenching, nitriding does not require that. This reduces potential problems such as component cracking or warping caused by rapid quenching. Additionally, nitriding can result in improvements not only in corrosion resistance but also in fatigue resistance, which is the resistance of a component to failure under repeated cyclic loading. (More about this later).

Flame and Induction Hardening

Medium-carbon steels can benefit from this Here, the component is heated preferentially at the surface using a high-frequency induction heater or a gas torch, followed by rapid quenching of the surface using water jets. Since the cooling rate at the surface is high and reduces with increased distance beneath the surface, martensite can be expected to form at the surface, below which bainite is formed, and then at greater depth, we find a tough pearlite-containing core. A laser or electron beam can also generate local surface heating instead.

Carbonitriding

Here, carbon and nitrogen are introduced to the steel surface using a gas mixture of carbon monoxide (or hydrocarbons) and ammonia.

8.6 Cast Iron

Cast iron composition starts from 2.14 wt% C and higher. Depending on the composition, heat treatment, and cooling rates, different cast irons could be produced. There are four major types of cast iron. **Gray Iron**, **Ductile Iron**, **White iron**, and **Malleable iron** (Fig. 8.11).

White Iron

When the content of Si drops to 0.5 wt% or less, cementite will become a stable compound as opposed to graphite. Additionally, even if enough silicon is present to stabilize graphite, cementite can still form if cooling is conducted rapidly, since cementite will not have enough time to dissociate. Cementite, a brittle white phase, will produce a fractured surface that appears silvery white, hence the name white iron. The cooling rate effect will also manifest in the size of the cast iron section. If it

Fig. 8.11 Classification of four major cast irons

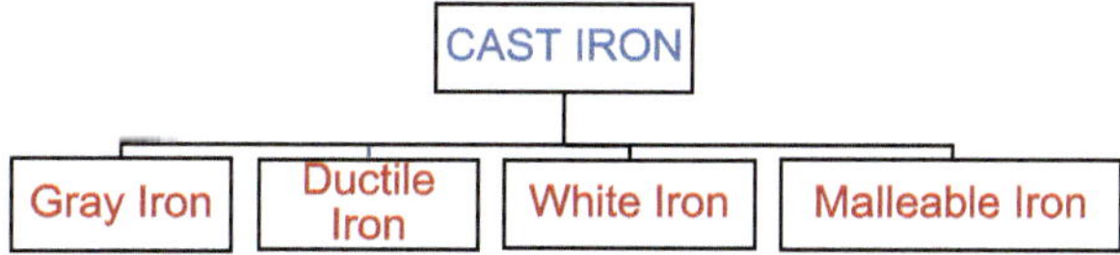

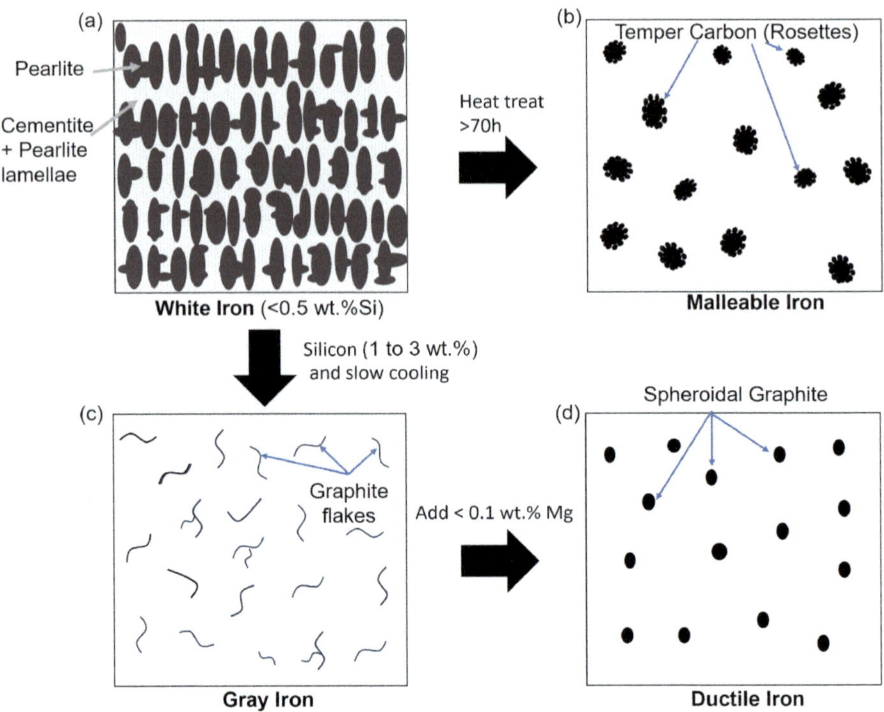

Fig. 8.12 Microstructures of four types of cast irons, **a** white iron, **b** malleable iron, **c** gray iron and **d** ductile iron (multiple sources)

is too large, then slower cooling rates result in gray cast iron, while very thin sections with faster cooling rates will produce white iron. Figure 8.12a shows a sketch of such a microstructure.

Gray Iron

These are cast irons that typically contain between 2.5 and 4.0 wt% C and between 1 and 3 wt% Si as major constituents, although Mn content can also vary between 0.1 and 1.2 wt%. Other impurities in low quantities may be present in the form of sulfur and phosphorus. Recall that Si is one of these elements that can destabilize Fe_3C and result in the formation of graphite. Hence, the main feature of the microstructure of this cast iron is graphite flakes, resembling potato chips or corn flakes within a metallic matrix (Fig. 8.12c), typically pearlite. Since the graphite flakes have pointed ends, they can act as stress raisers, reducing strength and toughness. These graphite flakes make the fracture surface appear gray, hence the name gray iron.

Reducing the silicon content or increasing the cooling rate can reduce the size of these flakes. If this is successfully done, a stronger gray iron can be produced.

ASTM A48 has classified gray irons with respect to their tensile strengths (Class 20–60). Class 20 has a minimum tensile strength of 140 MPa, and Class 60 has a minimum tensile strength of 410 MPa. Gray irons' properties include good thermal shock and wear resistance.

Mottled Iron

Suppose we have intermediate silicon contents (or cooling rates) between those needed to produce gray iron and white iron. In that case, a mixture of graphite and cementite in a pearlite matrix will result. This is called mottled iron, which is weak and brittle. Since the surface appears mottled, it is hence the name.

Malleable Iron

This iron results from the heat treatment of white iron. The cementite is replaced by a rosette form of graphite (temper carbon) as shown in Fig. 8.12b, yielding a cast iron with decent ductility.

Ductile (spheroidal) Iron

The flake morphology, with its pointed features that are determinantal to strength and ductility, can be replaced with more spherical graphite particles simply by the addition of very small amounts of magnesium (no more than 0.1 wt%), which promotes the nucleation of more spherical particles. Figure 8.12d shows a sketch of such a microstructure.

Mechanical Properties of Steels and Cast Iron

As mentioned earlier, the incredible versatility in changing composition and heat treatment approaches for steels has resulted in a plethora of steels with a wide range of properties. Figure 8.13 shows this for hot-rolled plain carbon steels.

We need to remember that ferrite is pure iron at 0.0 wt% C and has a very low solubility of carbon; out of the three phases, ferrite, pearlite, and cementite, ferrite is the softest. Conversely, cementite is a very hard compound, and pearlite is simply a microconstituent mixture of these two phases. In other words, if we wanted to rank these phases/microconstituents in order of strength, cementite would be the strongest, followed by pearlite and then ferrite. Also, ferrite will be the most ductile, followed by pearlite, then cementite, which is brittle. It then becomes common sense that as the microstructure contains more and more of a stronger phase, the overall strength of the steel will increase, and ductility will decrease. Recall that at 0.0 wt% C, the microstructure consists totally of ferrite (α), and as the carbon content is increased from 0.0 to 0.8 wt% C, more and more pearlite (P) is formed until the microstructure becomes 100% pearlite at 0.8 wt% C. Beyond that point, an increase in carbon content results in cementite (Fe_3C) formation at the expense of pearlite. For example, at a carbon content of 1.4 wt% C, the microstructure consists of 89.8 wt% pearlite and 10.2 wt% Fe_3C. This significant change in the microstructure with carbon content has a major effect on the mechanical behavior of plain carbon steels and,

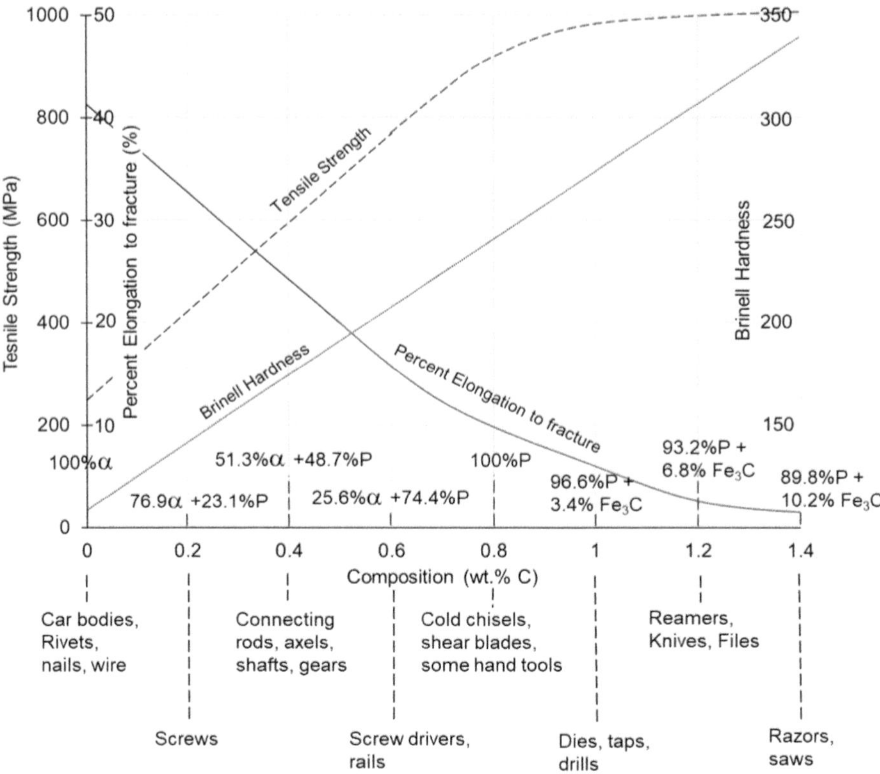

Fig. 8.13 Effect of carbon content on some mechanical properties of hot rolled carbon steels. The figure also shows the percent ferrite, pearlite (P) and cementite (Fe₃C) in the microstructure for different carbon contents (adapted and modified from R. A. Higgins, Properties of Engineering Materials, Hodder and Stoughton, London (1986))

thus, their applications. This can be seen in Fig. 8.13, where the increase in carbon content results in an increase in tensile strength and the Brinell hardness number, and simultaneously a reduction in ductility (percent elongation to fracture). The figure also shows how changing the carbon content can open the door to new applications.

Cast irons typically have moduli that are lower than those of steel and have far inferior impact toughness. However, cast iron, such as gray iron, has a much superior damping capacity than steel or any other cast iron. This has led to its use as a base for machinery to absorb vibrational energy. Contrary to some beliefs, not all cast irons are brittle; malleable and ductile cast irons have appreciable ductility. Table 8.6 shows examples of some of these properties. As can also be seen, Gray iron is classified into different classes according to their minimum tensile strengths in ksi; Class 20 has a tensile strength of 20 ksi, and Class 40 has a tensile strength of 40 ksi. In the table, SI units are used. All the gray and white irons listed have zero ductility. Another fact is that the compressive strengths of cast irons are significantly higher than their corresponding tensile strengths.

Table 8.6 Properties of some cast irons and steels

Material	Yield strength (MPa)	Tensile Strength (MPa)	Compressive strength (MPa)	Ductility[a] (%)	Modulus (GPa)	Brinell hardness
Gray Iron						
Class 20	–	138	655	0	83	175
Class 35	–	241	930	0	110	200
Class 40	–	276	930	0	124	210
Ductile Iron						
60-45-18	310	413	827	18	172	175
80-55-06	413	551	1102	6	172	235
Malleable Iron	220	345	1433	10	179	125
Grade 32510	413	551	1654	3	179	225
White Iron Grade A, Nitr-HiC	–	310	–	0	172	550
Steels 1040 plain carbon steel	600	750	–	17	200	235
8630 Low-Alloy Steel	680	800	–	22	–	220
304 Stainless Steel	205	515	–	40	193	–
410 Stainless Steel	700	800	–	22	200	250
L2 Tool Steel	1380	1550	–	12	–	–

[a] Percent elongation to fracture

Multiple sources: including K. G. Budinski and M. K. Budinski, Engineering Materials, Properties and Selection, 8th edition, Pearson, Prentice Hall (2005), and Metals Handbook, 8th edition vol.1

The Sphere
Layer 4

While you were busy reading and learning, my graduate student and I not only determined the exact composition of **sample 4, which represents layer 4 in the Sphere, but we also** *manufactured many Charpy specimens with the same composition. Moreover, we tested them to find out their DBTT. We needed this to confirm our suspicion regarding the origins of this sample 4, which we did. The composition of sample 4 was found to be iron, with the remaining elements being 0.2 wt% C, 0.46 wt% Mn, 0.044 wt% P, 0.067 wt% S, 0.017 wt% Si, 0.021 wt% Cu, 0.014 wt% O and 0.0031 wt% N.*

Now, let me tell you a story about something you thought you had already heard everything there is to know about. An earlier possible example of brittle fracture that had initially gone unnoticed was that of the **Titanic** (Fig. 8.14, in As A Matter of Fact). Everyone knows the story of this cruise liner, which sank on its maiden voyage.

After colliding with the iceberg, the Titanic was broken in two, as evidenced by the wreckage found in 1985 by Robert Ballard. After an expedition in 1996, the steel from its hull was retrieved and sent to the then University of Missouri-Rolla (now Missouri University of Science and Technology) for analysis. In a very interesting 1998 paper published in JOM (see bibliography), Felkins, Leighly, and Jankovic reported that the steel composition from which the hull was made *was 0.21 wt% C, 0.47 wt% Mn, 0.045 wt% P, 0.069 wt% S, 0.017 wt% Si, 0.024 wt% Cu, 0.013 wt% O and 0.0035 wt% N (with the balance being Fe). The composition of the steel used to build the Titanic contains relatively higher amounts of phosphorus, sulfur, and oxygen, which contributed to the embrittlement of that steel at low temperatures.*

Steel was used to build ships even during the early 1900s. One of those famous (or infamous) ships was the Titanic, as seen in Fig. 8.14.

Fig. 8.14 A ductile-to-brittle transition possibly contributed to the sinking of the Titanic (main image generated using ChatGPT 4.0 by K. Morsi)

*This composition looks close to **sample 4**. They even carried out impact tests on the material over various temperatures. Aren't you curious as to what they found? The material exhibited a ductile-to-brittle transition, and the DBTT temperature at an impact energy of 20 J was 32 °C. This means that, before 1912, when the Titanic builders tested the material at room temperature, unknown to them, the material was on its way down to more and more brittle behavior as the temperature went down. Now for the big question. Do you know what the water temperature was during this tragic collision? − 2 °C. We can confidently say that all the conditions favored a brittle fracture, which did not help the situation. Other researchers have also suggested a brittle failure of rivets as a contributing factor.*

*So, the answer to your question as to the origin of **sample 4**, well, I guess that it is from the Titanic. The **Sphere**, as we have by now established, is an incredible **time capsule**. Sample 4 is a perfect example of a material (arguably now a notorious material) that most probably contributed to one of the deadliest sea disasters in history and the riveting story that has captivated the minds and hearts of humans for over a century: the story of the Titanic.*

I like to look at the bigger picture with all these disasters, whether the Titanic, the Liberty ships, or even the Challenger. Hence, I draw major lessons from these disasters, especially the Titanic. The first is that whenever something wrong happens, an inquiry is launched, and a thorough investigation is conducted to determine why such a disaster happened. Measures are taken so that the disaster does not repeat itself. This is what an advanced nation does, which is one of the ways it advances, by learning from its mistakes. For example, following the Challenger disaster in 1986, the Committee on Science and Technology presented a report to Congress (House Report 99-1016, October 29, 1986) on the Challenger Disaster. The following are quotes from the report:

"From the outset, the focus of the Committee's investigation has been on understanding each of the following: What was the cause, or causes, of the Challenger accident? Are there other inherent hardware or management-related deficiencies that could cause additional accidents in the future? What must be done to correct all of these problems so that the Space Shuttle can be safely returned to flight status?….…..The Committee found that NASA's drive to achieve a launch schedule of 24 flights per year created pressure throughout the agency that directly contributed to unsafe launch operations…….The Committee, the Congress, and the Administration have played a contributing role in creating this pressure".

Look at how they eventually blame themselves. This is how we advance through diligence, integrity, and honesty despite forces that may want to hide the truth. That is why I have a zero-tolerance policy for cheating in my classes. Can we have an engineer who cheats? Or hides the truth? If that were the case, buildings would tumble, and planes would fall from the sky. Anyway, back to the Challenger. As quoted below, the committee draws even more universal lessons from the report"

"…..What we as a Committee, NASA as an agency, and the Nation as a whole, also must realize is that the lessons learned by the Challenger accident are universally applicable, not just for NASA but for governments, and for society. We hope that this report will serve this much larger purpose".

If you think this approach of investigating disasters and trying not to repeat them again is limited only to nations or organizations, you are dead wrong. This applies to you and me. Imagine a student who fails an exam; shouldn't that student investigate why he/she failed? was it his or her lack of studying? lack of understanding? lack of problem-solving practice? did he or she not have a good night's sleep before the exam? were other things going on in his/her life that made it difficult to focus? was the Professor a bad teacher? …. etc. Now, I have seen people quickly conclude that this happened because of this or that person or the professor (taking themselves completely out of the equation). It could be true that the Professor did a lousy job, but what if he/she didn't? then the student will keep failing exams since he/she did not identify the source of the problem and fix it. Depending on the identified cause, he/she may need to adopt a different solution. Believe me when I tell you there are still places today in which the same type of disaster keeps happening yearly because they don't effectively identify and solve the problems. Likewise, I say an advanced student, like an advanced nation, learns from mistakes after careful, honest, and <u>unbiased</u> analysis.

This makes a lot of sense. You said there was more than one lesson you could draw from this Titanic disaster; what else did you learn, Prof.?".

*When the Titanic was built, I believe there was a sense of arrogance, and some people boasted about the idea that they had built the **unsinkable** ship. Ironically, it was not on its second or third but on its first maiden voyage that the unsinkable ship was **sunk**. I learned that always seeking knowledge and doing your best to acquire it is essential. However, once you acquire it, you should remain humble. No one can claim to know everything; do you know all the information on Google? Of course not. Does Google know everything? Absolutely not. My friend, the knowledge we are given is very little indeed. Although such steel could be regarded as notorious today, the layer may also point to steel in general, one of the most impactful materials in history. Imagine a world without steel, no bridges, no tall buildings, no railway tracks…I can go on and on!*
*Now get some sleep, Alex; **Layer 5** awaits you tomorrow.*

Problems

8.1. What type of steel is formed when Ni, Cr, and Mn are added to a total content of 8 wt%?

8.2. What is the difference between monotectic, eutectic, and eutectoid reactions?

8.3. What is the maximum solubility of carbon in austenite, and at what temperature does it occur? Compare that to the maximum carbon solubility in ferrite, and explain the reason for the difference.

8.4. Describe pearlite and state the compositions of its constituents and their mass fractions using Fig. 8.2. Is pearlite regarded as a phase? If not, what is it?

8.5. With reference to Fig. 8.2, sketch the microstructure of a hypereutectoid plain carbon steel alloy of composition 1.2 wt% C at a temperature of 726 °C. What is the composition and mass fraction of pearlite?

8.6. With reference to Fig. 8.2, sketch the microstructure of a hypoeutectoid plain carbon steel alloy of composition 0.6 wt% C, at a temperature of 726 °C. What are the compositions and mass fractions of pearlite and proeutectoid ferrite?

8.7. With reference to Fig. 8.2, for a plain steel alloy of composition 0.55 wt% C at a temperature of 726 °C. What are the compositions and mass fractions of total ferrite and cementite?

8.8. With reference to Fig. 8.2, for a plain steel alloy of composition 0.50 wt% C at 726 °C. What is the composition and mass fraction of eutectoid ferrite?

8.9. With reference to Fig. 8.6. Describe the microstructure that would develop if we:

(a) Quench from 800 to 220 °C.
(b) Quench from 800 to − 60 °C
(c) Quench from 800 to − 60 °C, then heat to 300 °C and stay there for 2 h.

8.10. With reference to Fig. 8.6. Describe the microstructure that would develop if we:

(a) Quench from 800 to 600 °C, and stay at 600 °C until you reach the dashed 50% transformations line, then quench to − 60 °C.
(b) Quench from 800 to 520 °C and remain there for enough time to cross the red line.
(c) Quench from 800 to 300 °C and remain there for enough time to cross the red line.

8.11. What would be the final microstructure if pearlite is heated to 700 °C and left there for 90 h?

8.12. A customer requires steel with a bainite microstructure. You have plain carbon steel with a martensite microstructure. What procedures would you conduct to transform the martensite to bainite?

8.13. A customer requires steel with a pearlitic microstructure. You have plain carbon steel with a bainite microstructure. What procedures would you conduct to transform the bainite into pearlite?

8.14. Martensite is quite hard but highly brittle. What procedure would you carry out to improve its ductility while maintaining an appreciable level of strength? Sketch the resulting microstructure.

8.15. What is the difference between carbonitriding and nitriding?

8.16. What is the advantage of nitriding over carbonizing?

8.17. What is the difference between tempered martensite and spheroidite?

8.18. Discuss Austempering and Martempering.

8.19. State two elements considered ferrite (α) stabilizers and two austenite (γ) stabilizers.

8.20. How does increasing carbon content affect plain carbon steel's tensile strength and ductility? Explain the reasons for these effects in terms of the expected microstructures.

8.21. What is the difference between the mechanical properties of gray cast iron and ductile iron? Explain the reason for the difference in terms of their microstructures.

8.22. What is the difference between white iron and malleable iron?

8.23. What is mottled iron?

8.24. Explain the benefits of spheroidizing anneals.

Chapter 9
Strengthening and Corrosion of Metals and Alloys

9.1 Introduction

This chapter covers strengthening methods for increasing the strength of metals and alloys. It also discusses corrosion.

9.2 Strengthening Methods for Metals and Alloys

As has been mentioned so far, our job as materials scientists and engineers is first to understand the behavior of materials and how they react to external stimuli (whether stress, temperature, atmosphere, electrical potential, etc.), then use our knowledge and skills to engineer the material in a way that improves its properties, making it suitable for more and more applications. In this chapter, we discuss several ways by which metals and alloys can be strengthened, all of which depend on restricting the motion of dislocations. So far, we have learned that dislocations are primarily responsible for the observed ductility in metals and alloys. Their presence caused the observed lower strengths (orders of magnitude) than were first theoretically predicted by early scientists. It follows that when dislocations are allowed to move freely under a resolved shear stress, high ductility is expected in that material. We have seen before that during a tensile test, the resolved shear stress acting on a dislocation causes its motion when the shear stress reaches a critical value. However, what would happen if we made it difficult for the dislocation to move? Wouldn't we need a higher shear stress to get it moving again? Doesn't this translate to an increased yield strength? Hence, if we modify the material condition to make it difficult for dislocations to move, then the material will become stronger and harder (i.e., its yield strength and hardness will increase). Per the second golden rule in Chap. 1 of this book, if we do something to the metal/alloy to make it stronger, it will automatically make it less ductile, and vice versa. Hence, we also expect the ductility to decrease when we

K. Morsi, *An Engaging Approach to the Science and Engineering of Materials*,
https://doi.org/10.1007/978-3-032-06231-4_9

make it difficult for dislocations to move. It turns out that there are several ways in which we can oppose the motion of dislocations and, in turn, strengthen the material:

1. **Solid Solution Strengthening**: This method relies on the alloying of metals, where the alloying elements' atoms impede the dislocation motion.
2. **Precipitation Hardening** introduces new hard particles/precipitates into the material's microstructure, which impedes a dislocation's motion. These can be introduced using heat treatment or mixing the hard particles with the main material during manufacturing.
3. **Grain Size Reduction**: Reducing grain size increases the grain boundary density. Here, grain boundaries act as obstacles to dislocations.
4. **Strain (Work) Hardening**: This involves increasing the number of dislocations (i.e., dislocation density), which also hinders dislocation motion due to dislocation stress field repulsive interactions.

These strengthening methods are additive in nature, so we can employ all of them to increase the strength of materials, each making its contribution. Let us consider each of these strengthening methods in more detail.

9.2.1   Solid Solution Strengthening

We stated that pure metals are softer and more ductile than impure metals. The concept of alloying builds on this principle, in which we intentionally add impurities to a metal as alloying elements. But how do these impurity atoms cause strengthening? As mentioned earlier, they must somehow impede the motion of dislocations if the impurity atoms dissolve in the lattice of the metal (e.g., to form a substitutional solid solution). The dissolved impurity atoms can be slightly smaller or larger than the host metal atoms. This means that when they enter the crystal lattice, they will also cause some local lattice strains and stresses. Now, imagine a dislocation going about its business and trying to glide on a slip plane in response to an applied resolved shear stress. If this slip plane now contains impurity atoms that are, for example, larger than the parent metal atoms, the slip plane will be, in effect, rougher than a plane that only has host atoms. This roughness acts as an obstacle to the motion of dislocations. Logically, the more impurity atoms there are, the rougher the surface and, hence, the more formidable the opposition to dislocation motion. The effect of impurity concentration (X) and atomic size mismatch (M) on the yield strength (σ_y) is shown in Eq. 9.1:

$$\sigma_y \propto M^{3/2} \times X^{1/2} \tag{9.1}$$

From the equation, the higher the mismatch in atomic size between the impurity atoms and the host atoms, the higher the yield strength; and the higher the concentration (i.e., weight percent) of the impurity atoms, the higher the yield strength. Such a strengthening mechanism operates in the 5000 aluminum alloy series and many other alloys, including brass (an alloy of Cu and Zn). Depending on the phase diagram, some alloy compositions can produce a single phase at high temperatures but precipitate a secondary phase as they cool down and crosses the solubility limit line in the phase diagram. If a second phase is precipitated, the concentration of the impurity atoms in the original single-phase microstructure would decrease. Suppose we quench this alloy from the high-temperature single-phase field to room temperature. In that case, we can prevent the precipitation of the second phase and keep the impurity atoms still in solution within the single-phase microstructure. This is because, for the excess impurity atoms to leave the single phase, they need energy, which is not available at the low temperature when the material is quenched. Per Eq. 9.1, by quenching from the single-phase field, we would have successfully retained a high level of X impurities; hence, that alloy will be stronger for it. Since we prevented the impurity atoms from going out of solution, we have maintained a single phase that contains more impurity atoms than allowed by the equilibrium phase diagram, i.e., it is a **supersaturated** single phase.

Here, it is essential to highlight a few facts. Imagine if we were adding impurity atoms from metal B to metal A. Let's assume metal A has a higher yield strength than metal B. It does not matter which of these materials has a higher yield strength than the other. It is the basic principle of adding atoms of one material that are mismatched with other atoms. This means that if we keep adding element B atoms to the host metal A, the yield strength (and, for that matter, the tensile strength) will increase. But it is not without a limit. Once we add atoms of metal B to a concentration where atoms of element B become the majority, the situation is reversed. Element B becomes the host lattice, and element A would now represent the impurity atoms. This means that as we increase impurity atoms from element B to metal A, the yield strength will increase, but from the other end if we add atoms from metal A to metal B, the yield strength of metal B will also increase. This is shown in Fig. 9.1, where the maximum yield strength is obtained around a central composition. Note that the ductility will follow an opposite trend, such that adding B atoms to metal A or impurity atoms from metal A to metal B will decrease the ductility, and the ductility will be lowest around a central composition in the plot.

Adding impurity atoms (alloying elements) to metals can also influence their mechanical properties in another way. During solidification, these impurity atoms gather around dislocations to reduce the strain energy in that region. As previously discussed, these were called **Cottrell** or **solute atmospheres** and were responsible for the upper and lower yield points observed in mild steel.

9.2.2 Precipitation Hardening

In the previous section, we discussed the possibility of quenching an alloy from a single-phase field to low temperatures, and by doing so, we obtained a supersaturated solid solution, which contains more solute atoms than allowed by the equilibrium phase diagram. Although the excess solute atoms help the material to be stronger

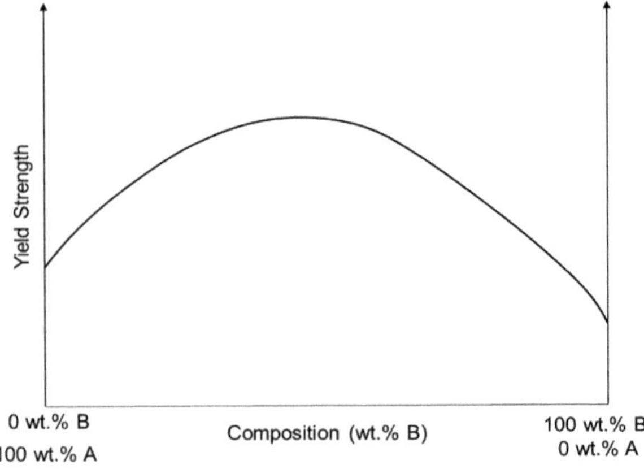

Fig. 9.1 Schematic of yield strength versus composition for a hypothetical A-B alloy system. Where the yield strength of metal A is higher than that of metal B. Here we see that the addition of impurity atoms to metal A or B increases the yield strength

via the solid solution strengthening mechanism, the material in this state is far from equilibrium. Hence, there is an innate urge to precipitate the excess solute atoms. If properly controlled, this urge can be guided to precipitate fine, coherent precipitates capable of effectively impeding dislocations. Although such an approach is used in many materials, we will look at aluminum alloys to explain the process further. The discovery of precipitation hardening (also called age hardening) in some aluminum alloys happened at the beginning of the 1900s. It was utilized in the First World War for components in military airplanes. A fundamental prerequisite to precipitation hardening is a phase diagram with a solubility limit that increases with an increase in temperature, like the one presented in Fig. 9.2, a rough schematic of a small part of the Al-Cu phase diagram. A well-known composition is Al-4wt Cu. The figure shows that at high temperatures above the solid solubility line, the alloy's microstructure consists of grains of single-phase α. Hence, the alloy is heated and left at that temperature for some time in a process called **solution treatment** or **solutioning**. This process aims to ensure that all the alloying elements' atoms are in solid solution, i.e., within a single phase, α. If we allow the alloy to cool and cross the solid solubility line, a second phase ($CuAl_2$) will precipitate. This phase is an *incoherent* precipitate and generates a phase boundary with the α lattice, making it less effective at impeding dislocations. Moreover, the phase is not only brittle but by taking copper atoms out of solution from the supersaturated α phase, the concentration of impurity atoms in the α phase will decrease, and hence the α phase will become softer (due to a decline in solid solution strengthening). The figure also shows the alloy's microstructure if we simply quench from the single-phase field to a temperature of 0 °C. The microstructure consists of grains of single-phase α, which is supersaturated. The maximum concentration of Cu atoms under equilibrium conditions that can dissolve in α is 0.2 wt% Cu. However, now the single phase is supersaturated because it contains 4 wt% Cu— these additional 3.8 wt% Cu will strengthen the alloy based on the solid solution strengthening principles mentioned in the previous section. However, as mentioned earlier, the material doesn't feel comfortable since it is in a non-equilibrium state. Suppose the quenched, supersaturated alloy is heated to a temperature below 180 °C.

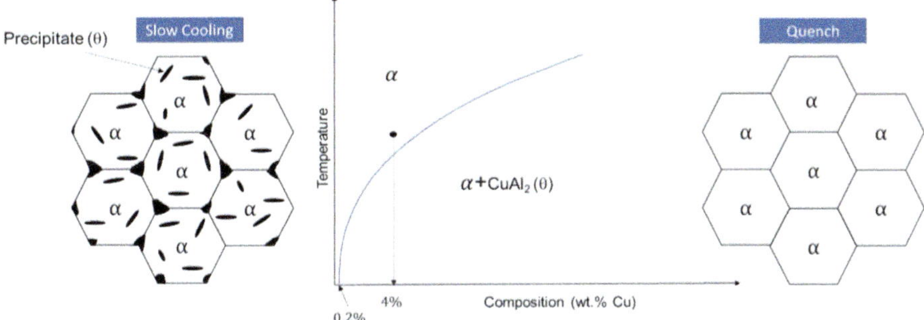

Fig. 9.2 Schematic part of the Al-Cu phase diagram, showing the effect of cooling rate (slow cooling and rapid quenching) on the microstructure for 4 wt% Cu alloy originally in the α single phase field

In that case, diffusion can be controlled, such that copper atoms exit the supersaturated phase and form very small discs on the order of 10 nm in diameter and 1 nm thick, which are called **GP zones** after **Guinier and Preston**, who discovered them. These discs are *coherent* with the aluminum lattice, showing coherency strains at their edges but no strain at the interface between the aluminum lattice and the disc's circular faces. The strains generated are enough to resist dislocation motion and strengthen the alloy. As the aging time increases, some of these GP zones coarsen to become bigger precipitates (i.e., larger discs, ~ 100 nm diameter, and 10 nm thickness) while other GP zones dissolve and, through diffusion, supply more copper to the θ'' to grow them. These θ'' precipitates are also *coherent*, exhibiting no strain at the interface between aluminum and the disc faces, but they also generate coherency strains at their edges. At even longer aging times, the coherency of the edges of the growing discs is lost (hence, no strains exist there), and since the disc faces are perfectly coherent, no strain is there either. At this stage, the coarser disc precipitates (~ 1 μm thickness) have an inter-disc spacing of about ~ 1 μm. Of course, the size of these precipitates depends on the aging temperature and the aging time. Finally, the precipitates lose all their coherency with the aluminum matrix and transform to the $CuAl_2$ equilibrium phase, which forms at grain boundaries and the interfaces between the θ ' and aluminum matrix. The total loss of coherency also results in the loss of the disc shape. Figure 9.3 shows these transitions' effect on the alloy's yield strength.

In summary, precipitation hardening or strengthening has several steps. The first is a solution treatment to produce a homogeneous single phase. This is followed by quenching (usually to room temperature) and then aging. Aging can occur either at room temperature, which is called **natural aging**, or at an elevated temperature, which is called **artificial aging**.

In general, dislocations interact with precipitates. The dislocations have two methods by which they can overcome precipitates; if the precipitates happen to be softer than the main material or very small in size, the dislocation can cut through the precipitate and continue its journey (Fig. 9.4). The stress needed to cut the precipitate

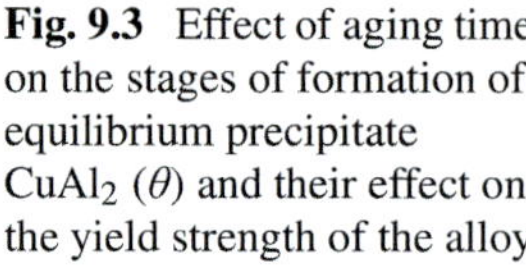

Fig. 9.3 Effect of aging time on the stages of formation of equilibrium precipitate CuAl$_2$ (θ) and their effect on the yield strength of the alloy

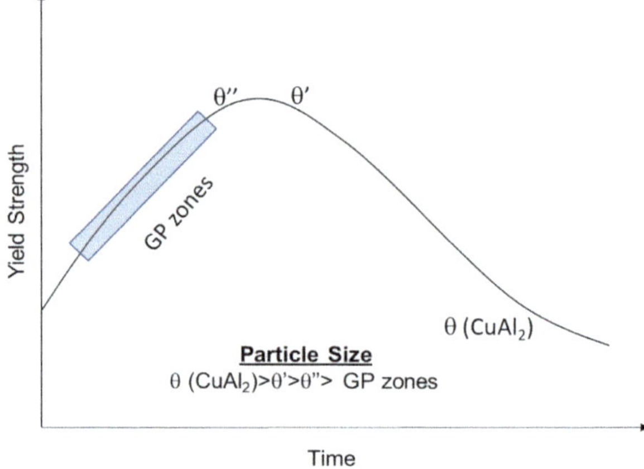

Fig. 9.4 Dislocation cutting particle

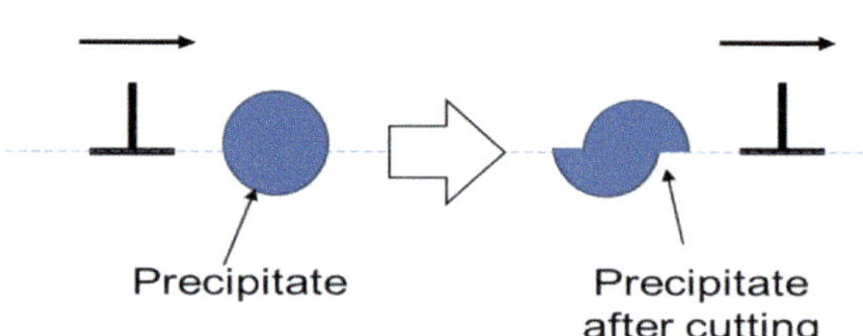

increases with an increase in precipitate size since, obviously, the shear strength of the precipitate poses resistance. This continues until a critical precipitate particle size is reached, where the bowing of dislocations past precipitates requires less stress. In other cases where the precipitates are larger and very strong, the dislocation has no other way than to bow around the precipitates. In doing so, it leaves behind what we call **dislocation loops**. This mechanism is called **Orowan looping**. Figure 9.5 shows schematically how the Orowan looping mechanism operates. Figure 9.5a shows a dislocation approaching two particles or radius r at a distance X apart. Figure 9.5b shows the dislocation bowing past the particles and bulging out. Figure 9.5c shows that the bulged dislocation reconnects with itself, forming again the dislocation but leaving behind a loop around the particles. Figure 9.5d shows a new dislocation approaching the particles surrounding the dislocation loops. The strain fields of the dislocation loops act to repel the incoming dislocation, which requires higher stress for it to bow between the particles. When it does pass, it will leave another dislocation loop so that each particle will have two dislocation loops around it.

The distance between the particles will significantly affect the stress needed for dislocations to bow around the particles. The smaller the distance, the higher the stress needed. One can imagine chewing bubble gum and trying to blow a bubble through one's mouth. If we make the opening between our lips smaller, it will take a much stronger blowing force to get the bubble out than when we open our lips a little. Equation 9.2 presents the relationship between the shear stress needed for

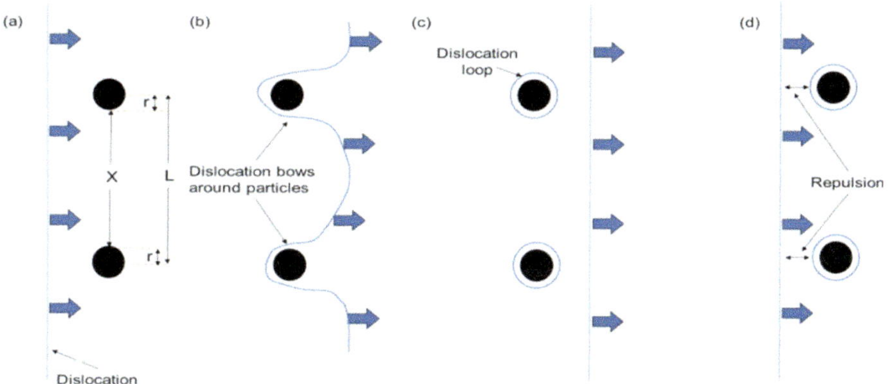

Fig. 9.5 Schematic of the Orowan looping mechanism: **a** dislocation approaching two particles, **b** dislocation bows in around particles, **c** dislocation passes particles and leaves behind a loop around particles, **d** incoming second dislocation encounters the particles and dislocation loop and is resisted (drawn with modification after T. H. Courtney, Mechanical Behavior of Materials, 2nd edition, McGraw-Hill (2000))

dislocations to bow between the particles and the distance between particles. The parameters are those shown in Fig. 9.5a.

$$\tau = \frac{bG}{L - 2r} = \frac{bG}{X} \tag{9.2}$$

where b is the burger's vector, G is the shear modulus, and X is the distance between particles (interparticle spacing). Decreasing distance X leads to an increase in the shear stress needed to bow past the particles.

One needs to realize that if we start with an alloy containing a certain volume fraction of precipitates and then anneal it at a high temperature, there will be a tendency for these precipitates to coarsen over time *while still maintaining their volume fraction*. A consequence of this will be that the inter-precipitate separation will increase as the precipitate size increases. Two effects will result from this: first, coarse particles will make it progressively more difficult for dislocations to cut through them, requiring higher and higher cutting stress. Second, the required bowing stress will progressively decrease since the distance between precipitates increases with precipitate coarsening. Therefore, there will come a point or a precipitate size of the growing precipitates when the stress needed to cut particles and bow around particles becomes equal. This can be seen in Fig. 9.6 at a critical precipitate radius of r^*.

Above this particle size, the precipitates are overcome by the bowing of dislocations and the generation of dislocation loops (since the stress needed to bow around precipitates will be lower than that needed to cut them), while at precipitate sizes less than this critical radius, particle cutting will be in operation (since the stress needed to cut precipitates will be lower than that needed to bow around them). This leads to

Fig. 9.6 Effect of
precipitate particle size on
the mechanism by which
dislocations overcome
precipitates (drawn with
modification after T. H.
Courtney, Mechanical
Behavior of Materials, 2nd
edition, McGraw-Hill
(2000))

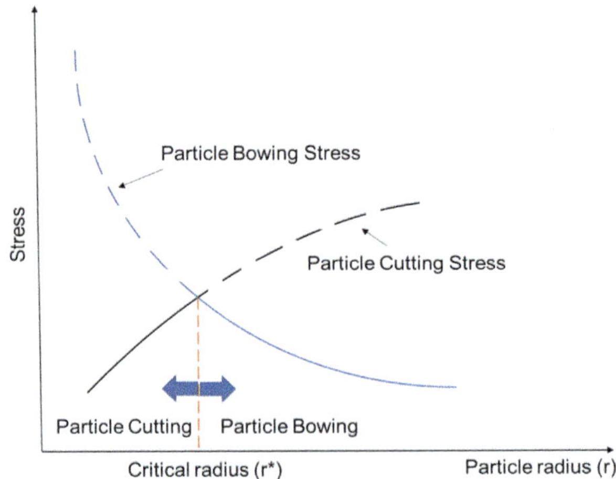

a material strength that increases with increased precipitate size to a maximum, then
a decline in strength or softening. This is seen in the solid lines (not dashed) on the
curves in Fig. 9.6.

9.2.3 Grain Size Reduction

This strengthening method effectively increases the materials' yield strength and
hardness. For a given material volume, a reduction in grain size leads to an increase
in grain boundaries and, hence, the grain boundary density. For dislocations to effectively impart ductility on a material, they need to be able to move (glide) within
a grain and cross the grain boundaries of the material on their journey across the
material. Here, a dislocation is faced with two hurdles. First, within the grain, the
dislocation glides on close-packed planes (to minimize lattice distortion) in which
atoms are highly ordered; however, when the dislocation meets the amorphous (and
much less packed) grain boundary region, it finds it difficult to cross. Suppose the
dislocation can overcome the grain boundary and arrive in the new grain to keep
gliding on the same glide plane within the new grain. In that case, it will need to
change direction (since the new grain is oriented differently than the initial grain),
which is again a hindrance to dislocation motion. This is shown schematically in
Fig. 9.7.

Therefore, the presence of more grain boundaries in the pathway of dislocations
hinders their movement, necessitating an increase in the applied stress to allow dislocations to travel across the whole material. This is equivalent to an increase in the
yield strength. The **Hall–Petch equation** (Eq. 9.3) provides a relationship between
the grain size (d) and the yield strength (σ_y) of a material.

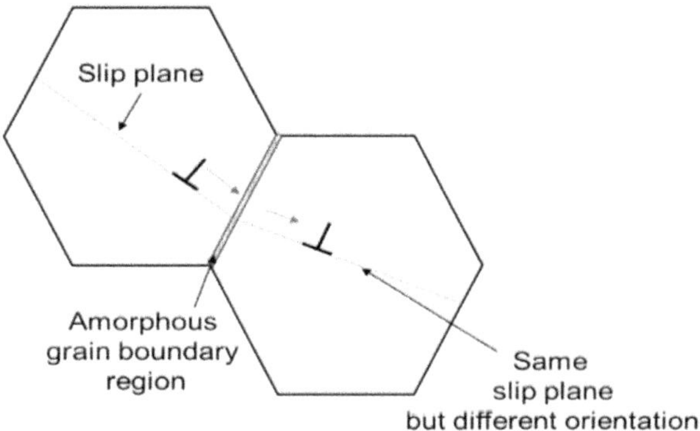

Fig. 9.7 Dislocation crossing grain boundary requiring an increase in applied stress

$$\sigma_y = \sigma_o + k\frac{1}{\sqrt{d}} \tag{9.3}$$

σ_o is the intrinsic lattice resistance (can also be taken as the yield strength of the coarse-grained metal), and k is a constant.

It is important to note that this equation is extremely valuable. However, it is invalid when the grain size is too large or below a critical size within the nano-range. This will be elaborated on in a future chapter. If we plot σ_y versus $\frac{1}{\sqrt{d}}$, we obtain a straight line with a slope of k and an intercept of σ_o. Noting that an increase in $\frac{1}{\sqrt{d}}$ signifies a decrease in grain size. Figure 9.8 shows that yield strength increases with a decrease in grain size.

Quite interestingly, ceramics also observe a Hall–Petch relation; many tests are, however, carried out through hardness testing. Hence, a plot of hardness versus 1/d yields a similar trend to that found in Fig. 9.8. The Hall–Petch equation becomes:

Fig. 9.8 The Hall–Petch relation: Effect of grain size (*d*) on the yield strength

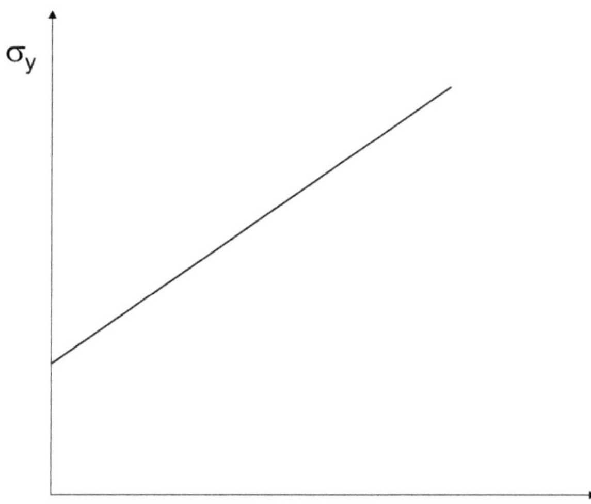

$$H = H_o + C\frac{1}{\sqrt{d}} \tag{9.4}$$

C and H_o are constants, and H is the hardness at a grain size d.

Example Problem 9.1
The yield strength of very coarse-grained aluminum in your warehouse is 100 MPa. A customer requires aluminum with a yield strength of 200 MPa. It is possible to refine the microstructure of this metal and reduce its grain size using some metallurgical processes. What grain size should the aluminum be for the yield strength to be 200 MPa? *Hint:* $k = 0.04$ MPa $\cdot$ m$^{1/2}$.

Solution

$$\sigma_y = \sigma_o + k\frac{1}{\sqrt{d}}$$

Substituting $\sigma_o = 100$ MPa, $\sigma_y = 200$ MPa and $k = 0.04$ MPa $\cdot$ m$^{1/2}$ into the Hall–Petch equation gives a value of d $= 0.16 \times 10^{-6}$ m or 0.16 μm. This means we need to somehow produce the same metal with a grain size of 0.16 μm.

9.2.4   *Work (Strain) Hardening*

Early in the scientific investigation of materials, scientists concluded that during plastic deformation, the number of dislocations must increase to allow the continuation of plastic deformation. In effect, there was a mechanism by which dislocations were generated. Such a mechanism, confirmed experimentally, is the **Frank–Read source**, where, in essence, dislocations give birth to new dislocations during plastic deformation. Things look pretty rosy if one can imagine a single dislocation traveling along a glide/slip plane with no obstacles! However, suppose its movement depends on the movement of other dislocations ahead of it. In that case, as soon as it gets close to another dislocation gliding on the same plane, a repulsive force will be experienced due to the strain fields of the dislocations (Fig. 9.9a). This repulsive force opposes the shear stress that was moving the dislocation in the first place, which means the shear stress now needs to be increased to continue the dislocation motion. More dislocations are formed during deformation, which requires higher stress for deformation to continue. This situation continues until the applied stress increases enough to break the sample altogether.

The increased shear stress requirement means the material is getting stronger as plastic deformation proceeds, referred to as **work hardening**. Fundamentally, work hardening results from increased dislocation density during plastic deformation.

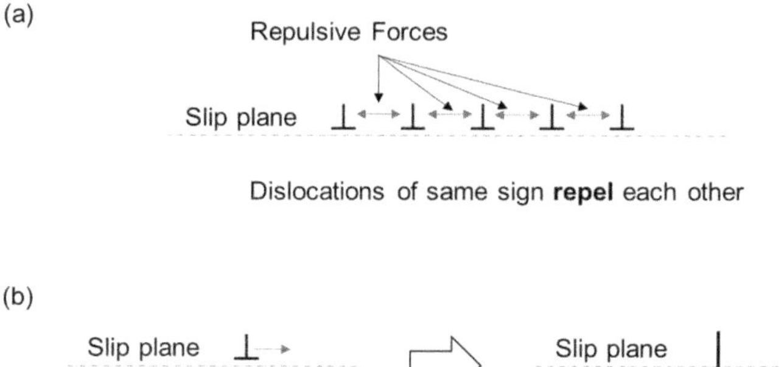

Fig. 9.9 Dislocation interaction when they meet on the same slip plane: **a** same sign, **b** opposite signs

Imagine an edge dislocation traveling on a slip plane; two things can happen if it meets another edge dislocation traveling on the same slip plane. If both dislocations are of the same sign, meaning the extra half-plane of atoms is on the same side of the slip plane, their strain fields will cause them to repel each other. However, if they lie on opposite sides of the slip plane (i.e., opposite signs), then when they meet, they will join and become a regular plane. In other words, there will be no more dislocations, and hence, the previous dislocations would be said to be **annihilated**. This is shown schematically in Fig. 9.9b.

It turns out that dislocations of equal signs vastly outnumber those of opposite signs, so although dislocation annihilation occurs, the main effect of increasing dislocation density with increased deformation is that same-sign dislocations are more able to resist further deformation. This concept is used to increase the strength of metals and alloys.

Bulk deformation processes such as rolling, forging, and extrusion can deform metals and alloys into useful shapes suited for engineering applications. Figure 9.10 shows a schematic of a cold rolling process (cold refers to a low temperature in relation to the material's melting point), where a plate of an initial thickness (t_o) and width (W_o) is passed between two rotating hardened steel rollers. The rollers are separated by a gap smaller than the plate's initial thickness. As the plate approaches the rollers, they bite onto it and pull it through the gap. This results in the plate being plastically deformed and will, therefore, exit having a final thickness (t_f) that is smaller than the initial thickness (t_o). We can assume for now that the width does not change since the biting of the rollers on the plate will resist sideways material flow due to frictional forces between the plate and the rollers. Some increase in width can occur in practice, but for the sake of simplicity, we will assume a constant width before and after rolling. However, since the plate before rolling had a finite volume, the deformed plate should have the same volume (constant volume assumption). So,

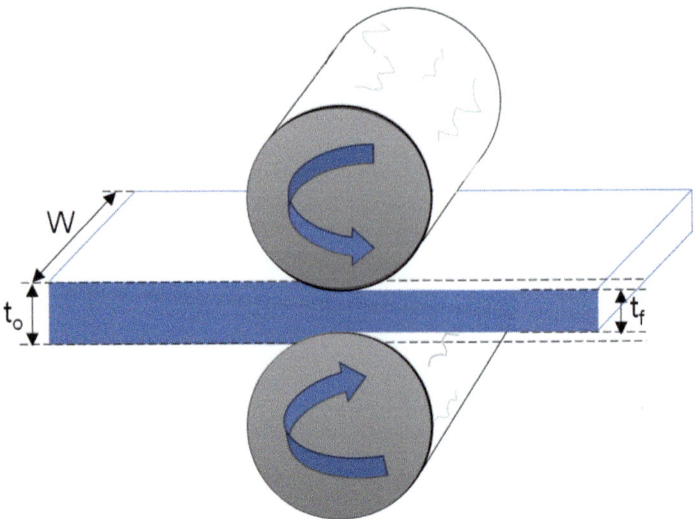

Fig. 9.10 Cold rolling of a metal plate

if the thickness decreases and the width stays the same, the length must increase. Hence, the deformed plate will take longer than the initial plate. Since the thickness of the final plate will be smaller, the cross-sectional area of the final plate ($W \times t_f$) will also be smaller than the cross-sectional area of the initial plate ($W \times t_o$). This reduction in cross-sectional area is permanent since it occurred through plastic deformation. Often, it is necessary to reduce the thickness of a plate to arrive at the required thickness for a component. If the required thickness is much smaller than the thickness of an available plate, it could be possible to cold roll the plate to the required thickness.

This typically cannot be done in one step. Instead, the plate is gradually reduced in thickness through a series of rolling operations, where the gap between the rollers is incrementally reduced with each operation until the final thickness is reached. Since plastic deformation occurs with each rolling operation, the **dislocation** density increases, making the plate stronger and harder. Plastic deformation increases the number of dislocations and other crystallographic defects, such as vacancies. Approximately 5% of the energy used in the cold rolling process will be transformed into these newly formed crystallographic defects, increasing the material's stored internal energy. The other 95% typically transforms into heat. Hence, when a metal deforms quickly, its temperature increases. This is a problem in metalworking processes such as hot extrusion, where, under certain conditions, metals can suffer from what is known as incipient melting, where a small region of the extruding metal heats up and melts. Upon the extruding metal exiting the die, that part of the metal that had melted will give way to cracking.

With increased strength due to cold working comes a reduction in ductility. It is prudent at this stage to introduce Percent Cold Work (%CW) as simply the percent reduction in the cross-sectional area after the plate passes through the rollers. This is shown in Eq. 9.5:

$$\%CW = \frac{A_o - A_f}{A_o} \times 100 \tag{9.5}$$

A_o is the plate's original cross-sectional area, and A_f is the final cross-sectional area of the cold-worked plate. For the condition where the plate does not change in width during cold rolling, Eq. 9.5 simplifies to Eq. 9.6, where the %CW can be found from measurements of the thickness only:

$$\%CW = \frac{Wt_o - Wt_f}{Wt_o} \times 100 = \frac{t_o - t_f}{t_o} \times 100 \tag{9.6}$$

Obviously, as the material experiences more %CW, the dislocation density will increase further, which results in the material increasing in stored energy. Moreover, the increase in dislocation density also affects the material's mechanical properties. Such an effect on hardness can be presented in a %CW versus yield strength, tensile strength, hardness, impact toughness, and ductility plots. Examples are shown in Fig. 9.11a and b. In Fig. 9.11a, we see that as the percent cold work (CW) increases, so do the metal or alloy's yield strength, tensile strength, and hardness. Note that tensile strength is always greater than the yield strength. Also, since hardness is related to strength, a similar trend is observed when hardness is measured. The increase in all three properties directly results from increased dislocation density with increased plastic deformation during cold working. Figure 9.11b shows that the ductility and impact toughness decrease as the percent cold work increases. Again, since dislocation density increases with an increase in the percentage of cold work, so will the resistance to the motion of dislocations, which will also reduce ductility and impact toughness. In essence, the material becomes increasingly brittle as cold working proceeds to higher and higher degrees.

Figure 9.12 shows a simple schematic of the effect of percent cold work on hardness for a hypothetical metal A and an alloy of the same metal A (for example, when metal A is alloyed with another metal, say B). As expected, the hardness of the pure metal increases with an increase in the percentage of cold work. However, we see that although the metal alloy also follows a similar trend, the hardness values are higher than those of the pure metal. This would be expected if solid solution strengthening operated as a strengthening mechanism in the alloy. The alloy is strengthened through cold working (like pure metal) and solid solution strengthening. This only emphasizes the additive nature of the strengthening mechanisms. The ductility would follow an opposite trend, decreasing with an increase in the percentage of cold work, but this time, the ductility values of the metal alloy would be lower than those of the pure metal. Since the added effect of solid solution strengthening is to increase hardness and strength, it would also reduce ductility.

Suppose the required final thickness is significantly smaller than the initial thickness. In that case, there may come a point during the consecutive rolling operations

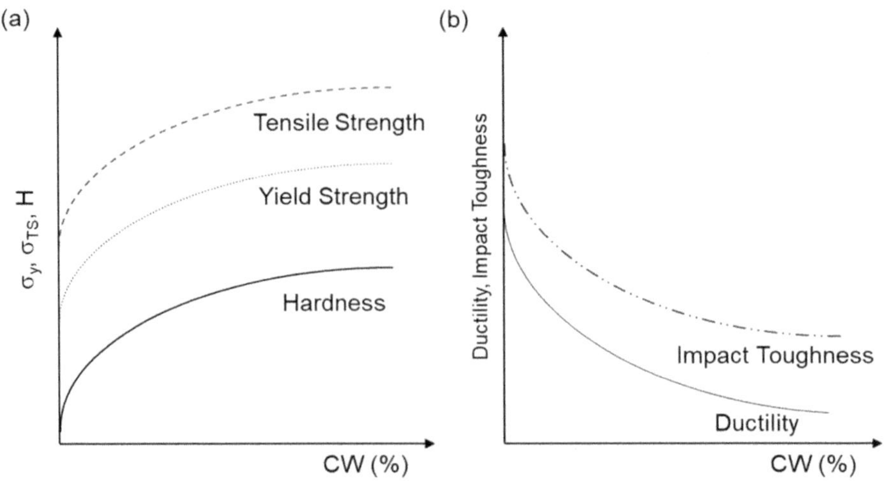

Fig. 9.11 Schematic showing the general effect of percent cold work on **a** the yield strength, tensile strength, and hardness, and **b** the ductility and impact toughness of metals and alloys

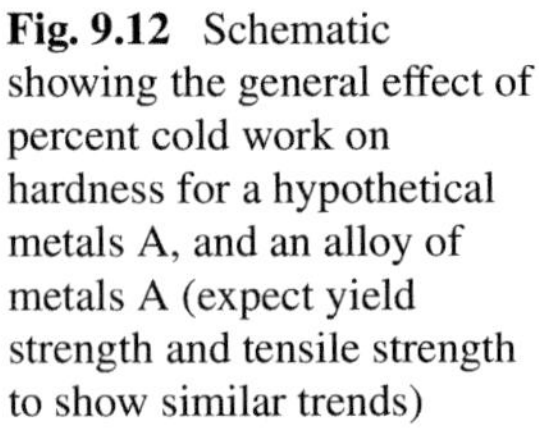

Fig. 9.12 Schematic showing the general effect of percent cold work on hardness for a hypothetical metals A, and an alloy of metals A (expect yield strength and tensile strength to show similar trends)

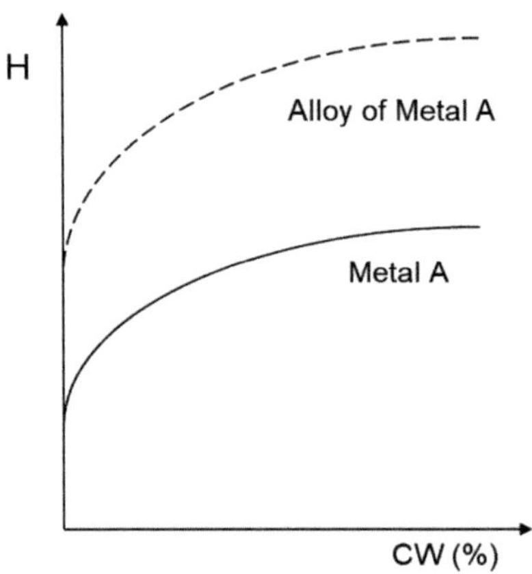

where the plate runs out of ductility before reaching the final thickness. Any attempt to cold roll the plate further will result in cracking. So then, what do we do? Since the addition of every dislocation results in an increase in strain energy within the plate, a significant increase in dislocation density results in a significant increase in excess energy within the plate. This is not a stable situation, and the material will be driven to reduce this excess energy in one way or another. The excess energy, therefore, becomes a driving force for material processes that can enable a reduction in this energy. One could imagine that the microstructure of a cold-rolled metal plate will involve elongated grains (along the direction of rolling). The dislocation density would be quite high within each of these grains, meaning that the microstructure has a high level of stored energy. This stored energy will act as a driving force for microstructural change if atomic diffusion is allowed to occur. At low temperatures, the material does not have the ability or means to enable such an energy reduction. However, if the plate is heated to a temperature where atomic diffusion is possible, then microstructural changes can occur in a way that reduces the plate's energy. Here, two things can occur. The first is a process called **recovery**. In this process, dislocations of opposite signs can annihilate each other, and the dislocations may also arrange themselves in a cell-like network structure. Both of which will result in an energy reduction. During isothermal (constant temperature) annealing, the recovery rate decreases with increasing time.

Alex, did you know that such cell-like dislocation networks have been exploited to produce nanostructured materials? In powder metallurgy, there is a process called mechanical milling. Here, metal powders like aluminum are placed under an argon atmosphere in a hardened steel vial/container with hard milling balls. The vial is then agitated at high speed, which allows the balls to impact the powders; in doing so, it plastically deforms the powders and increases the dislocation density (and hence increases the internal energy of the powders), and during the milling process, the dislocations arrange themselves in a cell-like structure. These cells have diameters typically on the nanoscale. The process, hence, results in milled micro-scale powders that possess a grain structure on the nanoscale. If such powders are then consolidated at relatively low temperatures, for example, through a process such as spark plasma extrusion, then bulk nanostructured materials can be produced.

It is pretty interesting, Prof. It is incredible how knowledge can be used to impact society at large practically.

Ok, now back to As a Matter of Fact.

Recovery is typically said to be followed by a process called **Recrystallization**. Although, in reality, they can overlap. During recrystallization, we see a significant change in the microstructure that ultimately reduces its stored energy. In effect, the original microstructure is replaced by a brand-new one. Unlike the recovery process, the rate of which decreases with annealing time, recrystallization is more of a nucleation and growth process. Hence, the recrystallization rate first increases with time up to a maximum, followed by a decline with time. Typically, the rate of recovery or recrystallization is determined experimentally by measuring the energy release rate during the process. While the older grains were elongated and possessed a high dislocation density, the new grains are equiaxed, containing the equilibrium number of dislocations (i.e., not the excess amount previously brought about by the cold working process). To achieve this, the process starts with the nucleation of strain-free grains at local areas of high energy; these include grain boundaries, slip line intersections, deformation twin intersections, and even foreign particles. The nuclei size must be at least a minimum critical size for the process to start. Figure 9.13a shows a schematic of the grain structure of a metal or alloy after it has been cold rolled or cold extruded. Notice the elongated grain structure along the rolling direction. At higher temperatures following initial recovery, recrystallization takes place within the microstructure, over time, gradually consuming the older microstructure and replacing it with strain-free equiaxed grains (i.e., which have not experienced any deformation and hence have the equilibrium number of dislocations). This is seen in Fig. 9.13b. The microstructure at this stage is called "partially" recrystallized. Eventually, the old microstructure will be completely replaced with strain-free equiaxed grains over time, resulting in a "fully" recrystallized microstructure. Since this microstructure no longer has excess dislocations, it will possess properties similar to those of the metal or alloy before cold working (assuming initially it was in an annealed condition). This microstructure is represented by the schematic in Fig. 9.13c.

A real-life example of recrystallization can be seen in Fig. 9.14a–c, which are EBSD inverse pole figure images of the microstructure of tantalum–tungsten alloy (Ta-5 wt% W) having been first cold rolled to 70% CW, then annealed at two different temperatures. At 1200 °C after 10 min, we see the emergence of some strain-free grains in parts of the now partially recrystallized microstructure, while at 1350 °C after 240 min, the microstructure becomes completely recrystallized. It is noteworthy that EBSD is an abbreviation for Electron Backscatter Diffraction, an analytical method used in electron microscopy, whereby not only an image is generated, but also the crystallographic orientation of grains, among other things. Figure 9.14c also shows different orientations of the recrystallized grains within the fully recrystallized microstructure as identified by color codes.

In recrystallization, the substitution of cold-worked microstructure by new strain-free grains does not happen instantaneously but via a nucleation and growth process over time. This can be illustrated in Fig. 9.15.

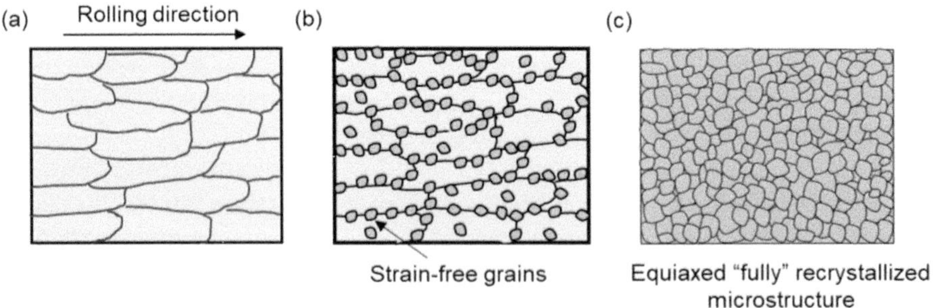

Fig. 9.13 Schematic of **a** cold worked microstructure, **b** partially recrystallized microstructure, **c** fully recrystallized microstructure

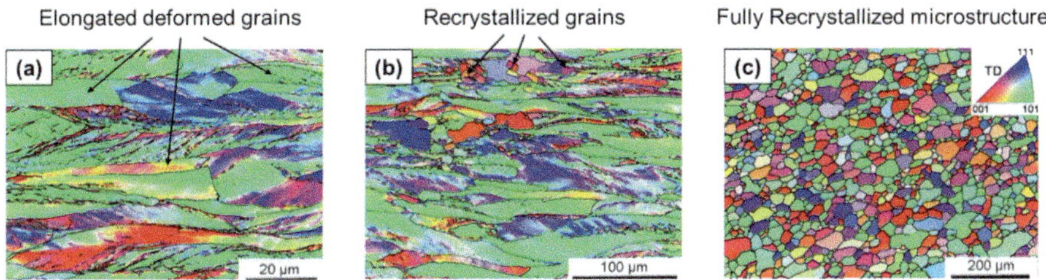

Fig. 9.14 EBSD inverse pole figure images for the microstructure of Ta-4% W alloy: **a** after cold rolling (70% CW), **b** after annealing the cold-rolled alloy at 1200 °C for 10 min, and **c** after annealing the cold-rolled alloy at 1350 °C for 240 min (image from graphical abstract reproduced and modified with permission from Guoqiang Ma, Qiongyao He, Xuan Luo, Guilin Wu, Qiang Chen, Effect of Recrystallization Annealing on Corrosion Behavior of Ta-4% W Alloy-Creative Commons Attribution (CC BY) license (http://creativecommons.org/licenses/by/4.0/)

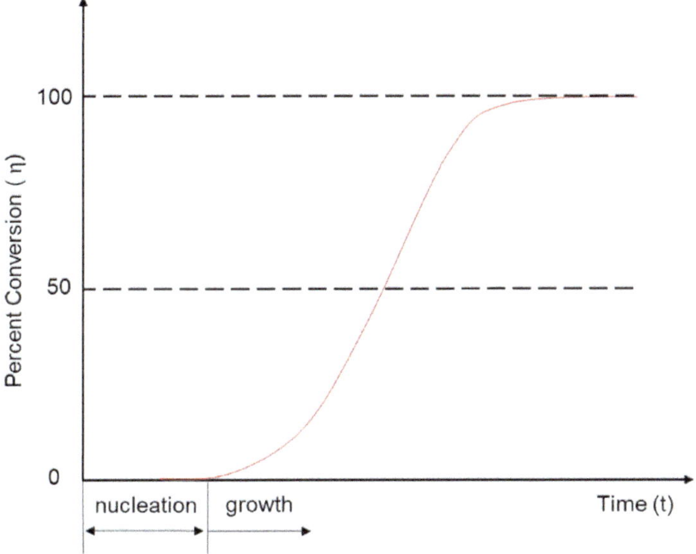

Fig. 9.15 Effect of annealing time on the percent microstructurethat has been recrystallized

The growth curve can be described by Eq. 9.7, also called the ***Avrami equation***.

$$Z = 1 - e^{-kt^n} \qquad (9.7)$$

In this equation, Z is the fraction of the microstructure that has been recrystallized, t is time, and k and n are time-independent constants. If the fraction is 0.5, for example, then "$t_{0.5}$" will be the time needed for 50% of the microstructure to be recrystallized. If the microstructure is fully recrystallized (i.e., $Z = 1$), then $t_{1.0}$ will be the time needed for the microstructure to be fully recrystallized.

If we take the reciprocal of any of these times (i.e., $1/t$), it will represent the **rate of recrystallization (r)**, as shown in Eq. 9.8. For example, $r_{0.5}$ is the rate of recrystallizing 50% of the microstructure, whereas $r_{1.0}$ is the rate of recrystallization for fully recrystallizing the microstructure.

$$r = \frac{1}{t} \tag{9.8}$$

The most important parameter that influences recrystallization has not been mentioned yet: the annealing temperature (T). Inherently, recrystallization is enabled by atomic diffusion, the rate of which we know is exponentially dependent on temperature. It then follows that the recrystallization rate should also be exponentially dependent on temperature. Equation 9.9 shows this dependence.

$$r = \frac{1}{t} = Ae^{-\frac{Q}{RT}} \tag{9.9}$$

A is a constant, R is the gas constant, T is the temperature in Kelvin, and Q is the activation energy for recrystallization. Note here that Q does not carry any precise physical meaning (since multiple processes could be responsible for recrystallization) and hence should be taken as merely an empirical constant. Moreover, the equation is not the only one to describe the effect of temperature on the rate of recrystallization. Other equations can be used for different metals. Hence, one can only consider the equation a rough relation between recrystallization time and temperature.

Both time and temperature (in addition to other factors) affect the degree of recrystallization. Traditionally, the **recrystallization temperature (T_r)** has been defined as *the temperature at which the microstructure is fully recrystallized after 1 h.* We can use this temperature to examine the influence of other factors on recrystallization. The fundamental driving force for recrystallization is the excess stored energy in the microstructure brought about by cold working. It then follows that since an increase in percent cold work increases the dislocation density and, therefore, the stored energy (depending on the %CW, could be on the order of 2 MJ/m^3), then the driving force for recrystallization increases with an increase in percent cold work. This has the effect of a decrease in recrystallization temperature with an increase in the percentage of cold work. Figure 9.16 shows such a dependence. Two things can be deduced from the figure. The first is that a minimum temperature must be reached before recrystallization can occur, no matter how high the %CW is. The second is that a minimum %CW value needs to be applied before recrystallization can occur. In other words, recrystallization will never occur below this minimum (critical) %CW value, as the driving force will not be large enough. This minimum critical %CW could have values of 5%, but different metals and different types of deformation (e.g., compression, tension, shear, torsion, rolling) can also influence the value. We

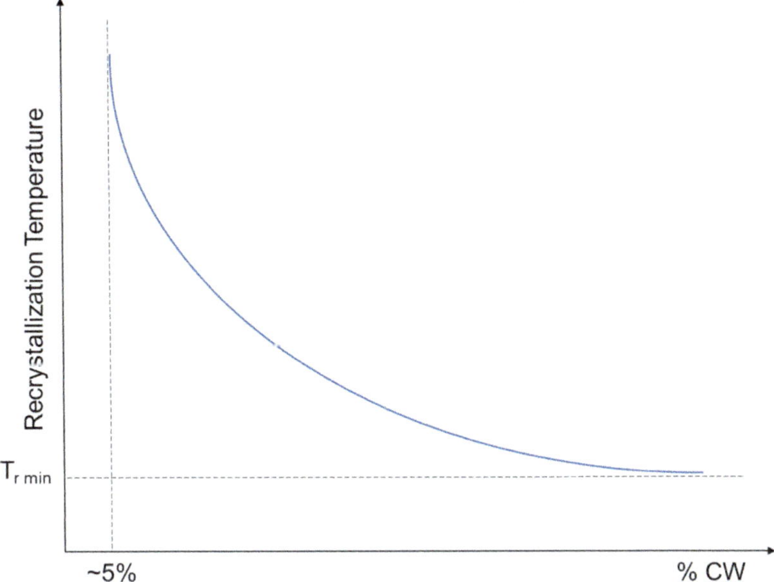

Fig. 9.16 Effect of %CW on the recrystallization temperature

could expect minimum critical %CW values ranging from 2% to even over 15%. The figure clearly shows a decline in the recrystallization temperature with an increase in %CW.

Alex, I guess you now understand the effect of % CW on the recrystallization temperature. However, what if two samples of the same metal were separately deformed to two different % CW but annealed at the same temperature? What will be different from one sample to the other?

Since the sample with a higher %CW will have a higher stored energy and, therefore, a higher driving force for recrystallization, I would expect it to recrystallize faster.

You got it, Alex!

It is essential to point out that recrystallization is often described as a nucleation and growth process. However, the nucleation process should not be understood in the conventional sense, where atoms collect together to form an embryo and nucleus. In recrystallization, nuclei are formed in microstructure regions with crystallographic orientations different from those of the surrounding regions.

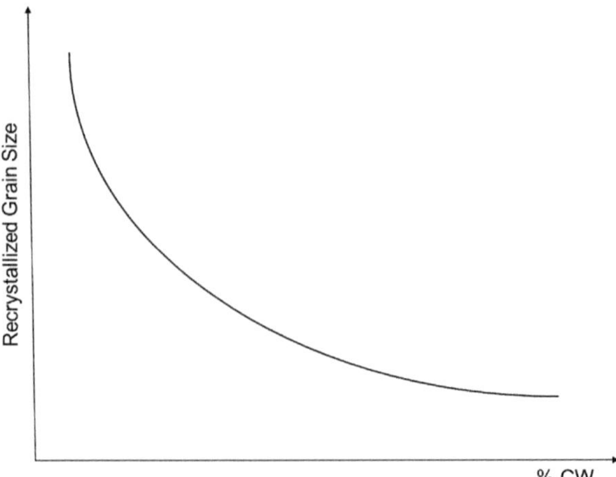

Fig. 9.17 Effect of %CW on the recrystallized grain size

The %CW affects the recrystallization temperature and the final recrystallized grain size. Recall that as %CW increases, the dislocation density increases, which provides more nucleation sites for new strain-free grains to form. The higher the number of nuclei, the less area they can grow before contacting another growing grain, resulting in a fully recrystallized microstructure with a smaller grain size. On the contrary, at low %CW, the dislocation density will be low, and hence, there will be fewer nucleation sites for grains, leaving more areas for them to grow into larger grains before they contact other growing grains. In other words, an increase in %CW will tend to result in a decline in the recrystallized grain size. This is again depicted in Fig. 9.17.

We now turn our attention to other factors that can influence recrystallization. The first is the *grain size of the initial metal* before cold working. When a metal is deformed, dislocations are generated within grains, then as they attempt to travel through the metal, they meet grain boundaries, at which dislocations can pile up like in a traffic jam. Imagine two samples of the same metal (sample A and sample B). Sample A has a smaller grain size than sample B. Suppose both samples were deformed to the same %CW. In that case, we expect sample A to have more grain boundaries (with their associated dislocation traffic jams and, therefore, a distorted lattice next to it) than sample B. This means sample A will have more nucleation sites than sample B, leading to a fully recrystallized microstructure with a smaller grain size.

Another factor is *metal purity*. The recrystallization process starts with the formation of a nucleus. The nucleus obviously will have a boundary with the surrounding (deformed) microstructure. The boundary must be mobile for this nucleus to grow and travel into the deformed microstructure. The presence of impurity atoms can preferentially segregate to this boundary and, in doing so, will slow it down. This makes recrystallization difficult, resulting in a decline in the rate of recrystallization and, ultimately, an increase in the recrystallization temperature. For example, cold-worked aluminum with 99.970% purity has a recrystallization temperature of almost 200 °C higher than aluminum with a higher purity (say 99.998%). This means the slightest of impurities can have a significant effect on the recrystallization temperature. Not only that, but the type of impurity atom can also be important. Consider pure copper with an impurity content of 0.01%. The impurity atom type at that impurity level will influence the recrystallization temperature. For example, although Ni impurity atoms do not increase the recrystallization temperature of pure copper, iron can increase it by 15 K, silver by 80 K, tin by 180 K, and tellurium by a whopping 240 K. It is no wonder that alloys typically have high recrystallization temperatures up to ~ 0.7 of the alloy's melting point in Kelvins, while metals can be around 0.3–0.5 of the melting point in Kelvins.

Let's say that at the recrystallization temperature, the metal is fully recrystallized after one hour of exposure. What happens if we leave it longer than one hour? To answer this question, let's put ourselves in the shoes of an energy-efficient material. We initially find ourselves with a fine-grained microstructure, and we discussed previously that the smaller the grain size, the more the grain boundaries and, hence, the grain boundary density. Every grain boundary has energy (i.e., grain boundary energy). This means even after getting rid of the excess dislocation through recrystallization, there remains stored energy in the form of grain boundary energy that the material can still try to get rid of. The material overcomes this through a process called **grain growth**. The process has also previously been referred to as *grain cannibalism*, where larger grains eat up smaller ones over time, increasing the average grain size and, hence, a lower number of grain boundaries (i.e., lower grain boundary energy). But what is the precise dependence of grain size on annealing time? Initial experiments in the last century were carried out on soap bubbles between two glass sheets. In effect, the behavior of 2D bubbles was monitored as time progressed. Several important conclusions were drawn. First, bubbles with convex surfaces tended to shrink with time, while those with concave surfaces tended to expand. Larger bubbles tended to grow at the expense of smaller neighboring ones. The process was governed by the diffusion of gas from the smaller bubbles to the larger ones, and with the progress of time, the smaller bubbles shrank until they disappeared while the larger bubbles increased in size. This has the effect of increasing the average bubble size with time. Assuming that in metals, atoms also diffuse from smaller grains to larger neighboring ones, then the same equation to describe the effect of time on the average grain size is valid. Equation 9.10 shows this general relation.

$$d^{1/n} - d_o^{1/n} = Kt \tag{9.10}$$

where d_o is the average grain size at the start of annealing, and d is the average grain size after annealing for time, t. k is a temperature-dependent material constant, and n is the grain growth exponent. If the starting grain size is much smaller than the final grain size, then it can be omitted from the equation, leading to Eq. 9.11:

$$d = Kt^n \tag{9.11}$$

If we include the effect of temperature, then we arrive at Eq. 9.11

$$d = k.\mathrm{e}^{-\frac{Q}{2RT}}.t^{n} \tag{9.12}$$

where $K = k.\mathrm{e}^{-\frac{Q}{2RT}}$

T is the temperature in kelvins, k is a constant, and R is the gas constant. Again, we see the Arrhenius equation depicting the temperature dependence. This is expected since grain growth depends on diffusion, and since diffusion depends exponentially on temperature, it follows that grain growth will be dependent exponentially on temperature, too. It is important to note that precipitates or pores present on grain boundaries hinder grain growth.

Now, we come to the grain growth exponent, n. In the bubble experiments, n was found to have a value of 0.5. For metals, n is dependent on a couple of factors. For example, the n value generally increases with an increase in annealing temperature to 0.5 (the value found for the bubbles). Pure metals can have an n value of 0.5, but as impurities are added, n decreases. Incidentally, for materials with grain sizes less than 100 nm (i.e., nanostructured materials), n values lower than 0.1 have been reported, signifying a resistance to grain growth. This was attributed to impurity atoms preferentially segregating at grain boundaries, reducing grain boundary mobility.

Recrystallization aims to replace the highly deformed microstructure with strain-free grains, refine the microstructure, and control the grain size. This is valuable as grain size plays a major role in the mechanical behavior of materials.

As mentioned, when a metal or alloy is mechanically deformed to change its shape, it is said to be worked. This work can be conducted at different temperatures. At low temperatures (in relation to the material's melting point in Kelvin), the material will initially have relatively high yield strength (compared with higher temperatures). Hence, it will generally require significant stress or work to deform it. We say the material has been **cold-worked** and will have generally clean surfaces. Alternatively, at high temperatures, the yield strength of the material will be low as the metal or alloy will generally be softer. Hence, the stress needed to deform the material will also be low. However, metals and alloys can suffer from environmental attacks at high temperatures, such as oxidation. Surface oxide scales can form, and the surfaces will not be as clean as those of cold-worked metals. When a metal is worked at high temperatures, it is said to be **hot-worked**. We see this in hot forging, hot extrusion, and hot rolling. Suppose we, however, decide to avoid the oxide scale formation accompanied by hot working as well as the high stresses encountered during cold working. In that case, we can compromise by working the material at a temperature that is not too high for scale formation and not too low for stresses to be too high. In this case, the material would be said to be undergoing **warm working**. There is no true consensus on which temperatures signify each of these working types, as they can change from material to material. One definition considers that hot working occurs when deformation is carried out above the material's recrystallization temperature.

Example Problem 9.2

Consider a metal plate 10 mm thick, 50 mm wide, and 10 cm long cold rolled to 50%CW. Assuming that the friction between the plate and the steel rollers is high enough to prevent an increase in plate width during rolling. What would be the final thickness and length of the cold-rolled plate? How do you think the cold rolling operation would have affected the strength and ductility of the plate?

Solution

Since the width (W) does not change after cold rolling, the cross-sectional area will be reduced solely because of the reduction in thickness (t). Hence it follows

$$\%\text{CW} = \frac{Wt_o - Wt_f}{Wt_o} = \frac{t_o - t_f}{t_o}$$

t_o is the initial thickness, and t_f is the final thickness after cold working.

$$0.5 = \frac{10 - t_f}{10}$$
$$t_f = 5 \text{ mm}$$

The length can be found by invoking the constant volume assumption. In other words, since we are not adding or subtracting material from the plate during cold rolling, the plate's volume before and after rolling is the same.

$$l_o W t_o = l_f W t_f$$
$$l_f = \frac{l_o t_o}{t_f} = \frac{100 \times 10}{5} = 200 \text{ mm}$$

This means this operation doubled the plate length from 10 to 20 cm. Since the cold working operation would have increased the dislocation density within the plate, the cold-rolled plate would be stronger and less ductile than the original plate.

Example Problem 9.3

You are an engineer in an industrial company. An order arrives to deliver a metal or alloy plate with a tensile strength of 430 MPa or greater. In your warehouse, you find plates of Metal A and Alloy B. Now, the task is to determine whether any or both of these metals/alloys are possible candidates, using Fig. 9.18.

Solution

The tensile strength-%CW plot (Fig. 9.18) shows how tensile strength depends on %CW. A horizontal cutoff line is drawn on the plot strictly at a tensile strength of

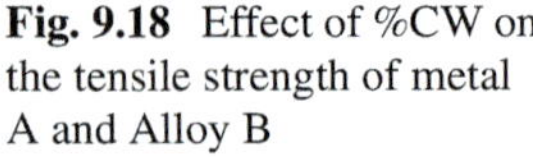

Fig. 9.18 Effect of %CW on the tensile strength of metal A and Alloy B

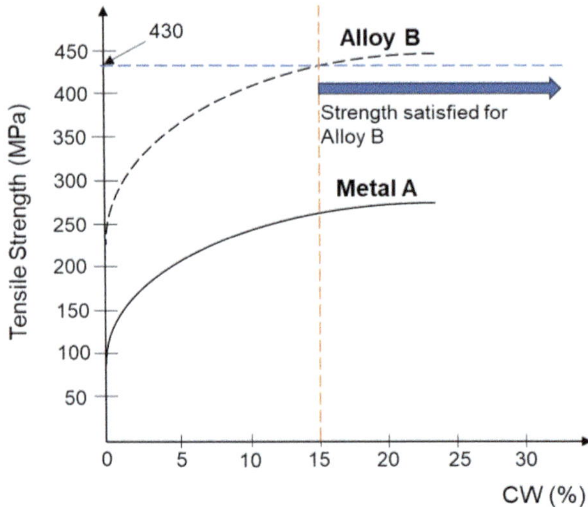

430 MPa. Any part of a curve above the line will satisfy the customer's request, while anything below will not.

The figure shows that only alloy B qualifies with a tensile strength equal to or greater than 430 MPa, but only after it is cold worked for 15% or greater. Metal A does not satisfy the strength requirement even if cold worked. Hence, the answer is to cold work alloy B for ≥ 15% and suggest that to the customer. Note that the customer, this time, did not place any restrictions on other plate properties, for example, ductility. So, the decision in this case is easy. The next problem tackles the situation when more than one property is requested.

Example Problem 9.4

You are an engineer in an industrial company, and an order arrives to deliver a particular metal (metal C) plate with a tensile strength of 250 MPa or more. Still, it should not have a ductility of less than 20%. In your warehouse, you find plates of Metal A that have previously been fully annealed. Now, the task is to determine whether you can deliver the customer's request.

Solution

Here, we must consider the %CW plots for tensile strength and ductility for metal C, as presented in Fig. 9.19. As far as the tensile strength goes, for the metal C plate to have a tensile strength of 250 MPa or greater, it is clear from Fig. 9.19a that it needs to be cold worked 10% or more. As for the ductility, it will be satisfied if we don't cold work the plate more than 15%, i.e., if the ductility would be independently satisfied if we did nothing to the plate (i.e., 0% CW) or worked it but not more than 20%, as seen in Fig. 9.19b. So, the result of this exercise is that we can indeed supply the customer with what he or she wants if we take the annealed plate of metal C and cold roll it anywhere between 10 and 15%, thus satisfying the strength and ductility

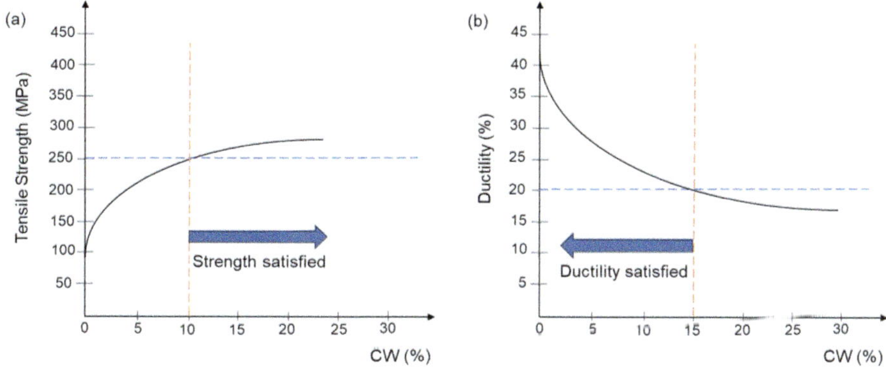

Fig. 9.19 Effect of %CW on **a** tensile strength and **b** ductility of metal C

requirements. The logical thing to do is to take the middle path, cold roll it for 12.5%, and then hand it to the customer and make some money ☺.

Example Problem 9.5

If the original plate of metal C in example problem 9.4 was 10 mm thick, what would be the final thickness of the plate handed to the customer?

Solution

Since we decided to cold work the plate 12.5%, we can use the cold working equation and simply input the %CW value of 12.5% and the original thickness of 10 mm to find the final thickness. We assume that the plate's width does not change during cold working.

$$\%CW = \frac{Wt_o - Wt_f}{Wt_o} \times 100 = \frac{t_o - t_f}{t_o} \times 100$$

$$\%CW = 12.5 = \frac{t_o - t_f}{t_o} \times 100$$

$$0.125 = \frac{t_o - t_f}{t_o}$$

$$0.125 = \frac{10 \times 10^{-3} - t_f}{10 \times 10^{-3}}$$

$$t_f = 8.75 \times 10^{-3} \text{ m}$$

In other words, the plate has been reduced in thickness from 10 mm to 8.75 mm.

9.3   Corrosion of Metals and Alloys

Nations bear significant costs because of corrosion. Despite some valuable benefits of corrosion, e.g., chemical etching, electrochemical machining, and electrolytic polishing (which rely on anodic oxidation), most corrosion impacts are detrimental. At the onset, we differentiate between **dry corrosion** and **wet corrosion**. Dry corrosion is when a material reacts with gases without a liquid, i.e., in a dry gaseous environment (for example, oxygen, hot dry steam, or even dry air). Another word for dry corrosion is **oxidation**. In wet corrosion (sometimes also called **aqueous corrosion**), the material is attacked via water as a medium. Both types of corrosion reactions typically take place at the material surface.

9.3.1   Dry Corrosion (Oxidation) of Metals and Alloys

Except for gold, all metals oxidize. A simple look at the metals found on our planet will reveal that many metals exist as oxides. As mentioned earlier, the stability of the metal oxide played a significant role in our ability to extract and later use pure metal. This is why it was relatively easy to smelt and reduce iron oxide, while it needed a more robust process to free aluminum from its oxide. The word black in Blacksmith refers to iron oxide (FeO) formed at the surface of iron when the iron is heated in air, precisely due to a reaction with oxygen in the gaseous atmosphere (air). Both electrons and a chemical reaction are involved in oxidation. Often, we hear the words oxidation and reduction, but what exactly do they mean? In **oxidation**, a material releases electrons and becomes oxidized, while in reduction, the material receives electrons and is said to be reduced. Counter to popular belief, oxidation is not only about forming an oxide; as long as a material loses electrons, it is said to be oxidized, even if it doesn't form an oxide per se. For example, nickel is oxidized at high temperatures in the presence of sulfurous gases and forms a sulfide scale. Nickel-base superalloy turbine blades oxidize while in service and react with combustion gaseous products such as SO_2 and H_2S.

Let us take iron as an example. A reaction with oxygen produces an oxide layer at the surface of the iron. The oxidation and reduction reactions are as follows.

$$Fe \rightarrow Fe^{++} + 2e^- \quad \text{(iron is oxidized)}$$

$$O + 2e^- \rightarrow O^{--} \quad \text{(oxygen is reduced)}$$

$$2Fe + O_2 \rightarrow 2FeO$$

e^- refers to an electron.

When an oxide layer is first formed, it is referred to as a film (typically of small thickness); thicker layers are called oxide scales. In forming the oxide film/scale, ions of both iron and oxygen are involved in addition to electrons. The buildup of

Table 9.1 Different oxide scale growth laws during oxidation

Law	Relation	Comments
Linear	$x \propto t$	Occurs initially before the oxide film growth hinders further oxidation
Logarithmic	$x \propto \ln(t)$	Occurs typically at low temperatures, e.g., Fe below 200 °C
Parabolic	$x \propto t^{1/2}$	Occurs at intermediate temperatures, e.g., Fe between 200–900 °C
Breakaway	Linear relation composed of repeated parabolic steps	Occurs due to oxide scale cracking or spalling

Source Sir Alan Cottrell, An Introduction to Metallurgy, second edition, The Institute of Materials, (1995)

the oxide film/scale depends on the ability of ions and electrons to diffuse through it.

With reference to the case where an oxide film/scale is formed on the surface of a metal or alloy, the oxide itself can take several forms, which can affect the progress or hinder further scale buildup. For example, the film/scale can be coherent with the metal surface or incoherent; it can be porous or continuous; it can shrink upon formation or even expand; it can adhere firmly to the surface, break up, spall off, or even evaporate. For example, sometimes high-temperature oxidation results in the formation of volatile oxides, such is the case for MoO_3 oxide formation on molybdenum. In this case, oxide formation and its subsequent volatilization expose the surface again and again to further oxidation and metal depletion. Depletion occurs since the oxide already picks up metal in the reaction. The ideal scenario is for the oxide film or scale to form and act as a barrier to further oxidation. The buildup of the oxide scale can occur at different rates, and different laws of growth can be observed depending on several factors. These are listed in Table 9.1 and shown in Fig. 9.20. Typically, a sample's mass change is measured with respect to time at the oxidation temperature. An increase in mass signifies an increase in oxide scale thickness (x). However, metals that form volatile oxides experience a mass loss. In the case of iron, oxygen must first be allowed access to the surface for the oxide scale to start forming. Oxygen molecules then chemosorb to the surface and react with the metal to form oxide nuclei, which grow into a film. These films (thick or thin) can generally be detected with the naked eye. Observing thick films is obvious; for example, during the hot working of iron alloys at high temperatures, the formation of thick oxide scales that break off during hot working is an example. As for thin film scales, at a temperature of 300 °C, steel forms a thin black–blue film that protects the surface. The process is called **bluing**. Not all oxide films are visible; for example, a 20 nm aluminum oxide film on aluminum is transparent. This film is quite strong, though, and effective in preventing further oxidation.

The parabolic law of oxidation, showing the dependence of scale thickness on time, can be expressed as shown in Eq. 9.13:

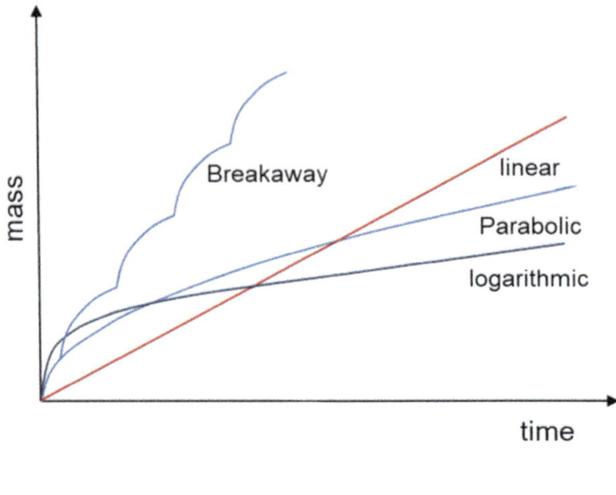

Fig. 9.20 Different oxide growth behavior during oxidation (after: Sir Alan Cottrell, An Introduction to Metallurgy, second edition, The Institute of Materials (1995))

$$x^2 = kt \tag{9.13}$$

where k is a temperature-dependent constant, many metals, typically at intermediate temperatures, observe this relation, including iron, nickel, copper, aluminum, and cobalt. Since diffusion is critical in the oxide scale formation process, one would expect that temperature significantly influences the rate of oxide scale buildup. Therefore, the effect of temperature is incorporated in constant k, where k is given by Eq. 9.14.

$$k = A.e^{\frac{-Q}{RT}} \tag{9.14}$$

A is a constant that includes the diffusion coefficient. It should be noted that the growth of the oxide depends not only on the transfer of electrons but also on ions. Oxygen anions (negatively charged ions) travel in the opposite direction to the metal cations. In some cases, the oxygen anions travel much faster than the metal cations, leading to oxide film/scale growth at the inner metal/oxide interface. Examples of materials that oxidize in this manner are zirconium, uranium, and titanium. Sometimes, the metal ions travel much faster than the anions to form additional oxide layers, but this time at the oxide/air interface. Since the metal anions leave the metal/oxide interface relatively quickly, they leave vacancies at the metal/oxide interface. Materials that behave in this manner include copper, chromium, and iron. This is seen in Fig. 9.21.

Due to the dependence of oxide film/scale growth on electron transfer and ionic diffusion, one would expect electrically insulating oxide films to be more stable and protect from further oxidation. At any given oxidation temperature, one would expect ionic diffusion coefficients in oxide films/scales of higher melting points to be slow compared with other oxides of relatively lower melting points. This would also hinder further oxidation. In this case, the oxide film will be quite protective, which is true for the alumina (Al_2O_3) film that forms on aluminum. It does a great job of passivating aluminum and protecting it from further oxidation.

Fig. 9.21 **a** Anions travel faster than cations leading to oxide buildup at the metal/oxide interface, **b** cations travel faster than the anions leading to oxide buildup at the oxide/air interface

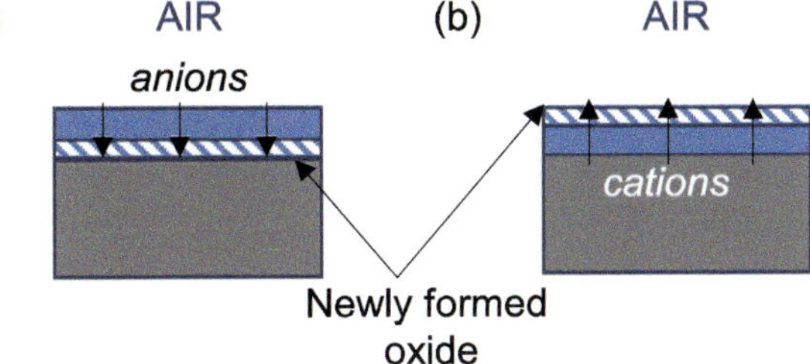

Example Problem 9.6

A silicon wafer is heated to 900 °C in air. If it is known that it takes 2 million hours to produce a 100 μm thick oxide scale on the substrate, how thick would the oxide film be after only 2000 h of exposure at the same temperature? Assume parabolic oxide film growth.

Solution

Since the parabolic law applies

$$x^2 = kt$$

Or

$$x = \sqrt{k}\sqrt{t}$$

$x_1 = 100 \times 10^{-6}$ m when $t_1 = 2 \times 10^6$ h $= 7.2 \times 10^9$ s

what is x_2 when $t_2 = 2000$ h $= 7.2 \times 10^6$ s

$$\frac{x_1}{x_2} = \frac{\sqrt{k}\sqrt{t_1}}{\sqrt{k}\sqrt{t_2}} = \sqrt{\frac{t_1}{t_2}}$$

$$\frac{x_1}{x_2} = \sqrt{\frac{7.2 \times 10^9}{7.2 \times 10^6}} = 31.62$$

Therefore,

$$x_2 = \frac{x_1}{31.62} = \frac{100 \times 10^{-6} \text{ m}}{31.62} = 3.16 \times 10^{-6} \text{ m} = 3.163 \text{ μm}$$

The Pilling–Bedworth Ratio (PBR)

An important property of the oxide film/scale being formed is its volume compared to the metal's. Suppose the formed oxide has a volume considerably different from the metal's. In that case, stress buildup is expected, which can cause specific structural changes to the oxide or even total detachment or cracking. For example, when the oxide volume is considerably greater than the metal's, it tends to spring away from the metal and detach at the metal/oxide interface. In other words, it flakes off. On the other hand, an oxide with a volume much less than that of the metal will compensate for the smaller volume by forming cracks and porosity. In both cases, the oxide scale or film will be ineffective at protecting the metal from further oxidation. An ideal situation is when the oxide has a volume similar to the metal's or even slightly greater; in this case, a protective oxide layer is formed. The Pilling–Bedworth ratio was introduced in 1923 to determine the integrity of oxide layers on metals and, ultimately, their effectiveness in protecting the metal from further oxidation, and is given by Eq. 9.15.

$$\text{PBR} = \frac{\text{Volume of 1 mol of } M_xO_y}{x \times \text{Volume of 1 mole of metal}} = \frac{M_o \times \rho_m}{x \times M_m \times \rho_o} \qquad (9.15)$$

M_o and M_o are the atomic or molecular masses of the oxide and metal, respectively. ρ_m and ρ_o are the densities of the oxide and metal, respectively. x is the number of mols of the metal in the oxide formula.

The PBR can reveal important insights into the type of oxide formed on the metal and whether it is protective. A value less than 1 means the oxide will be porous and contain cracks, hence not providing good protection. A value greater than 2 means the oxide will spall off and not provide any protection. Ideally, a PBR of 1–2 provides a dense protective oxide layer and protects the metal from further oxidation.

Example Problem 9.7

Aluminum is known to have a highly adherent alumina (Al_2O_3) film on its surface that protects it from oxidation. What is the Pilling–Bedworth ratio for aluminum/alumina? The density of alumina (ρ_o) and aluminum (ρ_m) is 3.987 g/cm^3 and 2.7 g/cm^3, respectively. The molecular mass of alumina (M_o) is 101.96 g/mol, and the atomic mass of aluminum (M_m) is 26.982 g/mol.

Solution

For Al_2O_3, x is 2.

$$\text{PBR} = \frac{M_o \times \rho_m}{x \times M_m \times \rho_o} = \frac{101.96 \times 2.7}{2 \times 26.982 \times 3.987} = 1.28$$

This value of 1.28 is between 1 and 2, and hence we expect a dense (non-porous and crack-free) adherent protective Al_2O_3 oxide film on aluminum.

Example Problem 9.8

What is the Pilling–Bedworth ratio for Mg/MgO? The density of MgO (ρ_o) and Mg (ρ_m) is 3.58 g/cm^3 and 1.74 g/cm^3, respectively. The molecular mass of MgO (M_o) is 40.304 g/mol, and the atomic mass of Mg (M_m) is 24.305 g/mol.

Solution

For MgO, x is 1.

$$\text{PBR} = \frac{M_o \times \rho_m}{x \times M_m \times \rho_o} = \frac{40.304 \times 1.74}{1 \times 24.305 \times 3.58} = 0.81$$

The value of 0.81 is less than 1; hence, we expect a porous cracked oxide film on magnesium that will not protect the metal from oxidation.

Alex, I promise you, you will not truly understand until you do it yourself. So now calculate the PBR for Ni/NiO and Fe/Fe$_2$O$_3$. Go through the same procedure and determine whether they make protective or non-protective films. I will give you the answers to make things easy for you, but you need to find them out by following the steps to gain confidence in problem-solving. (The PBR ratio Ni/ NiO is 1.65, and Fe/Fe$_2$O$_3$ is 2.14.)

9.3.2 Wet (or Aqueous) Corrosion

We start here with an interesting observation. We have all seen the reddish-brown rust formation on iron when left outside unprotected. Interestingly, iron would never rust if it were in contact with pure "oxygen-free" water. What causes the rust of iron is its presence in a moist atmosphere that contains both water and oxygen. So **rust**, ferric hydroxide (Fe(OH)$_3$), is formed via the reaction in Eq. 9.16.

$$4Fe + 6H_2O + 3O_2 \rightarrow 4Fe(OH)_3 \text{ (rust)} \tag{9.16}$$

Before we delve into wet corrosion, one must understand that, like in oxidation, in wet corrosion, a metal that loses electrons will corrode, i.e., it loses matter and generates a rust deposit. It then becomes evident that the ability of metals to hold on to their valence electrons is essential. It turns out that some metals can do this better than others. For example, noble metals such as gold, silver, and platinum are better at doing this than other metals. Metals are hence classified according to something called the **electrode potential**. The electrode potential of metals is measured relative to hydrogen, which is then given the value of zero electrode potential. This is shown in Table 9.2, also called the electrochemical series. The tendency of a metal to corrode increases going down the table. In this case, calcium has the highest tendency to corrode compared with the other listed metals. Let's consider two electrochemical cells to exemplify how these electrode potentials are measured, as shown in Fig. 9.22. The setups presented in the figure use hydrogen as the reference electrode. This is practically done by using a platinum tube and introducing hydrogen,

Table 9.2 Examples of electrode potentials (E) for some metals (25 °C)

Metal Ion	Electrode Potential (Volts)
Gold (Au^{+++})	+ 1.50 (**Cathodic**)
Platinum (Pt^{++++})	+ 0.86
Silver (Ag^{+})	+ 0.80
Copper (Cu^{+})	+ 0.52
Copper (Cu^{++})	+ 0.34
Hydrogen (H^{+})	0 (**Reference**)
Iron (Fe^{+++})	− 0.05
Lead (Pb^{++})	− 0.13
Tin (Sn^{++})	− 0.14
Nickel (Ni^{++})	− 0.25
Cadmium (Cd^{++})	− 0.40
Iron (Fe^{++})	− 0.44
Chromium (Cr^{+++})	− 0.74
Zinc (Zn^{++})	− 0.76
Aluminum (Al^{+++})	− 0.166
Magnesium (Mg^{++})	− 2.37
Calcium (Ca^{++})	− 2.87 (**Anodic**)

Source R. A. Higgins, Properties of Engineering Materials, Hodder and Stoughton (1986)

which dissolves interstitially in the platinum tube. A potentiometer measures the voltage between it and the other electrode. In Fig. 9.22a, we see copper as the other electrode; according to the electrochemical series (Table 9.2), copper has a higher electrode potential than hydrogen, making it cathodic relative to hydrogen. This means the hydrogen electrode will be the anode. This is reversed when we test iron instead of copper. Since iron is lower than hydrogen in Table 9.2, iron becomes the anode while hydrogen is the cathode. This means iron should be the one to corrode once a closed circuit is established between the electrodes (i.e., an electrical contact between the metals, including an electrolyte as part of the circuit). The electrolyte must be present. An electrolyte is simply a fluid that contains free-moving ions, which allows the passage of an electric current. Rust occurs in real life when water and oxygen are present. So, is water then a suitable electrolyte? It turns out that pure water has a very small concentration of dissociated H^{+} and OH^{-} ions, which is not enough to cause any significant corrosion. However, impurities in water, such as carbon dioxide or sulfur dioxide, will significantly increase the concentration of dissociated ions in water, making corrosion a genuine concern. Such impurities are common in industrial settings. So, then, what happens if an electric circuit is indeed established? Look at Fig. 9.23, in which such a setup is established. Here, we see the electrodes connected electrically at the top, and the circuit is completed through the electrolyte in the vessel. The difference in the electrode potential between copper and

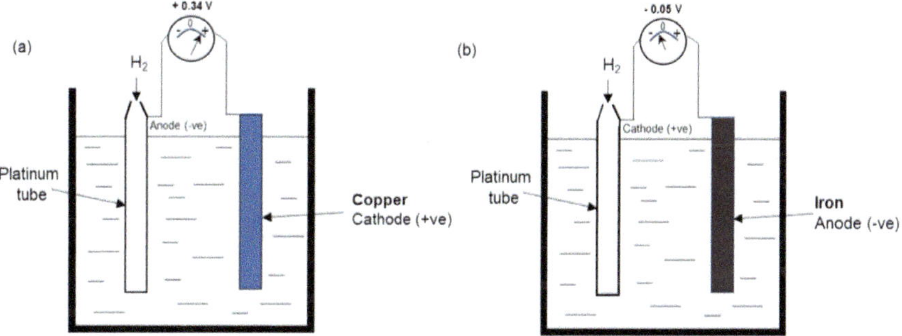

Fig. 9.22 Measurement of the electrode potential (E) of **a** copper and **b** iron (after R.A. Higgins, Properties of Engineering Materials, Hodder and Stoughton (1986)

iron results in the release of electrons from the iron electrode (anode) and the generation of iron ions, which enter the electrolyte. The released electrons then travel to the copper cathode through the wire connection, thus generating a negative charge at the cathode. The ionized positively charged hydrogen ions in the electrolyte are then attracted to the negative charge. The negatively charged OH$^-$ will also be attracted to the hydrogen ions, as seen in the figure. The electrons in the cathode combine with the H$^+$ to produce hydrogen (H) and, eventually, hydrogen gas, which is released in the form of bubbles. The reactions are as follows:

$$2H^+ + 2e^- \rightarrow 2H \rightarrow H_2$$

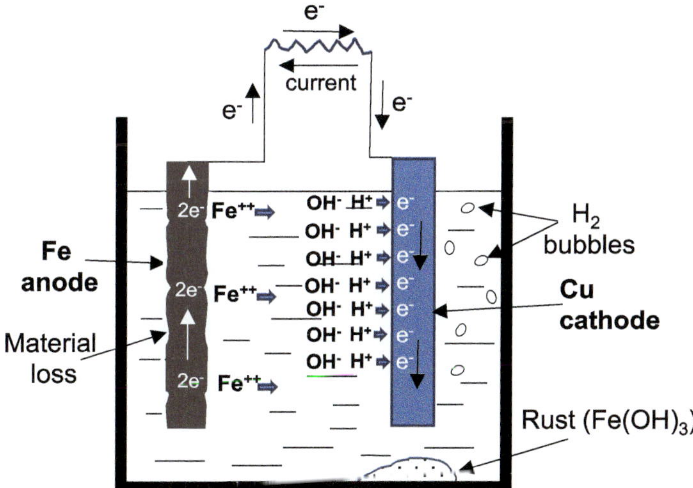

Fig. 9.23 Fe–Cu cell Measurement of the electrode potential of **a** copper and **b** iron (after R. A. Higgins, Properties of Engineering Materials, Hodder and Stoughton (1986))

The Fe^{++} ions from the anode travel to the cathode and react with OH^- to eventually form Fe $(OH)_3$ (rust) deposited at the container's bottom.

It is important to note that no corrosion would occur without an electrolyte containing ions to enable the formation and transfer of ions from the anode.

This difference between the electrode potential of different metals can be the basis of **galvanic corrosion** processes when two dissimilar metals are in contact (electrically connected) in an oxygen-containing aqueous medium. We hence see corrosion situations occurring in practice, for example, two pipes of different metals connected while passing water through them. In that case, the metal pipe with the lower electrode potential will assume the anode designation, and the other metal pipe will be the cathode. Here, ions of the anodic metal will be released into the water, and over time, the metal will be depleted. The released ions will travel, combine with ionized OH^- ions, and form rust $(Fe(OH)_3)$ at the cathode.

Corrosion is not only in pipes but has been seen in various situations. Consider manganese bronze propellers connected to a ship's steel hull; both are exposed to seawater (an electrolyte). Here, the steel becomes the anode, and corrosion is expected there. To remedy this situation, zinc plates are attached to the hull next to the propeller; these plates have a lower electrode potential than the hull and the propeller and hence will act as the anode and sacrificially corrode instead of the hull.

Corrosion can also occur on the microscopic level if the microstructure of a material is not homogeneous and contains different phases of different electrode potentials. An example would be steel with a pearlitic microstructure. Here, pearlite, which consists of cementite (Fe_3C) and ferrite (Fe), in the presence of water and oxygen, will establish a microscopic electrochemical cell between the cementite and ferrite, where the ferrite regions become the anode and cementite the cathode. As can be seen from Fig. 9.24, the ferrite preferentially corrodes through the release of Fe^{+++} ions. This mechanism can be extended to precipitates at grain boundaries, which will affect the local microscopic electrode potentials.

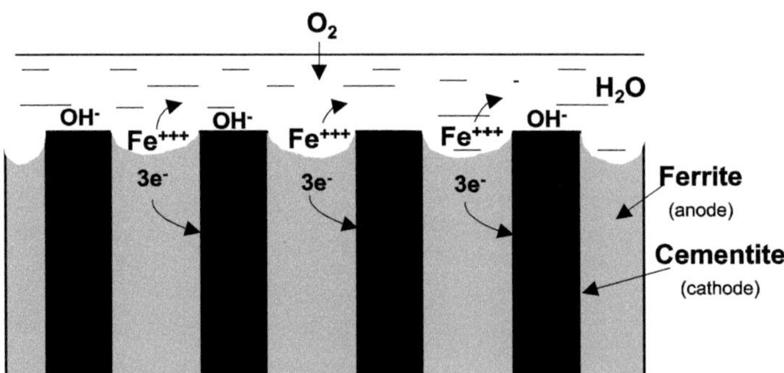

Fig. 9.24 Microscopic corrosion taking place in pearlite, since ferrite has a lower electrode potential than cementite, then corrosion takes place in the ferrite regions of the microstructure (after R. A. Higgins, Properties of Engineering Materials, Hodder and Stoughton (1986))

So far, the examples have focused on the contact of dissimilar materials with different electrode potentials in the presence of an electrolyte or even dissimilar microscopic phases within one material in the presence of an electrolyte. However, it is indeed possible for materials of the same type and composition to experience corrosion, even if their microstructure consists of a single homogeneous phase (i.e., not two phases as discussed before). Both the electrolyte composition and the condition of the material can play a significant role in corrosion. For example, if one part of a material has been cold rolled to a higher level than another, the electrode potential difference between the two parts will be affected. Nails are made by flattening one side of a thick wire using a hammer. This flattening operation involves considerable plastic deformation, increasing the dislocation density in that area and, hence, the stored energy there. This sets up a different electrode potential in the head compared to the rest of the nail, making the head the anode and a region further down the nail the cathode. If left outdoors and exposed to a wet environment, the head will rust more. We often hear **pitting** or **crevice corrosion**, but what do they mean?

9.3.3 Crevice Corrosion

Crevice corrosion is one type of corrosion that can occur in components of the same material. Consider two steel plates attached at one location. Figure 9.25 shows a schematic of such a situation, where initially, water seeps in between the plates via capillary force and enters a space between them. Although the outer water is exposed to the atmosphere, allowing oxygen to dissolve in the water, the water seeping in cannot take the oxygen with it. This leads to a difference in concentration between the water outside (to the left of the image) and that inside the interface between the plate. Under this compositional difference (despite the two steel plates having the same composition and process history), the region of the steel plate next to the water that is open to the atmosphere becomes cathodic, while a close region in between the plates becomes anodic where electrons are removed, iron ions are formed and transport through the capillary channel to react with the OH^- and form the rust as shown in the figure.

In the same light, imagine a droplet on the surface of a metal plate in air (Fig. 9.25 b). Near the outer boundaries of the droplet, the oxygen concentration is expected to be higher than in the mid-region of the droplet. This difference in concentration can also set up a corrosion situation.

9.3.4 Pitting Corrosion

This type of corrosion is more dangerous than crevice corrosion as it can often go unnoticed, leading to the failure of components, for example, the leakage of an underground steel pipe. Fatigue cracks can also initiate from pits. The addition of

(a)

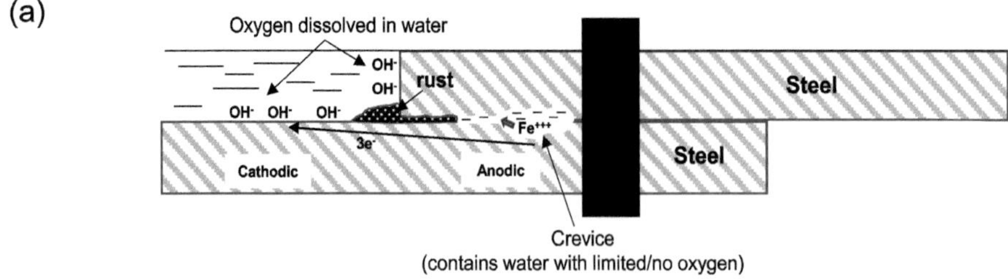

(b)

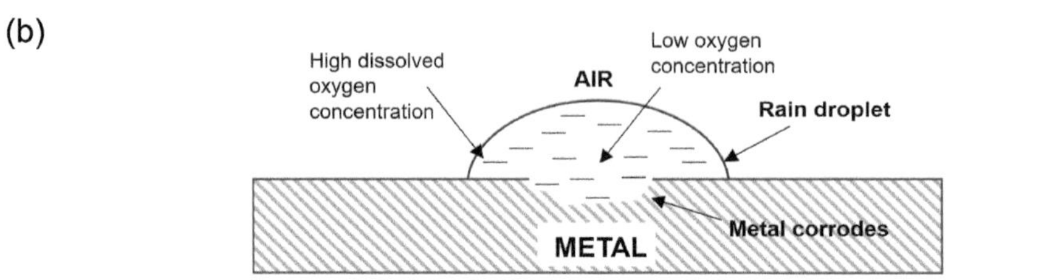

Fig. 9.25 a Crevice corrosion between two steel plates, **b** corrosion occurring due to oxygen concentration difference in rain drop (modified after R. A. Higgins, Properties of Engineering Materials, Hodder and Stoughton (1986))

alloying elements can limit pitting corrosion, such as chromium (~ 18 wt%) and nickel in stainless steel. Despite the advantage of adding chromium, one must still be careful when processing these materials. For example, in Fig. 9.26, pitting still occurs in 304 stainless steel at the welding junction and surrounds it in the heat-affected zone. Due to the high temperatures experienced in these locations, chromium precipitates are formed, further decreasing the concentration of chromium in the alloy, making it lower than 18 wt% (the recommended concentration to achieve corrosion resistance all over).

9.3.5 *Protecting Against Corrosion*

From the above examples, one can see that metals and alloys must be protected against corrosion. We know that materials with homogenous, single-phase microstructures are likely to cause fewer issues with corrosion. We also know that cold working can cause a deterioration of the corrosion properties of metals and alloys. The same can be said for multiphase materials and the oxygen concentration gradient within the electrolyte. There are several methods employed to counter corrosion. One way is to add **alloying elements** that tend to form a protective oxide layer, for example, chromium, which produces chromium oxide, or aluminum, which produces an aluminum oxide coat. Another is to coat the material with a protective layer, either an oxide or even a different metal, using established coating methods. Here, we must be careful since

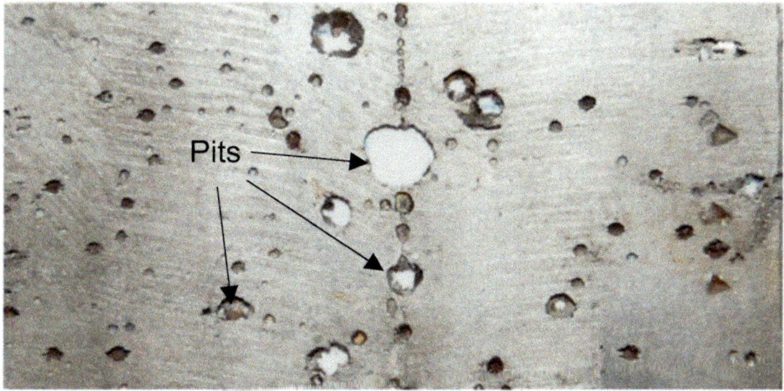

Fig. 9.26 Pitting corrosion at the welding and heat affected zones of cold rolled AISI 304 stainless steel (modified image reproduced with permission from MDPI under Creative Commons Attribution (CC BY) license http://creativecommons.org/licenses/by/4.0/). Francisco-Javier Cárcel-Carrasco, Manuel Pascual-Guillamón, Lorenzo Solano García, Fidel Salas Vicente, and Miguel-Angel Pérez-Puig, Pitting Corrosion in AISI 304 Rolled Stainless Steel Welding at Different Deformation Levels, Appl. Sci. 2019, 9, 3265)

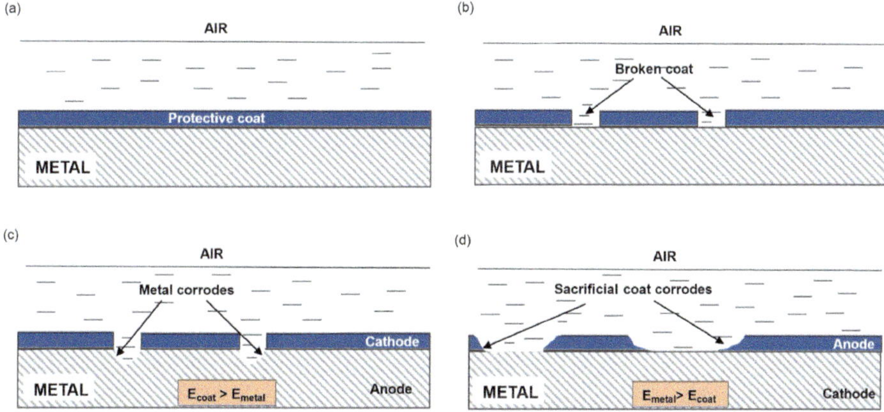

Fig. 9.27 **a** Coat covers metal entirely and provides protection **b** coat is damaged exposing the metal at two locations, **c** if electrode potential of coat is higher than that of metal, then metal becomes anode and the coat becomes the cathode, making the metal corrode, **d** when electrode potential of metal is higher than that of the coat, the coat corrodes and acts as a sacrificial coat that protects the metal (modified and expanded on after R. A. Higgins, Properties of Engineering Materials, Hodder and Stoughton (1986))

exposing the metal underneath will be highly problematic if any part of the coat is damaged. For example, when aluminum is used indoors, an oxide coat of ~ 5 nm thickness is sufficient to protect it; however, outdoors, aluminum products such as window frames need a much thicker coat to enhance their protection against the environment and possible damage. Here, processes such as anodization, which artificially increase the oxide thickness of aluminum to ~ 25 μm, are used. Still, if we are not careful and the coat is breached, we may end up setting up an electrochemical cell in the presence of an electrolyte (water and air). The protective coat can be a ceramic or a metal. Here, we must be careful and consider where the coat and the metal to be protected lie in the electrochemical series. Figure 9.27a shows a metal protected with a coat where the coat is adherent and completely covers the metal, which would hence provide good protection. Figure 9.27b shows the situation when the coat is somehow damaged, thus exposing the metal beneath it to the electrolyte. Here, two situations can arise: if the coat has an electrode potential (E_{coat}) greater than that of the metal (E_{metal}), then the metal will assume the role of the anode and, hence, start to lose ions and progressively corrode at the locations where the metal is exposed. This is seen in Fig. 9.27c. On the other hand, if $E_{metal} > E_{coat}$, then the coat will assume the anode role and the metal the cathode; hence, the coat will corrode this time. This could provide additional protection to the metal; if the coat is damaged, the metal will still be protected, and only the coat will be corroded sacrificially (Fig. 9.27d). Although this may prolong the life of the metal, ultimately, we will still have a corrosion problem down the road especially if the metal or alloy has a multiphase microstructure or for reasons explained before.

9.3.6 *Stress Corrosion or Stress Corrosion Cracking*

So far, we have been considering corrosion processes in the absence of any applied stress. However, in service, components are sometimes subjected to stress in a corrosive environment. Under such conditions, cracks can initiate and propagate due to a combined stress corrosion action, even when the mild corrosion environment would not cause issues. Stress in *stress corrosion* (also called *stress corrosion cracking (SCC)*) is essential; it can be an externally applied stress or an internal residual stress from deformation, etc. Cracks can propagate in a transgranular or intergranular manner. The stress at the tip of a crack is magnified to levels where plastic deformation can occur, resulting in the formation of a plastic zone. In the presence of a corrosive environment (even a mild one), the deformed metal at the crack's tip becomes anodic (Fig. 9.28), and hence, metal dissolution occurs at the tip, causing the crack propagation at a certain speed. For example, for austenitic stainless steel in a hot $MgCl_2$ solution, the crack can propagate at 1 mm per hour. Several metals and alloys suffer from stress corrosion cracking, including α-brass in ammonia environments, high-strength aluminum alloys in chloride solutions, and mild steel in caustic soda. Stress corrosion cracking can result in a brittle fracture in an otherwise ductile metal or alloy. Strategies to combat SCC include annealing to remove internal stresses and reduce external stresses.

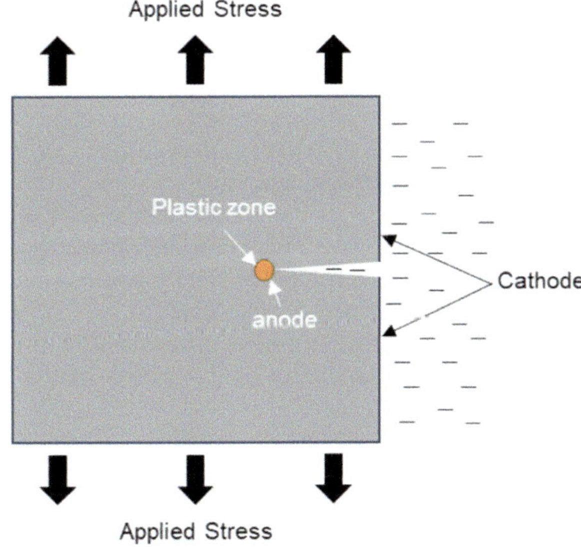

Fig. 9.28 Stress corrosion cracking metal dissolution occurs at the tip of the crack keeping the crack tip sharp (multiple sources)

9.3.7 Corrosion Fatigue

This occurs when the component is exposed to a corrosive environment (e.g., seawater or sea spray) while experiencing fatigue. The primary effect of this type of condition is a decrease in the component's fatigue strength. Figure 9.29 is an S–N curve that schematically shows this effect. As can be seen, the fatigue limit disappears under a corrosive environment, and so does the fatigue strength in general. The curve shifts to the left.

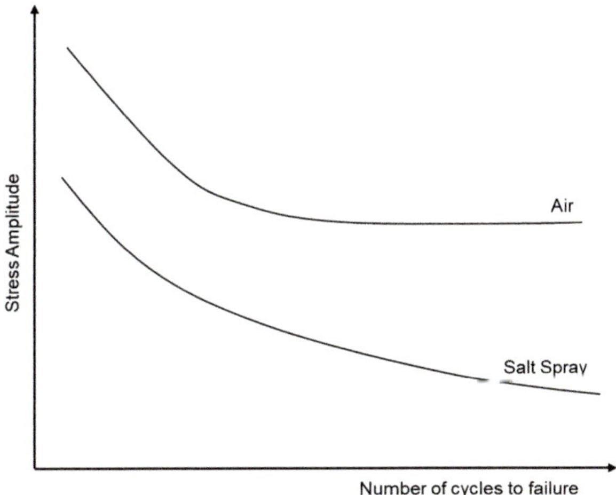

Fig. 9.29 S–N curve showing the effect of a corrosive salt spray environment on fatigue behavior. The fatigue strength is reduced, and fatigue limit eliminated. Redrawn and modified after A. Cottrell, An Introduction to Metallurgy, 2nd edition, Institute of Materials (1995)

Problems

9.1. Discuss four ways you could increase the strength of a metal.

9.2. Your company is considering alloying copper with 3 wt% solute atoms, either of zinc or nickel. Which of the two elements is expected to result in a higher yield strength? And why?

9.3. A copper alloy containing only 3 wt% of element X has a yield strength of 150 MPa. Assuming element X has complete solid solubility in copper and forms a substitutional solid solution, what would be the yield strength if 10 wt% of element X were added?

9.4. Describe the Orowan looping mechanism.

9.5. Under what conditions do dislocations prefer to cut precipitates rather than loop around them?

9.6. Grain size reduction is a strengthening method. What should be the grain size of aluminum to achieve a strength of 300 MPa? Note that the yield strength of very coarse aluminum is 100 MPa, and $k = 0.04 \text{ MPa} \cdot \text{m}^{1/2}$.

9.7. Why does cold working increase the tensile strength of a metal or alloy?

9.8. If cold rolling was carried out on a metal, what do you expect will happen to the yield strength and the ductility of the metal, and why?

9.9. What is the driving force for recrystallization? And what would happen to the strength and ductility of the cold-worked metal or alloy after recrystallization?

9.10. Schematically plot the effect of recrystallization time on the % recrystallized microstructure.

9.11. What is the recrystallization temperature? And how is it affected by the initial grain size of the metal to be deformed and the degree of cold work it has undergone?

9.12. A pure metal and an alloy based on that metal are both cold-worked to 20% CW and then recrystallized. Which one would have the higher recrystallization temperature?

9.13. With reference to Fig. 9.19, a customer requires a sample of metal C with a minimum tensile strength of 250 MPa and a minimum ductility of 30%. Is this possible? If so, how would you accomplish this, assuming you only had an annealed sample of metal C?

9.14. A 10 mm diameter copper specimen is cold worked by 40%.

   (a) What would be its final diameter?

   (b) What would happen to the strength and ductility of the sample after cold working?

   (c) If you wanted to maximize the ductility of the cold-worked copper specimen, what other procedure would you do?

   (d) If a brass alloy (an alloy of copper and zinc) was also cold worked by 40%. With everything else being the same, which cold-worked material (copper or brass) would be stronger and why?

9.15. Corrosion is a serious problem that is faced by all nations. What is the difference between wet corrosion and dry corrosion?

9.16. Consider a hypothetical metal A heated in air to 175 °C. If it takes 3 h to produce a 200 μm oxide scale on its surface, how long will it take to produce a 100 μm film? Assume a logarithmic growth of the film.

9.17. What is the Pilling–Bedworth ratio? Calculate it for Fe/FeO.

9.18. If copper is in contact with tin in the presence of an electrolyte. Which one will corrode and why?

9.19. Discuss two methods used to protect metals against corrosion.

9.20. What is stress corrosion?

9.21. What is corrosion fatigue?

Chapter 10
Failure of Metals and Alloys

10.1 Introduction

Material failure is an extremely important topic. Why occasionally do helicopters fall from the sky, bridges collapse, or large structures fail? Understanding such failure mechanisms is critical for designing materials that avoid such failures. This chapter deals with material failure and discusses several failure mechanisms, including fracture, creep, and fatigue.

10.2 Introduction to Material Failure

As mentioned previously, we are in an incredibly unique age in the history of mankind, when hundreds of thousands of materials are available to us to design and build new structures and products. These different materials provide us with different properties and attributes, including electrical and thermal conductivity, thermal stability, magnetic properties, optical properties, and even recently, shape memory behavior, where a component can be deformed and upon heating revert to its initial shape. The mechanical behavior of materials is something we discussed in a previous chapter when we defined yield and tensile strengths, modulus of elasticity, and ductility. In these situations, all we needed to worry about when placing a component in service is that it does not exceed its yield strength, for example. Otherwise, failure could occur if the component changes shape during service. In this chapter, we consider other types of failures that we need to protect the material against. For example, we have not considered what happens if a crack or defect is present in a material, which is a very likely possibility, especially for large structures such as bridges and ships. How would this crack's presence affect the material's mechanical behavior? What about if the load applied to the material was cyclic, an on–off type repeated cyclic load, for example? In this case, a material can fail even

K. Morsi, *An Engaging Approach to the Science and Engineering of Materials*,
https://doi.org/10.1007/978-3-032-06231-4_10

if the stress applied is relatively small (within the elastic region of the stress–strain curve).

In many situations, a component may have to carry a certain magnitude of load for extended periods, which, at face value, sounds like a doable proposition, especially if the load or stress only takes that material into its elastic region. However, what has been found is that if this is carried out at elevated temperatures, the material will start changing shape over time, eventually leading to failure. In other words, there is much more than yield strength when considering material failure. A prolonged load on a material at elevated temperatures can lead to it changing shape and dimensions. This does not have to be a bad thing; wouldn't we want to use the same conditions that cause the material to change shape, to manufacture it into a useful component? Hence, we need to ask ourselves what our objective is. Material survival during service? Or a material that is amenable to manufacturing? Earlier, we discussed cracks and whether they are good or bad for a component. At first glance, everyone would say, "Bad, of course". But do they know that by controlling the growth of cracks, they can warn us before a boiler explodes and, in doing so, save lives? We mentioned earlier bulletproof materials and how cracks absorb the energy of an impinging projectile, thus protecting people. It is fundamentally cracks that allow the manufacturing of most engineering ceramics. Most ceramic components start as powders. Such, powders likely would have been produced through mechanical comminution, in which ceramics were mechanically broken down by cracking larger particles and reducing them to finer powders, which are later shaped and sintered into final products. Mechanical comminution relies fundamentally on generating and propagating cracks in brittle materials, producing fine ceramic powders. So what could cause failure during service may be utilized to manufacture engineering materials.

We now consider failure under a sustained load at elevated temperatures, which is referred to as **Creep**.

10.3   Creep

I'm sure there was at least one time when you had to change an incandescent light bulb. Have you ever wondered why it no longer works? The filament in the light bulb is made of tungsten (W), electricity passes through it, and Joule heating heats the filament to a temperature in the region of 2000 °C (2273 K); and as long as this continues to happen, everyone is happy. The **homologous temperature** (T_H) can be defined as the material's service temperature in Kelvins (T_s) divided by the melting temperature (T_m) of the material, also in Kelvins, as shown in Eq. 10.1.

$$T_H = \frac{T_s}{T_m} \tag{10.1}$$

Since the melting point of tungsten is 3422 °C, i.e., 3695 K, during its operation, the filament experiences a homologous temperature of ~ 0.6, which means the atoms have considerable thermal energy. This thermal energy allows the process of atomic diffusion to take place. Over time, atoms diffuse within the filament, and under the filament's own weight, the filament sags to the point where it fractures, breaking the electrical circuit, at which point it is time to change the light bulb. Of course, such a process takes time since diffusion takes time. This process is called **creep**. We see the influence of creep all around us. Have you ever seen bookshelves bow over time due to them carrying heavy books? What about a plastic paper clip that widens over time, becoming ineffective, and loosens its grip on paper? Creep is also a central failure mechanism in engineering materials used for pressure piping, gas turbines, steam turbines, bolts. It should be noted that creep affects metals, ceramics, and polymers.

When creep is not operating, once a load is applied and a resulting strain is produced, that strain will remain at the same value indefinitely. However, if the same material is heated to a temperature where creep is operative, the strain will continue to increase with time until the material fails. We need to consider creep deformation at or above a certain homologous temperature, depending on the class of materials.

For metals: $T_H > 0.3$–0.4
For ceramics: $T_H > 0.4$–0.5.

Since creep inherently depends on atomic diffusion (and diffusion is exponentially dependent on temperature), the closer the material's operating temperature is to the material's melting point (i.e., as T_H increases), the greater the level of deformation caused by creep. Interestingly, for polymers, it is not the melting point that is used as a limit, but rather the glass transition temperature, as will be explained in a later chapter.

Like with any material behavior, to understand more about it, we need to conduct laboratory tests; ASTM E139-70 describes in detail the specifics of such tests. We place a tensile testing sample in a furnace that maintains a constant temperature, apply a constant uniaxial load, and observe what happens to the strain (or extension) over time. The tests can take months and even years to complete. For example, some tests have been reported to take 10 years to complete. The output of such a test will be a plot of strain (creep strain) versus time, called a **creep curve**. One example of a creep curve is shown schematically in Fig. 10.1. With reference to this creep curve, the specimen is typically heated to the test temperature. The constant load is applied once the temperature has been homogenized within the sample. As soon as this happens, the material responds with an instantaneous increase in strain, called the **instantaneous strain**. This strain is mainly elastic (as anelastic strain can also be present), and since it is instantaneous, it is not considered a creep strain (i.e., time-dependent strain). However, it must still be considered when calculating the total strain in any creep situation. After this instantaneous strain, the material elongates (and strains) with time. Initially, the strain increases at a relatively high rate, and over time, the strain *rate* continuously decreases until it becomes approximately constant with respect to time. This signifies the end of the first creep stage, called

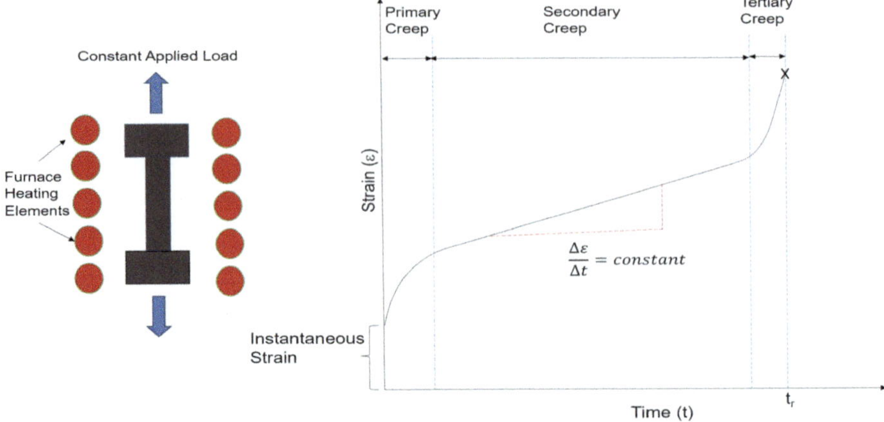

Fig. 10.1 Schematics of creep testing setup and resulting creep curve

primary creep. In primary creep, the decreasing strain rate can be explained by increasing strain hardening effects that counter the material recovery until, at the end of the primary creep stage, recovery and strain hardening balance out, giving rise to a strain rate that is constant with respect to time. This balance represents the condition during the long **secondary creep stage (also called steady-state creep)**. At this stage, the strain rate is constant, and engineers use it to design against creep. After secondary creep, the material enters the **tertiary creep** stage, represented by a strain rate that increases with time until rupture. Here, voids start to emerge, for example, at grain boundaries, and they grow until they link up and form a crack, at which point failure happens. In other words, the minimum creep rate throughout this test occurs in the secondary or steady-state creep stage. Depending on the application, this minimum creep rate can be used in design. For jet engine alloys, the stress at which the minimum strain rate of 2.8×10^{-10} s^{-1} is typically used, while for steam engines, the stress resulting in a creep rate of 2.8×10^{-11} s^{-1} is used. The time it takes for the sample to rupture is called the **rupture life**. The actual strain at which this happens is not as high as we may initially think; we are talking about a strain that could be less than 1%.

Professor, shouldn't we keep the stress and not the load constant during the creep test?

I think I know what you are getting at Alex. You are probably thinking that if we keep the load constant, and as the material creeps, the cross-sectional area of the sample should get smaller. Hence, for the same "so-called" constant load, the material will be effectively experiencing a continuously increasing stress (i.e., non-constant stress). Here is the answer, my friend; it turns out that a creep test under constant load is needed for all practical engineering purposes. Don't you think this would be the reality during service? Also, sometimes, when people refer to stress in these constant load tests, they typically refer to the initial stress before the cross section is reduced. Some people can conduct constant stress tests; such data will be more relevant to unravel the science behind creep, mechanisms, and so forth. Did you know that a constant stress test for the same material and conditions will result in the tertiary creep starting after much longer times than in a constant load test? This is straightforward since, for constant stress, increasing stress is not experienced, which would have otherwise shortened the material life.

The curve in Fig. 10.1 shows three distinct stages: *primary creep*, *secondary creep*, and *tertiary creep*. However, not all these stages are always distinct and clearly defined as presented. For example, at low stresses or low temperatures, only two stages are seen: primary and secondary creep. For example, only the primary stage is seen at room temperature for lead. So, the curve presented with the three distinct stages occurs at specific combinations of stresses and temperatures.

What Is the Effect of Increasing Stress or Temperature?

It turns out that increasing stress (or load) or temperature will have a very similar effect. What will happen to the instantaneous strain if we increase the load? It will increase, of course! Isn't this the natural relation between stress and strain? What about if we instead increase the temperature? Recall what happens to Young's modulus when we increase the temperature; it decreases, and the material becomes less stiff. This means that for the same load, if we increase temperature, the elastic strain will be larger (the material will stretch more), and hence, there will be a larger instantaneous strain on the curve. As will be explained later, stress also helps to drive diffusion, so it's not surprising that an increase in stress or temperature increases diffusion and hence results not only in the increase of the secondary (steady state) creep rate but also a reduction in the rupture life. In other words, the creep situation worsens with increased stress (load) or temperature. Figure 10.2 shows the effects of temperature and applied stress (or load) on the creep curve, as discussed.

As mentioned, the steady-state creep rate is constant over a long period, with the understanding that anything that increases this creep rate will enhance creep and shorten the material's life in service. We mentioned that increasing temperature and

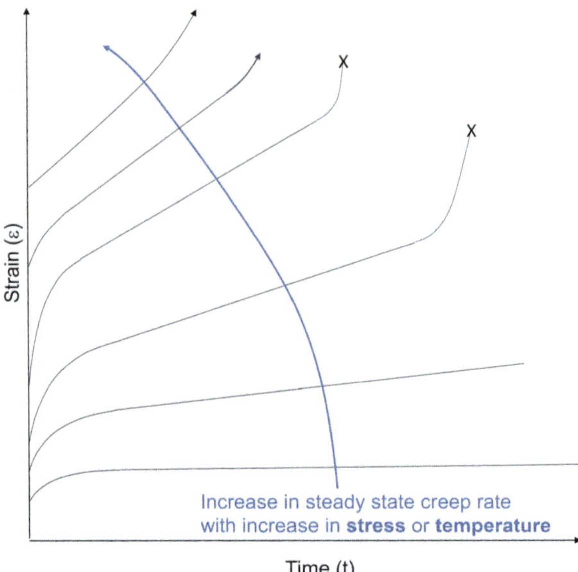

Fig. 10.2 Schematics of creep curves showing the effect of stress or temperature on the steady state creep rate

stress can increase the steady-state creep rate. To determine the relationship between the steady state creep rate and stress we can keep the temperature constant and vary stress. Equation 10.2 depicts the resulting relation.

$$\dot{\varepsilon}_{ss} = A.\sigma^{n} \tag{10.2}$$

$\dot{\varepsilon}_{ss}$ is the steady-state creep rate (i.e., the slope of the secondary creep stage in the creep curve).

A is a constant, σ is the stress, and n is the creep exponent.

If we take the log of both sides, we obtain Eq. 10.3.

$$\log(\dot{\varepsilon}_{ss}) = \log(A) + n\log(\sigma) \tag{10.3}$$

This is an equation of a straight line where log (A) is the intercept and n is the slope.

Figure 10.3 shows a schematic of a typical output of log $(\dot{\varepsilon}_{ss})$ versus log(σ). There are two slopes, one at low-stress values, where n equals 1, and at higher stress values, $n = 3$–8. The different slopes represent different diffusion mechanisms responsible for creep deformation operating at high-stress values, and the other operating at low-stress values.

A similar experiment can be conducted; this time, we keep the stress constant and only vary the temperature to determine the effect of temperature on the steady-state creep rate. Equation 10.4 shows the resulting relation.

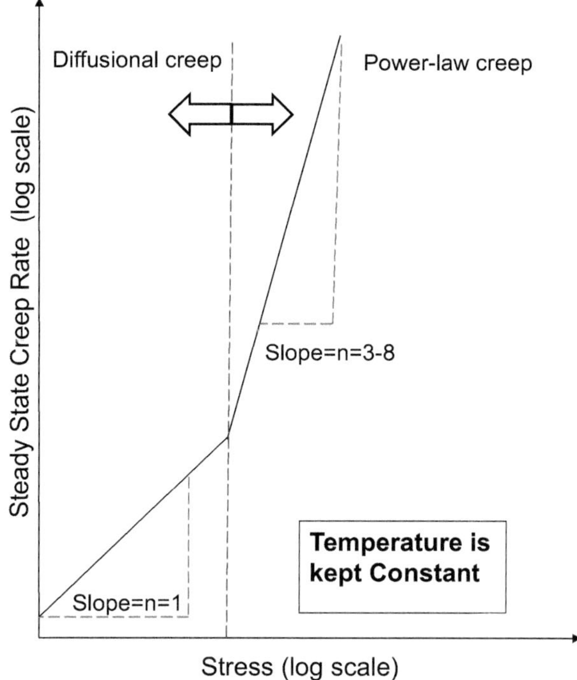

Fig. 10.3 Schematics of steady state creeprate- stress relation (axes on log scale) (after M. F. Ashby and D. R. Jones, Engineering Materials 1, An introduction to Properties, applications and Design, 4th edition, Butterworth-Heinemann, 2009)

$$\dot{\varepsilon}_{ss} = B.e^{\frac{-Q}{RT}} \tag{10.4a}$$

where B is a constant, R is the universal gas constant, T is the temperature in kelvins, and Q is the activation energy for creep (which depends on the creep mechanism)..

The **activation energy (Q)** for creep is typically the amount of energy needed for one mol of atoms to diffuse. Sometimes, Eq. 10.4b is displayed as shown below:

$$\dot{\varepsilon}_{ss} = B.e^{\frac{-Q}{kT}} \tag{10.4b}$$

In other words, if R is replaced with Boltzmann's constant k, the activation energy will be defined as the amount of energy needed for one atom to diffuse. The activation energy can be treated as a constant within a certain temperature range, as long as the dominant diffusion mechanism does not change.

It should be no surprise that we see an exponential dependence of steady-state creep rate on temperature. Isn't creep dependent on diffusion? And how is diffusion dependent on temperature? Exponentially, of course. So, the Arrhenius-type equation appears again.

Now we can take the natural log of both sides to obtain Eq. 10.5.

$$\ln(\dot{\varepsilon}_{ss}) = \ln(B) - \frac{Q}{RT} \tag{10.5}$$

Fig. 10.4 Schematics of $\ln(\dot{\varepsilon}_{ss})$ versus $1/T$ (after M. F. Ashby and D. R. Jones, Engineering Materials 1, An introduction to Properties, applications and Design, 4th edition, Butterworth-Heinemann, 2009)

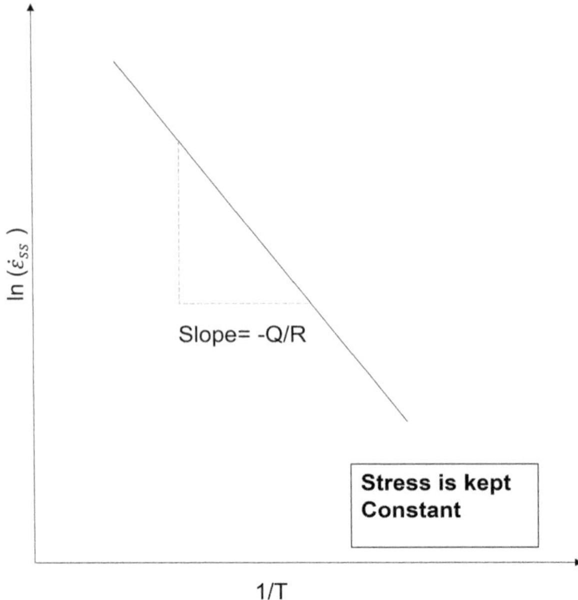

The equation is again that of a straight line (if we plot $\ln(\dot{\varepsilon}_{ss})$ vs. $\frac{1}{T}$), where $\ln(B)$ is the intercept and $(-Q/R)$ is the slope. A plot of $\ln(\dot{\varepsilon}_{ss})$ versus $\frac{1}{T}$ results in Fig. 10.4.

How Is Creep Failure Defined?

Here, we discuss what it means for a component to fail due to creep. At first glance, one would say that the specimen fractures. This is true, but only for specific components that do not depend on dimensional accuracy for their performance. For example, imagine a pressure pipe; it is said to have failed if it fractures. However, there are other components in which dimensional accuracy and tolerancing are important; imagine turbine blades in a steam or gas turbine. In these instances, we must know how the dimensions of such components change during service, as they cannot exceed the clearance allowed. Another application that is not too evident at the start, which is also affected by creep, is the use of bolts in engineering structures. When bolts are tightened, they impose a certain level of stress that holds things together. Ideally, this stress needs to be maintained throughout service. This is a situation where the strain automatically remains constant during service, but if creep occurs, it will be accommodated by a **relaxation** (a reduction) **of the stress**. Once the stress is reduced to a specific value, the effectiveness of the bolt will be compromised; in this case, the bolt needs to be retightened. So, creep failure means different things depending on the application. The stress-rupture test is the best test if we consider failure due to rupture. Here, stresses higher than those used in creep tests are used, and unlike in creep tests, failure strains in **stress-rupture tests** can be much higher, in the region of 50%, compared to, say, 0.5% in a creep test. Due to the higher stress levels, the tests can typically be completed in less than 1000 h (~ 42 days). Stress rupture-time

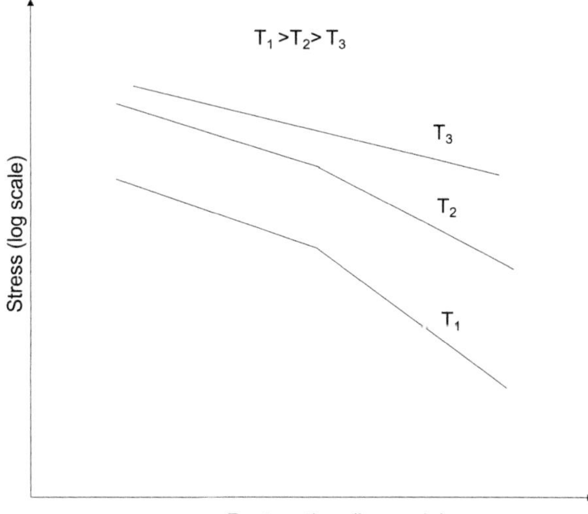

Fig. 10.5 Schematics of stress versus rupture life from a stress-rupture test conducted at three different temperatures

data can be plotted for materials to help in design. Figure 10.5 shows a schematic plot of tests conducted at three different temperatures. It is clear that for the same stress level, an increase in temperature results in a reduction in time to rupture; this is, of course, expected due to an increase in atomic diffusion at higher temperatures. Also, at any given temperature, a reduction in stress level increases the time to rupture. Notice in the figure that the relationship between stress and rupture time may change its slope after a certain amount of time; this can be caused by specific changes that occur in the material, for example, recrystallization, grain growth, a change in crack propagation from transgranular (across the grain) to intergranular (along grain boundaries), or even oxidation. Hence, designers must be aware of such potential changes. Otherwise, decisions made on the simple extrapolation of data may result in serious issues.

What Are the Mechanisms of Creep?

We have been saying for a while now that atomic diffusion is the reason behind creep, but how exactly does creep occur? It turns out that the applied stress can influence how atomic diffusion enables creep. At low stresses, what we refer to as diffusional creep can take place, while at higher stresses, dislocation climb (or power-law creep) takes over. Let's discuss these diffusion mechanisms in greater detail.

Diffusional Creep

We have already established that creep occurs when the temperature is at a significant homologous temperature (> 0.3–$0.4 T_H$ for metals). However, what process occurs in the material to enable a strain that increases with time? We will answer this question using Fig. 10.6, a schematic of part of the microstructure of a polycrystalline material. The applied stress, like the concentration gradient, is a driving force for atomic

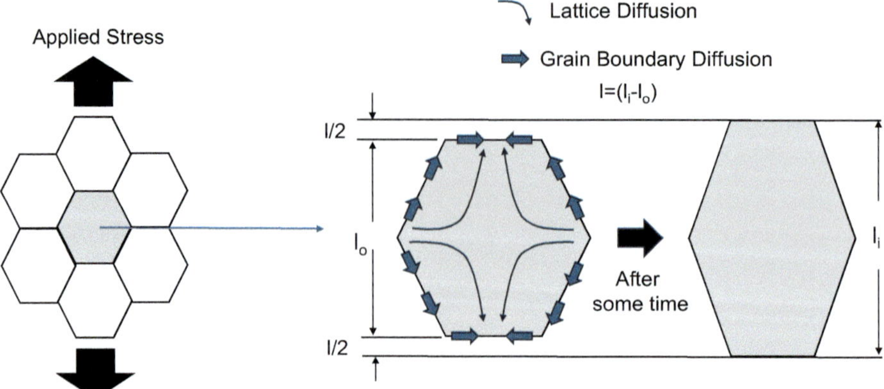

Fig. 10.6 Schematics of diffusional creep mechanism, showing grain boundary diffusion and lattice diffusion. The grain is shaded grey only for identification

diffusion. At low stresses, atoms travel from grain boundaries, as seen in the figure, to grain boundaries that are oriented close to perpendicular to the applied stress. Over time, the grain elongates along the loading direction and shrinks in the regions where the diffusing atoms originated. Since all grains have neighboring grains, to avoid any voids forming between grains, sliding of these grains (**grain sliding**) also occurs along the loading direction, generating a relatively smaller strain in that direction. All these strains are permanent (plastic) strains. We have established that the fundamental mechanism is that atoms diffuse to grain boundaries perpendicular to the applied stress. However, how do these atoms travel? If the atoms diffuse along grain boundaries, this is referred to as **Coble creep**; if they travel within the grain body, i.e., through the crystal lattice, we say it happens through lattice diffusion. If that is the case, the type of diffusional creep will be called **Nabarro–Herring creep**. We must remember that it's not like the atoms decide whether to diffuse through grain boundaries or the lattice; both will continue simultaneously. The question is, which one is more dominant? For the atoms to travel through the lattice, they must do so through vacancy diffusion, which means they will not be dominant at low temperatures since the concentration of vacancies will be small. In this case, grain boundary diffusion will dominate, so Coble creep is in play. Nabarro–Herring creep takes over at higher temperatures, as lattice (bulk) diffusion is dominant. Whether coble creep or Nabarro–Herring creep is in operation will influence the steady-state creep rate. This can be seen in Eqs. 10.6a and 10.6b.

Coble Creep:

$$\dot{\varepsilon}_{ss} = \frac{50\sigma b^4 D_{GB}}{kTd^3} \tag{10.6a}$$

Nabarro–Herring Creep:

$$\dot{\varepsilon}_{ss} = \frac{14\sigma b^3 D_L}{kTd^2} \tag{10.6b}$$

where σ is the applied stress, b is the burger's vector, k is Boltzmann's constant, T is the temperature in Kelvins, d is the grain size, D_L is the diffusion coefficient for lattice diffusion, and D_{GB} is the diffusion coefficient for grain boundary diffusion. Note that the stress exponent appearing in Eqs. 10.6a and 10.6b, is 1. (i.e., $n = 1$ in Eq. 10.1)

While both creep mechanisms linearly depend on stress, what is immediately apparent is that the dependence on grain size is different. Coble creep depends more on grain size than Nabarro–Herring creep, d^3 versus d^2. We also note that the temperature effect is incorporated within the diffusion coefficient, $D = D_o e^{-Q/RT}$.

Dislocation (Power-Law) Creep

At higher stress levels, what is known as dislocation (or power-law) creep occurs. In this case, the steady-state creep rate is determined using the power-law equation 10.7.

$$\dot{\varepsilon}_{ss} = \frac{ADbG}{kT} \left(\frac{\sigma}{G}\right)^n \tag{10.7}$$

where G is the shear modulus, D is the diffusion coefficient, and n is the creep exponent with a value ranging from 3 to 8. Typically, a value of 5 is observed.

The question then arises: how does dislocation creep work? Situations can arise where some obstacle, such as a precipitate in the microstructure, obstructs a gliding dislocation. If creep is not occurring, the dislocation will remain trapped and not contribute to any permanent strain. However, under moderate to high stress levels, the dislocation can climb over the obstacle and continue its glide until it meets another obstacle, and the process repeats. So, what exactly is dislocation climb? This is illustrated in Fig. 10.7, where a dislocation initially trapped or pinned by a precipitate obstacle manages to climb over the precipitate by the diffusion away of atoms within the dislocations that were in direct contact with the obstacle. This process frees the dislocation and allows it to continue its journey of glide! The dislocation motion or glide causes the creep strain, but the diffusion of atoms away from the obstacle (which frees the dislocation) determines the creep rate. This diffusion can happen in two ways. At high temperatures, enough vacancies are available in the lattice for such diffusion to take place through the lattice. However, at lower temperatures, dislocation core diffusion becomes the easier pathway. Here, the atoms simply diffuse through the core of the dislocation, which is a **short-circuit diffusion** pathway, as discussed in a previous chapter. In both cases, the driving force for such diffusion is a force (stress) imposed on the dislocation by the obstacle. Remember, stress can have a similar effect as a concentration gradient to drive diffusion.

Professor M. F. Ashby introduced the concept of a deformation map that shows which deformation/creep mechanism operates at which combination of temperature and stress. Here, the stress is normalized by dividing it by the shear modulus (G), and the temperature in Kelvins is normalized by dividing it by the melting temperature

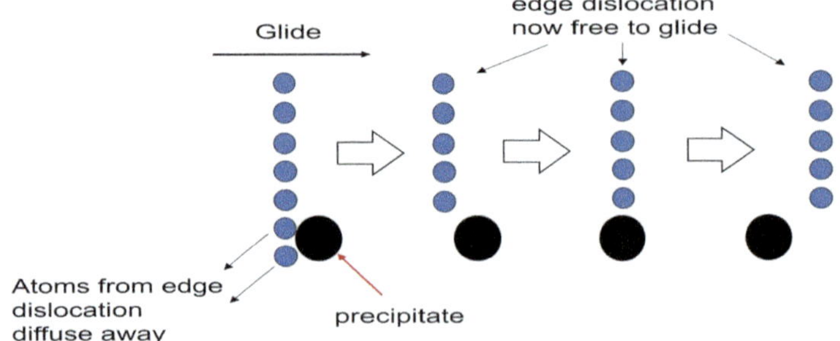

Fig. 10.7 An illustration of how an edge dislocation can climb over an obstacle, with the help of diffusion

(also in Kelvins); i.e., what is plotted on the x-axis is the homologous temperature. Figure 10.8 shows an example of such a map.

How Do We Protect Against Creep?

First, we need to determine which creep mechanism would be dominant during the service of a component; which is where the Ashby maps can come into play. In the case of diffusional creep, with reference to Fig. 10.6, it is evident that it would be more difficult for the atoms to diffuse to the top and bottom of the grain if the grain were large (coarse) and not small (fine), since longer distances need to be traveled by the atoms to elongate the grain. This would ultimately result in a reduced creep rate. To go even further, one may not rely on equiaxed grains but somewhat elongated grains

Fig. 10.8 An illustration of a deformation map (redrawn with some modification, based on M. F. Ashby and D. R. Jones, Engineering Materials 1, An introduction to Properties, applications, and Design, 4th edition, Butterworth-Heinemann, 2009)

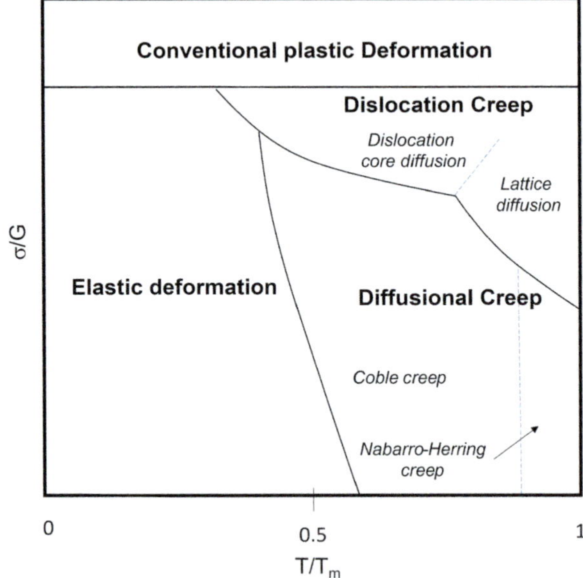

(which can be produced using directional solidification), where the longest dimension of the grain is aligned parallel to the stress application axis. This means even longer distances for atoms to travel and even lower creep rates than for coarse equiaxed grains. An even more extreme scenario is not to use polycrystalline materials in the first place, since grain boundaries are the source of this problem. In that case, a component can be made out of a single crystal, thus eliminating grain boundaries altogether. This would result in the best creep resistance. Of course, the choice would depend on cost, among other factors. Interestingly, all these designs have been utilized for high-temperature turbine blades. Since grain sliding is part of the mechanism, one could also restrict it by adding alloying elements that preferentially precipitate phases at the grain boundary, making the sliding more difficult.

On the other hand, if the power-law creep is dominant, since dislocation climb is at the heart of its success, we would simply try to add more obstacles in the pathway of dislocations. This will again be carried out through alloying to produce more precipitate.

We have so far been looking at creep in a negative light, as if it were our enemy and that we need to defeat it. This is the case if our objective is to preserve the component during service. However, creep deformation is so fundamental in many manufacturing processes, for example, in bulk deformation processes such as hot extrusion, hot forging, and hot rolling, and in powder consolidation processes such as hot pressing and hot isostatic pressing. The latter powder-based processes are not only used to produce metal products but are fundamental (together with solid state and liquid phase sintering) in producing ceramic products. In these cases, creep is our friend.

10.4 Fracture

In tensile testing, impact testing, and creep, we have seen that the ultimate failure of a component can occur through fracture. As it happens, crack initiation and propagation are at the heart of a very important field called **Fracture Mechanics**.

In Great Britain, the desire to make ships bigger and longer led designers in 1903 to use the H. M. S. Wolf, a destroyer, to carry out some essential experiments and measurements to ascertain its ability to resist fracture in the (sometimes) treacherous and violent seas. At the time, strain gages were placed in many locations on the ship, leading to experimental stress measurements in many locations. Understandably, although the hull is surrounded by water on the open seas, the water does not fully support the weights inside the ship, including the weight of fuel and engines. This leaves the hull experiencing some degree of bending. The Wolf experiments were conducted in a dry-dock, where the ship was supported mechanically (using evenly distributed keel blocks), water was gradually introduced in a controlled way, and stresses were gradually measured. The experiments went further by picking the worst sea conditions to launch the Wolf, expecting severe wave forces to impact the ship's hull. Stresses were measured there, too. Even on the open, vigorous seas, stresses measured did not exceed 80 MPa. This is far below the strength of the steel making up the hull, around 400 MPa. So, things looked assured. Yet ships were still sometimes breaking in two. What is significant about ships is that they contain many openings, hatchways, windows, and discontinuities in their structure. This means that large load-bearing structures have areas with no material to carry the load. Designers had initially considered such issues, but in a proportional way. This means that if a hole is introduced that takes away one-fourth of the material, the designers would say that only three-quarters of the material is available to carry the same load, which means the stress would be magnified by a factor of 4/3, and so on. A 1913 paper by **Sir Charles Inglis** shed critically important light on the problem. He refuted the idea of proportionality in calculating the stress, as designers claimed. He mathematically treated the problem in his paper and concluded that the stresses can be considerably greater than previously thought. Of course, few believed him then, but this was the birth of what we know today as **stress concentration**. The impact his work has had on the design of large structures cannot be overstated.

Professor, this is a great story, but why didn't the designers catch these high stresses when experimenting with the H. M. S. Wolf destroyer? I mean, they were experimentally taking strain (and therefore stress) measurements, weren't they?

Great question, Alex. As fate would have it, they never placed these strain gages near any discontinuity, such as next to a hole or a hatchway. So, they missed it!

The Concept of a Stress Concentration

To bring the concept of stress concentration closer, imagine a plate of material subjected to an applied tensile stress (σ_a), Fig. 10.9a, which keeps it in the elastic region. This plate is composed of atoms held together by spring-like bonds. Each set of atoms carries its share of the load or stress. Figure 10.9a shows a plate with some stress trajectory lines; all atoms along these trajectories, as in other locations in the plate carry their share of the load. Now, imagine an elliptical cavity of length 2a introduced in the center of the plate. In effect, the cavity replaces atoms that would have otherwise carried their share of the load. This leads to atoms on both sides of the cavity chipping in and carrying the share of their previous neighbor atoms that the cavity has replaced. They do the good neighborly thing; the closer atoms to the edge of the cavity will carry more of the share compared with further and further atoms (or neighbors). At a far distance from the cavity, the stress carried is just the applied stress. Therefore, the cavity's presence results in the material close to the tip of the cavity carrying much more than the material at a greater distance away from the cavity. In other words, the stress is magnified at locations close to the cavity tip, resulting in a stress level that is called the **magnified stress (σ_m)** in that region (Fig. 10.9b) for the situation where the dimensions of the cavity are much smaller than the dimensions of the plate. A **stress concentration factor** is defined as the magnified stress divided by the applied stress, as shown in Eq. 10.8.

$$K = \frac{\sigma_m}{\sigma_a} = 1 + 2\sqrt{\left(\frac{a}{\rho}\right)} \tag{10.8}$$

where K is the stress concentration factor, a is half of a central crack or cavity. Note if the crack only appears at the surface, i.e., a surface or edge crack, then "a" represents the total length of that crack. ρ is the radius of curvature of the cavity or crack. For an elliptical crack, $\rho = b^2/a$ (Fig. 10.10), where b is have the crack width.

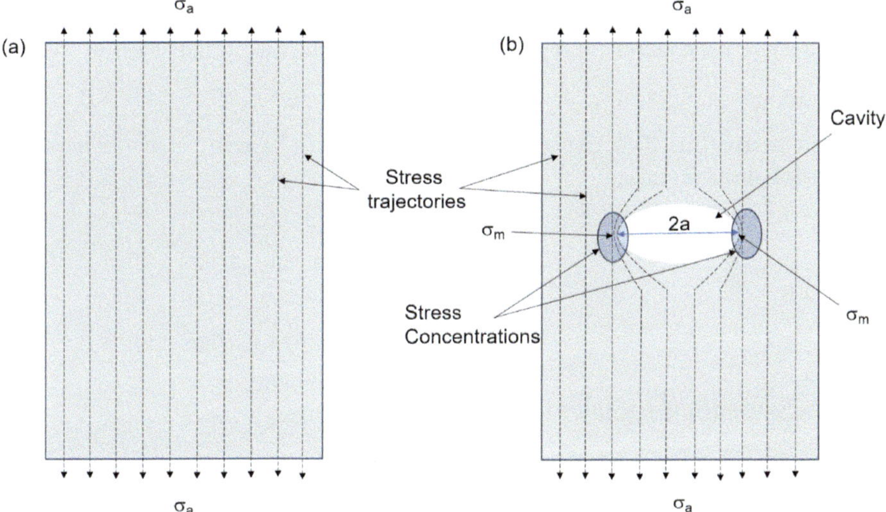

Fig. 10.9 Plate loaded in tension within the elastic region, **a** plate with no cavity, **b** plate with central cavity of length 2a

Fig. 10.10 Schematic of an internal cavity of length 2a and width 2b

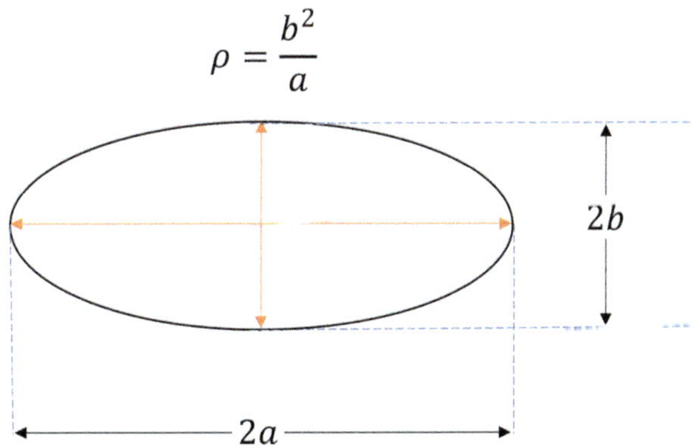

$$\rho = \frac{b^2}{a}$$

The stress concentration factor is highly dependent on the shape of the cavity and its size in relation to the structure, as can also be seen in the radius of curvature, meaning that sharper cracks with smaller radii of curvature will have a higher stress concentration factor. It should be mentioned that any discontinuity in a structure, whether a cavity, a window, a hatchway, or anything else, will cause a stress concentration close to it, where the stress is magnified. This means we need to be extra careful when designing such structures by considering stress concentrations when calculating the stress distributions within a structure.

Example Problem 10.1

Two very large plates must be subjected to a tensile stress of 100 MPa. One plate (plate A) has a small central elliptical cavity of length 10 mm and width 4 mm, while the other (plate B) has a small circular cavity of 10 mm diameter. Determine the stress concentration factor for each of the plates and the magnified stresses.

Solution

We need to use Eq. 10.8. Since the cavity is a central cavity, a in the equation represents half its length.

$$K = \frac{\sigma_m}{\sigma_a} = 1 + 2\sqrt{\left(\frac{a}{\rho}\right)}$$

Plate A:

First, we determine the radius of curvature for the cavity.

$$\rho = \frac{b^2}{a} = \frac{\left(2 \times 10^{-3}\right)^2}{5 \times 10^{-3}} = 0.8 \times 10^{-3}$$

$$K = 1 + 2\sqrt{\left(\frac{5 \times 10^{-3}}{0.8 \times 10^{-3}}\right)} = 6$$

This means that if the applied stress is 100 MPa, then the magnified stress will be 6 times greater, i.e., 600 MPa.

Plate B:

Here, we have a simple circle. This means $b = a$ in Fig. 10.10, making the radius of curvature simply the radius of the circle, which is 10 mm/2 = 5 mm or 5×10^{-3} m. Hence,

$$K = 1 + 2\sqrt{\left(\frac{5 \times 10^{-3}}{5 \times 10^{-3}}\right)} = 3$$

This means that the magnified stress is 3 times the applied stress, 300 MPa.

This simple exercise shows that the stress concentration of the circle is less than that of an ellipse. Now, as long as the size of the circular hole is much smaller than the size of the plate, the actual dimensions of the circular hole will not affect the stress concentration calculation. However, cavities and discontinuities in structures come in different shapes and sizes, and sometimes, we cannot assume the infinite plate simplification.

Example Problem 10.2

A large plate is expected to be subjected to a tensile stress of 100 MPa. The plate has a relatively small surface semi-elliptical flaw of length 10 mm and width 4 mm; determine the stress concentration factor and the magnified stress.

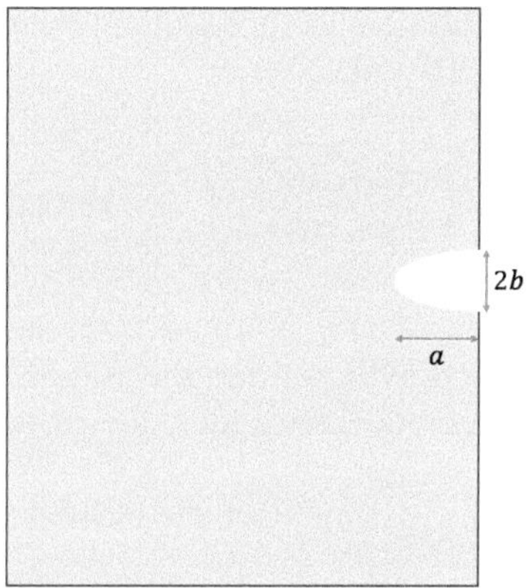

Solution

First, we determine the radius of curvature for the cavity.

$$\rho = \frac{b^2}{a} = \frac{\left(2 \times 10^{-3}\right)^2}{10 \times 10^{-3}} = 0.4 \times 10^{-3}$$

However, in the stress equation, we use 'a' as the total length of the surface cavity, i.e., 10×10^{-3} m.

$$K = 1 + 2\sqrt{\left(\frac{10 \times 10^{-3}}{0.4 \times 10^{-3}}\right)} = 11$$

Here, the stress concentration factor is higher than the central elliptical cavity discussed in the previous example problem. Therefore, the magnified stress is 11 times the 100 MPa applied stress, i.e., 1100 MPa.

Stress concentration and Y factors have been determined for different structures and cavity configurations, which are very helpful in structural design.

Example Problem 10.3
A large plate must be subjected to a tensile stress of 100 MPa. The plate has a relatively small surface "sharp" crack of length 10 mm and width 4 nm; determine the stress concentration factor and the magnified stress.

Solution

$$\rho = \frac{b^2}{a} = \frac{\left(2 \times 10^{-9}\right)^2}{10 \times 10^{-3}} = 0.4 \times 10^{-15} \text{ m}$$

However, in the stress concentration factor equation, we use "a" as the total length of the surface cavity, i.e., 10×10^{-3} m.

$$K = 1 + 2\sqrt{\left(\frac{10 \times 10^{-3}}{0.4 \times 10^{-15}}\right)} = \sim 10{,}000{,}000$$

Here, the stress concentrating factor is found to be astronomically high. The magnified stress is, therefore, 10 million times the 100 MPa applied stress, which would exceed the material's theoretical strength at the current stress conditions.

This last example points to an essential fact that Griffith picked up on, as pointed out below.

The work of **Sir Inglis** was groundbreaking and has had a tremendous impact on the design of structures today. It also had at least an equally significant impact on understanding the fracture of materials. A young man by the name of **A. A. Griffith** in 1920 revolutionized how we view fracture. Griffith wondered why brittle materials can be theoretically determined to have high strengths yet fail when tested at much lower stresses. The work of Sir Inglis regarding stress concentrations for large structures made Griffith wonder whether minute flaws were present inside these brittle materials, where stress is magnified at their tips. As seen from the previous example, hairline cracks or sharp cracks, according to Inglis's equation, can result in extremely large stress concentrations. Griffith should be highly credited for his insight in applying Inglis's ideas to the minute scale of flaws within materials. He carried out a set of groundbreaking experiments on the simplest brittle material he could think of, which was glass. Glass is an amorphous material, so no need to worry about the influence of grain boundaries, for example. He produced a series of glass fibers of different diameters down to 2.5 μm, with his then assistance **Sir Ben Lockspeiser**. We need to realize that during these days, such inquisitive scientists were not too appealing, as it was always expected to work on more practical issues and materials such as steel or wood. Hence, Griffith and his assistant worked somewhat behind the scenes until a fire broke out because Sir Lockspeiser had left the Bunsen burner on, which they used to heat and form the glass. I am mentioning this story as an example of the desire and passion of the scientific figures who shaped our understanding today.

Griffith conducted a great experiment. He took a glass rod with a 1 mm diameter and tested it in tension, which resulted in a strength of 170 MPa. Then he heated the middle part of the rod and pulled it so that the heated part of the glass was drawn (and thinned) down in diameter. He did this to produce glass with different diameters down to 2.5 μm and tested them in tension.

Griffith then plotted the tensile strengths of fibers versus their diameters, and the plot looked something like Fig. 10.11. A clear correlation was observed between diameter and strength. The smaller the diameter, the higher the strength, such that at a diameter of 2.5 μm, the strength was extremely high, ~ 6000 MPa. A simple extrapolation of the data can show that at even smaller diameters, the theoretical strength can be reached (~ 11,000 MPa). It should be noted that these strength values were for fibers that had just been drawn; however, left only a few hours, the same 6000 MPa strong fiber now had a strength of 3500 MPa. Obviously, some vulnerability of the fiber to damage was evident.

Griffith strongly suspected cracks in the glass and that the cracks somehow decreased with a decrease in the glass rod diameter. It turns out that glass is very sensitive to damage from its surroundings, and this sensitivity decreases with a decline in diameter. Very thin fibers are easily bent, and this may be a mechanism that helps protect them from damage. The most important outcome was that crack-free glass can approach the theoretical strength. Later, this was also confirmed for other materials produced as whiskers grown from solution. Whiskers typically can have diameters

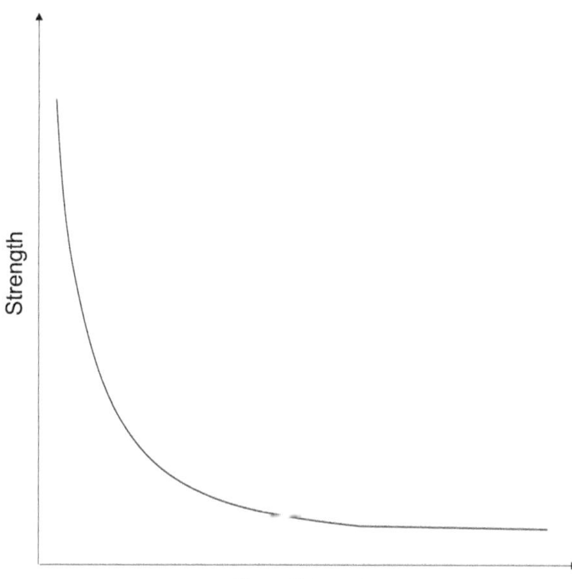

Fig. 10.11 Schematic of glass fiber diameter versus strength

in the region of a few micrometers. A similar trend to the Griffith glass strength vs. diameter was observed for these materials, which was attributed to step-like features at the whiskers' surface due to their fabrication method.

Griffith made another groundbreaking contribution that shaped our current understanding of crack propagation. He proposed that a crack will propagate only if it reduces energy, and that there must be a mechanism by which it can grow. Figure 10.12 shows a plate of *unit thickness* loaded elastically using stress; at this stage, the plate possesses strain energy of $\frac{1}{2}\,\sigma^2/E$, where E is Young's modulus of the plate, if a crack of length "a" is introduced, approximately two triangular regions of material behind the crack (identified in blue) will be relaxed and, as such, will release their strain energy. The area of these regions amounts to approximately a^2. Here, we must do some accounting of the energies involved. A crack of length "a" will have two new surfaces; each surface will have an energy equivalent to $a.\gamma$, where is the surface energy. Hence, the total energy of the crack, because it has two surfaces, will be $2a\gamma$. This energy has to come from somewhere, which is the released strain energy. Hence, when the released strain energy equals the energy needed to form a crack, crack propagation will occur, leading to the following Eq. 10.9.

$$\sigma = \sqrt{\frac{2\gamma E}{\pi a}} \tag{10.9}$$

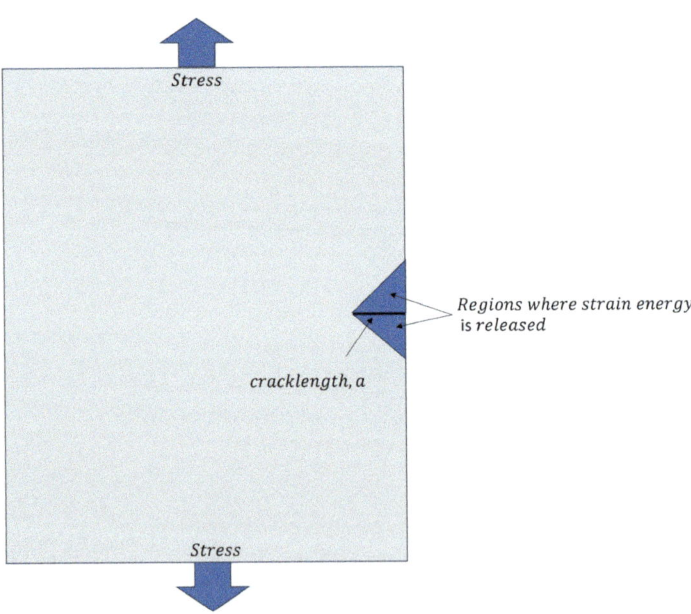

Fig. 10.12 Schematic of loaded glass plate with edge crack

Recall, this equation is valid only at the point of fracture. Assuming, γ and E are constant for a particular material, the equation shows that the stress at fracture is inversely proportional to the square root of the crack size. In this equation, a is the total length of an edge or surface crack and half the length of a central (or internal) crack.

In other words, the smaller the flaw size, the higher will be the stress needed to break the plate. Also, conversely, a plate with a long crack will only need a small stress to break it. Here, we can use this equation in two ways: if we are wondering whether or not to subject a plate to a given (known) stress level. We could use this equation to find out what crack size will satisfy the equation by solving for a. The resulting 'a' value will be called the **critical flaw size**. Alternatively, if we knew the longest flaw size in the plate, we could input this value along with the Young's modulus and surface energy to obtain the stress that satisfied the equation. This stress will be called **the critical stress**.

Example Problem 10.4

You intend to subject a glass plate to a tensile stress of 40 MPa. Should you go ahead with it? The surface energy of glass is 1 J/m^2, and its Young's modulus is 70 GPa.

Solution

Equation 10.9 can be rearranged in terms of the critical flaw size, a.

$$a = \frac{2E.\gamma}{\pi \sigma^2}$$

Inputting the stress values, Young's modulus, and surface energy results in a critical flaw size of 27.9 μm. With typical crack detection techniques limited to detecting cracks no less than 1 mm, I guess we would be taking a risk by assuming we do not have a 27.9 μm crack because we cannot detect it. Most non-destructive testing techniques (NDT) can only tell us if we have a crack of ~ 1 mm length or greater. They will not detect a smaller crack. So, we cannot be assured. I would avoid applying the stress; otherwise, the plate would fracture.

Dislocations can also lead to the initiation of a crack. In metallic materials, during deformation, dislocations glide on a glide plane and meet a grain boundary (which can block their motion); at some point, we can have a pileup of many dislocations at a grain boundary, which will lead to a situation like that shown in Fig. 10.13. As can be seen, beneath the half-plane of atoms will be a gap, referred to as a crack "embryo", and hence can be an initiation site for a crack. Similar situations can result when dislocations moving on different slip planes suddenly intersect within the crystal, resulting in a crack.

Equation (10.10) is a valuable equation typically used to determine conditions for fast fracture. It introduces a new property, the stress intensity factor, K_I.

$$K_I = Y\sigma\sqrt{\pi a} \tag{10.10}$$

Fig. 10.13 Schematic showing how a dislocation pileup can result in a gap at grain boundaries (crack embryo) that can initiate a crack

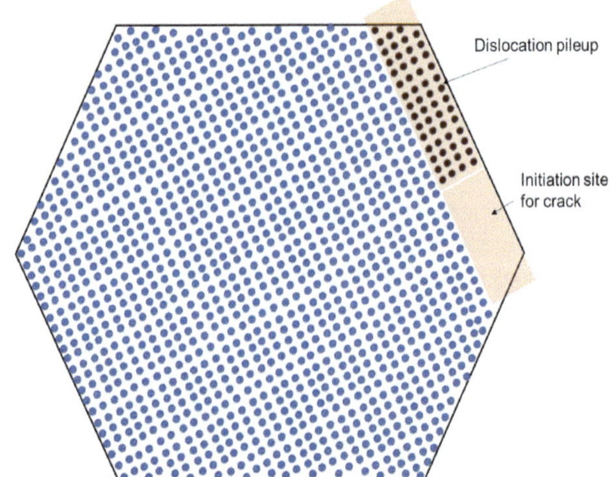

Y is a dimensionless parameter that depends on the crack/material geometry. A small crack present in what could be considered as an infinite plate will result in a Y value of 1, σ is the applied stress, a is the flaw size (total length of a surface crack or half the length of a central or internal crack). As one can imagine, if we don't apply any stress, then the stress intensity factor will be zero. As the applied stress increases, the value of the stress intensity factor also increases until a specific value is reached, at which point fracture occurs. At this value, the stress intensity factor will be called the **critical stress intensity factor**. This value is highly dependent on the thickness of the component, decreasing with increasing thickness until a certain thickness is reached, beyond which the critical stress intensity factor remains constant and is purely a material property after then. This is called the **plane strain fracture toughness (K_{IC})**, while at thicknesses lower than that critical value, the material will be under plane stress conditions. The plane strain fracture toughness, K_{IC}, is the value usually quoted for materials (as it is a conservative value); it is a material property that indicates the material's resistance to fast fracture. A high value of K_{IC} means the material is resistant to fast fracture; metals are examples of materials with high fracture toughness values. Low fracture toughness values mean the material is poor at resisting fast fracture, like ceramics. Since plane strain fracture toughness is a material property, it can be measured in the lab, just like other properties such as tensile strength, yield strength, and ductility.

Table 10.1 lists the fracture toughness values of some materials. As can be seen, metals typically have high fracture toughness, while ceramics and polymers have low fracture toughness values. For some materials, ranges of fracture toughness values are listed, which reflect either different compositions or heat treatments carried out on the materials. An example can be seen in titanium. Titanium has many compositions and, as such, different alloys. Commercially pure titanium (CP-Ti) and Ti6Al4V (sometimes referred to as Ti64) are but two examples. Ti64 is mainly used both as a biomaterial and an aerospace material. The alloy is referred to as a two-phase alloy

Table 10.1 Fracture toughness values of some common materials

Material	Fracture toughness K_{IC} (MPam$^{1/2}$)	Material	Fracture toughness K_{IC} (MPam$^{1/2}$)
Metals		LDPE	1
Pure ductile metals (Cu, Al, Ag, Ni)	100–350	HDPE	2
Al alloys (high strength-low strength)	23–45	Epoxy	0.3–0.5
Mild Steel	140	**Ceramics and glasses**	
Medium carbon steel	51	Alumina	3–5
High strength steels	50–154	Silicon carbide	3
cast iron	6–20	Silicon nitride	4–5
Ti–6Al–4V	55–115	Magnesia	3
Polymers		Soda glass	0.7–0.8
Wood (crack parallel to grain)	0.5–1	**Composites**	
PMMA	0.9–1.4	Steel reinforced cement	10–15
Polystyrene	2.0	WC–Co (cermets)	14–16
Polyester	0.5	Glass fiber reinforced epoxy (fiberglass)	42–60
Polycarbonate	1.0–2.6	Carbon fiber reinforced plastics	32–45
Nylon	3.0	Steel reinforced cement	10–15

Source M. F. Ashby, D. R. H. Jones, Engineering Materials 1, An introduction to their properties and Applications, Butterworth and Heinemann (1996)

where the α-Ti phase (HCP) and β-Ti (FCC) comprise the microstructure. The material has exceptional strength and toughness and is biocompatible and corrosion resistant. β-Ti is, in fact, a Ti phase that is usually stable at high temperatures. However, adding vanadium at (4 wt%) stabilizes the phase and allows it to exist at room temperature. Al is an α stabilizer, which helps stabilize the room temperature hcp phase α. One notices from the table that pure metals can have exceptionally high fracture toughness values, mainly due to the high ductility displayed by these materials. When considering structural materials, i.e., materials used in load-bearing applications, while strength is important, a more substantial property is fracture toughness. It can sometimes be said that a material with a high fracture toughness is forgiving, while a material with low fracture toughness is unforgiving. Ice is a material not listed in the table; when tested at subzero temperatures, it displays a fracture toughness of ~ 0.5 MPa m$^{1/2}$. Many efforts have been expended on improving the fracture toughness of ceramics; an example is the doping of zirconia (ZrO$_2$) ceramic with 2–3 mol% Yttria (Y$_2$O$_3$), which produces tetragonal zirconia polycrystals (TZP), with improved mechanical properties; for example, fracture toughness values can be raised to ~ 10 MPam$^{1/2}$. Additionally, magnesia partially stabilized zirconia ceramic can have fracture toughness values of 15 MPam$^{1/2}$. This kind of ceramic has been referred to as "ceramic steel", due to its improved mechanical properties. It is yet another example of how we can engineer materials to achieve greater and greater properties—more about ceramics in a later chapter.

Example Problem 10.6

A mild steel component is expected to carry a tensile stress of 500 MPa. It was found that it contains a surface crack of 1 mm in length. The material has a plain strain fracture toughness of 140 MPam$^{1/2}$. Would the component fail once the stress is applied? You can assume that $Y = 1$.

Solution

We can solve this problem in three ways.

Method 1

Substitute the values of a (= 1×10^{-3} mm), Y (= 1), and K_{IC} (140 MPam$^{1/2}$) in the K_{IC} equation and find out the value of the critical stress. If the value is higher than the expected applied stress (500 MPa), then the material will not fail once we apply the stress (as long as it is not a cyclic stress; more about this later). Remember that "a" in the equation is the total length of the surface crack. Alternatively, if the critical stress value lies below the intended applied stress, that means the applied stress exceeds the critical stress, and hence, failure will occur.

$$K_{Ic} = Y\sigma\sqrt{\pi a}$$

$$\sigma_c = \frac{K_{IC}}{Y\sqrt{\pi a}} = \frac{140 \times 10^6}{1 \times \sqrt{\pi} \times 1 \times 10^{-3}} = 2.5 \text{ GPa}$$

According to method 1, the critical stress far exceeds the applied stress. Hence, the component will not fail.

Method 2

Substitute in the K_{IC} equation the values of σ (= 500 MPa), Y (= 1), and K_{IC} (140 MPam$^{1/2}$) and find out what is the value of the critical flaw size (a). If the size of the critical flaw is greater than that of the flaw found in the component, it means it hasn't reached the critical size yet, and the component will not fail upon applying the stress. If, however, the critical flaw size happens to be lower than the size of the flaw present in the component, that means the flaw in the component exceeds the critical size, and the component will fail upon stress application

$$K_{Ic} = Y\sigma\sqrt{\pi a}$$

$$a_c = \frac{1}{\pi} \times \left(\frac{K_{IC}}{Y\sigma}\right)^2 = \frac{1}{\pi} \times \left(\frac{140 \times 10^6}{1 \times 500 \times 10^6}\right)^2 = 25 \text{ mm.}$$

The critical flaw size is much greater than the flaw size of 1 mm, hence, fracture will not occur.

Method 3

Here, we calculate the stress intensity factor using all the values given for the other parameters. If the stress intensity factor is below the fracture toughness value, then fracture will not occur. If it is equal to or greater, then fracture will occur. So, let's input the values and see what happens.

$$K_I = 1 \times 500 \times 10^6 \times \sqrt{\left(\pi \times 1 \times 10^{-3}\right)}$$
$$= 28 \, \text{MPa} \, \text{m}^{1/2}$$

This is less than the fracture toughness value of 140 MPa m$^{1/2}$. Hence, a fracture will not occur.

10.5 Fatigue

We learned in the previous section about the influence of cracks on the failure of components. However, sometimes all calculations are conducted appropriately, and components are deemed safe, yet components still fail after a specific time during service. This can happen if the component is subjected to a cyclic (repeated) stress, even if this stress falls within the elastic region of the stress–strain curve and is low compared to the material's yield strength. This stress can arise from machine vibrations, fluctuating wind blowing on a bridge, the vibrations of the wings of an aircraft, or even a key turning time and time again in a keyhole. In this situation, the material is said to have been **fatigued**. The first person to notice this was **Sir William Fairbairn**, who, in 1861, found that while a wrought iron beam was able to withstand a 12-ton dead weight, it only took a 3-ton weight that was raised and lowered on the beam about 3 million times to result in the beam fracturing. He also discovered a very unusual thing: the beam never fractured when he cycled a weight of less than 3 tons. These three tons today would be called a fatigue limit, as discussed later.

Normally, we may have designed the component well in terms of the expected level of static stress, considering the fracture toughness, calculating the critical flaw size, and ensuring that the size of the biggest flaw is much smaller than the critical flaw size. However, we may also need to consider the influence of the cyclic nature of the stress, if a cyclic stress is applied to the component. It turns out that if fatigue is occurring, then with each cycle of stress, the initial small (i.e., deemed safe) crack size increases incrementally, such that after a certain number of cycles (i.e., time), the crack finally reaches the critical flaw size we had previously calculated, and the material then fractures. Although we can initially assume the absence of flaws in smaller components like ball races, gear teeth, and axles, large structures like bridges, pressure vessels, and ships almost always have flaws. If a material does not have a pre-existing crack, it does not mean it is immune to fatigue failure. It may take a certain number of cycles to initiate a crack, followed by a number of cycles that propagate the crack to the critical flaw size. We will be putting ourselves in a vulnerable situation if we design components with corners and discontinuities, where stress becomes concentrated. I'm sure these days, we will not see a square window in an airplane, where fatigue cracks can initiate at the corners due to vibrations during flight. That would be suicidal. So, typically, rounded edges and less abrupt transitions in structures are a safer way. In large structures that contain pre-existing cracks, no cycles are needed to initiate a crack (since it is already there), but rather, the stress cycles directly affect the propagation of the crack. Since fatigue cracks typically originate at the surface (since it is the most vulnerable part of the component in contact with the surroundings), surface modification has been used as a means to combat fatigue failure, for example, by imposing residual compressive stress through processes such as shot peening, where very tiny balls are shot at the surface of a metal component.

Apart from the apparent presence of a surface or internal flaw in a material that can propagate during fatigue, fatigue cracks can also initiate from certain microstructural features within a material that initially did not contain any flaws/cracks. An internal pore, for example, could present a stress concentration and become an initiation site for a fatigue crack. Even though stress during cyclic fatigue is mostly lower than the yield strength of the metal, some plastic deformation can still locally take place, and slip bands have been observed at the surface of the material, which result in either extrusions or intrusions on the surface of the material experiencing cyclic stress. This is shown in Fig. 10.14, which notes that the intrusion can be an initiation site for fatigue cracks. These intrusions can typically be on the scale of about 1 μm.

As one can imagine, fatigue stress cycles can come in different magnitudes and shapes. One of these cycles is what we call a **random stress cycle**. The figure shows the positive (tensile stress) and the negative (compressive stress) on the axis. The

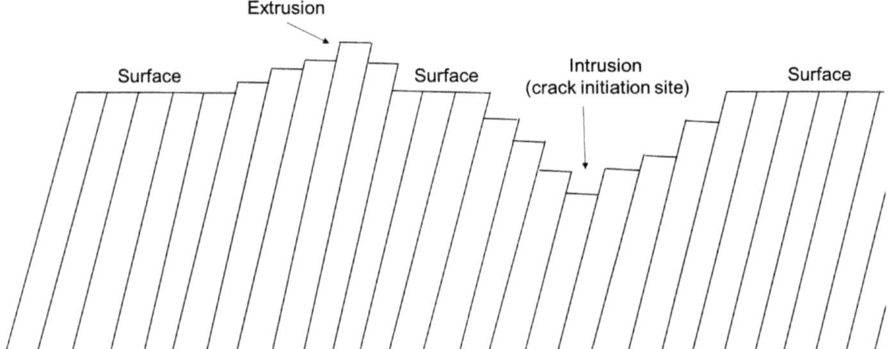

Fig. 10.14 Schematic showing localize slip that can occur during cyclic loading resulting in extrusions and intrusion of the surface of a metal during cyclic loading. Intrusions can be initiation sites for fatigue cracks (after R. A. Higgins, Properties of Engineering Materials, Hodder and Stoughton, 1986)

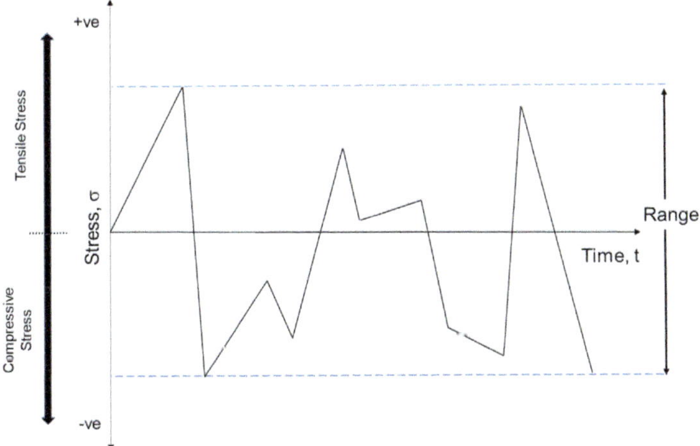

Fig. 10.15 Schematic of random stress cycles

figure is just one example of a possible random stress cycle, but more complex cycles can be observed. In essence, anything goes as long as it is not regular (Fig. 10.15).

I guess, Alex, this would be your favorite cycle on an exam; if I asked you to draw a random stress cycle, you could almost draw anything, and you would be correct.

Professor, are you saying anything goes? I guess I can even sign my name, huh?

Not really.

Why is that?

If you sign your name, the likelihood is that you would have to drag the pen backward on the plot. Just try writing the letter "e" and you'll know what I mean. Unless you are Michael J. Fox, I don't think you can go back in time ☺, and you will get a zero on that question!

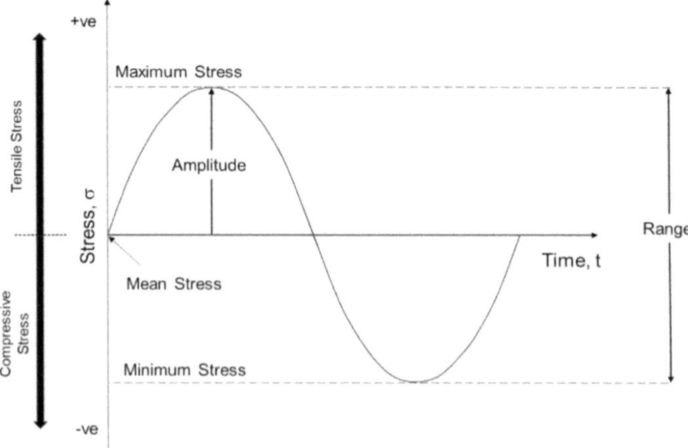

Fig. 10.16 Schematic of a reversed stress cycle

The simplest of these stress cycles is the **reversed stress cycle**, shown in Fig. 10.16. The figure shows the stress being applied as a sine curve, showing the stress amplitude, the stress range, and the mean stress.

Not all applications will subject the component to a reversed stress cycle, where the magnitude of the minimum and maximum stress in the cycle is the same. If we shift the cycle in Fig. 10.16, up or down the y-axis, we will obtain a cycle that looks like it. However, the stress range and stress amplitude will be the same. However, its maximum and minimum stresses will differ in magnitude. It also follows that the

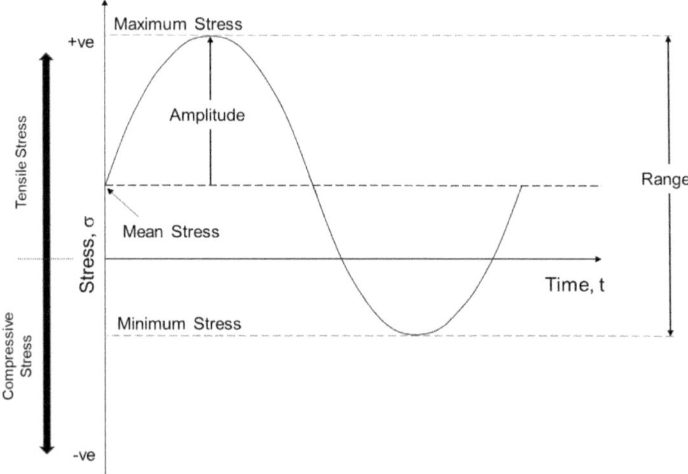

Fig. 10.17 Schematic of a repeated stress cycle

mean stress will not be zero. Such a stress cycle is called a **repeated stress cycle**, as shown in Fig. 10.17.

Several parameters are important to note when considering fatigue cycles. The first is the maximum and minimum applied stress during the stress cycle. The difference between them is the **stress range**, $\Delta\sigma_r$, as shown in Eq. 10.11.

$$\Delta\sigma_r = \sigma_{\max} - \sigma_{\min} \tag{10.11}$$

We need to be careful here. By convention, any compressive stress is considered negative.

Example Problem 10.7

A component is subjected to a repeated stress cycle, having a maximum tensile stress of 500 MPa and a minimum compressive stress of 50 MPa. What is the stress range?

Solution

The maximum stress is tensile, hence positive. The minimum stress is compressive; hence, it is negative. The stress range will then be:

$$\Delta\sigma_r = 500 \times 10^6 - \left(-50 \times 10^6\right) = 550 \times 10^6 \text{ Pa}$$

Another critical parameter is the stress amplitude, σ_a, which is defined as half of the stress range, as shown in Eq. 10.12.

$$\sigma_a = \frac{\Delta\sigma_r}{2} \tag{10.12}$$

Yet another extremely important parameter is the **mean stress**, defined as the average stress in the cycle, considering the sign convention for stress. Equation 10.11 is used to calculate the mean stress.

$$\sigma_m = \frac{\sigma_{\max} + \sigma_{\min}}{2} \tag{10.13}$$

The mean stress can strongly affect fatigue failure.

Example Problem 10.8

A component is subjected to a repeated stress cycle, having a maximum tensile stress of 250 MPa and a minimum compressive stress of 100 MPa. What is the stress range, stress amplitude, and the mean stress?

Solution

$$\Delta\sigma_r = 250 \times 10^6 - \left(-100 \times 10^6\right) = 350 \times 10^6 \text{ Pa}$$

The stress amplitude is half the stress range, 175×10^6 Pa.
The mean stress is:

$$\sigma_m = \frac{250 \times 10^6 + \left(-100 \times 10^6\right)}{2} = \frac{150 \times 10^6}{2} = 75 \times 10^6 \text{ Pa}$$

The **stress ratio** is another parameter used to characterize fatigue stress cycles. The stress ratio is given by Eq. 10.15:

$$R = \frac{\sigma_{min}}{\sigma_{max}} \tag{10.15}$$

Example Problem 10.9

A steel component is subjected to a reversed stress fatigue cycle. What is the stress ratio?

Solution

Since the stress is reversed, the minimum and maximum stress will have the same stress magnitude but opposite signs, since the maximum stress is tensile, and the minimum is compressive. This means that the stress ratio will be $- 1$.

Like all other properties, we can investigate the fatigue resistance of materials in laboratory tests. Here, too, we follow standards that dictate the geometry of the sample to be tested and the method of testing. The stress cycles discussed previously can be simulated on the sample. For example, if a material is subjected to a particular stress cycle, we wait until it fails (breaks). Then, we note the number of cycles to failure that coincide with the stress amplitude of each stress cycle investigated. We do this for different stress amplitudes; as you can imagine, the higher the stress amplitude, the smaller the number of cycles to failure. Conversely, the smaller the stress amplitude, the higher the number of cycles to failure. The output is a plot of the stress versus the number of cycles to failure; the stress can be stress amplitude (a), the maximum stress in the cycle (max), or even the minimum stress (σ_{min}). These are called **S–N curves** (stress-number of cycles to failure curves). Here, two types of S–N curves can be produced. The first (Fig. 10.18) shows a curve where the number of cycles to failure continuously increases as the stress is reduced. If a material exhibits this kind of S–N curve, such as aluminum, we can design structural components while considering fatigue failure in two ways. The first is to identify an acceptable number of cycles to failure, say 100 million cycles, and read off the curve the stress amplitude that would coincide with it on the curve. This stress level will be called the **fatigue strength (or the endurance limit).** Using a safety factor, we operate the component at a fraction of that stress amplitude. The other approach is to know what stress level will be applied and see how many cycles to failure on the S–N curve will coincide with this stress. The number of cycles to failure is referred to as the **fatigue life**. Let's, for example, assume the fatigue life for a stress amplitude of 100 MPa is 10,000,000 cycles. We would need to replace the component with a much smaller number of cycles, say 100,000 cycles, for example. Of course, probability of failure calculations will also need to be conducted. The designers ultimately will determine the specifics of the safety factors.

The other type of S–N curve is one where, initially, the number of cycles to failure increases with a decline in stress amplitude until a certain stress level is reached, at which point the curve plateaus out. This plateau occurs at a stress level called the fatigue limit, typically around 50% of the tensile strength of steels. Figure 10.19 shows such an S–N curve exhibited by steels and titanium.

Effect of Mean Stress on Fatigue Failure

Most fatigue tests use the reversed stress cycle; however, the mean stress may not be zero in engineering, i.e., the reversed stress cycle may not be operative. From the previous discussions, a reversed stress cycle can have the same stress amplitude as a repeated stress cycle while both the maximum and minimum stress are tensile. Logically, we cannot expect a component subjected to these two different cycles to fail after the same number of cycles, even if the stress amplitude is the same. The reversed stress cycle has a mean stress of zero, whereas the repeated stress cycle would have a tensile mean stress which would be more damaging, and cause the component to fail after a lower number of cycles.

Miner's Rule for Cumulative Damage

So far, we have assumed that a given component will be subjected to the same stress cycle until it fails. This may not happen, and a component could be subjected to

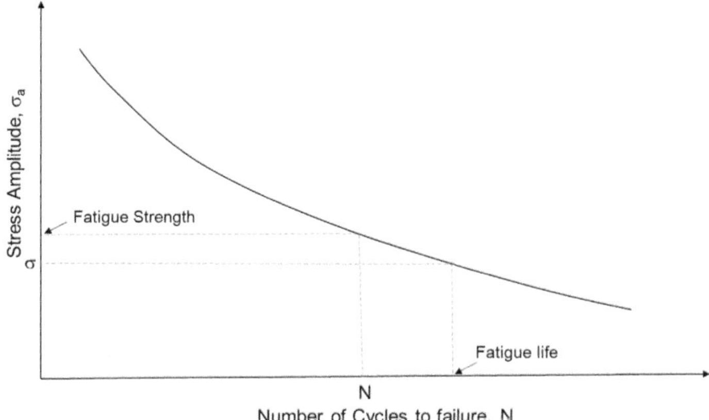

Fig. 10.18 S–N curve with continuously increasing number of cycles to failure with a decline in stress amplitude

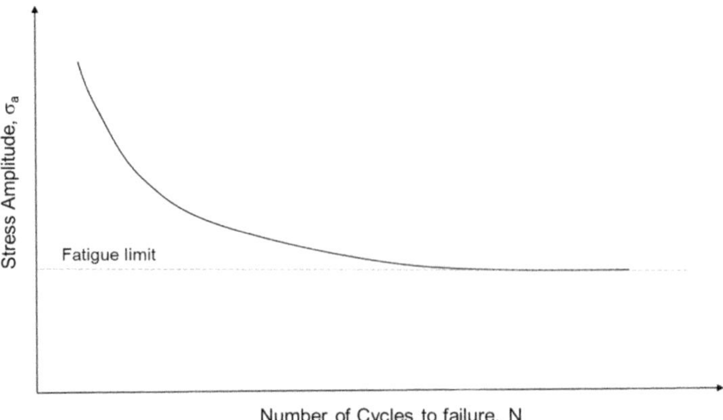

Fig. 10.19 S–N curve with a fatigue limit

a combination of different stress cycles, for example, with different mean stresses. Alternatively, it can sometimes be the case where a component is subjected to stress cycles of different stress amplitudes but the same mean stress. In any of these cases, how can we predict the number of cycles to failure? The fatigue damage will be a cumulative effect of the different types of cycles experienced by the material. Figure 10.20 shows a stress amplitude versus time plot, where a component has been subjected to a sequence of random cycles (1–7), each having a different mean stress. Cycle 4 (having n_4 number of cycles) has the highest mean stress, followed by cycle 1 (having n_1 number of cycles) and then cycle 5 (having n_5 number of cycles), and so on. It is logical that if cycle 4 were the only type of cycle the component experiences, it would fail after the smallest number of cycles (N_{f4}) compared with the other cycle types. If the same was considered for cycle 5, the number of cycles to failure (N_{f5}) is expected to be greater than that of cycle 4 (N_{f5}) since cycle 5 is less severe.

Miner's rule (an empirical rule) takes this into account by taking each type of cycle and dividing the number of cycles by the number of cycles to failure had that cycle type been the only cycle the component was experiencing. This will thus result in a fraction. The same will be repeated with the other types of cycles. Finally, all the fractions must add up to 1 if failure is to occur. Equation 10.16 depicts this result.

$$\sum_i \frac{n_i}{N_{fi}} = 1 \tag{10.16}$$

where n_i is the number of cycles in cycle type i, and N_{fi} is the number of cycles that would cause failure under cycle type i. Although Miner's rule is used in design against fatigue, critical components require additional tests since Miner's rule, at the end of the day, is still an empirical rule.

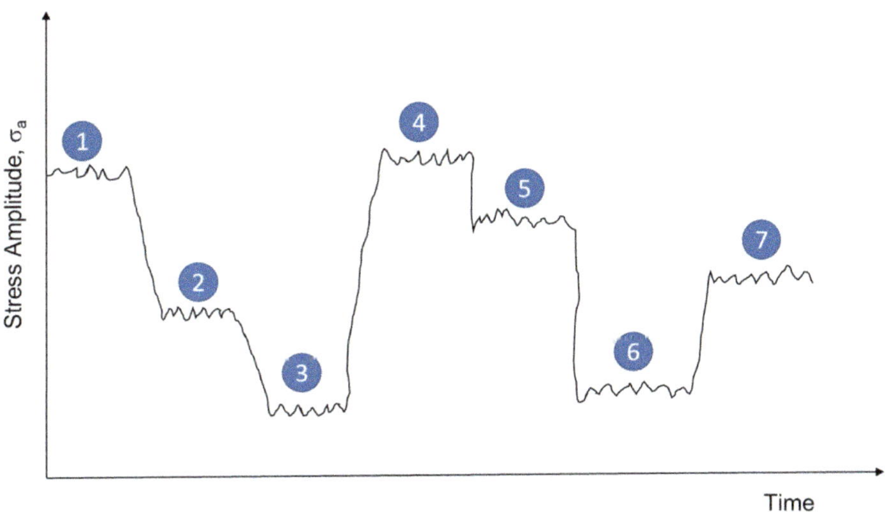

Fig. 10.20 Stress amplitude versus time showing a material that has been subjected to varying types of random cycles of different mean stresses over time

Example Problem 10.10

A non-critical component will be subjected to 3 different types of fatigue stress cycles. The first cycle type involved 100,000 cycles, and normally, the number of cycles to failure is 1,000,000. The second type of cycles involved 2,000,000 cycles, with the expected number of cycles to failure for that cycle type being 10,000,000 cycles, and finally, the third cycle type involved 300,000 cycles, with the number of cycles to failure for that cycle type is 500,000 cycles. Will this component fail?

Solution

Since different stress cycle types will be applied to the component, we can use Miner's rule, as shown below:

$$\frac{100,000}{1000000} + \frac{2,000,000}{10,000,000} + \frac{300,000}{500,000} = 0.1 + 0.2 + 0.6 = 0.9$$

Since the total is 0.9, which is less than 1, then the component should not fail.

What are the Signs of Fatigue Failure?

The field of fractography, where we examine the fractured surface of a component, is very important in identifying the reason for failure. It is usually part of a more significant investigation, called a material's post-mortem investigation. The telltale signs of a fatigue failure include the presence of **beachmarks, striations, a crack initiation site**, and typically a region in the fracture surface (where the fast fracture occurred) that looks different from the rest of the surface. Often, crack propagation during a fatigue cycle that involves tensile and compressive stresses as part of the stress cycle, the crack surfaces rub against each other (during the compressive part), resulting in a somewhat polished surface, which looks different than the region in the fracture surface where fast fracture finally occurs. These are identified in Fig. 10.21, showing a crack initiation site and the direction of crack propagation until the crack reaches the critical flaw size, where fast fracture occurs (also identified in the figure). Large, dashed contours are shown (beachmarks), and much finer markings are seen (striations) between them. Each **striation** represents one fatigue cycle; hence, if we counted the total number of striations from the crack initiation point until the crack reaches the critical flaw size, we would have counted the number of cycles to failure. So, if striations represent fatigue cycles, then what are beachmarks? Since components are usually part of a given machine or structure. Sometimes, a machine is switched on for a certain period, then switched off during the weekend, and then on Monday, it gets switched on again. Every time the machine is switched on and off, a beachmark is formed if fatigue is occurring. Hence, the total number of beachmarks represents the total number of times the machine in which the component is located is switched off and on.

Example Problem 10.11

During a post-mortem investigation of a failed component, the investigators reported finding 77 beachmarks and 10,000 striations. What do these numbers tell you?

Solution

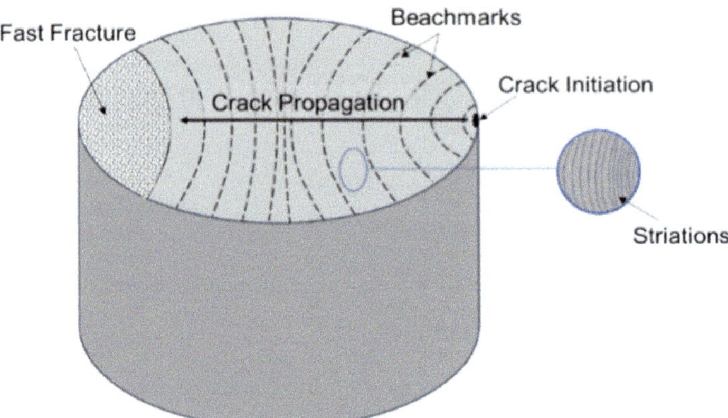

Fig. 10.21 Schematic of the fracture surface of a component that had failed by fatigue. The image shows, where the crack initiated, the direction of crack propagation, beachmarks, striations and a region where fast fracture occurred

First, the component failed because of fatigue, since all the telltale signs were there. Second, it took 10,000 stress cycles to propagate the crack to the critical size. Finally, the component was possibly part of some machine that switched on and off 77 times during its life.

In this chapter, we learned about fracture, creep, and fatigue. There are some conditions where both creep and fatigue must be considered, such as turbine blades operating at high temperatures. In this case, the treatment is more complex. It is important to note that when it comes to critical components, experimental simulations are typically required in addition to analyses to make sure nothing is missed. Finally, you may come across the terms **low-** and **high-cycle fatigue**. In low-cycle fatigue, the material fails under 10^4 or 10^5 cycles, and this is due to high-stress levels being applied. At the same time, high-cycle fatigue occurs for low applied stresses, where the material survives for much longer cycles.

The material is usually tested for 5×10^8 cycles for low-cycle fatigue.

Problems

10.1. What is the difference between creep and fatigue?

10.2. At what homologous temperatures should we worry about creep in ceramics and metals?

10.3. Sketch a creep curve and identify on the curve the instantaneous strain, primary creep, secondary creep, and tertiary creep regions on the creep curve. Also, identify the rupture life. Make sure to label the axes.

10.4. How does increasing temperature or stress affect the steady-state creep rate?

10.5. With the aid of sketches, discuss the Nabarro–Herring creep mechanism. How does it differ from the Coble creep mechanism? What is the value of the creep exponent for each mechanism?

10.6. With the aid of sketches, discuss the dislocation climb creep mechanism. Why is it sometimes called power-law creep?

10.7. Why isn't it good practice to have square windows on airplanes?

10.8. Determine the stress concentration factor of an internal/central elliptical cavity of length 100 mm in an infinite plate. Assume the radius of curvature to be 1 mm.

10.9. What would be the stress concentration factor if the cavity in problem 10.8 were an edge/surface cavity or notch of the same length and radius of curvature?

10.10. A material has a plane strain fracture toughness of 25 MPam$^{1/2}$. Calculate the central critical flaw size if a tensile stress of 300 MPa is applied. Assume $Y = 1$.

10.11. A component has a surface crack of 1 mm in length. Would the component fracture if a 700 MPa tensile stress is applied? Why or why not? The fracture toughness of the component is 50 MPam$^{1/2}$, and $Y - 1$.

10.12. A component is expected to be loaded in tension at a stress level of 100 MPa. You are considering two materials to make the component out of, a copper alloy and a steel. Which one of those materials has the higher critical central flaw size? Assume $Y = 1$, K_{IC} for copper alloy is 50 MPam$^{1/2}$, and that of steel is 25 MPam$^{1/2}$.

10.13. Metals are known to be forgiving materials. Unlike ceramics, they tend to deform plastically rather than shatter or experience fast fracture. You are provided with a metal. If the yield strength of this metal is 100 MPa, would it experience fast fracture or yield plastically if it contains a 100 mm surface crack?

10.14. What is the difference between high-cycle fatigue and low-cycle fatigue?

10.15. Sketch and discuss three types of fatigue stress cycles. Make sure to label all axes.

10.16. A fatigue cycle has a maximum stress of 500 MPa and a minimum stress of 100 MPa. What is the stress range and the stress amplitude?

10.17. What is the stress ratio of a fatigue cycle with a stress range of 300 MPa and a mean stress of 100 MPa?

10.18. Sketch the S–N curves for a material with a fatigue limit and one without. Label all axes.

10.19. What is meant by fatigue life and fatigue strength?

10.20. What would happen if a material were subjected to a fatigue cycle that has a stress amplitude that lies below the material's fatigue limit?

10.21. A component was subjected to a fatigue cycle having a 300 MPa stress amplitude for two hundred cycles (at this stress level, the material is expected to fail after 1000 cycles). The same component was subjected again to another stress amplitude of 50 MPa for 20,000 cycles (at this stress level, the material is expected to fail after 130,000 cycles). Would this component fail after applying both cycles? Show analysis.

10.22. The fracture surface of a component that failed due to fatigue was examined. It was found to contain 77 beachmarks and 200,000 striations. What do these numbers mean?

10.23. Describe a process that can improve the fatigue life of a component.

Part IV
Layer 5: Save Me

The Sphere

Layer 5
"Save me."

Professor, the color of sample 5 is black.

Alex, based on my many years of experience, I highly suspect that layer 5 is probably a polymer, but what type? I don't know. I also would not be too hung up on the color. Polymers are easily colored.

I want to say that it does feel like a plastic..

Please read the chapter "Introduction to Polymers" in As a Matter of Fact to gain the knowledge needed to unravel this material layer's identity. Then I want you to measure its density, tensile strength, ductility, and modulus, then let's talk again.

I'm on it, Prof.!

Chapter 11
Introduction to Polymers

11.1 Introduction

Polymers play a significant role in today's society. This chapter introduces their structure and properties.

11.2 The Structure of Polymers

So far, we have discussed metals and ceramics composed of atoms as their basic building blocks. These atoms form unit cells for crystalline materials that, when repeated, provide long-range atomic order and crystallinity within the material. In amorphous materials, such unit cells do not exist; instead, a disordered collection of atoms dictates the structure. However, other important materials rely on building blocks in the form of macromolecules instead of unit cells. These materials are called polymers. They can be subdivided into **natural polymers** (which have existed since the beginning of time) or **synthetic (man-made) polymers**.

Natural polymers include wood, protein, natural rubber, and even DNA. One of the earlier driving forces to produce *synthetic polymers* was related to the games of billiards, pool, and snooker. Initially, ivory was used to make the balls, which unfortunately relied on killing elephants to obtain their tusks. Apart from the obvious humane reasons we have today, which prevent us from doing so, at the time, this posed a problem in terms of supply and expense, which prompted the consideration of other cheaper and more available options. These included ivory dust bonded with a natural polymer, shellac (a resin secreted by the female lac bug on trees), and collodion. This was later replaced with cellulose nitrate and camphor (from the camphor tree), which were also used to make dolls at the time. The material was probably the first man-made polymer, which was ultimately named **celluloid**. However, its flammability led to its downfall. The invention of Bakelite in 1907 (a Phenol–Formaldehyde polymer) saw

K. Morsi, *An Engaging Approach to the Science and Engineering of Materials*,
https://doi.org/10.1007/978-3-032-06231-4_11

the proper arrival of polymers, after which it was used on a large scale, especially as an electrical insulator in addition to billiard balls. Today, we hear of many polymers used in our daily lives. Everyone must have heard the names PVC, polyethylene, Teflon®, polycarbonate, nylon, polyester, Saran wrap®, and ABS. Some select polymers and their applications are listed in Table 11.1. A polymer like polyethylene is one of the simplest and most widely used polymers today. Depending on its molecular structure, molecular arrangements, and the lengths of its molecules, a variety of polyethylene is available to us today, including low-density polyethylene (LDPE), linear low-density polyethylene (LLDPE), high-density polyethylene (HDPE), and molecular weight polyethylene (UHMWPE) with highly long polymer molecules. UHMWPE is used as a biomaterial when it is crosslinked (meaning its molecules are made to connect at specific locations permanently). In that state, the material has good wear resistance, a low coefficient of friction, and self-lubricating properties. These properties have prompted its use as an acetabular cup in hip joint replacement. Polyethylene comes in different densities, for example, low-density polyethylene (~ 0.92 g/cm^3) and high-density polyethylene (~ 0.96 g/cm^3). The difference in molecular architecture of the same polymer can result in different properties; this is seen in Table 11.1. Interestingly, HDPE can be boiled in water for extended periods, which is helpful for sterilization.

The wide variety of structures and compositions found in polymers results in materials with a wide range of properties, which opens the door to many applications, as seen from a few examples of synthetic polymers in Table 11.2.

Let us first break down the word "polymer". Mer is from the Greek "meros", which means "part", and poly means "many". In fact, polymers are made up of molecules, each composed of many repeated parts called mers.

When all these mers are the same and repeated to produce a polymer molecule, we call the resulting polymer a **homopolymer**. If different types of mers combine to produce a polymer molecule, we call the polymer a **copolymer** (more about this later). For now, let's start with one of the simplest and most popular man-made

Table 11.1 Selected properties of LDPE and HDPE

Property	LDPE	HDPE
Density (g/cm^3)	0.92	0.96
Crystallinity (v/o)	~ 0	~ 50
Thermal expansion (°C^{-1})	180 × 10–6	120 × 10–6
Thermal Conductivity (W/m^2)(°C/m)	0.34	0.52
Tensile Strength (MPa)	5–15	20–40
Young's Modulus (MPa)	100–250	400–1200
Heat resistance for continuous use (°C)	55–80	80–120
10 min temperature exposure (°C)	80–85	120–125

Source L. H. Van Vlack, Elements of Materials Science and Engineering, 6th edition, Addison Wesley (1989)

Table 11.2 Some common polymers and their uses

Polymer	Uses
Polyethylene (PE)	Supermarket bags (HDPE), clear packaging film (LDPE), food container, textile fibers
Polypropylene (PP)	Molded bottles, chemical tanks, packaging
Nylon	Gears, cams, coatings on dishwasher racks, appliance wheels, rope, clothing, textiles, carpeting
Polyester	Boil in bag containers, beverage containers, fibers, photographic films
Polycarbonate (PC)	Safety shields, Football helmets, compact discs, electrical and appliance housing, glazing for solar collectors
Polymethyl methacrylate (PMMA) e.g. Plexiglas®	Camera lenses, flashlights, safety glasses, windows, skylights, glazing for aircrafts and boats, safety shields
Polytetrafluoroethylene (PTFE), e.g. Teflon®, Fluon®	Moldings, films, anti-corrosive seals, high temperature electronic parts, chemical pipes and valves
Acrylonitrile Butadiene Styrene (ABS)	Lego bricks, popular 3D printing filament, pipes fitting, computer and telephone housings
Polyetheretherketone (PEEK)	High temperature electrical insulation and coatings
Styrene-Butadene rubber (SBR)	Tire treads
Polyvinylchloride (PVC)	Plasticized PVC: Imitation leather, flexible tubing, wire coatings, rigid PVC (without plasticizer): low cost piping, guards, ducts, fume hoods
Polyvinylidene chloride (PVDC)	Saran Wrap® and similar products
Polystyrene (PS)	Toys, battery cases, appliance housing, plastic tableware, food containers

Multiple Sources including: K. G. Budinski, M. K. Budinski, Engineering Materials: Properties and selection, eighth edition, Pearson-Prentice Hall, 2005

homopolymers out there today: "poly"ethylene (sometimes referred to as polythene). We will go through the process of its production, and by doing so, we will learn more about the structure of this important polymer. Specifically, we will discuss a **polymerization** process called **free radical addition polymerization**. The process can be divided into three stages: *initiation*, *propagation*, and *termination*. It all starts with the molecule ethylene (C_2H_4), which we will call a **monomer**. The ethylene molecules form a gas, and Fig. 11.1 shows the chemical structure of this molecule. As can be seen, ethylene has a double covalent bond between its two carbon atoms. Such molecules with a double bond (or even triple bond) are referred to as **unsaturated**. This means that they have a driving force to become **saturated** and, hence, can react with other molecules to do so, and in the process, the molecule grows in length. It is essential to point out that monomers that can grow into polymer molecules must be able to react with other molecules at more than one location within their structure. Otherwise, any molecule produced from the first reaction will not be able to grow further. We call these monomers **bifunctional**, if two sites are available for reaction, **trifunctional**, if three sites are available for reaction, or generally **poly (multi)-functional** if more than two sites are available for reaction. The growing polymer molecules can take different shapes and architectures depending on such functionality. For example, a monomer that can react on its left and right sides should only grow linearly. Suppose another monomer is multifunctional and it can grow on its left and right sides in addition to other locations. In that case, the likelihood is that the growing molecule will also grow branches and may also link up and bond with other molecules, to such an extent that it produces a 3D bonded polymer. Hence, the molecular structure would be different in all these resulting polymers, so we expect different properties.

Through thermal decomposition, a free radical can be produced; an example is a peroxide, such as Benzoyl peroxide, which eventually breaks down into phenyl radicals and carbon dioxide. These phenyl radicals can then be introduced to ethylene gas under certain conditions of heat and pressure. The radicals, in essence, act as a catalyst by attracting one of the electrons in the double bond and linking up to the ethylene molecule. In doing so, the bonding in the other carbon atom becomes incomplete; hence, the carbon atom acts as a free radical, ready to react with more ethylene molecules. The monomer is now said to be activated. This is the *initiation stage*. Step by step, the polyethylene molecule grows by this process (Fig. 11.1), i.e., *the propagation stage*. As one can imagine, this is happening in many locations in the reaction chamber, and an extremely large number of molecules are being grown simultaneously (not all having the same length). We have control over when to stop

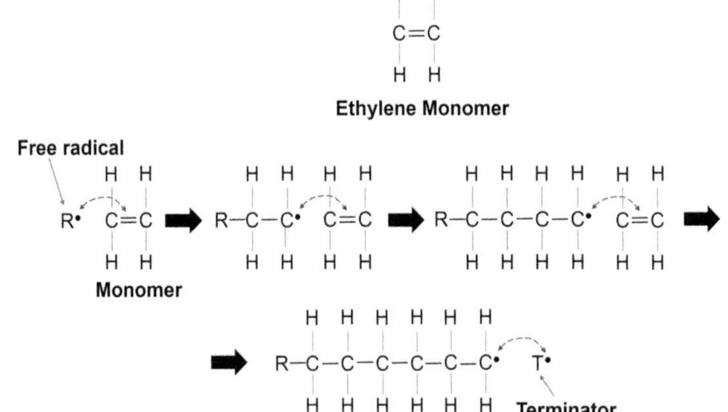

Fig. 11.1 Schematic of polymerization process starting with free radical reacting with the monomer and ending with a terminator radical

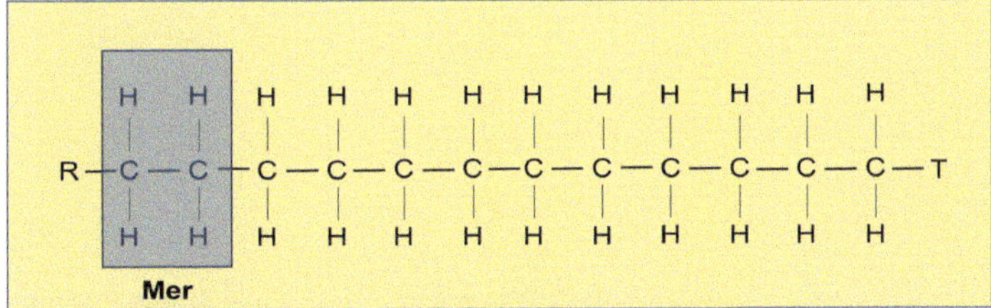

Fig. 11.2 Schematic of terminated polyethylene polymer molecule with 6 mers

the reaction and, hence, decide the average size of the synthesized polymer molecules. Why is this important? This is because the length of the polymer molecules can also affect the final properties of the same polymer. To do this, either the free radical on the growing molecule bonds with another free radical on another growing molecule, and in doing so, the combined molecule ceases to react further, or a terminator molecule is introduced, such as an OH group, whereby the growth of the growing molecule is terminated. This is shown in Fig. 11.1, where a terminator molecule is introduced to stop the growth of the molecule while it is still small in size. This is the *termination stage*. Figure 11.2 shows a longer polyethylene polymer molecule after termination. Note that the shaded part in the figure represents one of the repeated *mer* units, C_2H_4, which make up the polymer molecule. In this case, the polymer molecule is made of 6 mers.

Another way to represent such a polymer is shown in Fig. 11.3.

The repeated mer is placed between the brackets, and 6 is the number of mers in the polymer molecule. If one of the hydrogen atoms between the brackets is replaced with a different atom or several atoms/side groups, we obtain a *vinyl polymer*. Alternatively, if two hydrogen atoms attached to a carbon atom are replaced with a different atom or side group, the resulting polymer will be called a *vinylidene polymer*. These are summarized in Fig. 11.4. Such polymers are also called **Ethenic** polymers, meaning they depend or build on the C=C backbone of the polyethylene monomer.

We also refer to these polymers as **linear polymers**, pretty much like spaghetti (Fig. 11.5a). It is possible that linear polymer chains can fold into **crystalline** regions (also called crystallites). These regions will still be surrounded by amorphous structures of the spaghetti-like random entangled arrangements of molecules (Fig. 11.5b). Hence, this type of polymer is referred to as **semi-crystalline**. These crystallites can range from 5 to 95% in polymers. Polyethylene is a semi-crystalline polymer.

Fig. 11.3 An alternative representation of polyethylene having 6 mers in its chains

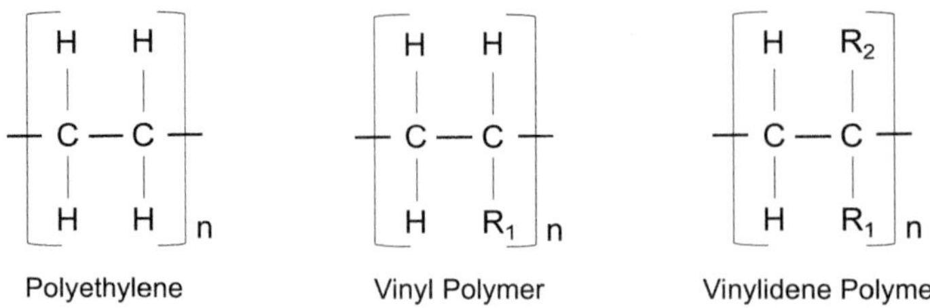

Fig. 11.4 Mer structures of polyethylene, Vinyl and Vinylidene polymers

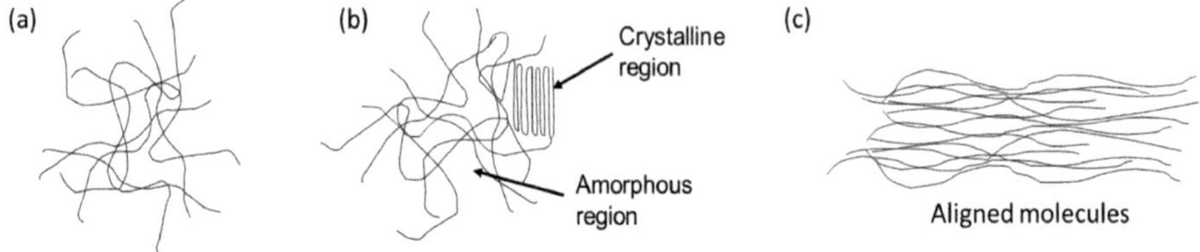

Fig. 11.5 **a** amorphous linear polymer, **b** semi-crystalline polymer showing both crystalline and amorphous regions in its microstructure, **c** amorphous polymer molecules aligned in a particular direction due to the application of deformation, for example extrusion of the polymer in that direction

Figure 11.5c shows how the molecules in an amorphous (also called **glassy**) linear polymer can be aligned in one direction. This can occur if the polymer is extruded or deformed in tension, which allows the linear molecules to align in the extrusion direction, resulting in increased strength and stiffness along that direction.

Many polymers can be formed using this **addition polymerization process**.

Table 11.3 lists several polymers, each with its chemical structure, melting point, and molecular weight of the mer. As can be seen, some of the mers have side groups that are simple atoms, such as PE and PVC, while some have more cumbersome side groups, such as PMMA and PVA. All these differences will affect the properties of the final polymer. Whether or not a linear polymer molecule can fold and produce crystalline regions will logically depend on factors such as the bulkiness of the side groups. For example, while polyethylene can achieve such crystalline structures, PMMA cannot. A quick look at Table 11.3, which shows the chemical structure of the two polymers, will quickly reveal the difference; PE has a simple H-side atom, while PMMA has a very bulky one.

Another approach to producing polymers is called **condensation polymerization**. Important polymers are produced this way. These include Nylon 6,6 and polyethylene terephthalate (PET), a polyester. Here, two different monomers are typically reacted to produce a larger molecule, and a byproduct is produced in the process. This byproduct, depending on the reacting monomers, can be H_2O (hence the name condensation) or Methanol, etc.

As mentioned, we can control the size of the polymer chains during the synthesis of polymers. Of course, not all the chains will come out the same size, but a distribution

Table 11.3 Several polymers with their chemical structure, molecular weight of the individual mer and the melting point of the polymer

Polymer	Chemical structure	Molecular weight of Mer (g/mol)	Melting point (°C)
Polyethylene (PE)	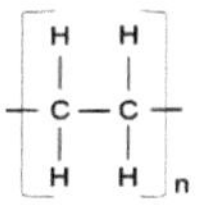	28	110–137
Polypropylene (PP)	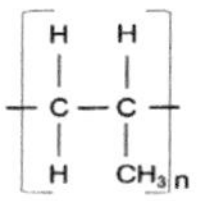	42	165–177
Polyvinyl chloride (PVC)	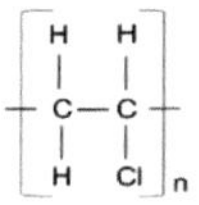	62.5	− 204
Polystyrene (PS)	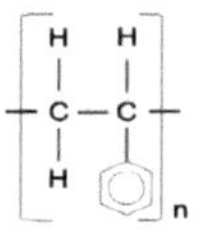	104	150–243
Polyacrylonitrile (PAN)	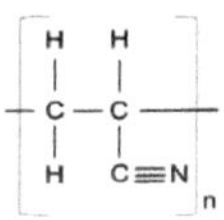	53	Does not melt
Polyvinyl acetate (PVA)	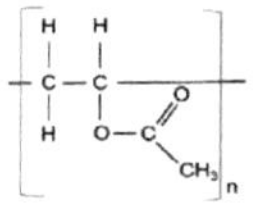	86	177
Polymethyl methacrylate (PMMA)	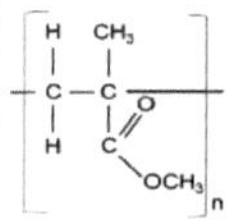	100	160

Source (except molecular weight of mer): W. F. Smith, J. Hashemi, Foundations of Materials Science and Engineering, 5th edition, McGraw Hill, 2006

Note that the phenyl group in polystyrene is represented by a hexagon with a circle inside it, as shown in Fig. 11.6

of chain lengths will be produced, and then an average is usually quoted. The longer the average chain length, the higher the melting or softening temperature of the polymer. The **average molecular weight** of the polymer is simply a representation of the chain length, in the sense that the longer an average chain is, the more atoms it has and hence weighs more. The average molecular weight has a bearing on the

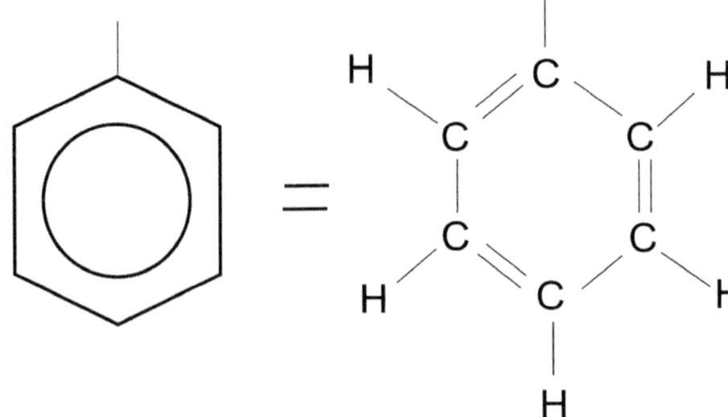

Fig. 11.6 Geometric representations of Phenol side group

polymer's state; as the molecular weight increases, the matter can change from gas to liquid to solid. When the polymer is in a solid state, it is called a **high polymer**.

Even the transparency of the polymer is dependent on its molecular weight, and an increase in molecular weight can change a polymer from being transparent to translucent to opaque.

Two average molecular weights can be quoted: a **number-average molecular weight** and a **mass-average molecular weight**. In the case of *number average molecular weight*, $\overline{M}_n$, the fraction of molecules within an interval of molecular weights is plotted for different intervals. For example, if a polymer has molecules with molecular weights ranging from 10,000 to 100,000 g/mol. We can divide this range into 10 intervals: 10,000–20,000, 20,000–30,000, 30,000–40,000, 40,000–50,000, ..., 90,000–100,000. We then determine the fraction of the total molecules in the polymer with different specified intervals. Then, we obtain a number-average molecular weight according to Eq. 11.1.

$$M_{\mathrm{Na}} = \sum x_{\mathrm{ni}} M_i \tag{11.1}$$

M_{Na} is the number-average molecular weight of the polymer and x_{ni} is the number fraction of molecules (out of the total) with a molecular weight M_i. M_i is the molecular weight value in the middle of the interval range. For example, for an interval range of 10,000–20,000 g/mol, M_i is 15,000 g/mol.

Similarly, the *mass-average molecular weight* of the polymer is given by Eq. 11.2.

$$M_{\mathrm{ma}} = \sum x_{\mathrm{mi}} M_i \tag{11.2}$$

M_{ma} is the mass average molecular weight of the polymer and x_{mi} is the mass fraction of molecules (out of the total mass of molecules) with a molecular weight M_i. M_i is the molecular weight value in the middle of the interval range. For example, for an interval range of 20,000–30,000 g/mol, M_i is 25,000 g/mol.

M_{ma} is almost always greater than M_{na}, for the simple fact that although a high-molecular-weight molecule and a low-molecular-weight molecule will contribute the same to the calculation of the number fraction, they do not when considering a weight fraction. The high-molecular-weight molecule will weigh more than a low-molecular-weight molecule, contributing more to the weight fraction calculation. This ends up biasing the average molecular weight to higher values when weight fractions are considered, as opposed to number fractions.

Example Problem 11.1

The molecular weight distribution of a linear polymer is shown in Table 11.4. Calculate the number average molecular weight and this polymer's average molecular weight.

Solution

Table E11.1 lists all the molecular weight intervals of the polymer, the mass fraction (x_{mi}), and the number fraction (x_{ni}) of each interval. It calculates the mid molecular weight value (M_i), $x_{mi}M_i$ and $x_{ni}M_i$ for each interval. In accordance with Eqs. 10.1 and 10.2, the mass average molecular weight (M_{ma}) and the number average molecular weight (M_{na}) are calculated by summing all the $x_{mi}M_i$ and $x_{ni}M_i$ values respectively. This yields $M_{ma} = 27{,}500$ and $M_{na} = 23{,}000$. Note that $M_{ma} > M_{na}$.

Another important parameter is the degree of polymerization (DP), which is given by Eq. 11.3 and basically shows how many mers are in an average polymer molecule.

$$\mathrm{DP} = \frac{M_{na}}{M_{mer}} \text{ or } \frac{M_{ma}}{M_{mer}} \tag{11.3}$$

where M_{mer} is the molecular weight of the mer (i.e., the repeated unit). Obviously, when quoting the DP it is important to also state which average molecular weight values you used, number or weight average.

Table E11.1 Molecular weight intervals of a linear polymer with their corresponding mass and number fractions, with relevant calculated parameters

Molecular weight (MW) interval (g/mol)	Mid value of MW interval (M_1)	Mass fraction (X_{mi})	$X_{mi}\,M_i$ (g/mol)	Number fraction (X_{ni})	$X_{ni}\,M_i$ (g/mol)
1000–11,000	6000	0.10	600	0.20	1200
11,000–21,000	16,000	0.20	3200	0.25	4000
21,000–31,000	26,000	0.30	7800	0.28	7280
31,000–41,000	36,000	0.20	7200	0.15	5400
41,000–51,000	41,000	0.15	6150	0.10	4100
51,000–61,000	51,000	0.05	2550	0.02	1020
Total	176,000	1	27,500 M_{ma}	1	23,000 M_{na}

Example Problem 11.2
Polyethylene, polystyrene, and PMMA samples, each with a mass average molecular weight of 50,000 g/mol, are available. What is the degree of polymerization in each one?

Solution:

Equation 11.3 shows the degree of polymerization as

$$DP = \frac{M_{ma}}{M_{mer}}$$

Table 11.3 shows that the M_{mer} for PE, PS, and PMMA is 28, 104, and 100 g/mol, respectively.

Hence for PE, $DP = \frac{50000\,g/mol}{28\,g/mol} = 1786$.

The same calculation for PS and PMMS results in DP values of 476 and 500 respectively.

A closer look at M_{na} and M_{ma} will reveal that M_{na} gives voice to molecules of low molecular weight and treats them equally to ones of higher molecular weight, whereas M_{wa} favors more molecules of higher molecular weight. As such, the broadness of the molecular weight distribution in a polymer can be described in terms of what is termed the **polydispersity index** (PDI), (sometimes also called **dispersity** or **heterogeneity index**), which is given by Eq. 11.4. This index would have a value approaching 1 for mono (or homo) dispersed polymers, that is polymers of more uniform polymer molecule lengths, and a value greater than 1 for broader distributions, for example, a value of 5. Typically, commercial polymers can have values of 4 or 5.

$$PDI = \frac{M_{ma}}{M_{na}} \tag{11.4}$$

Example Problem 11.3
In Example Problem 11.1, it was determined that the mass average molecular weight (M_{ma}) is 27500 g/mol and the number average molecular weight (M_{na}) is 23,000 g/mol. What is the polydispersity index?

Solution:

$$PDI = \frac{M_{ma}}{M_{na}} = \frac{27500}{23000} = 1.2$$

This value seems closer to uniform molecular lengths than the typical larger values of 5 for commercial products.

It is important to note that so far, we have been assuming that each molecule will stretch out and simply be a straight line. If we assume a molecule of PE has 1000 mers, since the length of the C–C bond is 0.154 nm, and each mer has an equivalent of two bond lengths, then for the 1000 mers, we have 2000 bonds, each with a length of 0.154. This means the total length of the molecule will be 308 nm. This is not true primarily since the carbon–carbon backbone will be more like a zigzag, as shown in Fig. 11.7, which will shorten the molecule. The zigzag nature of the molecule will shorten the molecule to a length of 251.5 nm (i.e., $308 \times \sin(109.5/2)$). This still does not necessarily mean that the end-to-end distance is 251.5 nm. However, this length represents the maximum possible end-to-end distance the molecule can attain. Second, C–C single bonds are typically free to rotate in many polymers. It is also possible for a polymer molecule to kink, twist, and coil, which would decrease its length, even to the extent that the front and end parts of the molecule can be in contact with one another. The root mean square length (average) will be some length between these extremities. This length is given by Eq. 11.5 and represented in Fig. 11.8.

$$\bar{L} = l\sqrt{x} \tag{11.5}$$

l is the length of the C–C bond, and x is the number of bonds in the molecule. If we take the example presented previously regarding the short PE molecule with 1000 mers. This molecule contains 2000 C–C bonds with an individual C–C bond length of 0.154 nm. Hence $\bar{L} = 0.154 \sqrt{2000} = 6.9$ nm.

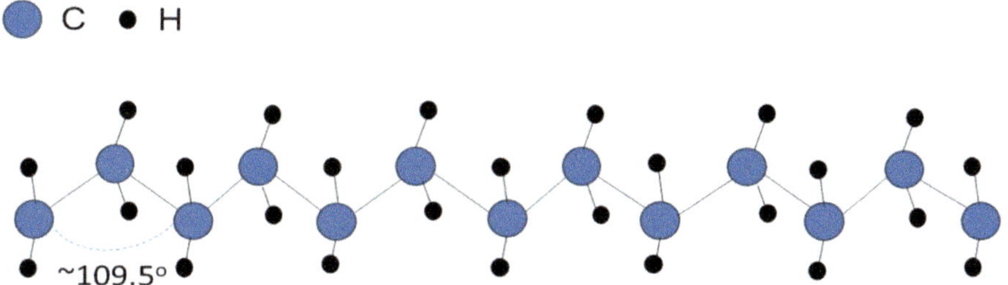

Fig. 11.7 Polyethylene molecule showing the zigzag-like nature of the molecule

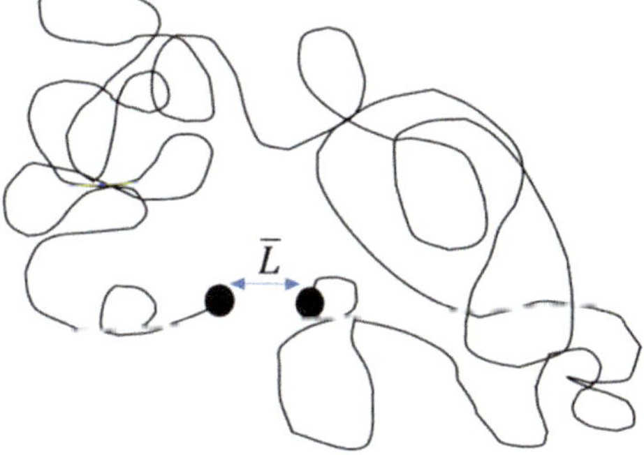

Fig. 11.8 Average root mean square length (end-to-end distance) between the start and end of an amorphous linear molecule

Example Problem 11.4

What is the end-to-end distance for a PVC polymer molecule having a molecular weight of 100,000 g/mol? How much would that distance change if the molecular weight was 500,000 g/mol instead?

Solution:

Table 11.3 informs us that the molecular weight of the PVC mer (repeat unit) is 62.5 g/mol. The degree of polymerization (DP) will give us how many mers there are in this molecule, as follows:

$$DP = \frac{100000}{62.5} = 1600$$

This means we have 1600 mers, and in each mer, we have 2 C–C bonds. Hence, the molecules contain 3200 C–C bonds, each with a length of 0.154 nm. The end-to-end distance will be given by Eq. 11.5

$$\overline{L} = 0.154\sqrt{3200} = 8.7\,\text{nm}$$

For a 500,000 g/mol molecule, DP would be 8000, giving 16,000 C–C bonds. Hence,

$$\overline{L} = 0.154\sqrt{16000} = 19.5\,\text{nm}$$

Therefore, a 5-time increase in the molecular weight of the molecule will more than double the end-to-end distance.

It is worth noting that we have so far represented the polymer molecule as a simple repetition of the mer in perfect order. Several variations can occur for the same polymer, resulting in **Structural isomerism** and **Configurational isomerism**. In *structural isomerism*, the chemical composition of the monomer may remain, but the structure could be different. An example is shown in Fig. 11.9 for polystyrene and poly (*p*-xylylene), two isomers with the same chemical composition but different structures and, therefore, properties.

Fig. 11.9 Isomers of polystyrene, having the same chemical composition but differ in the structure

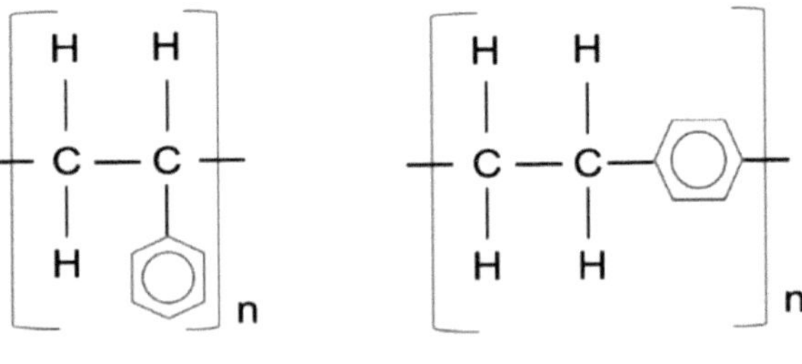

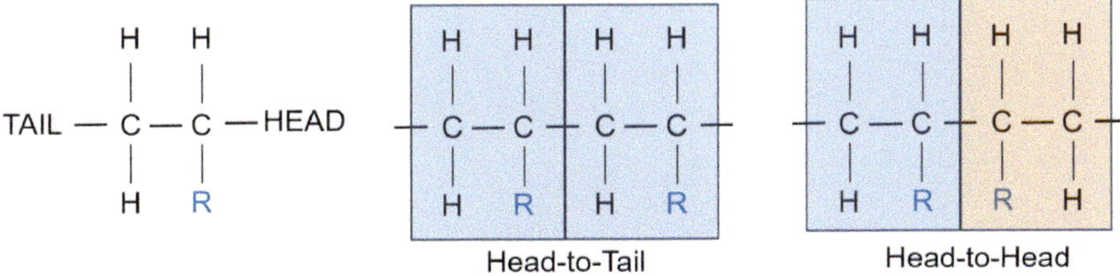

Fig. 11.10   Structural variations, head-to-tail, and head-to-head

Another variation relates to how the monomers attach to one another. A monomer normally attaches to another in a way called *head-to-tail*, see Fig. 11.10. Occasionally, head-to-head attachment can be realized, but predominately the head-to-tail prevails.

In **Configurational isomerism**, a polymer molecule may experience variations in the positioning of the side groups; for example, they may switch their position with a hydrogen atom. This is called **stereospecificity, stereoisomerism, or tacticity**. For example, if we consider a linear polymer with a side group R. Figure 11.11 shows three ways the side groups can be located on the main polymer chain to produce three different isomers. The **Isotactic** configuration is the regular arrangement where the side groups are all on one side of the chain, the **syndiotactic** configuration is when the side groups are at the same position along the length of the chain but alternate from one side to the other. Finally, **atactic** is a random positioning of the side groups on the chain. Each of the three isomers will result in different properties; remember, structure affects properties (one of the golden rules). Polymer tacticity can be estimated using nuclear magnetic resonance spectroscopy.

The third type of isomers is called **Geometric Isomers**; these can occur in polymers having double bonds in place of the single C–C bonds, which are the backbone of the polymer molecule. Let's look at the isoprene mer. One form of this mer is shown on the left in Fig. 11.12, which is the building block of natural rubber called *cis-polyisoprene*; here, at the double bond, the methyl (CH_3) group is on the same side as hydrogen. A different isomer is shown to the right, producing a *trans-polyisoprene* natural polymer called gutta-percha, which is used today, including in dentistry. Trans-polyisoprene has the methyl group on the opposite side to the hydrogen as seen in the figure. The properties of both isomers are different.

Now that we have shown the different variations that can occur in polymer chains, we need to consider other even more pronounced variations in the structure of polymers. Here, polymers can be divided into linear, branched, cross-linked, and network polymers, which are discussed below.

How the polymer molecules reside within a polymer in terms of arrangement and bonding with other polymer chains can result in different polymers. These are presented in Fig. 11.13 and are discussed below.

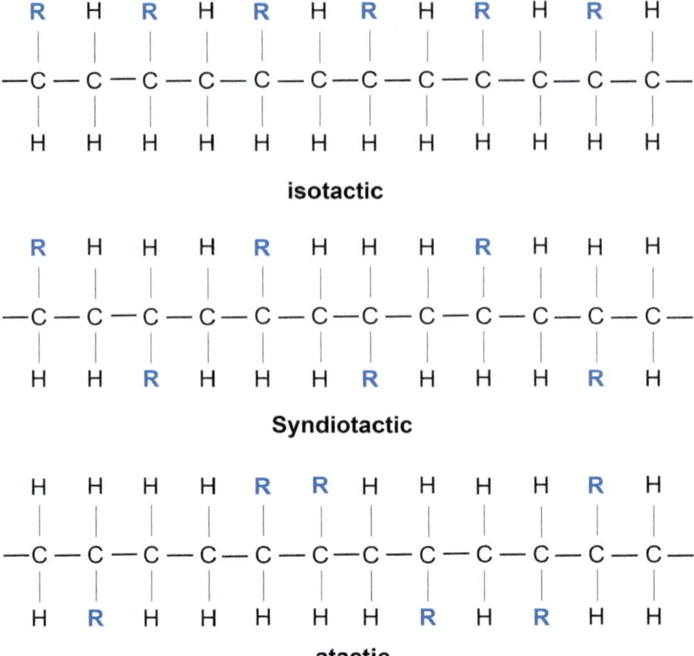

isotactic

Syndiotactic

atactic

Fig. 11.11 Three ways side groups can be located on the main polymer chain to produce three different isomers

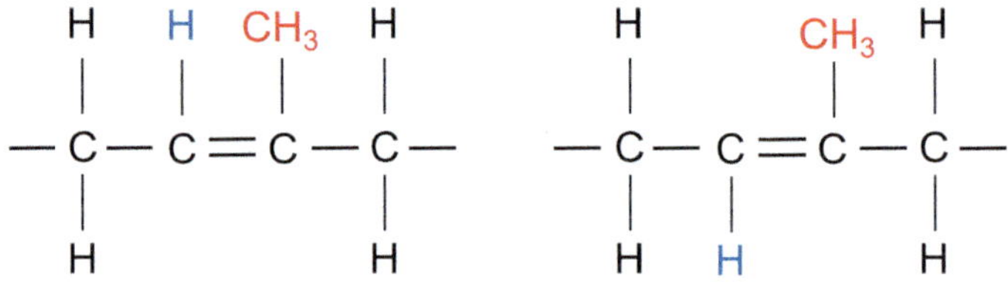

Mer for *cis*-polyisoprene
(natural rubber)

Mer for *trans*-polyisoprene
(gutta percha)

Fig. 11.12 Geometric isomers of isoprene

11.2.1 Linear polymers

Linear polymers are those made up of linear molecules, see Fig. 11.13a, examples of this type of polymer are polyethylene, polystyrene, polypropylene, and PMMA. Many of them make up what we call **thermoplastics**, referring to the thermomechanical behavior of such polymers, such that they can be mechanically deformed and shaped when heated. The intramolecular bonds within the chain are covalent, which

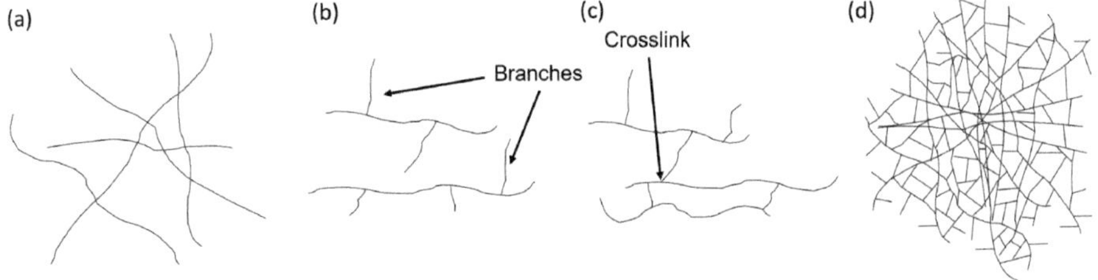

Fig. 11.13 Different types of polymers; **a** linear, **b** branched, **c** crosslinked, and **d** network (2 D representation of a 3D structure)

makes the individual polymer chains very strong. However, the different molecules are held together by weak secondary bonds (van der Waals bonds) which can be overcome by the application of heat due to the thermal vibrations of the molecules.

At relatively high temperatures, these linear molecules can easily slide past one another under applied stress, resulting in plastic deformation. Another good feature about plastics is that they can melt and be reshaped multiple times. Let us revisit the concept of crystallites. As mentioned earlier, there are many semi-crystalline polymers. Polymers with crystalline regions are stronger and stiffer than purely amorphous ones, but what are the factors that favor crystallite formation? It's logical: the more uniform a polymer chain that excludes bulky side groups is, the more likely the polymer will show crystallinity. We see this clearly in polyethylene, which is semi-crystalline. A polymer such as PMMA with quite bulky and non-uniform mers is amorphous.

We often talk about the melting point of polymers, but what does that mean? Such a temperature is clearly defined for metals, but it needs further explanation for polymers. If we consider a semi-crystalline polymer, its microstructure consists of crystallites (folded molecules) surrounded by amorphous regions, but the crystallites will come in different sizes, small and larger. As the polymer is heated, the crystallites will gradually become amorphous, with the smaller ones first and then the larger ones. This leads to a gradual transition to a truly amorphous phase once all crystallites have been dismantled due to the thermal vibrations. The temperature at which this happens is called the **melting temperature (T_m)** of the polymer. Hence, the term *melting temperature* refers to the *crystalline regions*. This melting temperature depends on the thickness of the crystallites; thicker crystallites result in higher melting temperatures. Melting is also highly dependent on the heating rates, if the heating rate is increased, it will give less time for the folded molecules to unfold, thus increasing the melting temperature.

Specific volume (volume/mass), which is really the reciprocal of density, is a property that can be measured and used to follow these transitions, noting that crystallites are better packed than amorphous regions and hence denser; they occupy less volume. Hence, a transition from a crystalline to an amorphous structure comes with increased specific volume. Figure 11.14a shows such a transition for crystalline polymeric structures. As can be seen at the lower temperatures, there is initially a gradual increase in specific volume when the smaller crystallites become amorphous, and at the higher temperature, much more of the process is happening throughout the polymer, leading to a more pronounced rise in the specific volume. The melting point marks the end of this process. The polymers do also have a considerable amount of amorphous regions, where molecules are randomly configured. Figure 11.14b shows how the specific volume changes as the temperature is reduced for an amorphous polymer or region. At higher temperatures, the molecules vibrate due to thermal energy that overcomes the Van der Waals forces holding the molecules together; as the temperature declines, so does the thermal energy, giving rise to an increase in the effectiveness of the Van der Waals forces, which result in a decline in the specific volume. This process continues until a particular temperature is reached; the **glass transition temperature (T_g)**, above this temperature, and the amorphous polymer behaves in a leathery then rubbery fashion. Below this temperature, molecular motion is restricted, and only atomic vibration is in effect. The polymer below the glass transition temperature behaves like glass; it is brittle and lacks toughness, but it is also considerably stronger and stiffer. T_g is similar to the ductile-to-brittle transition temperature we saw for some metals. The glass transition temperature hence relates to amorphous polymers or amorphous regions of a polymeric microstructure, and melting temperature relates to crystalline polymers or crystalline regions in the microstructure. Therefore, a semi-crystalline polymer will experience both temperatures, although it is more difficult to detect such glass transition in these polymers than fully amorphous ones. Another important point is that the glass transition temperature depends significantly on the cooling rate; slower cooling rates result in lower T_g values. Table 11.4 shows T_g and T_m values for several polymers.

Another way to look at the glass transition temperature is the concept of **free volume**, which is the empty space not occupied by molecules in a polymer. As one can imagine, in molten polymers, there is a lot of space and free volume, which allows polymer molecules to move easily. As the temperature is reduced, however, the free volume is also reduced, until a temperature where polymer molecules are

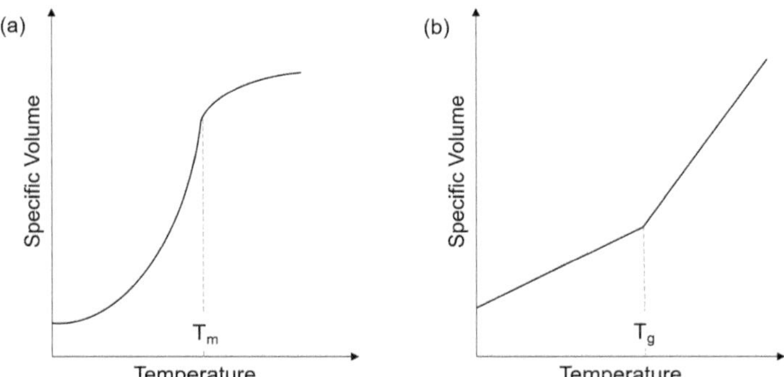

Fig. 11.14 Effect of temperature on specific volume of **a** a crystalline polymer, **b** an amorphous polymer (after R. A. Higgins, Mechanical Properties of Engineering Materials, Hodder and Stoughton (1986))

Table 11.4 Melting and glass transition temperatures for selected polymers

Polymer	T_m (°C)	T_g (°C)
Polyethylene		− 120
Polyethylene (50% crystalline)	120	
Polyethylene (80% crystalline)	135	
Polypropylene	176	− 27
Polyvinyl chloride (PVC)	212	87
Polytetrafluoroethylene (PTFE)	327	126
Natural rubber	28	
Polystyrene		100
Polyvinylidene chloride	198	− 17

Source R. A. Higgins, Mechanical Properties of Engineering Materials, Hodder and Stoughton (1986)

no longer able to rotate or move; this is the glass transition temperature. Polymer molecules become frozen in their place below this temperature, and the polymer becomes stiff. Figure 11.15 shows a schematic depicting the free volume below and above T_g. Many polymers are indeed used in this glassy state. It does lose its stiffness though if heated above the glass transition temperature.

So let us imagine a glassy polymer heated from a temperature below T_g; as soon as T_g is reached, the polymers start to gain mobility, and it becomes freer to rotate and move. Any feature of the polymer molecules that hinders this motion will raise T_g. As such, adding bulky side groups to otherwise simple and flexible polymers such as polyethylene will do this. If the side groups were polar, they would have a more significant effect than non-polar side groups since such polar interactions between molecules end up restricting the flexibility of the polymers. This can be seen in the higher T_g of polyvinyl chloride compared with polypropylene. As mentioned, any situation that reduces the flexibility of polymers will increase T_g, which is also the

Fig. 11.15 Effect of temperature on free volume

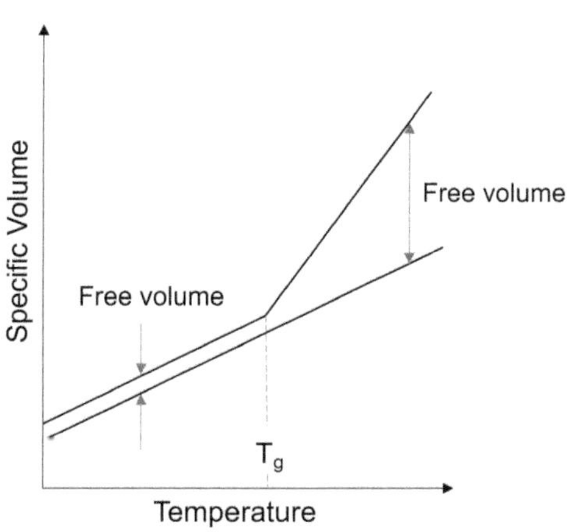

case for linear polymers that possess a lot of side branches (discussed next); these side branches restrict molecular motion. By the way, the melting temperature (T_m) is also affected similarly by polymer chain flexibility, the less flexible polymers are, the higher T_m and T_g. In fact, for homopolymers that display T_g and T_m, T_g is typically between 0.5 and 0.8 T_m. Hence, it would be impossible to change the T_g of these polymers without changing the T_m. However, we can do this for copolymers. Copolymers will be discussed later.

Some things increase T_g and some decrease it, I'm still not comfortable with what does it really mean to increase T_g? I mean, what is going on in the polymer?

Let's simplify it a little, Alex. Imagine two polymer blocks at absolute zero (0 K) one consisting of linear and flexible polymer chains and the other with linear chains that contain bulky side groups (i.e., stiffer). At absolute zero temperature, both polymers are glassy. If we increase the temperature and consequently the thermal energy there will come a temperature above which the polymer chains will be able to move and rotate. This is the glass transition temperature. You and I agree that the polymer with the inherently flexible chains may be able to gain that mobility first, and it would require a higher temperature (and higher thermal energy) to realize the same effect in the polymer with the stiff chains containing bulky side groups. In other words, the polymer with the stiffer polymer molecules (with bulky side groups) will have a higher T_g. I hope this helps!

Thanks. Makes sense!

Really? Ok, so tell me, if you are required to shape a polymer component out of a linear polymer and you need to be able to conduct this shaping process at room temperature. Would you want a polymer with a T_g higher or lower than room temperature?

Easy, Prof.! Lower, of course. I want my material to be above its T_g when I deform it at room temperature so that its molecules can flow and enable the deformation and shape change. This means I need a polymer with a lower T_g than room temperature (the temperature of operation of the polymer), such that room temperature will be above its T_g

Professor, it seems clear that the concept of free volume can help us understand how T_g works and, hence, predict how it will be affected by structural or chemical changes in the polymer. Yet, I'm not sure if the molecular weight of a polymer affects T_g. Do you have any insights?

First, you are correct in sticking to the concept of free volume. A low molecular weight polymer means that the polymer has short molecules, and a higher molecular weight polymer has longer molecules. Don't you think there will be some space or free volume at the ends of each of these molecules? That is how you should look at it; if we look at two polymer blocks of the same size, the one with a low molecular weight will have a relatively large number of short polymer chains (with many chain endings and hence additional free volume). In contrast, the polymer with a high molecular weight will have a lower number of longer polymer chains (with fewer endings and hence less free volume) and more restricted chain movement. This means T_g increases with an increase in molecular weight. I would go as far as saying that if we introduce a few side branches to the molecules of a linear polymer, we will increase its free volume because of the additional branch endings with their additional free volume provided by the branches. This will ultimately decrease T_g. Of course, if you add many branches, then the freedom of movement of the molecules will be restricted, increasing T_g. Differential scanning calorimetry (DSC) can experimentally determine polymers' glass transition and melting temperatures

11.2.2   Branched polymers

Sometimes, linear polymer molecules have side branches (Fig. 11.13b); these could be intentionally or unintentionally produced. It is logical that if a polymer comprises such molecules, they will not be able to pack efficiently, as do purely linear polymers. This has a few consequences. The first is that such materials will have a lower density compared with the same material (having the same molecular weight) that has a purely linear structure to its molecules. The branching will also obstruct the alignment of chains into more densely packed crystallites; hence, crystallinity is greatly reduced. An example of a branched polymer is low-density polyethylene (LDPE). The branches can also interfere with the sliding of molecules past each other as a deformation mechanism for the polymer and, hence, may cause an inherent increase in resistance to molecular sliding. When branches, however, physically separate molecules, it hinders the Van der Waal bonding forces, which reduces the strength of the polymer. Strength is also affected by other factors, including crystallite amount and polymer density. A decrease in density and crystallite amounts can decrease strength significantly. Low density polyethylene (a branched polymer) is, therefore, weaker than high-density polyethylene due to its lower density and lower crystallite content.

11.2.3   Cross-linked Polymers

While a branched polymer molecule contains side branches just like a tree branch contains twigs, cross-linked polymers have side branches that act as strong mechanical bridges between different molecules (Fig. 11.13c). This means two molecules will be strongly connected by these branches. The number of these cross-links will have a profound effect on the properties of polymers. For example, light cross-linking can produce polymers that are referred to as **elastomers**, which have distinct properties compared with branched and purely linear (unbranched) polymers. Here, we are talking about a crosslink every 100 or more mers. An excessive amount of crosslinking will take away any flexibility polymer chains may have had and result in rigid polymers, which we call **network polymers**. Examples of lightly crosslinked polymers include synthetic rubber, which is an elastomer.

11.2.4   Network Polymers

When the degree of cross-linking becomes excessive, say, every member has a cross-link or two, then a 3D network of connections is established. This is possible when the monomer is multifunctional. Examples of these kinds of network polymers include Bakelite (phenol–formaldehyde), the first synthetic polymer (or copolymer; more about this later), epoxy, and polyurethanes. A 2D representation is shown in Fig. 11.13d.

11.2.5   A Closer Look at Crystallinity in Polymers

As was mentioned in the previous section, many polymers crystallize when they are produced or even when they are deformed. Crystallinity, when it comes to polymers, basically means the folding or alignment of molecules to form a crystalline region (i.e., an ordered region). These regions are called **crystallites**. An example schematic of a crystallite is shown in Fig. 11.16, which shows a lamellae structure, sometimes referred to as platelets, the thickness of which is relatively small in the region of 10 nm, with a length in the region of micrometers. Each lamella can be viewed as a polymer chain folded on itself; some parts are seen leaving the crystalline region and then returning, and other molecules are dangling. Hence, the crystallites are not perfectly crystalline but crystalline enough to diffract X-rays. Hence, XRD can be used to characterize the structure of these regions, which are relatively complicated; for example, an orthorhombic structure can be observed in polyethylene.

   When a crystallizable polymer is cooled from a viscous melt, crystallization typically occurs by forming what are referred to as **spherulites**. In essence, crystallites grow into ribbons that can also twist. It can be imagined that these ribbons nucleate

Fig. 11.16 Simplified schematic of a crystallite consisting of lamellae of folded polymer chains

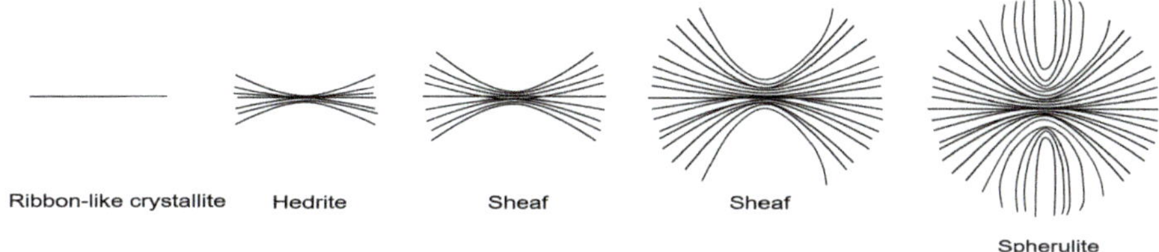

Fig. 11.17 Steps in the formation of a spherulite starting with a crystallite. (after: Gránásy L, Pusztai T, Börzsöni T, Tóth GI, Tegze G, Warren JA, Douglas JF. Polycrystalline patterns in far-from-equilibrium freezing: a phase field study. Philos Mag 2006; 86:3757–78)

from a nucleus and grow outwards to form something resembling a dandelion, with an overall spherical geometry. Such structures can evolve (which has also been observed experimentally) starting from a ribbon-like crystallite; then, other crystallites form parallel to the initial crystallite and splay outwards. The process is repeated until the newly formed and splayed crystallite curves onto itself. This signifies the end of the spherulite formation if they do not first impinge on other growing spherulites. This process is seen in Fig. 11.17, which shows the transition from a crystallite to a **Hedrite** to a **sheaf** and then a spherulite. It is important to note that, in between the crystallites, amorphous regions still exist, so the spherulite is not 100% crystalline. Even some molecules from one crystallite can be connected to other crystallites. In general, the crystallinity of polymers can lie between 5 and 95%, i.e., obtaining a 100% crystalline polymer is impossible. The cooling rate used to crystallize polymers from the melt is important; a slow cooling rate allows crystallites to form and hence favors crystallization. Figure 11.18 shows an electron micrograph of Hedrites.

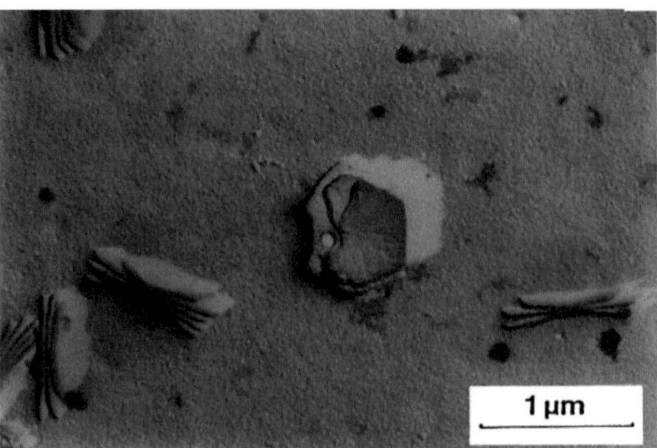

Fig. 11.18 Electron micrograph showing hedrites source: (Fig. 1b) in A.S. Vaughan, On morphology and polymer blends: polystyrene and polyethylene, Polymer, vol. 33, 12, 1992, pp. 2513–2521. Reproduced with modification with permission from Elsevier

When spherulites grow extensively, they end up impinging on one another; at this point, their spherical geometry is not maintained, and what looks like boundaries (which can be viewed as equivalent to grain boundaries in metals and ceramics) develop between spherulites. When crystallized from the melt, such spherulitic structures can be found in polymers such as polyethylene, nylon, PTFE, PVC, and polypropylene. Figure 11.19a is an image of polypropylene obtained using cross-polarized light, which shows a characteristic Maltese cross pattern. Figure 11.19b shows how the size of the spherulites can be refined and controlled with the addition of graphite nanosheets.

As mentioned, many polymers are semi-crystalline, with crystalline and amorphous regions making up their microstructures. Since crystalline regions consist of molecules that are better packed than amorphous regions, they are hence denser. It is, therefore, possible to obtain the degree of crystallinity of a polymer with knowledge of the density of the sample (ρ), amorphous region (ρ_{am}), and crystalline region (ρ_{cr}) using Eq. 11.6.

$$\text{Percent Crystallinity} = \frac{\rho_{cr}(\rho - \rho_{am})}{\rho(\rho_{cr} - \rho_{am})} \times 100 \tag{11.6}$$

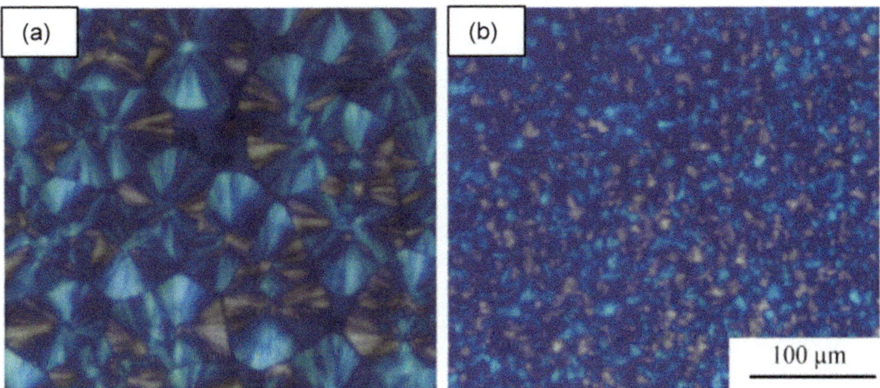

Fig. 11.19 Spherulites in **a** polypropylene, **b** in polypropylene and graphite nanosheet causing refinement in the size of the spherulites. *Source* Ya Liu, Yanjin Guan, Jiqiang Zhai, Lei Zhang, Fengjiao Chen and Jun Lin, Refined Spherulites of PP Induced by Supercritical N_2 and Graphite Nanosheet and Foaming Performance, Polymers, Performance. Polymers 2023, 15, 1204. https://doi.org/. Reproduced and modified under license: Creative Commons Attribution (CC BY) license (https:// creativecommons.org/licenses/by/ 4.0/)

Example Problem 11.5

A semi-crystalline polyethylene sample of density 0.96 g/cm^3 contains crystallites with an orthorhombic crystal structure, such that the crystallite average density is 0.9972 g/cm^3. If the percent crystallinity is 80%, what would be the average density of the amorphous region?

Solution:

In this problem, we have been given the values of all the terms in Eq. 11.6, except for the density of the amorphous region. By inputting the given values, we can then solve for ρ_{am}.

$$\text{Percent Crystallinity} = \frac{\rho_{cr}(\rho - \rho_{am})}{\rho(\rho_{cr} - \rho_{am})} \times 100$$

$$80 = \frac{0.9972(0.96 - \rho_{am})}{0.96(0.9972 - \rho_{am})} \times 100$$

$$0.80 = \frac{0.9972(0.96 - \rho_{am})}{0.96(0.9972 - \rho_{am})}$$

$$0.80 \times 0.96(0.9972 - \rho_{am}) = 0.9972(0.96 - \rho_{am})$$

$$0.766 - 0.768\rho_{am} = 0.957 - 0.997\rho_{am}$$

$$0.229\rho_{am} = 0.191$$

$$\rho_{am} = 0.83 \text{ g/cm}^3$$

Hence, the average density of the amorphous regions is 0.83 g/cm^3.

The degree of crystallinity can also be determined through analytical techniques such as differential scanning calorimetry (DSC) and quantitative X-ray diffraction. Crystallization can occur, as mentioned above, through melt processing, solution processing, and even mechanical deformation (where polymeric chains can align in the direction of a uniaxial stress).

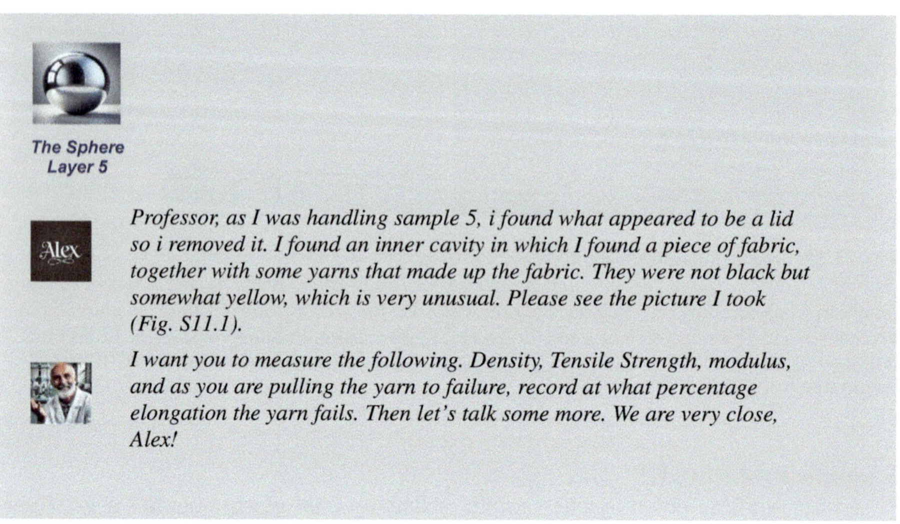

Professor, as I was handling sample 5, i found what appeared to be a lid so i removed it. I found an inner cavity in which I found a piece of fabric, together with some yarns that made up the fabric. They were not black but somewhat yellow, which is very unusual. Please see the picture I took (Fig. S11.1).

I want you to measure the following. Density, Tensile Strength, modulus, and as you are pulling the yarn to failure, record at what percentage elongation the yarn fails. Then let's talk some more. We are very close, Alex!

Fig. S11.1 Fabric sample present inside the cavity of sample 5 (photography by K.Morsi)

11.3 Copolymers

What we have been discussing so far are **homopolymers**, where polymer chains are produced from the repetition of one type of mer. It is, however, possible to design very useful polymers by using two or more different mers as building blocks for the polymer chains. One can imagine by doing so we can come up with different combinations of these mers to produce different types of copolymers. These are outlined in Fig. 11.20; the red and black solid circles represent different types of mers. Of course, the polymer chains, in reality, are not straight as drawn, but the molecules are drawn as such for simplicity. An **alternating copolymer** is when we have alternating red and blue mers in the chain. A **random copolymer** is when the red and blue mers are randomly positioned along the polymer chain. An example would be vinylidene chloride-vinyl chloride copolymer, SBR (Styrene Butadiene Rubber) copolymer, and Nitrile rubber (NBR), which is an acrylonitrile-Butadiene copolymer. Another way to arrange the mers is to produce what is known as a **block copolymer** in which blocks of the different mers appear on the molecule. An example of a block copolymer is styrene-butadiene, also called **impact-modified polystyrene**. Block copolymers display the properties of the homopolymers of both mers. Finally, it is also possible to produce what is referred to as a **graft copolymer**, in which its molecules can be viewed as linear polymer molecules made up of one type of mer while the branch is made up of a different type of mer. An example of this type of polymer is ABS (acrylonitrile–butadiene–styrene), where butadiene grafts are attached to the main linear copolymer molecule of styrene-acrylonitrile. This can be seen in Fig. 11.21. The composition of copolymers can be determined using infrared spectroscopy.

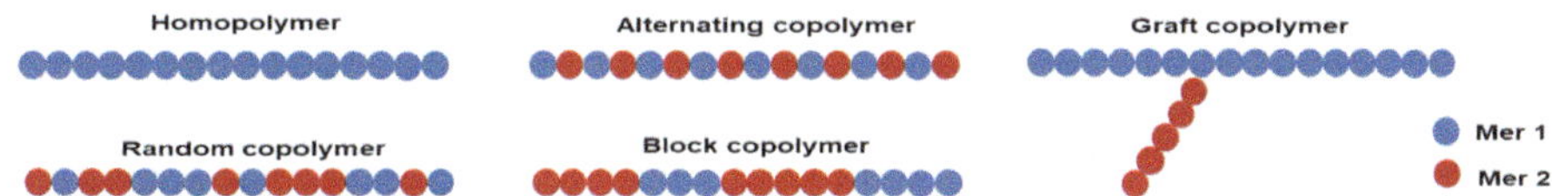

Fig. 11.20 Schematic of homopolymer, alternating copolymer, random copolymer, block copolymer and graft copolymer

Fig. 11.21 Schematic of a graft terpolymer of acrylonitrile-styrene-butadiene

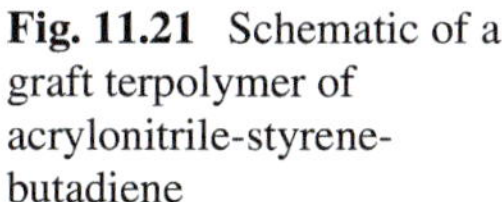

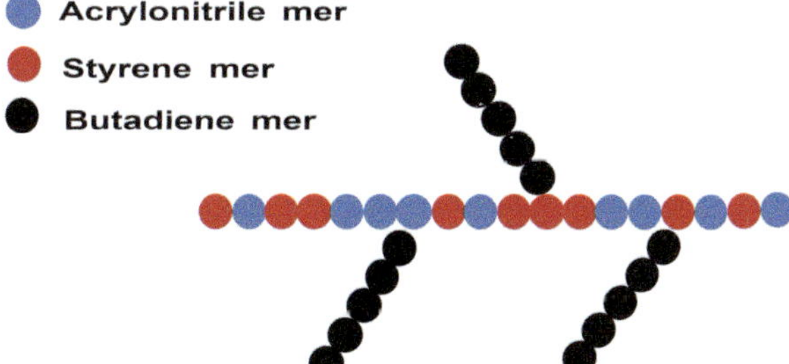

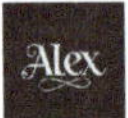

Professor, I understand the different types of copolymers, but how can people practically execute such a design? How do they do it?

*Ok, let's take a graft copolymer as an example. It consists of a homopolymer main chain, onto which side branches of another homopolymer are attached. If we can remove one of the hydrogen side atoms and create a free radical there, we will make the condition for attachment of a side branch. This can be done typically using irradiation. Here, both the long molecule and the new monomers are irradiated. Another approach is to add a free radical and a new monomer to the homopolymer; the free radical will remove the hydrogen atom and allow the attachment of the new monomer. The process is called **Transfer grafting**. To give you an example, PMMA molecules can be added as branches onto polystyrene molecules in this way, using a peroxide as a catalyst/free radical.*

11.4   Polymer Blends

The development of new polymers can be quite costly. Alternatively, a much cheaper option is the mixing or blending of different polymer chains, which can result in a **blended polymer** with new properties. In the same way that alloys can be produced with the mixing of different atoms to produce a single-phase solid solution or when sugar molecules and water molecules mix to produce a single-phase syrup, polymers (although having much longer molecules) can do the same. Here, two conditions can emerge. The first is that if the polymers are **miscible** in one another, in this case, an equivalent of a single-phase solid solution is produced with its unique properties. For example, if the two polymers had different glass transition temperatures, the new blended homogenous polymer would have its unique glass transition temperature. For a polymer blend made from two homopolymers (1 and 2), the new T_g for the blend can be determined using the **Fox equation**, as shown in Eq. 11.7.

$$\frac{1}{T_g} = \frac{X_1}{T_{g1}} + \frac{X_2}{T_{g2}} \tag{11.7}$$

where T_g, T_{g1}, and T_{g2} are the glass transition temperatures of the polymer blend, polymer 1, and polymer 2, respectively, and X_1 and X_2 are the mass fractions of polymers 1 and 2, respectively.

If, however, the two polymers are immiscible, then two phases will be produced within the blended polymer microstructure, i.e., an inhomogeneous microstructure will be produced. In this case if each polymer had a glass transition temperature, the new blended polymer would display two glass transition temperatures. Figure 11.22 describes these types of blends. An example of such a blend is polyvinylidene

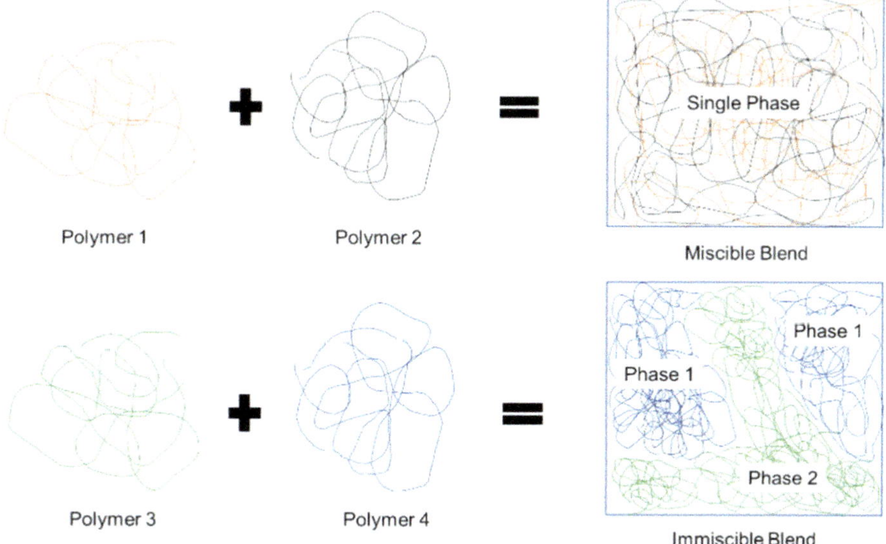

Fig. 11.22 Miscible and immiscible polymer blends

chloride-polyvinyl chloride (which acts as the matrix) and polystyrene-polymethyl methacrylate, which act as dispersions in the matrix (i.e., a separate phase). Both copolymers (phases) are amorphous.

Many copolymers can also produce inhomogeneous blends in a different way than presented so far. Consider the block copolymer styrene-butadiene, shown in Fig. 11.23a. In Fig. 11.23b, the molecules line up in the polymer so that the block of a particular mer lines up with the same mer block in another molecule. This results in regions within the polymer that identify the different polymers.

Fig. 11.23 **a** A Styrene-Butadiene copolymer, **b** inhomogeneous polymer blend produced by the alignment of block polymer molecules, showing different colored regions (phases) in the microstructure

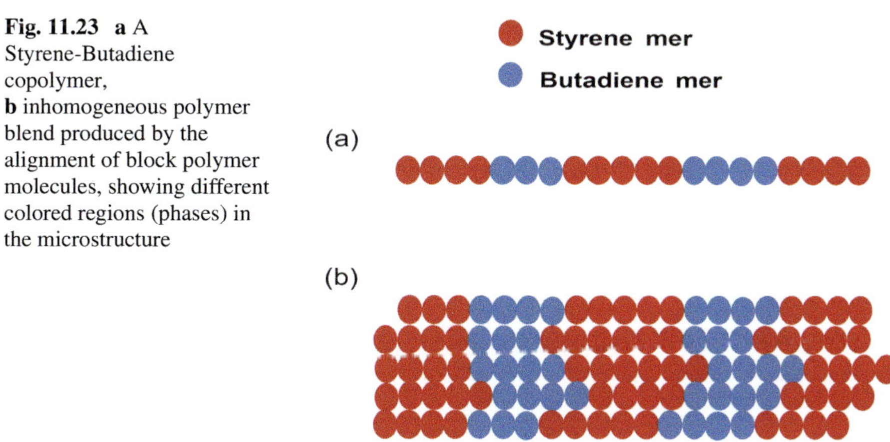

Liquid Crystal Polymers (LCPs)

Unlike semi-crystalline and amorphous polymers, liquid crystal polymers (LCP) consist of rod-like rigid molecules. Their alignment gives rise to crystalline regions that remain in the melt state, even though the separations between rods are greater than in the solid state. These polymers can be solid or liquid at room temperature. The liquid types are called **Cholesteric**, while some **nematic** LCPs are solid at room temperature. An example of liquid crystal polymers is Kevlar®. Different types of LCPs can be transparent and flame-resistant and have low coefficients of thermal expansion, excellent thermal stability, high impact strength, and chemical inertness when exposed to different solvents and acids. Uses include liquid crystal displays (LCDs) used in flat panel computer monitors and televisions, digital watches and electronic displays, interconnect devices, capacitor housing, photocopiers, and medical equipment.

11.5   Additives

It would be misleading to assume that all polymers can be used without any additional additives. Polymers can be susceptible to environmental and chemical attacks. For example, sunlight can disintegrate a polymer. Some polymers can be susceptible to UV attack, and oxidation can also be problematic. The glass transition temperature can also be lowered by adding low-molecular-weight polymers. What are these additives, and what exactly do they offer? This will be discussed in this section.

Plasticizers

Plasticizers are additives that can improve the flexibility and toughness of polymers. They can have a low molecular weight (for example 300 g/cm^3) or a high molecular weight. For example, PVC is plasticized using high molecular weight phthalate esters. Without plasticizers, PVC products would be quite brittle; imagine the impact it would have on shower curtains, vinyl flooring, and clothing.

The choice of plasticizer and its compatibility with the polymer system is critical. Plasticizers can be polar or non-polar; in both cases, the function is to reduce the Van der Waals attractive forces between the polymer molecules by physically separating molecules and/or eliminating the polar attraction between the polymer molecules to allow easier flow. They also reduce the T_g of polymers and their melt viscosity. In semi-crystalline polymers, the plasticizer usually dissolves in the amorphous regions where molecules are less tightly packed than crystallites. However, these additives still need to be carefully selected. The following points are considered in their selection:

1. Plasticizers must be nontoxic and soluble in the polymer.
2. Plasticizers should have high boiling points and low vapor pressures to prevent their evaporation, which could render the polymer brittle again. The drying of paint is accompanied by the evaporation of plasticizers, which consequently causes cross-linking due to the exposure to oxygen.

3. They should have more affinity for the plasticized polymer molecules than for another plasticizer molecule.
4. It should not migrate to the surface of the polymer.
5. Too much plasticizer addition could turn the polymer into a liquid.
6. The plasticizer should not be soluble in other solvents, which could lead to plasticizer depletion and embrittlement of the polymer. For example, petroleum products can dissolve different types of polymer molecules.

Figure 11.24 shows a schematic of how plasticizer molecules can prevent molecules from approaching each other and bonding. Figure 11.24a shows areas of secondary bonding, and Fig. 11.24b shows how the plasticizer acts as a separator and hence decreases the Van der Waals attractive forces.

Fillers

These are typically powders of inexpensive non-polymeric materials. One of the primary reasons for adding these fillers is to reduce the overall cost of the polymer. Carbon black is a known filler for rubber. It is known that Bakelite typically contains fillers such as mica and wood flour, among others. Apart from the obvious cost savings, these additives act as reinforcements to the polymer, which can improve some of their properties, including strength by resisting molecular motion and improving dimensional stability. An additive like graphite can reduce friction, and asbestos can enhance temperature resistance. Asbestos, as a material, has health risks.

Other additives

The susceptibility of polymers to sunlight (UV exposures) and oxygen requires the addition of *ultraviolet absorbers* (e.g., carbon black) to protect the polymer from UV radiation and *antioxidants* to protect against oxidation. Moreover, color can be imparted to the polymer using either pigments (particles) or dyes. Finally, many polymers are flammable, apart from polymers such as PVC and PTFE, which contain considerable amounts of chlorine and fluorine, thus reducing flammability. Flame retardants are warranted for halogen-free polymers, and chlorine-, fluorine-, bromine-, and phosphorous-containing additives are used as flame retardants. Polyethylene is a polymer that benefits from the addition of such flame retardants.

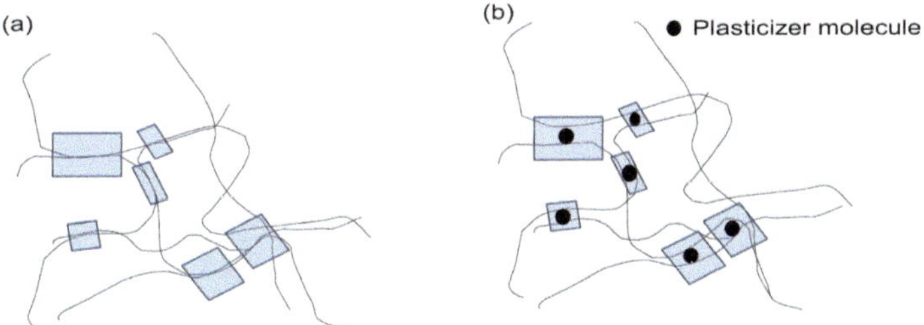

Fig. 11.24 a Polymer molecules bonded via Van der Waals bonding in certain locations, **b** effectiveness of plasticizers in separating these molecules

All additives need to mix efficiently with the polymer. If they are not mixed correctly, additives could be depleted in some areas of the microstructure, resulting in potential issues at these locations.

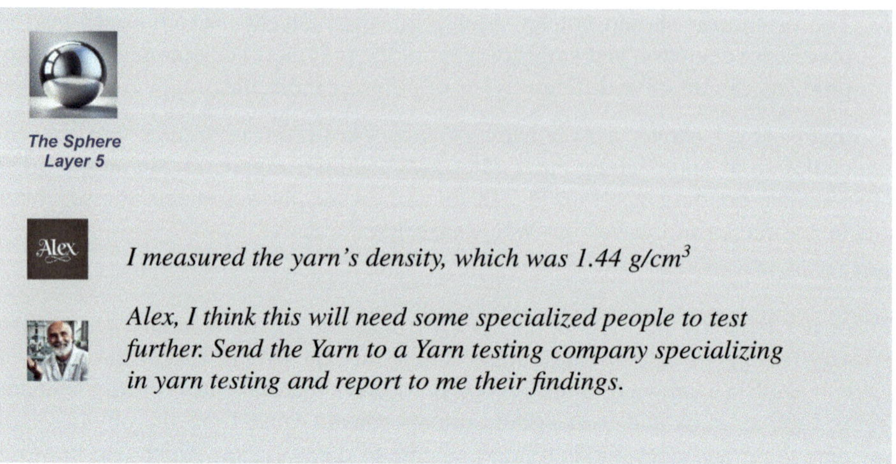

I measured the yarn's density, which was 1.44 g/cm³

Alex, I think this will need some specialized people to test further. Send the Yarn to a Yarn testing company specializing in yarn testing and report to me their findings.

11.6 A Closer Look at Some Popular Polymers

Now that we have understood many aspects of polymers, let us discuss in more detail several commercial polymers that have had and are still having major impacts on our lives today. In doing so, we will try to capture examples covering the different types of polymers discussed so far: linear, branched, crosslinked, network, copolymers, and polymer blends.

Nylons (Polyamides)

With tensile strengths that can be on par with some soft aluminums, Nylons (also referred to as polyamides) stand as some of the most important engineering polymers. The development of Nylon 6 in 1930 saw its use in the Second World War as a replacement for silk used in parachutes. Nylons have the advantage of being easily processed through melt processing approaches. With properties such as high strength, good chemical resistance, low friction, and the ability to carry mechanical stress at elevated temperatures, it is not surprising that it has found considerable use in today's market. Today, we see its use in many engineering applications such as gears for windshield wipers and speedometers, electrical parts (due to their good dielectric strengths, e.g., Nylon 6,6 has a dielectric strength of ~ 15 kV/mm) that are exposed to relatively elevated temperatures, bearings, and antifriction parts. This is in addition to the numerous other applications that include ropes, clothing, tire reinforcements,

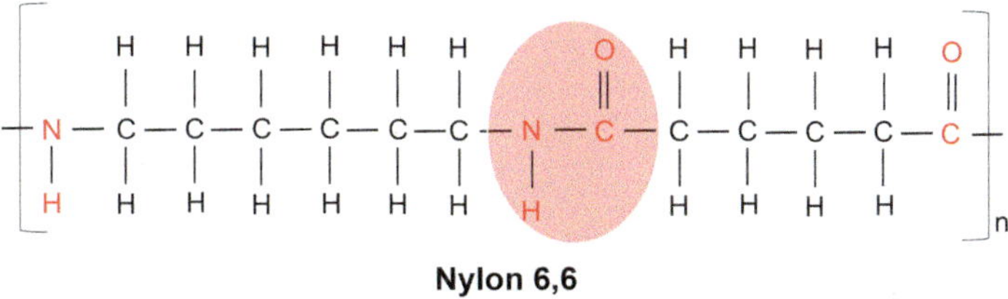

Fig. 11.25   Nylon 6,6 mer, also highlighting the amide link in pink

coatings on dishwasher racks, and carpeting. This is a semi-crystalline linear polymer, and when formed from the melt, it can form spherulites.

Nylons include Nylon 6, Nylon 6,6, Nylon 11, and Nylon 12. However, Nylon 6 and Nylon 6,6 have been occupying most of the market in the US compared with other Nylons. The molecular structure of Nylon 6,6 is presented in Fig. 11.24.

What does the number and its format after the word "Nylon" mean, Prof.? I see different numbers. Sometimes, only one number appears, sometimes two numbers are present with a comma between them, and sometimes even no comma.

Great question, Alex. The numbers after the word Nylon, refer to the number of carbon atoms in each of the monomers used to form the Nylon. Let me explain further. In the case of Nylon 6,6, this polymer was produced through a condensation polymerization reaction, as shown in Fig. 11.26. *(in As a Matter of Fact). The figure shows the repeated mer unit of Nylon* 6,6, *highlighting the amide link. Here, Hexamethylene diamine (having six carbon atoms) and adipic acid (also having six carbon atoms) react, and water is produced as a byproduct; the process continues to produce longpolymer chains. A similar process can be carried out, for example, to produce Nylon 6,9, Nylon 6,10, and Nylon 6,12 by reacting hexamethylene diamine (6 carbon atoms) with azelaic acid (9 carbon atoms), sebacic acid (10 carbon atoms) and dodecanedioic acid (12 carbon atoms) respectively. The Nylon 6, however, is produced using a different process called* **Ring-opening polymerization.** *This is shown in* Fig. 11.27. *(in As a Matter of Fact), where the repeat mer is shown. Note that if the mer is repeated head-to-tail, the amide link (OCNH) will be formed. .Now continue reading 'As a Matter of Fact.'*

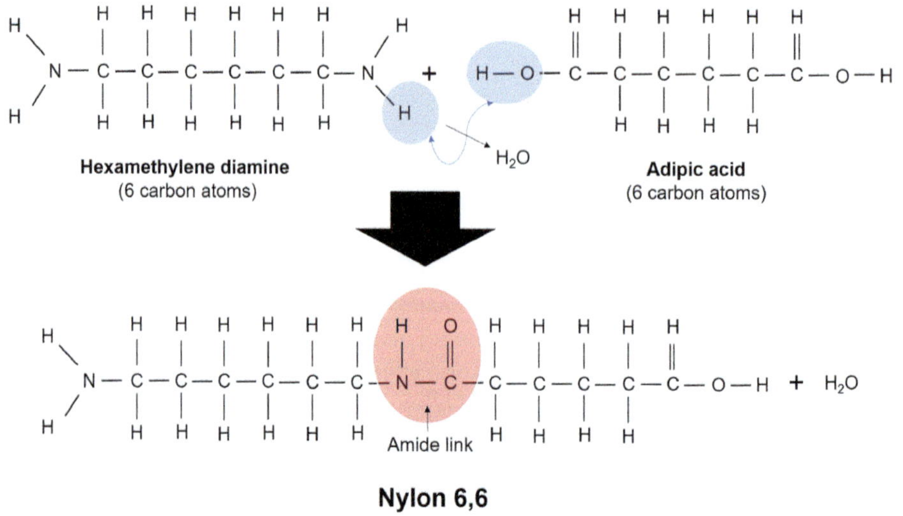

Fig. 11.26 Condensation reaction to produce Nylon 6,6, the amide link is common to all Nylons. (Modified from: D. R. Askeland, W. J. Wright, Essentials of Materials Science and Engineering (SI edition, D. K. Bhattacharya), Third edition, Cengage learning, 2014)

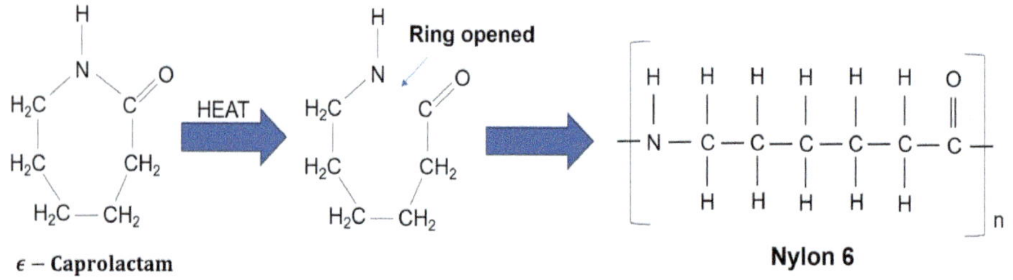

Fig. 11.27 Ring opening polymerization to produce Nylon 6. (Modified from W. F. Smith, J. Hashemi, Foundations of Materials Science and Engineering, 5th edition, McGraw Hill, 2006)

The amide link is at the heart of all Nylons, and in fact, it is the reason why they have high strengths. This happens due to hydrogen bonding between chains enabled by the amide link, which, due to its polarity and hydrogen bonding, also results in water absorption (moisture). Hence Nylons have a significant disadvantage in that they are affected by moisture. Nylon 6 can increase in weight by up to 10 at 100% relative humidity and 20 °C due to moisture absorption. This has major implications for the weight of a component, and its dimensions can significantly change during service because of it. This should, hence, be accounted for in the design of such components. There are applications where dimensional accuracy is not important, though, for example, in the Nylon coatings on dishwasher racks. From a mechanical property standpoint, though absorbed moisture can lead to a reduction in strength and stiffness of the Nylon but an increase in toughness. The moisture molecules act in a similar way to a plasticizer. Fortunately, there are Nylons that are more moisture resistant, for example, Nylon 11 and Nylon 12. These polymers have long chains between the amide links, resulting in less moisture pick up.

Natural Rubber

Natural rubber is produced from the Hevea Brasiliensis latex, typically known as the rubber tree, originally grown in the Amazon rainforest. Today, Southeast Asia is growing them for commercial use. The latex is the white substance that emerges from the tree when it is injured to protect it against insects. It is a suspension of polymer particles in water and other components, including proteins, lipids, ash, and inorganic salts. The mixture is perfectly designed to coagulate and harden after it leaves the tree and comes into contact with air. The latex from the trees is treated, and other ingredients are added. To prevent coagulation and to set the basis for rubber-like behavior, sulfur is added, as will be explained later. However, before we delve into this, let us start by visualizing the molecular structure of natural rubber. As mentioned previously, the chemical structure of the mer of natural rubber is *cis*-1,4 isoprene, as shown in Fig. 11.28. The reference to 1 and 4 in the name has to do with the first and last carbon atoms in the mer, in that these are where they link up with other mers. *Cis* means the methyl (CH_3) group, together with the hydrogen atom, is on the same side of the carbon double bond, providing some bulkiness and inflexibility on that side. However, it allows the molecule to coil toward the other side, which gives natural rubber molecules flexibility. The *trans*-isoprene mer, on the other hand, has the methyl group and hydrogen on opposite sides of the carbon double bond, which takes away the flexibility that *cis-isoprene* had. Hence, its properties are vastly different from those of natural rubber. Natural rubber displays elasticity and plasticity. The molecules are coiled, and any applied stress to the coiled molecules will elastically stretch them, such that when the load is released, they coil back, and the strain is recovered. There could be some entanglements between different molecules that also help this behavior. At sustained loads, the stretched molecules can slowly pass by one another, such that the molecules now occupy different relative positions to one another; thus, when the load is removed, elastic recovery happens, but some permanent strain remains. More about this behavior later.

In 1839, **Charles Goodyear** invented a process called **vulcanization**, where natural rubbers (and now synthetic rubber) can gain the elasticity we take for granted

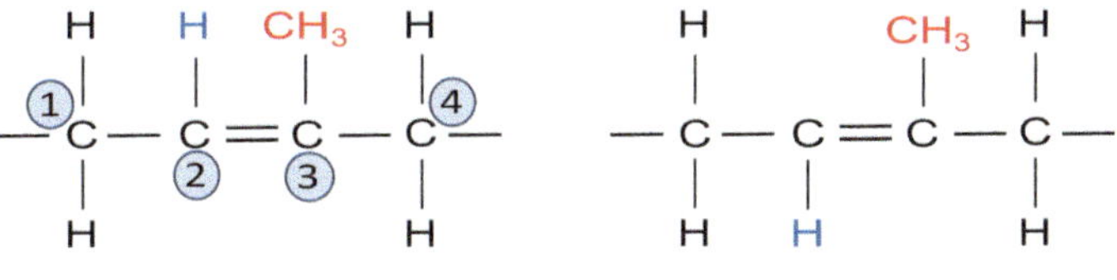

Fig. 11.28 Chemical structure of cis 1,4 isoprene and trans isoprene

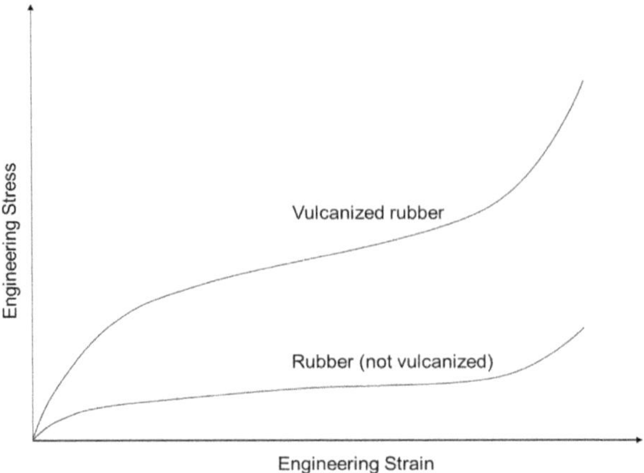

Fig. 11.29 Engineering stress–strain curves for rubber that has not been vulcanized and one that has been, both tested above the glass transition temperature. Note the higher modulus and strength or the vulcanized rubber

today when we stretch an elastic band. The molecules in natural rubber are not connected by strong covalent bonds. However, numerous unsaturated double bonds in each polymer chain can provide locations where reactions can occur, and links can be made between different molecules. Goodyear found that when sulfur is added and after curing the sulfur-containing rubber to elevated temperatures (typically less or equal to 200 °C), the resulting rubber not only possessed hyper elasticities (elastic strains of 500–1000%), but the vulcanized rubber is also significantly stronger and stiffer than natural rubber that is not vulcanized. This can be seen schematically in Fig. 11.29 for rubber tested above the glass transition temperature. Below the glass transition temperature, rubber is glassy and brittle, displaying linear elastic behavior.

Figure 11.30 shows that sulfur atoms bridge and crosslink between different molecules. The more sulfur added, the more double-bond regions react with sulfur to form crosslinks. If this process is continued, such that all the double bonds were reacted and crosslinks formed there, the resulting material would be quite hard and rigid. When this occurred, materials such as *vulcanite* and *ebonite* were produced and used at the time. These have, however, been replaced by other polymers. Therefore, to yield the hyperelastic properties sought, crosslinking must only happen in a few double bonds. For example, a sulfur content that would react with only 5% of the available double bonds has been used for automobile tires. This, however, leaves a vast number of unsaturated double bonds remaining in the rubber, which makes the rubber vulnerable to processes such as oxidation or ultraviolet (UV) light. For example, exposure to oxygen or ozone can result in the cross-linking of the oxygen atoms between double bonds in the same way as sulfur, which results in the deterioration of the properties of the rubber. The addition of antioxidants overcomes this. Similarly, UV light attacks the double bonds, so UV absorbers such as carbon black are added to rubber to counter this. Hence, the rubber we see today has many added

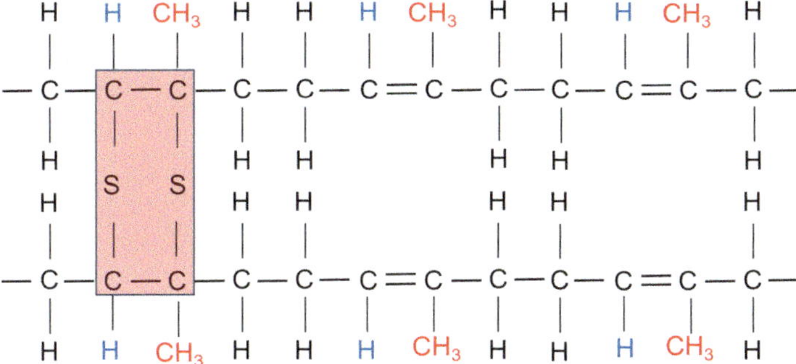

Fig. 11.30 Vulcanization of two polyisoprene polymer chains, where sulfur atoms attach to the carbon atoms which previously had a double bond. Note, the carbon bond is now a single bond in that location

components to combat environmental effects. In addition, fillers are added, which help reduce the material cost of the rubber product. Contact with solvents such as petroleum products is also another concern.

Styrene-Butadiene Rubber (SBR)

This is a copolymer of styrene and butadiene monomers. It is the most widely used synthetic rubber, and cheaper than natural rubber. The chemical structure of this rubber is shown in Fig. 11.31. It can be seen from the figure that the polybutadiene mer contains a double bond. Hence, it is understandable that vulcanization by adding sulfur can take place through these double bonds in the same way that it occurs for natural rubber.

EPOXY

This is a rigid network polymer with properties that include good adhesion to other materials, high strength, good chemical resistance, and good electrical insulation. Initially, they were developed for the aerospace industry. However, the material has gained a broader scope for application, for example, as electrical switches, as an

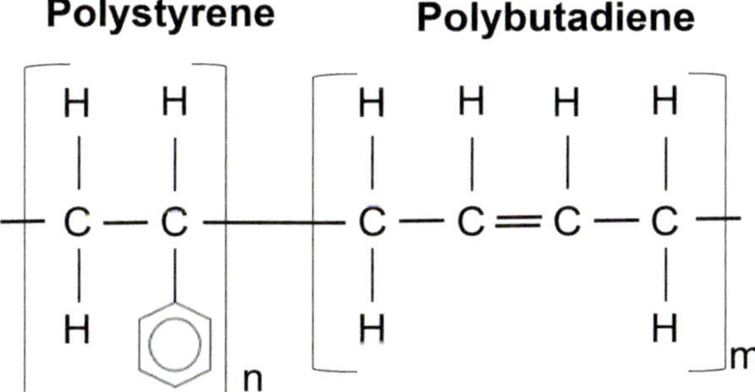

Fig. 11.31 Chemical structure of Polystyrene Butadiene copolymer

(a)

$$CH_2 - CH -$$ (epoxy group)

(b)

$$CH_2 - CH - CH_2 - \left[O - C_6H_4 - \underset{CH_3}{\overset{CH_3}{C}} - C_6H_4 - O - CH_2 - \underset{OH}{CH} - CH_2 \right]_n - O - C_6H_4 - \underset{CH_3}{\overset{CH_3}{C}} - C_6H_4 - O - CH_2 - CH - CH_2 -$$

Fig. 11.32 **a** Epoxy group, **b** general chemical structure of epoxy resin

encapsulating material for transistors, and as surface coatings. They are also used as a matrix material for fiberglass (glass fiber-reinforced epoxy composite). The epoxy group contained in the polymer is shown in Fig. 11.32 a and b shows a general chemical structure of epoxy resin.

The **n** in Fig. 11.32b is two or more if the resin is a solid and less than one for liquid resins. For the resin to gain some rigid state, it needs to be cross-linked extensively; this is done by adding a cross-linking agent (typically referred to as a hardener). The hardener (e.g., amines like diethylene triamine and ethylene tetramine) is intimately mixed with the resin. Over time, cross-linking typically proceeds at the chains' epoxy groups and the OH. The epoxy structure shown in Fig. 11.32 is typical, but other structures are also available. Epoxy does not become malleable when heated; heat would eventually decompose it. The material also has poor thermal conductivity. Aside from being able to cure epoxy and harden it at room temperature, one of the advantages of epoxy is that before the hardener takes full effect, the epoxy resin is in the liquid state and hence can easily be poured and shaped. The other is that no byproducts are produced when the curing takes place, which results in low shrinkage after curing.

11.7 Mechanical Behavior of Polymers

As with materials in general, the structure of polymers significantly affects their mechanical behavior. We will start by introducing/or reintroducing some terms to describe polymers we are already acquainted with. The first is **thermoplastic**. This is a term that describes a polymer that, upon the application of heat and stress, can be plastically deformed. These polymers typically include polymers with linear or slightly branched polymer chains, like polyethylene. Another type of polymer is a **thermoset**; these are polymers with polymer molecules that are practically "set" in their place because of a high degree of crosslinking; examples would be the network polymers Bakelite and epoxy. **Plastics** can typically refer to either a thermoplastic or a thermoset. The third polymer type is an **elastomer**; polymers with slightly

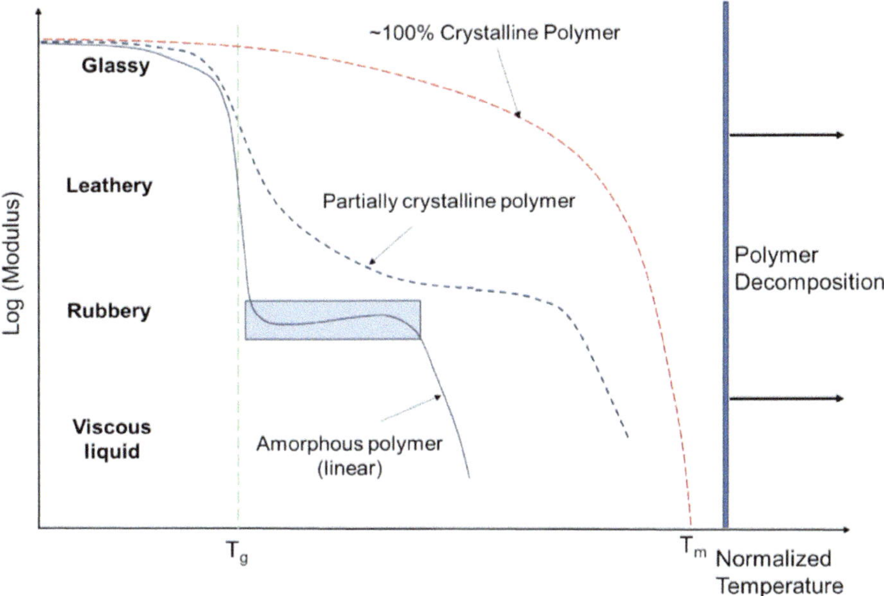

Fig. 11.33 Effect of normalized temperature on the relaxation modulus (based on multiple sources including M. F. Ashby, and D. R. H. Jones, Engineering Materials 2, second edition Butterworth Heinemann, (1998))

crosslinked polymer chains are ideal. The crosslinks anchor the molecules to each other, such that the sliding of one molecule relative to the other can enable large elastic strains, which are then easily restored upon removing the stress (mainly due to these anchoring crosslinks). Too many crosslinks between molecules will generate an inflexible and rigid thermoset. As we discussed, examples of elastomers are natural rubber and SBR, which are used in automobile tires.

Now, when we consider the mechanical response of a polymer to an applied stress, we note that although polymers contain macromolecules, they also contain covalent bonds within these macromolecules. Hence, the mechanical response of the polymer will depend on the resistance of the chemical bonds (covalent) and the macromolecules, which are entities that contribute to the mechanical behavior. Before we discuss the stress–strain curves of some polymers, let us first consider the effect of temperature on the elastic modulus of polymers. Unlike metals and ceramics, polymers are highly sensitive to temperature, molecular weight, and loading rate. In fact, an increase in temperature can result in a 1000 times difference in a polymer's modulus or strength values. Moreover, a property such as a modulus will depend on the loading time for polymers. This means that if we apply stress to a polymer, the strain will not stay constant but instead will increase with time, resulting in a decline in modulus with time. This modulus is also called the **relaxation modulus**. So, when dealing with polymers, one must be cautious of this aspect. This time-dependence of strain is not the case for metals and ceramics; it can only appear if the metal or ceramic is undergoing creep deformation at temperatures equal to or greater than 0.4 times the absolute temperature. Since the modulus of polymers can vary by three orders of magnitude when tested at different temperatures, it makes sense to plot the modulus data on a log scale for polymers with respect to temperature. Here we must pause. In fact, instead of simply plotting temperature, it is more appropriate to plot a **normalized temperature** (T_N) for the polymer in question, which is the testing or service temperature (T_s) divided by the glass transition temperature (T_g), (i.e., $T_N = T_s/T_g$). If the log of the modulus (at a particular loading time, let's say after 1 s of loading time) is plotted with respect to normalized temperature, the polymer can be seen to go through several structural and behavioral changes (what are known as **mechanical states**) as outlined in Fig. 11.33.

The **glassy state** exists from 0 K to just before T_g. In this state, the polymer, irrespective of the type of polymer, is brittle. Within this regime, the modulus declines similarly to metals and ceramics with an increase in temperature. Figure 11.34 shows a schematic of the stress–strain curve for a polymer within the glassy state.

We also observe some effect on the modulus value if we consider different loading times. The decline in modulus with temperature within the glassy regime is partly caused by thermal expansion during heating, which reduces intermolecular forces of attraction. As seen in Fig. 11.33, around T_g, there is a more significant abrupt drop in modulus, and the polymer assumes a leathery state within this region and gains some flexibility. This drop in modulus is a little less dramatic for thermosets, such as epoxy, than for thermoplastics, for example. However, at an even higher normalized temperature, a **rubbery state** for all types of polymers exists, and the modulus slightly increases with an increase in temperature within this regime. Unless the polymer is an elastomer, the material, although rubbery, does not display the high degree of elasticity observed in regular elastomers (which contain physical cross-links). Following this rubbery state, the polymer enters a **viscous state** where it can flow. Finally, at a temperature considerably higher than T_g, the material decomposes, as the C–C bonds are gradually destroyed. It is important to point out that the main curve in Fig. 11.33 refers to an amorphous linear polymer; however, if the polymer contains crystalline regions, the curve shifts to the right, and the rubbery state shifts upwards. The figure also shows the curve for a ~ 100% crystalline polymer where the abrupt drop in modulus and the leathery state do not exist. Instead, at a higher temperature, the polymer observes a drop in the modulus that ends with the polymer melting at T_m. Recall, amorphous polymers do not melt, but crystalline ones do.

Professor, what about the effect of crosslinking of amorphous polymers on the modulus-temperature curve?

Cross-linking has a similar effect to crystallization in regard to shifting the curve to the right and the rubbery state upward. However, the curves will not dip down and melt as in crystalline materials. Instead, a heavily cross-linked polymer will degrade when it reaches the decomposition temperature.

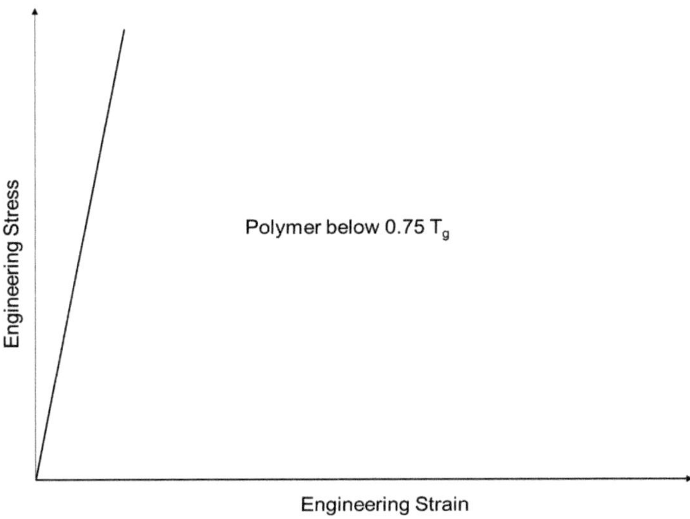

Fig. 11.34 Engineering stress–strain curves for polymers at a temperature below 0.75 T_g

We have so far seen the stress–strain curves for elastomers, thermosets and also all polymers at temperatures much lower than T_g. However, what does a typical stress–strain curve look like for a thermoplastic? Fig. 11.35 shows this curve for a thermoplastic above its T_g. Initially, the polymer displays a linear elastic region up to point X. At stresses above point X, the curve behaves nonlinearly till point Y. The stress level at point Y is taken as the **yield strength**. Beyond this point, necking starts, and for amorphous regions, molecular disentanglement occurs, whereas for crystalline regions, the crystallites start to unfold; in the plateau region, further stretching and unfolding of chains of the crystallites occur, and the molecules become aligned inside the necked region. The material in that necked region is stronger than anywhere else in the gage length because the molecules are aligned and brought closer, increasing the intermolecular bonding. Due to the strength of the necked region, any further deformation instead occurs in the softer neighboring regions of the gage length, which leads to what is seen as a spreading of the necked region within the gage length. The curve then shows a rapid increase in stress with further strain until the point of fracture. The stress at which fracture occurs is taken as the **tensile strength** of the polymer. The increase in stress up to that point occurs due to the improved alignment of molecules and their proximity to one another. It is a region where the material becomes continuously hardened.

Polymer recycling is an important topic. The only polymers that can be recycled are thermoplastics since thermosets and elastomers tend to degrade and decompose at high temperatures and therefore are not suitable for recycling. However, an exception is a material known as a **thermoplastic elastomer**. This material consists of copolymer molecules where the ends are made of a polymer with a higher T_g than the inner portion. This is seen in Fig. 11.36 for a styrene–butadiene–styrene

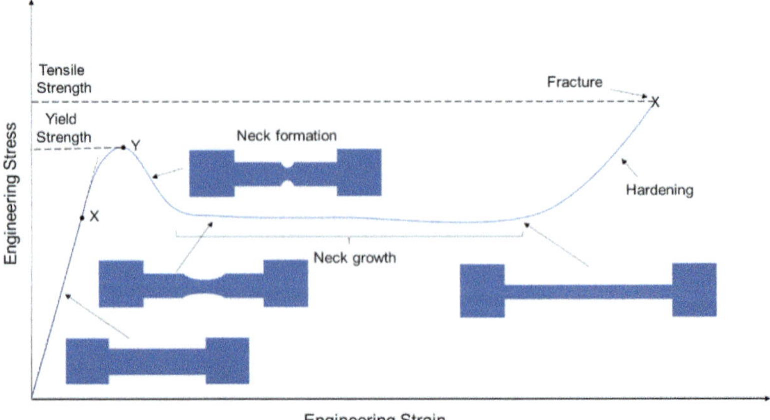

Fig. 11.35 Engineering stress-engineering strain curve for e thermoplastic above its T_g. (after multiple sources including D. R. Askeland, W. J. Wright, Essentials of Materials Science and Engineering, SI edition, Cengage Learning, (2014))

copolymer. The figure shows the polymer molecule and the molecular arrangements within the polymer, where the blue styrene regions are aligned (and hence effectively bond well) in a way where they act as an effective crosslink. As long as the overall polymer is at a temperature above the T_g of butadiene and below that of styrene, the material behaves as an elastomer. However, above the T_g of styrene, the material loses its elastomeric properties and acts like a thermoplastic and hence can be recycled.

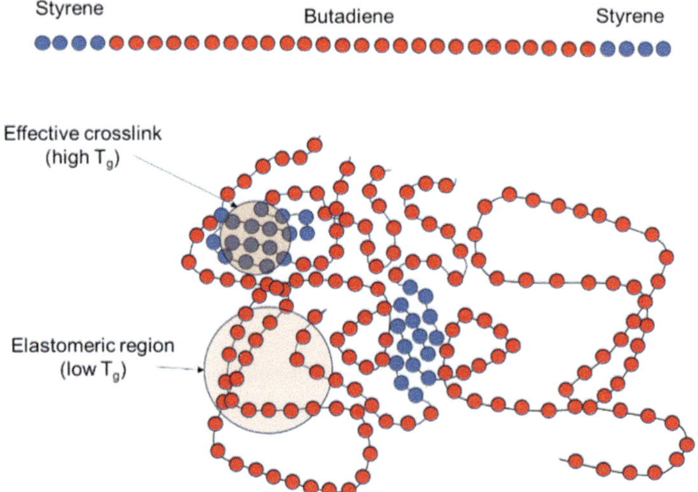

Fig. 11.36 Thermoplastic elastomer consisting of styrene–butadiene–styrene polymer chains

**The Sphere
Layer 5**

*The results are back, Prof. I can't say, I understand them, though!
Anyway, they are as follows:*

Denier $=$ 1500

Density 1.44 g/cm^3.

Moisture $=$ 5.5%

Elongation to break $=$ 3.55%

Tensile Modulus $=$ 70 GPa

Breaking Tenacity $=$ 2.87 GPa

Thermal Conductivity $=$ 0.038 W/m K

Specific heat at 25 °C $=$ 1390 J/kg K

Decomposition temperature $=$ 450 °C.

Heat of Combustion $=$ 3.6 $\times$ 10^7 J/kg

*You must be wondering what all these names mean. Let me
relieve you. Many of those terms are specific to fibers. For
example, a denier refers to the weight of 9000 m of yarn,
typically given in grams. The breaking tenacity is the
equivalent of the tensile strength. Finally, the heat of
combustion is the amount of heat released when the material is
burned. With all this information. I know what the material is
now. But one last test. I want you to place the fabric you found
on a piece of wood and then, using the sharpest kitchen knife
you have, stab it as quickly and hard as possible to pierce it*

*I'm sorry. I tried my best, but no matter how hard I try, I can't
pierce it.*

*Ok, Alex. From all the information provided, this material is **Kevlar® 29**. This is yet another material that came into existence only by accident. Its story is an interesting one. A scientist named **Stephanie Kwolek** was searching for a strong polymeric fiber material to replace steel in tires. Since polymers are lighter than steel, they would primarily help with weight savings. She directed her attention to the aromatic polyamide family of polymers, specifically poly-para-phenylene terephthalamide. She suddenly discovered a liquid crystal material while trying to spin it into fibers. Due to the cloudiness of the solution, she was met with stiff resistance when she wanted to spin it. After many tests and persistence, she managed to spin the solution eventually. The resulting fibers were spun into yarns, which could then be converted into fabric. Of course, with the different ways by which to weave it. The material is not only much lighter than steel (1.44 g/cm^3 vs. 7.8 g/cm^3), but it has a strength multiple times that of steel. Kevlar is now used in many applications, including knife- and bullet-proof vests, protective helmets, ropes, and so many more. When a bullet impacts a series of stacked Kevlar fabrics, they tend to capture the bullet, absorbing energy in the process. They can also be used for thermal protection. Can you imagine how many lives this invention has saved? No wonder she received the National Medal of Technology and Innovation award in Chemistry in 1996 and was inducted into the National Inventors Hall of Fame*

Alex, let's tackle Layer 6. I can't wait, and I'm sure neither, can you.

Problems

11.1. Give two examples of a natural polymer and two examples of a synthetic polymer.

11.2. What is meant by a bifunctional polymer?

11.3. Discuss the Ring-opening polymerization process.

11.4. Consider a linear polyethylene and a branched polyethylene. Which one has the lower density and why?

11.5. What is an Ethenic polymer?

11.6.   What is the polydispersity index of a polymer with a mass average molecular weight of 32,500 g/mol and a number average molecular weight of 19,100 g/mol? What does the value tell you about the polymer?

11.7.   What is the end-to-end distance for a polystyrene polymer molecule having a molecular weight of 125,00 g/mol?

11.8.   What are geometric isomers?

11.9.   What is the difference between an atactic and a syndiotactic polymer? Provide a real polymer example for both.

11.10.   Discuss the difference between a glass transition temperature and a melting temperature as they relate to polymers.

11.11.   Sketch the relationship between specific volume and temperature for a crystalline and an amorphous polymer. Identify on the sketch the melting point and glass transition temperature as appropriate.

11.12.   What is a sheaf?

11.13.   A semi-crystalline polymer is 60% crystalline. The density of the polymer is 0.9 g/cm^3. If the density of the amorphous region is 0.7 g/cm^3, what is the density of the crystalline region?

11.14.   What is the difference between a homopolymer and a copolymer?

11.15.   With the aid of sketches, describe the difference between an alternating copolymer and a block copolymer.

11.16.   Describe the molecular structure of an elastomer.

11.17.   Give an example of a network polymer. Does a network polymer have a melting point? Explain your answer.

11.18.   What is the effect of vulcanization on the modulus and strength of an elastomer?

11.19.   Draw two stress–strain curves, one for a thermoplastic above its glass transition temperature and one below its glass transition temperature. Indicate on the plots as appropriate, the yield strength and the tensile strength.

11.20.   Using sketches, describe a thermoplastic elastomer and example how it acts as an elastomer and at the same time a thermoplastic. Is it recyclable?

Part V
Layer 6: Communication is Key

The Sphere

Layer 6
"Communication is Key"

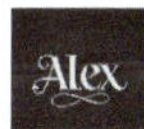

Professor, I carried out a few tests on sample 6. This is what I have found so far:
Density: 2.33 g/cm^3
Hardness: 1000 HV
Color: Dark grey
I also noticed that during Vickers macro-hardness testing, cracks sometimes formed; they were seen to emerge from the corners of the indent. I don't know what that means.

Alex, based on your findings, this material seems brittle, judging from your reported crack formation. It is also hard and lightweight. I'm sure the color may come in handy later on. Great job so far. Since it is brittle, why don't you start learning about some famously (or infamously) brittle materials? Ceramics. I would direct you to Chap. 12: Ceramics. Hopefully, you pick up some helpful knowledge. Let's talk some more later.

Chapter 12
Ceramics

12.1 Introduction

Ceramics play an important role in today's society. Unlike traditional ceramics, such as coffee mugs and dinner plates, engineering ceramics have been engineered to enable their use in engineering applications. This chapter introduces ceramics, classifies them, and discusses their structure and mechanical properties.

12.2 Introduction to Ceramics

Ceramic, as a word, stems from the Greek word "Keramos", which is very roughly translated as a burned thing or stuff. So, what exactly can be regarded as a ceramic? This is a topic in which there is no universal agreement. However, here we will use the basic definition put forward by the father of ceramic engineering, W. D. Kingery, in his classic book *Introduction to Ceramics*, to quote him:" The *art and science of making and using solid articles which have as their essential component, and are composed in large part of, inorganic, nonmetallic materials*". Under this definition, we are talking about a very broad spectrum of materials labeled as ceramics. These include glass, concrete, cement, table salt, pottery, and engineering ceramics such as alumina (Al_2O_3), zirconia (ZrO_2), and silicon carbide (SiC). Even ice can be referred to as a ceramic under this definition. Basically, everything that is not a metal or a polymer (since polymers are organic). Of course, this definition is imperfect, since we would have to define silicon as a ceramic, although today it is typically referred to as a semiconductor. However, despite its slight deficiencies, we will use this definition for the lack of a better one. Ceramics have been previously classified as either traditional or technical/advanced/engineering. The former refers to ceramics such as pottery or the mugs in which we pour our coffee, while the latter includes ceramics more suited for engineering applications. Figure 12.1 provides a more elaborate classification

© The Author(s), under exclusive license to Springer Nature Switzerland AG 2026
K. Morsi, *An Engaging Approach to the Science and Engineering of Materials*,
https://doi.org/10.1007/978-3-032-06231-4_12

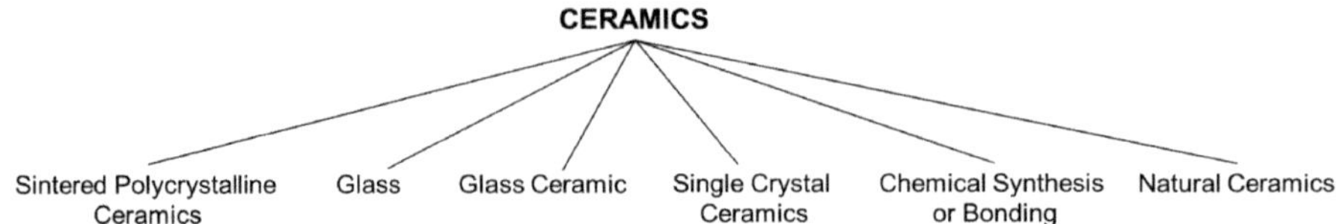

Fig. 12.1 Classification of ceramics (following the classification of D. W. Richerson in Modern Ceramic Engineering, 3rd edition, CRC (2006))

(classified in D. W. Richerson, Modern Ceramic Engineering, 3rd edition, CRC (2006)). We will be adopting this definition.

We will now describe the different ceramics categories presented in Fig. 12.1, expanding on W. D. Richerson's treatment.

12.2.1 Sintered Polycrystalline Ceramics

These ceramics start as powders, which are then compacted or shaped into a green compact or green body, respectively, followed by high-temperature sintering. Solid-state sintering is a high-temperature particle bonding process. At the sintering temperature, atomic diffusion occurs, and the contact points between particles grow to produce grain boundaries. Ideally, the space between particles is eliminated, resulting in high-density products. For ceramics, it is very difficult to eliminate all the porosity by simply sintering in a furnace, and hence, other powder densification processes are employed to produce fully dense products. These processes include hot pressing, hot isostatic pressing, recently spark plasma, and flash sintering. Figure 12.2 shows a simplified schematic of a solid-state sintering process and how pores are reduced with sintering time.

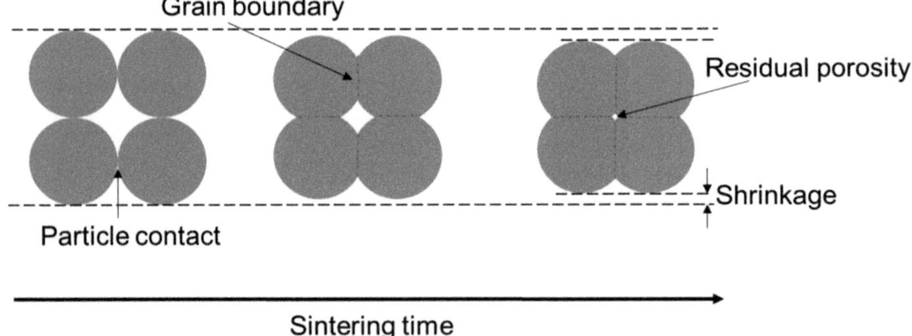

Fig. 12.2 Simplified schematic showing particle bonding and shrinkage with sintering time during an isothermal sintering process for four particles

12.2.2 Glass

Today, glass products are all around us, but even in the Stone Age, mankind used obsidian glass to make tools. Glass is currently used to produce lenses for microscopes, telescopes, windows, and even drinking and storage containers, not to mention fiber optics. The one ingredient of glass that is almost always present in significantly large quantities is silica (SiO_2). Another type of glass that does not depend on silica instead depends on P_2O_5 as the main ingredient. In any case, glass has been previously categorized into four categories (other ways to classify it also exist). One is **sheet glass**, which is used to make windows found in buildings and cars (auto glass). Auto glass is typically tempered (side windows) and laminated (windshields) for safety. While soda-lime glass has been used for centuries, recently, window glass has been modified in terms of composition, including the addition of magnesia (MgO), which prevents glass from crystallizing. **Container Glass** *is used for general containers, including drinkware, storage jars, bowls,* etc. This is also typically based on soda-lime glass with varied compositions, including adding MgO and, in some cases, K_2O. Although glass is transparent, it can also be colored by the addition of certain oxides. For example, copper oxide produces glass of a turquoise color, while iron (II) oxide produces a greenish-blue color glass. Even copper, when added, can generate a dark red color. At least from the aesthetic standpoint, glass is quite versatile. **Glass fibers** can be used for thermal insulation (glass wool, for example, which is made of soda lime glass fibers) or as reinforcements in composites. Cost-effective E-glass is an example of composite reinforcement. Its composition is SiO_2 (54 wt.%), Na_2O (1 wt.%), CaO (17 wt.%), Al_2O_3 (15 wt.%) MgO (4 wt.%) B_2O_3 (9 wt.%). *E* refers to the suitability of *E*-glass as an electrical insulator. **Advanced glass**, which includes optoelectronics, bioimplants, and glass used in telecommunication. Notably, calcium phosphate glass has been used as a biomaterial; one of its advantages is its ability to resorb into the human body. *Phosphate glass* is also known for its superior optical range, high coefficient of thermal expansion, and low melting point compared to silicate glass. The low melting point may facilitate easier forming. It can be used to produce composites for hard tissue regeneration, therapeutic ion release, drug/gene delivery, soft tissue engineering, and bioresorbable photonic devices. In bone fracture healing, Bioglass (bioactive glass) (Invented by **Professor Larry Hench**) is found to bond very well with soft and hard tissue and allow for bone regeneration. The composition of the first Bioglass (45S5) was SiO_2 (45 wt.%), CaO (24.5 wt.%), Na_2O (24.5 wt.%), P_2O_5 (6 wt.%). **Optical glass** can also be classified under advanced glasses; here, various glasses are produced with varying refractive indices, allowing their use in different applications such as eyeglasses, microscopes, and even telescopes. *Crown glass* is optical glass with a low refractive index, while *Flint glass* has a high refractive index.

Most glass is based on the soda-lime-silica glass, with varying compositions. Lime (CaO) improves fluidity, while soda (Na_2O) improves stability. Glass is an amorphous transparent material, amorphous meaning that it does not possess long-range atomic order. Imagine a liquid with no long-range atomic order, and as we cool it to produce a solid, two things can happen. Either the liquid crystallizes, as in the case of metals and ceramics, or it remains an amorphous liquid. If it remains a liquid, its viscosity will exponentially increase as the temperature is reduced. This is what happens to glass. Crystallization does not occur, which would have produced a polycrystalline material with grains containing atoms orderly arranged. On the contrary, the amorphous nature of the glass remains as the temperature is reduced, even at room temperature. At room temperature, glass is in a metastable state, and the glass would wish to crystallize if it could. However, due to its high viscosity, limited thermal energy, and limited atomic diffusion at low temperatures, this does not happen. Even at room temperature, it can take hundreds of millions of years for glass to devitrify (change from amorphous to crystalline). Hence, the reaction is astronomically sluggish. Table 12.1 shows compositions and some properties of some common glass. The composition of the glass is chosen to fulfill a particular application. For example, for light bulbs, the glass is thin, and a composition of SiO_2 (73 wt.%), Na_2O (17 wt.%), CaO (5 wt.%), Al_2O_3 (1 wt.%), and MgO (4 wt.%) is used. Whereas for lab glassware, resistance to chemical attack and temperature fluctuations is required. Here, *Pyrex* and *Vycor* both contain boric oxide (B_2O_3), which reduces the thermal expansion coefficient of glass. They also contain a high amount of silica, which also has a low thermal expansion coefficient. A low thermal expansion coefficient reduces thermal stresses, which limits cracking. Also, the refractive index of glass plays a role in the choice of glass compositions for fibers used in fiber optics. Before we delve into the data provided in Table 12.1, it is important to define a few terms. The density is an obvious property, but for glass, it depends on the cooling rate from the melt. A higher cooling rate results in less dense glass, giving less time for the glass to reorganize its atoms into more dense arrangements. Table 12.1 shows examples of some common glasses and some of their relevant properties. For example, the **softening point** for glass, which is the temperature at which the glass can slump under its own weight, is highest for fused silica. The softening temperature would, therefore, dictate the maximum service temperature of the glass. Regarding Young's modulus, all glasses presented in the table have a similar value of ~ 69 GPa. In contrast, glass fiber has a modulus of 87 GPa, making it very suitable for stiffening polymers, hence its use in fiber-reinforced polymer composites.

Most materials expand upon heating and contract upon cooling, to varying degrees. The **thermal expansion coefficient** quantifies this effect and is essential to know when using products at high or fluctuating temperatures. For a material being heated, for example, it describes the strain per degree centigrade or degree Kelvin. The linear thermal expansion coefficient (α) (for small changes in temperatures where α does not change significantly over the temperature range) can be approximated by the following expression in Eq. 12.1:

$$\alpha = \frac{\Delta l}{L} \times \frac{1}{\Delta T} = \frac{\varepsilon}{\Delta T} \tag{12.1}$$

In other words, it is the strain (which is unitless) per degree centigrade or degree Kelvin. For thermally isotropic materials (i.e., that expand (or contract) to the same

Table 12.1 Properties of some common glasses

Glass	Composition (wt.%)					Density (g/cm^3)	Young's modulus (GPa)	Softening point (°C)	TE ($\times$ 10^{-7}/°C)
	SiO_2	$Na_2O + K_2O$	$CaO + MgO$	B_2O_3	Al_2O_3				
Soda-lime (plate)	71–73	12–14	11–16	–	0.5–1.5	2.5	69	735	87
Soda-lime (containers)	70–74	13–16	10–13	–	0.5–1.5	2.5	69	730	85
Borosilicate	81	4	–	13	2	2.23	68	820	32
Fused Silica	100	–	–	–	–	2.2	69	1667	5.5
Fiber glass	54.5	0.5	22	8.5	14.5	2.57	87	830	60

Source Glass Engineering Handbook, 2nd ed. McGraw-Hill, 1958, p. 17, Ceramics and Glasses, volume 4, ASM, 1991

degree in all directions), the volume thermal expansion coefficient is 3α. Moreover, it happens that the thermal expansion coefficient is also related to Young's modulus according to Eq. 12.2:

$$\alpha = \frac{\rho C_\mathrm{p} \gamma_\mathrm{G}}{3E} \tag{12.2}$$

ρ is the density, C_p is the heat capacity at constant pressure, γ_G is Gruneisen's constant, which approximately has a value of 1 for most solids. The thermal expansion coefficient is inversely proportional to Young's modulus. This gives us some general understanding, as ceramics generally have higher modulus values than metals or polymers. As such, we see that the thermal expansion coefficients of ceramics are typically lower than those of metals and polymers. Since Young's modulus is also related to the melting point of a material, per the simplified Eq. 12.3, it follows that the thermal expansion coefficient can also be related to the material's melting point according to Eq. 12.4. Hence, materials with high melting points typically have low thermal expansion coefficients.

$$E \approx \frac{100kT_\mathrm{m}}{\Omega} \tag{12.3}$$

$$\alpha = \frac{\gamma_\mathrm{G}}{100T_\mathrm{m}} \tag{12.4}$$

K is the Boltzmann constant (1.380649×10^{-23} J/K), and Ω is the atomic volume.

Most materials will have a positive value for the thermal expansion coefficient, i.e., they will expand upon heating (and contract upon cooling), the degree of which will depend on the magnitude of the thermal expansion coefficient. Metals, for example, generally expand more than glass.

Some materials have almost zero thermal expansion coefficient. Let's imagine the following: a hot material object is suddenly quenched in a much colder environment, such as water. The region of the object closest to the surface will cool much faster than the material's interior. Hence, the object's surface will want to contract, while its interior will not, and instead pose a resistance to surface contraction. This generates stresses at the surface, which, if high enough, will cause the object to fracture (if brittle) or warp (if ductile and thin). Here, we define **thermal shock resistance** as the temperature change that can be tolerated without failure during the quenching of a material. The higher the thermal shock resistance, the safer the product is when it is subjected to a sudden change in temperature. Since the stresses were brought about by the outer material trying to contract, they are linked to the thermal expansion coefficient. The closer the thermal expansion coefficient is to zero, the less stress is generated during quenching or sudden heating. For example, if we place regular glass on a hot stove, it will likely shatter, but if we place fused silica instead, it will not because it has a low thermal expansion coefficient, as shown in Table 12.1. Let's discuss this further. Imagine an initially hot ceramic plate suddenly submerged in a much cooler environment, for example, water at room temperature (25 °C). The surface will cool quickly and reach the water temperature almost instantaneously, leaving the plate's interior hot. If the material at the surface were free to react to the reduction in temperature, it would shrink. However, the hotter interior is in no position to shrink (it is still hot). Hence, the interior restricts the contraction of the outer material layer, thereby generating a thermal tensile stress at the surface. Due to what is known as stress equilibrium, the plate develops a compressive stress internally to balance out the tensile stress. The highest surface tensile stress, which occurs at the surface in this situation, can be determined using Eq. 12.5.

$$\sigma = \frac{\alpha E}{1 - v}(\Delta T) \tag{12.5}$$

where σ is the stress, α is the thermal expansion coefficient, ΔT is the temperature difference between the surface and the interior, and *v is Poisson's ratio.* The equation can be applied to infinite slab shapes, long cylinders (hollow or solid), and spheres (hollow or solid). Since different materials can have different Poisson's ratios, elastic moduli, and thermal expansion coefficients, different materials are expected to experience different thermal stress levels for the exact temperature difference. Table 12.2 shows the calculated thermal stress for different materials for a temperature drop of 100 °C at the surface. The values for the moduli, Poisson's ratios, and thermal expansion coefficients come from different sources. The listed strength value is a lower value of the range of strengths if a range of strengths was reported in the original data source. While the table's highest calculated thermal stress value is for titanium diboride (TiB_2), the strength of titanium diboride is higher than the calculated thermal stress. Hence, we don't expect TiB_2 to fracture. However, the second highest thermal stress belongs to alumina (406 MPa), higher than its reported strength value of 276 MPa. Hence, at a first approximation, alumina may cause problems in terms of fracture. The equation is somewhat of a simplified depiction of thermal stress. Of course, for more accurate results, we need to consider other factors such as thermal conductivity, internal porosity, fracture toughness of the ceramic, and the distribution and size of microcracks, if present, etc. This simplified exercise nevertheless shows that both borosilicate glass (Pyrex) and silicon carbide should not fail. The thermal shock resistance can be calculated by solving for ΔT in Eq. 12.5 and substituting σ with the reported *strength* of the ceramic.

12.2.3 Ceramic Glass

It was mentioned earlier that when a glass is cooled, it will exist in a metastable state. This means it would achieve a more stable state if it could crystallize. We can indeed help the glass to do just that by adding some nucleating agents on which atoms of the glass can deposit during solidification, thereby forming a polycrystalline material. The ceramic glass can also be cooled and then heated again in a separate step

Table 12.2 Thermal stresses (σ) produced for a 100 °C temperature difference

Ceramic	E (GPa)	v	α ($\times 10^{-6}$)/°C	σ (MPa)	Strength (MPa)
Alumina	380	0.26	7.9	406	276
Titanium diboride	544	0.11	8.1	495	700
Silicon carbide (α)	345	0.19	5	213	621
Borosilicate glass (Pyrex)	70	0.2	4.6	40	69

E, v, α and strength data from multiple sources including Engineered Materials Handbook, Volume 4: Glasses and Ceramics, ASM (1991), page 30 & D. W. Richerson, Modern Ceramic Engineering, 3rd Edition, CRC Taylor and Francis, (2006)

to grow, in a controlled way, the crystals around the nucleating agent (TiO_2 is an example of a nucleating agent). This typically involves an elaborate heat treatment schedule where atoms can gain energy and move to enable crystallization, ultimately producing a polycrystalline material. Such material will have unique properties and features as opposed to a typical glass or ceramic. For example, the crystalline structure improves the mechanical properties of the parent glass. Additionally, unlike polycrystalline ceramics, ceramic glass can be produced with no porosity. This results from some residual glassy/amorphous phase that flows and fills the pores during high-temperature processing. Recently, the definition of a glass ceramic has been revised by Deubener et al. (2018) to be *"Glass-ceramics are inorganic, nonmetallic materials prepared by controlled crystallization of glasses via different processing methods. They contain at least one type of functional crystalline phase and a residual glass. The volume fraction crystallized may vary from ppm to almost 100%"*. Glass Ceramics (also referred to as ceramic glasses) typically have very low thermal expansion coefficients; some even have negative ones. One of their advantages is the ability to be initially shaped like regular glass. Moreover, they can be opaque or transparent, which opens the door to different applications. The material can be used in range tops, cookware, and other applications such as dental and biomedical. Examples of glass ceramics are compositions based on the $Li_2O–Al_2O_3–SiO_2$, $MgO–Al_2O_3–SiO_2$, and $Na_2O–BaO–Al_2O_3–SiO_2$ systems.

12.2.4 Single Crystal Ceramics

Single crystals can be found in nature or grown artificially. Examples of single crystals are sapphire, synthetic diamond, and ruby. Note that Ruby has the same structure as Sapphire (corundum) but contains chromium as an impurity, which renders its color red. On the other hand, Sapphire, although popularly known to be blue, includes all colors other than red. Metallic single crystals have limited applications; one notable example is single crystal turbine blades, which are needed for creep resistance at high temperatures. Single crystal Ceramics, on the other hand, have numerous applications. Table 12.3 lists examples of these applications spanning various fields. Single-crystal ceramics have also been used as reinforcements for metals. For example, TiB_w single-crystal whiskers have been grown in titanium matrices to produce titanium-TiB_w composites with significantly improved properties over commercially pure titanium.

Table 12.3 Examples of ceramic single crystal applications

Application	Single crystal
Semiconductor devices	
• Diodes	Si, Ge
• Photodiodes	Si, GaAs, $Cd_xHg_{1-x}Te$
• Transistors	Si, GaAs, SiC
• Photoconductive devices	$Cd_xHg_{1-x}TeSi$
• Integrated circuits	Si, GaAs
• Light-emitting diodes	GaAs, GaN, SiC
• Radiation detectors	Si, Ge, CdTe, YAG
• Strain gauges	Si
• Hall effect magnetometers	InSb
Mechanical components	
• Abrasives and cutting tools	SiC, Al_2O_3
• Substrates	Diamond, Al_2O_3
Piezoelectric devices	
• Resonant bulk wave devices	SiO_2, $LiTaO_3$, AlN
• Surface wave devices	SiO_2, $LiNbO_3$, AlN
Jewelry	Diamond, cubic zirconia
Magnetic devices	
• Transformer cores	Ferrites
• Electric motors	Ferrites
• Tape heads	Ferrites
• Microwave circulators	Garnets
Optical devices	
• Windows	Al_2O_3
• Lenses	CaF_2
• Polarizers	$CaCO_3$
• Laser hosts	YAG (Yttrium–aluminum-garnet), Al_2O_3

From: Chapter 29, Growing Single Crystals. In: Ceramic Materials. Springer, New York, NY (2007). https://doi.org/10.1007/978-0-387-46271-4_29

12.2.5 Chemical Bonding or Synthesis

Under this category, we find ceramics such as plaster and Portland cement. These materials are initially in powder form, and when mixed with water at room temperature, they react and form a ceramic. When gravel and sand are added to the cement–water mix, **concrete** is formed, which is a major construction material. Here, the cement bonds with the gravel/sand aggregates.

12.2.6 Natural Ceramics

These include bone, teeth, quartz, rock, stone, etc. Quartz, for example, has displayed piezoelectric properties, and the natural version of quartz (Quartz extracted from sandstone) is still used in oscillators and other devices. Our planet's crust is made mainly from raw materials suited to produce traditional ceramics, including silicates and aluminum silicates. These form the basis of traditional ceramics, including pottery (earthenware, stoneware, and China) and porcelain. They are based on naturally occurring clays (such as Kaolinite, $Al_2Si_2O_5$ $(OH)_4$) which are fine particles of hydrous aluminum silicates (layered structures), in addition to silica and feldspar (an aluminum silicate combined with either potassium, calcium, sodium, or barium, e.g., $KAlSi_3O_8$, $NaAlSiO_8$, $CaAl_2Si_2O_8$, or $BaAl_2Si_2O_8$). Unlike earthenware, stoneware has higher densities due to it being fired (heated) at higher temperatures, and hence, it can homogeneously distribute heat; as such, our coffee mug is stoneware. China is even more dense. Other natural ceramics include marble (primarily calcium carbonate) and ice. We only have to look at the ice cap in Antarctica, where the majority of the ice on our Earth resides.

12.3   Crystal Structures of Ceramics

Single-element ceramics such as diamond can be covalently bonded. Likewise, ceramic compounds can be covalently bonded; an example is silicon carbide (SiC). Most ceramics, however, are compounds between *metals and non-metals*. Hence, the crystal structure will involve not only two or more types of atoms, or more appropriately, ions, but also ionic bonding will be prevalent. Considering ionic bonding, the ceramic structure will involve the arrangement of anions (negatively charged ions) and cations (positively charged ions) while maintaining the crystal structure's charge neutrality. Typically, the anion forms the primary crystal structure (e.g., cubic, hexagonal, etc.), and the cations fill all or a fraction of the available interstitial sites. The relative sizes of the anions and cations are also important in determining the crystal structure. The radius of an ion will depend on several factors, including the

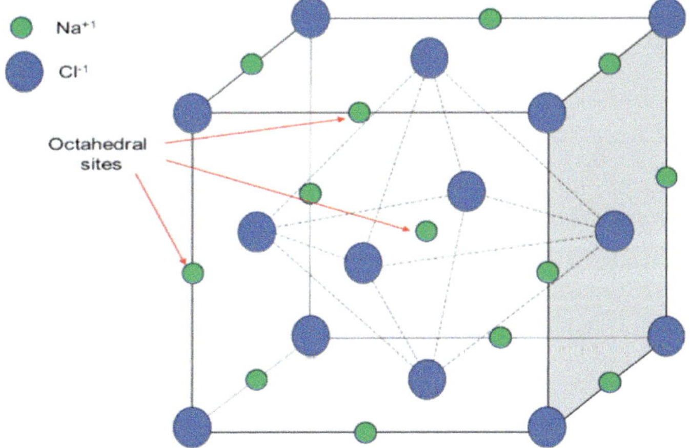

Fig. 12.3 Rock salt structure, showing anions occupying FCC structure and cations occupying all the octahedral interstitial sites

sign and magnitude of its charge, the type of bonding it is involved in, and the coordination number within the structure it occupies. Typically, anions are larger than cations, as explained in an earlier chapter. The type of crystal structure a ceramic will adopt will depend on several factors, including the ionic size ratio between the cation and the anion (i.e., radius of cation over the radius of the anion), the charge of the ion, and the degree of bonding (i.e., how much of the bonding is covalent and how much ionic). When all these factors are considered, ceramics can fall under several formulas, which can be related to the structure. AX, AX_2, A_2X_3…etc. A is an anion, and X is the cation (often oxygen for ceramics). As such, under the AX formula, we see ceramics such as MgO, NaCl, AgCl, and SiC; under the AX_2 formula, we see SiO_2, ZrO_2, and TiO_2; and under A_2X_3, we see one of the most popular engineering ceramics, which is Al_2O_3 (alumina) and also Fe_2O_3. For tertiary compounds, examples include ABX_3 (perovskites), in which A and B are two types of cations, and X is the anion. Examples would be $CaTiO_3$, $BaTiO_3$, and $MgTiO_3$. Table 12.4 summarizes some of these ceramics and their structures.

We see various structures under the AX structure classification, including Rock salt, CsCl, and zinc blende. Although the **Rock salt structure** is based on NaCl, several other ceramics adopt it, such as MgO, CaO, FeO, MnO, NiO, and AgCl. Here, bonding is predominantly ionic for monovalent ions such as Na^{1+} and Cl^{1-}. In this structure, the anions occupy the positions of an FCC unit cell. In contrast, the smaller cations occupy all the interstitial octahedral sites, as seen in Fig. 12.3. In the zinc blende structure, the anions take on an FCC structure. In contrast, the cations occupy *half* of the interstitial tetrahedral sites in the structure. This structure has a significant covalent component that contributes to the bonding. In fact, it resembles the structure of a diamond (which is, however, totally covalent). For Zinc Sulfide, the zinc ion has a valence of $+$ 2 and a sulfur valence of -2. There are a total of 4 anions of sulfur that reside inside the FCC unit cell borders, and likewise, 4 zinc

Table 12.4 Structures of different ceramics

Class/structure	Radius ratio (cation/anion)	Anion lattice	Cation occupation	Examples
AX				
Rock salt	0.41–0.73	Cubic close-packed	All Octahedral sites	NaCl, MgO, CaO, FeO, MnO, NiO, AgCl
Nickle arsenide	0.41–0.73	Hexagonal close-packed	All Octahedral sites	NiAs, FeS, VS, CoS
Wurtzite	0.22–0.41	Hexagonal close-packed	1/2 Octahedral sites	ZnS, SiC, ZnO, GaN
Zinc blende	0.22–0.41	Cubic close-packed	1/2 Tetrahedral sites	SiC, ZnS, BeO
Cesium chloride	> 0.73	Simple cubic	All cubic sites	CsCl, CsI, CsBr
AX$_2$				
Fluorite	> 0.73	Simple Cubic	½ cubic sites	ZrO$_2$, ZrF$_2$, HfO$_2$, ThO$_2$, CaF$_2$
Rutile	0.41–0.73	Tetragonal Close-packed	1/2 Octahedral sites	TiO$_2$, SnO$_2$, PbO$_2$, NbO$_2$
Silica	~ 0.33	Connected tetrahedra	–	SiO$_2$, GeO$_2$
A$_2$X$_3$				
Corundum		Hexagonal close-packed	2/3 Octahedral sites	Al$_2$O$_3$, Fe$_2$O$_3$, V$_2$O$_3$, Ti$_2$O$_3$, Cr$_2$O$_3$, Ga$_2$O$_3$
ABX$_3$				
Perovskite		Cubic close-packed	1/4 Octahedral sites	BaTiO$_3$, CaTiO$_3$, PbZrO$_3$, LaMnO$_3$
Ilmenite		Hexagonal close-packed	2/3 Octahedral sites	MgTiO$_3$, NiTiO$_3$, CoTiO$_3$
AB$_2$O$_4$				
Spinel		Cubic close packed	1/8 tetrahedral (*B*) sites, 1/2 octahedral (*A*, *B*) sites	MnFe$_2$O$_4$, MgAl$_2$O$_4$, FeAl$_2$O$_4$, CoAl$_2$O$_4$

Source D. W. Richerson, Modern Ceramic Engineering, CRC, 2006, p. 128, and Ceramics and Glasses, volume 4, ASM, 1991, page 880

anions reside also inside the unit cell. This equal number of anions and cations (of equal and opposite charge) results in the charges canceling out, giving rise to charge neutrality of the unit cell. Note that rock salt, zinc blende, and others, such as the wurtzite, nickel arsenide, and CsCl structures, fall under the **AX** classification. They can differ in how the ions pack, their relative coordination numbers, and the cation-to-anion radius ratios.

In cases where the cation (A) has twice the charge of the anion (X), to guarantee charge neutrality of the unit cell, the structure **AX$_2$** is preferred. Under this structure classification, we find the Fluorite, Rutile, and Silica structures. Named after CaF$_2$, the fluorite structure has relatively large cations (cation/anion radius ratio greater than 0.73). Due to this relatively large cation size, they occupy the body center position of a simple cubic arrangement of anions. However, only half of the body-centered sites are occupied to maintain charge neutrality (Figs. 12.4 and 12.5).

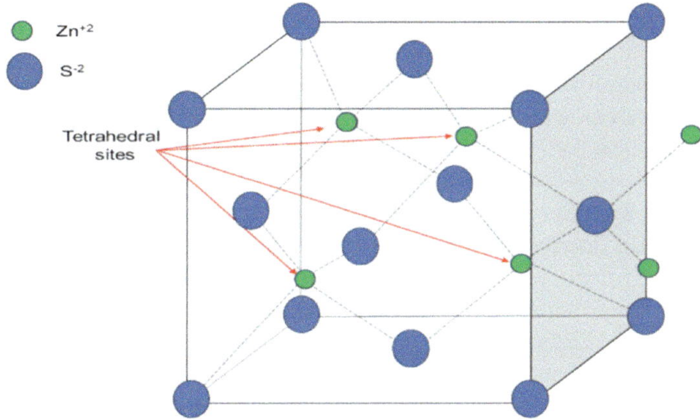

Fig. 12.4 Zinc Blende structure, showing anions occupying FCC structure and cations occupying half the tetrahedral interstitial sites

Fig. 12.5 Fluorite structure, showing anions occupying simple cubic structure and cations occupying half the body centered sites, example for Thoria (ThO$_2$)

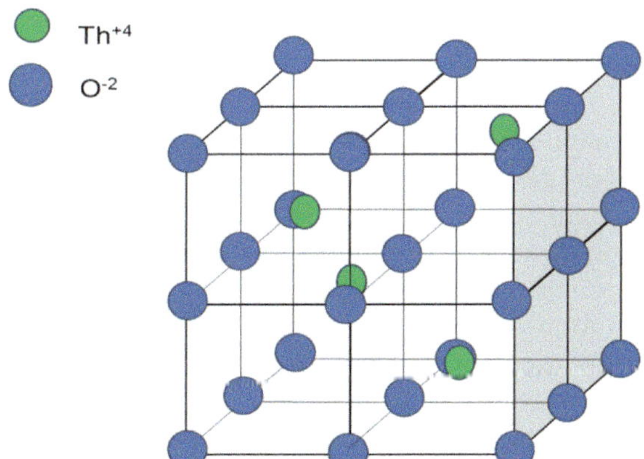

12.4 Mechanical Behavior of Ceramics

12.4.1 Strength

Table 12.1 shows one of the mechanical properties of glass: Young's modulus. Glasses have moduli values greater than polymers, similar to metals such as aluminum (~ 70 GPa) and higher than metals such as magnesium (~ 45 GPa). However, other ceramics can have exceedingly higher Young's modulus values. For example, the moduli of some popular ceramics are as follows: alumina (hexagonal crystal structure) is 380 GPa, TiC (cubic crystal structure) is 430 GPa, TiB_2 (hexagonal crystal structure) is 514–574 GPa, silicon carbide (hexagonal crystal structure) is 207–483 GPa, CVD-SiC (cubic crystal structure) is 415–451 GPa, and Si_3N_4 (hexagonal crystal structure) is 304 GPa. These relatively high moduli values are primarily due to the strong bonding observed in these materials.

One of the main problems with ceramics is that they typically do not exhibit a single strength value but rather a range of strengths for the same material. For metals, the average strength values can have a standard deviation of only a few percent, which makes designing structural components from these materials relatively easy. This is not the case for ceramics with an appreciable variation in strength for the same sample, inherently due to a population of flaws that can exist in a ceramic. These flaws can arise during manufacturing or service. Such inherent flaws come in different flaw size distributions and orientations within the volume of the ceramic, making the determination of a strength value for a ceramic challenging.

For metals and polymers, it is very easy to manufacture a tensile testing sample. For example, machining a metal to produce a dog-bone specimen is typically straightforward. We can still manufacture a tensile testing specimen out of a ceramic. Still, since it is very brittle, machining costs would be high, which traditionally require diamond tools and expensive polishing operations. Processes such as electro-discharge machining (EDM) have provided means to machine ceramics in a non-contact way. Still, even here, typically only electrically conductive ceramics such as ZrB_2, TiB_2, TiN, and ZrN are suitable candidates. Some of the most famous ceramics, including alumina, silicon carbide, silicon nitride, and zirconia, are nonconducting. Hence, other specialized modifications of the EDM process are used to enable their machining. Most researchers and engineers have historically resorted to other, much simpler methods for testing ceramics, which are 4-point and 3-point flexural (or bend) strength testing.

Here, several points need to be clarified. In 3-point and 4-point testing, the ceramic is produced in the form of a bar or pin and then is placed onto two hard supporting pins (positioned at a specified distance apart in accordance with standards). Another pin (or two pins separated by a specified distance for 4-point testing) comes down on the bar from the top, as shown in Fig. 12.6, and applies a force using a universal testing machine. The ceramic pin or bar can have a circular cross-section, a square cross-section, or a rectangular cross-section. Figure 12.6 shows the 3-point and 4-point strength testing setups for a ceramic bar with a rectangular cross-section and some dimensional parameters. The precise values of these parameters may vary and are dictated by standards. During the test, the universal testing machine's crosshead is moved down at a prescribed speed, and the load continuously increases until the bar fractures. The load at which this happens is then recorded and entered into Eqs. 12.6a and b, depending on whether the test was a 3-point or 4-point strength test, to obtain the fracture strength (σ_f), also known as the **modulus of rupture (MOR)**.

3-point bending test:

$$\sigma_{f\,3\text{pt}} = \text{MOR} = \frac{3FL}{2ab^2} \tag{12.6a}$$

4-point bending test:

$$\sigma_{f\,4\text{pt}} = \text{MOR} = \frac{3Fx}{ab^2} \tag{12.6b}$$

Example Problem 12.1

Two sets of ceramic bars with the same cross-section ($a = 4$ mm and $b = 3$ mm) were tested using 3-point and 4-point methods, per ASTM standard C 1161. In both cases, the distance (span) between the lower support pins was 40 mm (i.e., $L = 40$ mm). In the 4-point strength testing configuration, $X = 10$ mm. The average fracture force for the 3-point and 4-point bend strength tests was 20 N and 10 N, respectively. Calculate the modulus of rupture for both cases.

Solution:

In the case of the 3-point strength configuration:

$$\sigma_{f\,3\text{pt}} = \text{MOR} = \frac{3 \times 20 \times 40 \times 10^{-3}}{2 \times 4 \times 10^{-3} \times \left(3 \times 10^{-3}\right)^2} = 33.33\,\text{MPa}$$

In the case of the 4-point strength configuration:

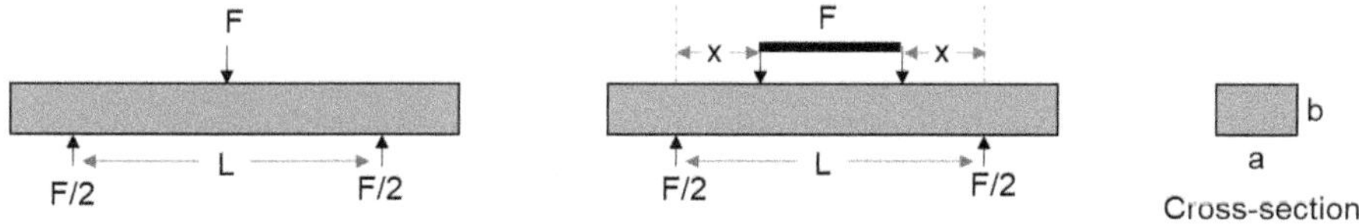

Fig. 12.6 **a** 3-point strength testing setup, **b** 4-point strength testing setup

$$\sigma_{f\,4\mathrm{pt}} = \mathrm{MOR} = \frac{3 \times 10 \times 10 \times 10^{-3}}{4 \times 10^{-3} \times \left(3 \times 10^{-3}\right)^{2}} = 8.33\,\mathrm{MPa}$$

Note that the 4-point bend MOR is lower than the 3-point bend one. To understand this, we must consider both configurations' stress state. During load application, the bottom surface of the ceramic bar will be subjected to a tensile force directed outwards toward the two support pins, and the top surface will be subjected to a state of compression. In Fig. 12.7a and b, the stress at the bottom surface is shown, as can be seen, for the 3-point strength testing, the stress is maximum only along a line on the bottom surface, coinciding with the position of the top pin that applies the force. The stress level decreases linearly on both sides of the line to a value of zero at the bottom supporting pin locations. On the other hand, in 4-point testing, the stress is maximum over an area (not a line) on the bottom surface, coinciding with the separation between the two pins on the top surface. Again, on both sides of the area, the stress magnitude is reduced until it becomes zero at the support pin locations. In Fig. 12.7c, it is clear that tensile testing promotes maximum stress in the reduced section of the sample, which is not an area or a line, but rather a volume. These geometrical differences in where the maximum stress lies between these three testing configurations can significantly influence the measured strength value of a ceramic. To understand this, we must realize that many ceramics contain inherent flaws (micro-cracks) within their volume. These flaws will be randomly distributed and have varying lengths and orientations with respect to the applied tensile stress on the bottom surface of the bend strength samples. In 3-point bend strength testing, we see that the maximum tensile stress occurs along a line on the bottom surface of the ceramic bar sample. Now we have to ask, what is the probability that the largest flaw is present along that line? Also, is it oriented favorably to be easily opened, i.e., does it lie perpendicular to the applied tensile stress acting on the bottom surface? It becomes clear that this probability will be low for the case of 3-point testing. Figure 12.8a shows how the largest flaw on the bottom surface of a ceramic bar being tested can be missed by the maximum stress line in 3-point bend testing. In this case, the strength value will be high, but definitely misleadingly high.

On the other hand, in 4-point strength testing, the same flaw will lie within the bigger area of maximum stress, resulting in a lower but relatively more realistic strength value. This larger area is shown in red in Fig. 12.8b. Tensile testing can catch the biggest flaw better than 4-point or 3-point bend strength testing, since now a much larger volume is subjected to the maximum stress and not an area or a line. Thus, strength values for the same ceramic will be highest for 3-point, followed by 4-point, and the lowest and most reliable will be the value obtained from tensile testing. When the probability of finding the largest flaw is low, it poses another problem. Let's assume we have a ceramic disk that was manufactured by hot pressing of ceramic powders in a graphite die under certain process conditions of temperature and pressure. The consolidated high-density disk has been subjected to the same manufacturing conditions everywhere. Yet there will be a random distribution of flaws within the disc. So, if we were to cut the disk into rectangular bars for testing,

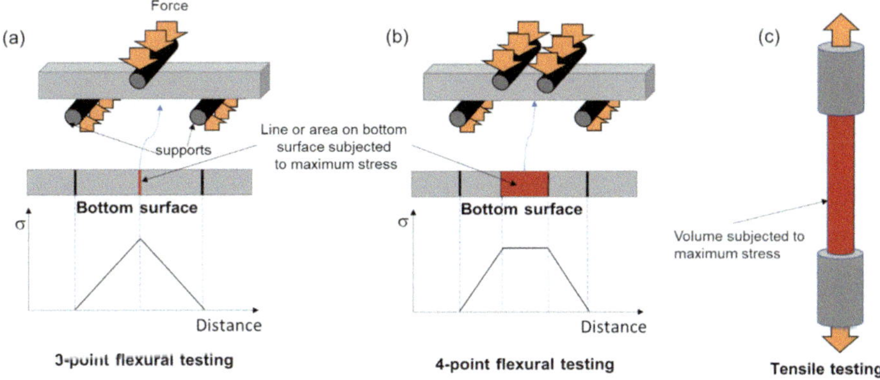

Fig. 12.7 Schematics of **a** 3-point, **b** 4-point, and **c** tensile testing

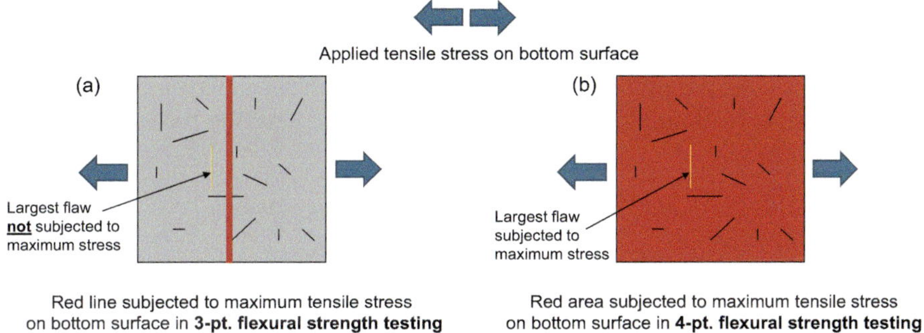

Fig. 12.8 Schematics of **a** maximum stress line (red) on bottom surface in 3-pt. testing, **b** maximum stress area (red) on bottom surface in 4-pt. testing

each bar could break at a different stress value than the other, just because it contains a different flaw size distribution or orientation than the next bar. This results in a range of strength values obtained for the same disk material when tested. Since 3-point bend testing strength values can be misleading, one of the top scientific journals, early on, banned researchers from submitting papers that report strength results obtained through 3-point bend testing. 4-point strength testing remains the most convenient, cheapest, and standard method for ceramic strength testing.

Did you know, Alex, that the concept of 3-point and 4-point flexural (or bend) testing is somewhat intuitive? If I gave you a pencil and told you to break it, I'm sure you would apply the 4-point bend principle using your thumbs and index fingers of both hands. What if I gave you a long, slender stick and asked you to break it? You would probably use your knee and two hands to break it in 3-point. You, see? It's not rocket science

Wow, I didn't think of it like that.

At this point, it should have become clear that when considering the strength of engineering ceramics, we may have to resort to probability analysis. This is where the **Weibull distribution** and the **Weibull modulus** are quite relevant. Weibull put forward what is referred to as the *weakest-link theory*, in that the variation of strengths of a ceramic is a result of a population of random flaws of varying lengths and orientations, and that the strength will be determined by the largest, most critically stressed flaw. For tensile loading, Eq. 12.7a was put forward to describe the probability of failure:

$$F = 1 - e^{-\int_v \left(\frac{\sigma - \sigma_u}{\sigma_0}\right)^m dV}$$

(12.7a)

F is the probability of failure, σ_0 is a characteristic strength value (usually assumed to be the stress resulting in a probability of failure of 0.632), and σ_u is the threshold stress below which no failure will occur (typically taken as zero). **m** is the **Weibull modulus**, a dimensionless parameter for which a high value represents a low scatter in strengths and more confidence in the strength value. Typically, pottery, whitewares, and brick may have m values of 3–5 or less, which is low. An m value greater than 10 is good, and an m value greater than or equal to 30 is exceptional, resulting in consistent strength values. We see that those materials, such as aluminum, extruded magnesium alloy, and steel, can have m values, for instance, between 90 and 100. In contrast, engineering ceramics such as alumina, silicon carbide, and silicon nitride may have values between 5 and 10. Magnesium-based *glassy* alloys can have values between 5 and 41. Hence, the Weibull modulus plays a vital role in determining the level of scatter that can be present in the strength of a material and, hence, will, in turn, affect our design approach.

For a specimen experiencing uniform tensile stress, Eq. 12.7a can be simplified to Eq. 12.7b:

$$F = 1 - e^{\left[-V\left(\frac{\sigma - \sigma_u}{\sigma_0}\right)^m\right]}$$

(12.7b)

If σ_u is taken to be zero, then the equation further simplifies to Eq. 12.7c:

$$F = 1 - e^{-V\left(\frac{\sigma}{\sigma_0}\right)^m} \tag{12.7c}$$

or

$$e^{-V\left(\frac{\sigma}{\sigma_0}\right)^m} = 1 - F$$

Taking the natural log of both sides gives

$$V_E\left(\frac{\sigma}{\sigma_0}\right)^m = -\ln(1-F) = \ln(1-F)^{-1} = \ln\left(\frac{1}{1-F)}\right)$$

If we take the natural log again, we get

$$\ln(V_E) - m\ln(\sigma_0) + m\ln(\sigma) = \ln\left(\ln\frac{1}{(1-F)}\right)$$

$\ln(V_E) - m\ln(\sigma_0)$ is a constant, let's call it C. Hence, the equation becomes

$$\ln\left(\ln\frac{1}{(1-F)}\right) = m\ln(\sigma) + \ln(V_E) - m\ln(\sigma_0) = m\ln(\sigma) + C$$

If we then plot $\ln\left(\ln\frac{1}{(1-F)}\right)$ versus $\ln(\sigma)$, it will yield a straight line with the slope being m, and the intercept as C. If fact this is how the data is plotted. An example schematic is shown in Fig. 12.9.

The strength of a ceramic depends on several factors. As we have seen, one of them is the type of test used to determine strength. However, the porosity in the

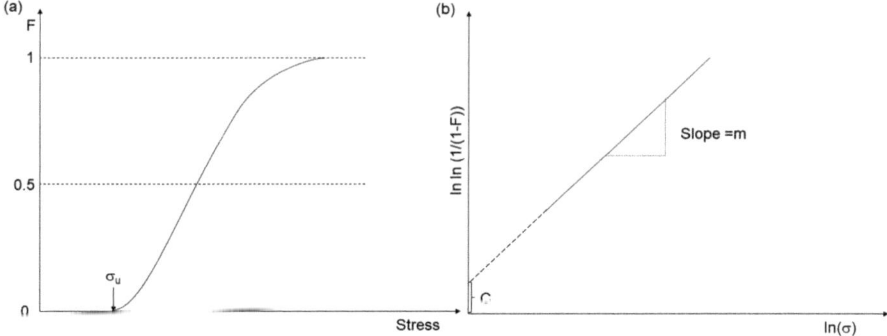

Fig. 12.9 **a** Probability of failure versus stress, **b** ln (ln(1/1 − F)) versus ln (stress)

microstructure will also cause a decline in strength. Noting that obtaining fully dense ceramics (0% porosity) is a challenge. Hence, when quoting the strength of ceramics, it is essential to state the testing method used and the material's residual porosity. Another factor that affects strength is grain size. Similar to the Hall–Petch equation, a grain size decline increases strength.

Table 12.5 lists MOR and tensile strength values for some ceramics. **W. D. Richerson** has compiled these strength values from various sources using different testing configurations (3-point/4-point bend strength for MOR and tensile testing data) and different porosity levels and grain sizes. Hence, the data presented only provides a general idea of the strengths of certain ceramics. By looking at the table, it becomes evident that, in general, for the same material, the tensile strengths (obtained from tensile tests) are lower than their MOR counterparts. Also, an increase in porosity results in a decline in strength, as seen in the strength of silicon nitride (Si_3N_4). Of course, the manufacturing process and conditions under which a ceramic is produced also have an effect due to their impact on the resulting microstructure.

Additionally, the strength of ceramics is highly sensitive to the surface condition of the specimen to be tested. Typically, the bottom face of the 3-point or 4-point bend strength tested bar is polished. This is important since the grinding operations (for example, using a flat-bed grinder) used to manufacture the ceramic bar will leave the surface with grinding-induced marks and flaws, which can easily be greater than any inherent flaw inside the ceramic. Thus, polishing is the best way to remove such large flaws, which would otherwise result in a false low-strength reading. Polishing, carried out only on the bottom surface (since it will experience tension during the test), also removes residual stresses that can affect the outcome of the strength test. Even the long edges/corners along the bottom surface are also chamfered (usually by diamond

Table 12.5 General strength values of some ceramics

Ceramic	MOR (MPa)	Tensile strength (MPa)
Sapphire (single crystal alumina)	620	–
Alumina (0–2% porosity)	350–580	200–310
Sintered SiC (~ 2% porosity)	450–520	–
Fused SiO_2	110	69
Stabilized ZrO_2 (< 5% porosity)	138–240	138
Spinel (< 5% porosity)	80–220	–
Mullite (< 5% porosity)	175	100
WC (< 2% porosity)	790–825	–
Si_3N_4 (< 1% porosity)	620–965	350–580
Si_3N_4 (15–25% porosity)	200–350	100–200
TiC (< 2% porosity)	275–450	240–275
Insulating fire brick	0.28	–
Graphite (grade ATJ)	28	12

D. W. Richerson, Modern Ceramic Engineering, Taylor and Francis, 2006

Table 12.6 Effect of flaw size on strength in ceramics for different ceramic materials

Flaw size	Material	Strength
nm	Pristine optical fibers, whiskers	10 GPa
	Ultrafine grained ceramics	
		1 GPa
μm		
	As handled glasses, single crystals, fine-grain polycrystalline ceramics	
		100 MPa
	Coarse-grain polycrystalline ceramics, composites	
	Rocks	
mm		
	Refractories	10 MPa
	Concretes	
m		
	Earth's crust, glaciers	1 MPa

With modifications after Brian Lawn, Fracture of Brittle Solids, Cambridge Solid State Science Series, Cambridge University Press, p. 308, 1993.

polishing) to blunt their sharpness, which could potentially be a location where failure could occur. To summarize, surface preparation is crucial in ceramic strength testing to reveal the actual effect of the inherent flaw population on the strength of ceramics and not some artificially imposed surface flaw or stress concentration in the form of sharp edges.

Table 12.6 shows different flaw sizes typical of various brittle materials and expected strength ranges to exemplify the sensitivity of strength to flaw size in ceramics.

12.4.2 *Fracture Toughness of Ceramics*

It is intuitive to think of ceramics as having very low fracture toughness values since they are known to be brittle. We all have broken a mug or plate at some point. This is true; however, over the years, some very important and, in some cases, ingenious methods have been developed to increase the fracture toughness of ceramics through careful microstructural design. In this section, we will discuss a number of these methods, all of which are approaches that make it difficult for a crack to propagate in a ceramic.

(a) **Crack Bridging**

The method restricts a crack from opening and extending by having microstructural features bridging the two crack surfaces behind the crack tip. If the crack

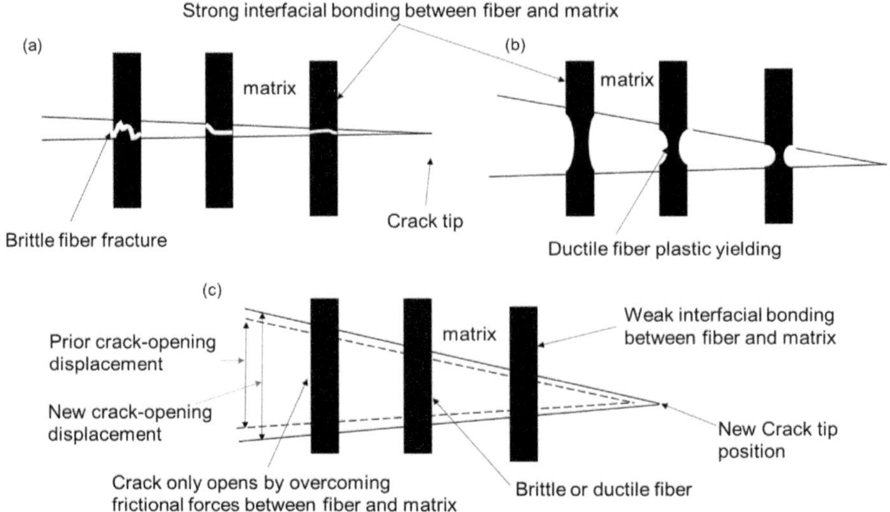

Fig. 12.10 **a** Crack bridging of brittle fibers where interface between fiber and matrix is very strong, **b** Crack bridging of ductile fiber where interface between fiber and matrix is very strong, **c** Pullout of brittle (or ductile) fibers when interface between fiber and matrix is weak. (multiple sources)

tries to advance further, it will need to open, which in turn will be resisted by these microstructural features, which act like anchors behind the crack tip. These microstructural features can be brittle or ductile, as shown in Fig. 12.10.

One notable example of applying this toughening mechanism is in coarse-grained alumina (Al_2O_3). This method is based on the principle that if a crack is to propagate and grow, its faces will need to open or separate more and, in turn, produce what is referred to as a **crack-opening displacement** between the two crack faces. The crack-bridging mechanism limits the crack opening displacement, making it difficult for the crack faces to open and consequently difficult for the crack to grow in length. Scientifically speaking, if the crack is allowed to open, the stress intensity at the tip would increase, thus allowing its extension, and by limiting the displacement, we limit the stress intensity at the tip and hinder the ability of the crack to grow. Crack bridging in alumina relies on the crack taking an intergranular pathway around grains as it advances, and once it passes a grain, the grain will be left behind the crack tip but still bridging the two crack faces. This means that if the crack faces wanted to open, at least one face would have to pull away from the grain partially by overcoming frictional forces between the grain and the surrounding microstructure in that location. This friction generates a crack closure force. As the crack advances, more and more grains will be left behind the crack tip and available for bridging. Crack bridging results in a unique outcome: as the crack initially propagates, the material gets tougher and tougher, i.e., the fracture toughness increases with increased crack length. Since the resistance (R) to crack propagation increases with increased crack length, this behavior is called **R-curve behavior**. This crack bridging mechanism and the resulting R-curve behavior are shown in Fig. 12.11. As the grain size

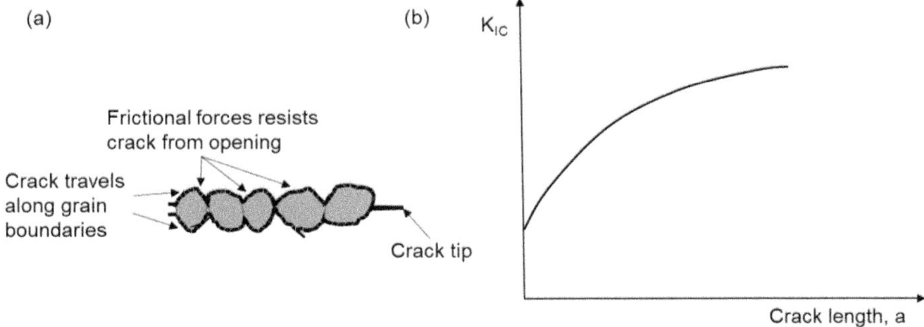

Fig. 12.11 a Schematic of crack bridging mechanism, **b** Resulting R-curve behavior (i.e., an increase in fracture toughness with an increase in crack growth)

increases, the effect of crack bridging increases since there will be a larger interfacial area between the bridging grain and the surrounding microstructure, thus imposing resistance to crack opening through increased friction. Similarly, *R*-curve behavior can also be found in fiber-reinforced ceramics, as shown in Fig. 12.10.

(b) **Crack Deflection**

Another approach to increasing the fracture toughness of ceramics is through the crack deflection mechanism. If a crack takes a straight path, it produces two new surfaces with their accompanying surface energy. This surface energy, which is energy per unit area, is the material's resistance to crack propagation. Suppose we force the crack to propagate instead in a somewhat serrated way. In that case, the crack will cover a larger surface area, and hence it will be required to supply a larger surface energy, which poses a higher resistance to crack propagation. There is also the added benefit that, as the crack is somewhat zigzagging while extending, it moves in different directions, which ends up reducing the stress intensity factor at the crack tip (which would otherwise be highest for a straight crack moving in a direction perpendicular to the applied load). Crack deflection can be observed in ceramic matrix composites where a ceramic is reinforced with ceramic particles of different thermal expansion coefficients. After sintering or high-temperature consolidation, and as the composite is cooling, the matrix and the reinforcements shrink at different rates; this causes internal stresses to build up. If controlled, these stresses can force the crack to either avoid the reinforcements and deflect around them or be attracted to them and even forced to intersect them. Figure 12.12a shows a schematic of how a crack can deflect around a spherical particle due to favorable internal stresses, while Fig. 12.12b shows the same but for elongated particles, forcing the crack to take a longer path and hence absorb more energy. Figure 12.12c and d is micrographs of a glass composite where thoria particles have been added to a different glass with different thermal expansion coefficients. By controlling the residual internal stresses, the crack is seen to deflect around the thoria particle in one micrograph, where the matrix is Corning 7740 Pyrex, and intersect through them in another, where the

(a)   (b)

(c)   (d)

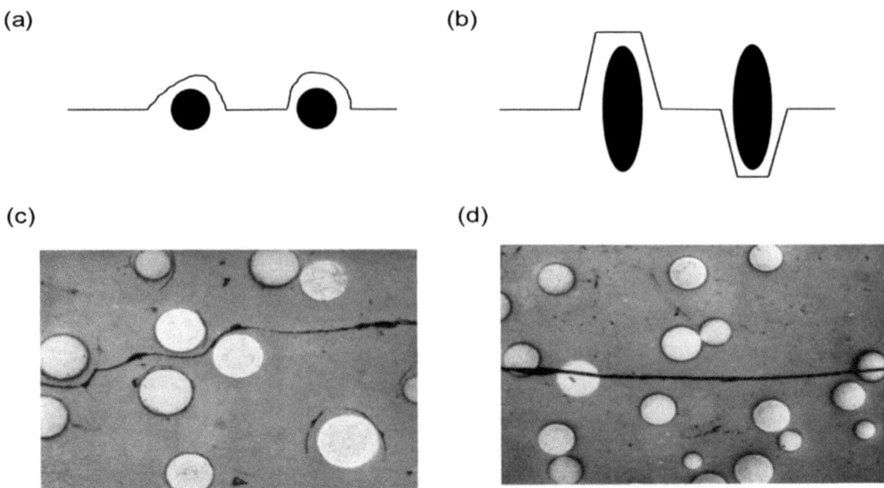

Fig. 12.12 Schematic of crack deflection mechanism around, **a** spherical particles, **b** elongated particles, and micrographs showing, **c** crack deflection around Thoria spherical particle in Pyrex glass matrix, **d** crack attracted toward Thoria spherical particles and intersecting them (matrix is soda lime glass). **a** and **b** based on multiple sources, (**c** and **d**) from R. W. Davidge, T. J. Green, The Strength of Two-phase Ceramic-Glass Materials", Journal of Materials Science 3 (1968), pp. 629–634. Reproduced with modification with permission from Springer Nature

matrix is GEX X8 Soda lime glass. This type of control of crack behavior during fracture can be used to improve the fracture toughness of ceramics.

(c) **Transformation Toughening**

This is probably the most ingenious mechanism of them all. This is seen explicitly in either zirconia or ceramics reinforced with zirconia particles. Discovered by **Garvie and co-workers** in 1975, the toughening effect of zirconia has been studied extensively. Before we discuss these mechanisms, we must first discuss the three polymorphs of zirconia (ZrO_2). Zirconia has a monoclinic crystal structure at room temperature up to 1170 °C, above which it transforms to the tetragonal crystal structure, which remains up to 2370 °C, above which it transforms to the cubic crystal structure, which is stable until the material melts at 2680 °C. Specifically, the tetragonal to monoclinic transformation occurs upon cooling below 1170 °C, from which the real benefit comes. This transformation results in a 3–5% volume expansion of the zirconia, which plays an important role in two important toughening mechanisms, as outlined below.

Microcracking

If zirconia particles are introduced as reinforcements in a matrix of alumina or cubic zirconia, they will transform from tetragonal to monoclinic crystal structure upon cooling.

If the zirconia particle size is large enough, the tangential stresses from the expansion below 1170 °C will result in microcracking in the surrounding matrix material (Fig. 12.13). Now, imagine an advancing crack accompanied by its surrounding tensile stress field. As it gets close to the microcracked particle, the microcracks can become influenced by the tensile stress field surrounding the crack and grow slightly under its influence, thus reducing the stress field of the major incoming crack. Also, if the crack intersects the particle, the crack can be deflected through the growth of the microcracks at different angles than the long growing crack (i.e., bifurcation). We must be very careful that when we allow these microcracks to form around the particles initially, we avoid them being too long so as to weaken the material. The size of microcracks will be related to the size of the initial tetragonal particle. If the particle size is too small, the tetragonal particle will not transform; if it's too large, it will transform but result in large microcracks that are not good for strength. The best size is somewhere in between these extremes.

Fig. 12.13 Schematic of microcracking in matrix surrounding zirconia particle upon its transformation from tetragonal to monoclinic (based on multiple sources)

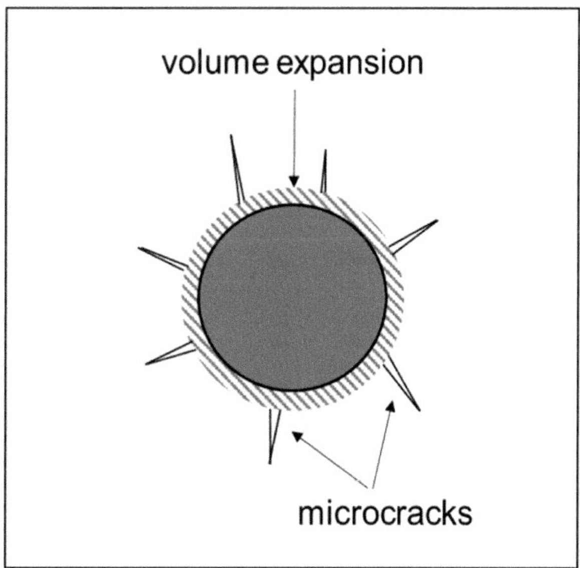

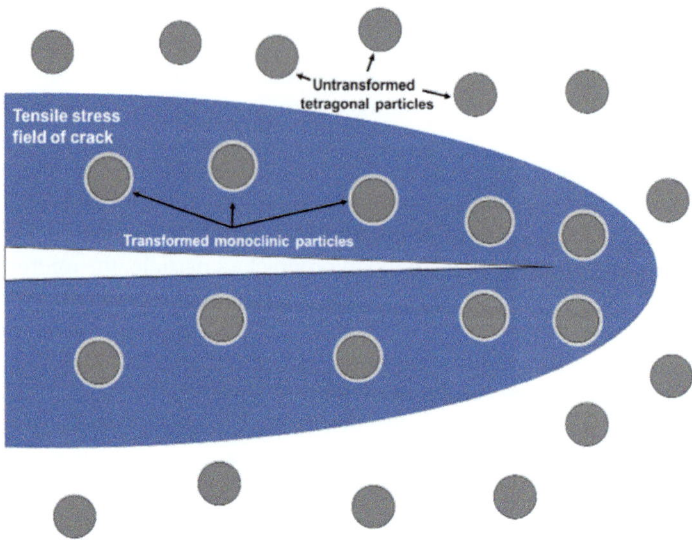

Fig. 12.14 Schematic of Stress-Induced Transformation Toughening mechanism (multiple sources)

Stress-Induced Transformation Toughening

This mechanism is probably one of the most ingenious mechanisms for toughening ceramics. It also builds on the tetragonal-to-monoclinic martensitic transformation with its accompanying expansion. Here, tetragonal particles are distributed in a ceramic matrix, constraining the tetragonal from expanding upon cooling. This leaves a matrix with metastable tetragonal reinforcements. As the crack advances, its accompanying tensile stress field can be enough to remove the matrix's constraint on the tetragonal particles, thereby promoting the tetragonal to monoclinic martensitic transformation, with its accompanying volume expansion of the particles. Due to the transformed particles' 3–5% volume expansion, the regions close to the particles will be under compression (Fig. 12.14). This reduces the energy of the advancing crack, promoting an increase in fracture toughness.

12.5 Fracture Toughness Measurement Techniques

The fracture toughness of ceramics has been measured using a variety of techniques. Some involve long cracks, some short cracks, and even indentation techniques have been used. This section discusses some of these techniques.

Long Crack Fracture Toughness Methods

This method involves introducing a long crack in a specimen of a specific geometry. One of these geometries is the **double cantilever beam** geometry, shown in

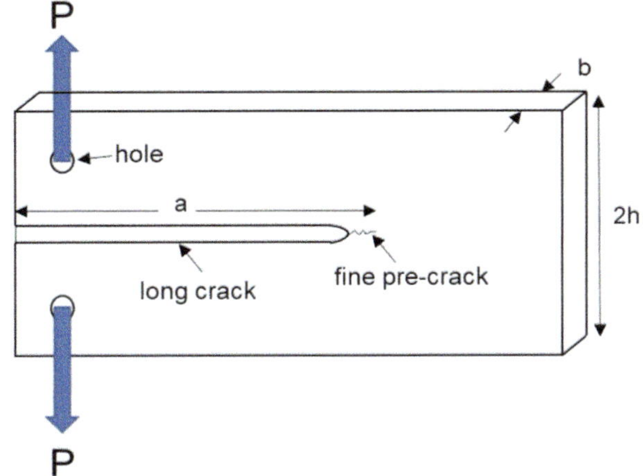

Fig. 12.15 Schematic of stress induced transformation toughening mechanism

Fig. 12.15. The long crack is typically introduced using a diamond or diamond-impregnated wire saw. These can produce long cracks, but since the blades of these saws typically have a thickness between 150 μm and 300 μm, the long crack will have a relatively blunt tip. If the sample is tested as is, the test would not be valid. This is because the crack is required to be a sharp hairline crack. This can be achieved through **pre-cracking**, where a fine crack is introduced at the tip of the long, blunt crack. Pre-cracking is highly challenging, to say the least, for ceramics; often, samples can fracture during this process. Additionally, the fine crack introduced must also have a straight through-thickness crack front. The sample is then tested in tension by attaching the sample to a universal testing machine adapter using pins that go through the sample holes and the adapters (Fig. 12.15). These holes are drilled using a core drill under water.

Equation 12.8 can be used to calculate the fracture toughness (K_{IC}) of a double cantilever specimen from the specimen dimensions and the fracture load (P), as seen in Fig. 12.15.

$$K_{IC} = 3.45 \frac{Pa}{b\sqrt{h}} \left[1 + 0.7 \left(\frac{h}{a} \right) \right] \tag{12.8}$$

Indentation Fracture Toughness

Historically, during the Vickers Hardness testing of brittle materials, sometimes cracks emerged at the corners of the indent at some critical load. Initially, this was regarded as a testing defect, and the hardness resulting from this situation would not be accepted. Over the years, the size of these cracks has been used in semi-empirical expressions to measure the fracture toughness of materials. Notably, the work of **Palmqvist** (1957) and **Lawn and coworkers** (1975) provided the basis for the study of indentation cracks and, ultimately, their relevance in determining the fracture toughness of materials, the method we know today as *Indentation Fracture Toughness*. In indentation fracture toughness testing, typical loads applied range from 10 to 300 N. In comparison to long-crack specimens for fracture toughness determination, this method has several advantages, which include.

1. Simplicity and low cost.
2. No pre-cracking required
3. Small sample size (this becomes even crucial for materials not available in large sizes or quantities).
4. Using conventional Vickers hardness testers, which are abundantly available. The surface to be indented needs to be polished, not only to remove potential residual compressive stresses but also to allow for easy measurement of the crack size.

At a particular load, cracks emerge from the corners of the impression of the indentation. Two types of crack systems can emerge; the first is the **Palmqvist crack system**, where the cracks start from the corners of the indentation impression and extend below the surface as shown in Fig. 12.16b; in this case, the crack size is given as "l" (measured from the surface as seen in Fig. 12.16a) and half of the indentation diagonal is "a". The other crack system is the **radial crack system**, as seen in Fig. 12.16c, where the crack extends from the center of the indentation impression radially and below the surface, as seen in Fig. 12.16a and c. This crack is sometimes referred to as a *half-penny crack*. In this case, the radial crack size shown is "c". Figure 12.16a shows how both cracks appear from the top view of the indent, which would enable the measurement of the appropriate parameters. Depending on the crack system, a different expression calculates the fracture toughness (K_{IC}). So, we must first identify the crack system in operation by simply polishing the surface indentation. As material is removed, if we see the radial crack separating from the corner of the indent, it is a Palmqvist crack system. Equation 12.9a is the expression used to calculate the fracture toughness of ceramics that show the Palmqvist crack system (from C. Ponton, R. Rawlings, "Vickers Indentation fracture toughness test: part 1, Materials Science and Technology 5 pp. 865–872, (1989)).

$$K_{IC} = 0.0089 \left(\frac{E}{H}\right)^{0.4} \frac{P}{a.l^{0.5}} \tag{12.9a}$$

E is the elastic modulus, *H* is the hardness, *l* is the crack size, and "a" is half of the indentation diagonal, as shown in Fig. 12.16b. *P* is the applied load, and K_{IC} is the fracture toughness.

For the radial crack system, fracture toughness can be obtained using the following expression (Eq. 12.9b) (from M. Sakai, R. Bradt, Fracture toughness testing of brittle materials", International Materials Review, vol. 38, no. 2, p. 66 (1993).

$$K_{IC} = A \left(\frac{E}{H}\right)^{n} \frac{P}{c^{1.5}} \tag{12.9b}$$

"A" typically has a value between 0.016 and 0.018, and n = 0.5.

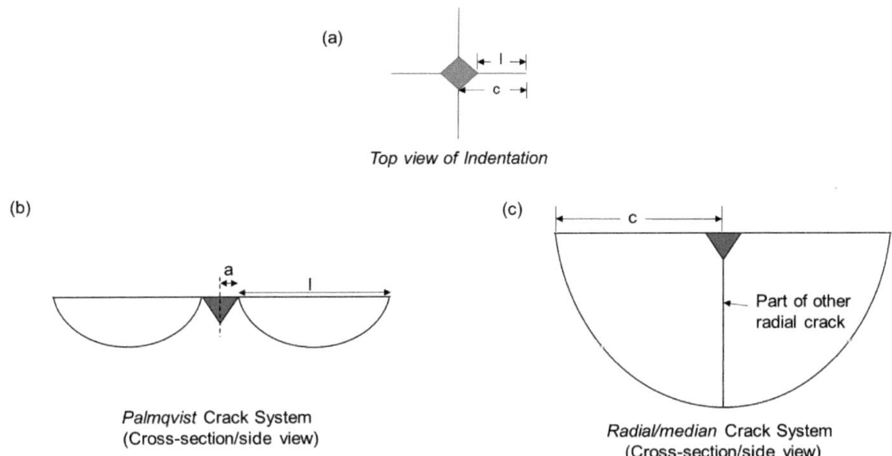

Fig. 12.16 **a** Top view of indentation showing cracks emerging and relevant parameters for Radial/Median and Palmqvist cracks, **b** Side view (cross-section) of Palmqvist crack system, **c** Side view (cross-section) of radial/median crack system

Indentation Strength in _Bending_ Method

This is another indentation-based fracture toughness measurement technique; however, this one does not need the measurement of any indentation crack. The technique involves indenting a ceramic bar midway on the bottom surface with an indentation load P, generating a radial crack, and then testing the bar under 4-point loading to obtain the indentation strength of the indented bar. Fracture toughness can then be calculated using Eq. 12.10:

$$K_{\mathrm{IC}} = 0.88\sigma^{3/4}P^{1/4} \tag{12.10}$$

where P is the indentation load and σ is the indentation strength (i.e., the 4-point bend strength of an indented bar), however, we must confirm that failure was initiated at the indentation cracks.

12.6 Fatigue of Ceramics

Ceramics' fatigue behavior can be measured following standards such as ASTM C1361-10 (2019), which provides standard practice for testing ceramics under constant amplitude, axial, and tension-tension fatigue situations. Flexural cyclic loading can also be used for fatigue testing of tensile samples. Of course, the larger the volume exposed to fatigue conditions, the better because of the ceramics' flaw size distribution.

An interesting, simplified indentation method for evaluating the *relative* fatigue properties of ceramics was proposed in the 1980s. The method was called *Fatigue by repeated indentation* . Although it is not like the standardized methods mentioned above, it highlights some interesting characteristics when ceramics are indented, which is discussed here. In this method, a sample is indented with a Vickers "macro"-indenter, and if the load is large enough, a lateral crack is formed underneath the indenter. The subsurface view of the indentation presented in Fig. 12.17 is somewhat incomplete. As the material is indented with an indenter, a plastic zone is developed beneath the indentation impression due to plastic deformation/flow of the ceramic underneath the indenter. In addition to the radial/median cracks shown in Fig. 12.17, a lateral crack underneath the indenter is also formed. If the material is indented in the indentation location using a lower load than that used to form the lateral crack in the first place. With each indentation, the lateral crack will grow, and after a certain number of indentation cycles, the crack will reach a critical size where it spalls and connects to the surface. At this point, the original indentation no longer becomes fully visible. Curves can be generate similar to the S-N curves by conducting these same procedures, using the same initial load to form the lateral crack, but different subsequent lower loads. The number of indentations until spalling occurs increases with a decline in the magnitude of the subsequent loads. Subsequent load vs. number of indents until spalling, resembles the S-N curves. Note although the indenter applies

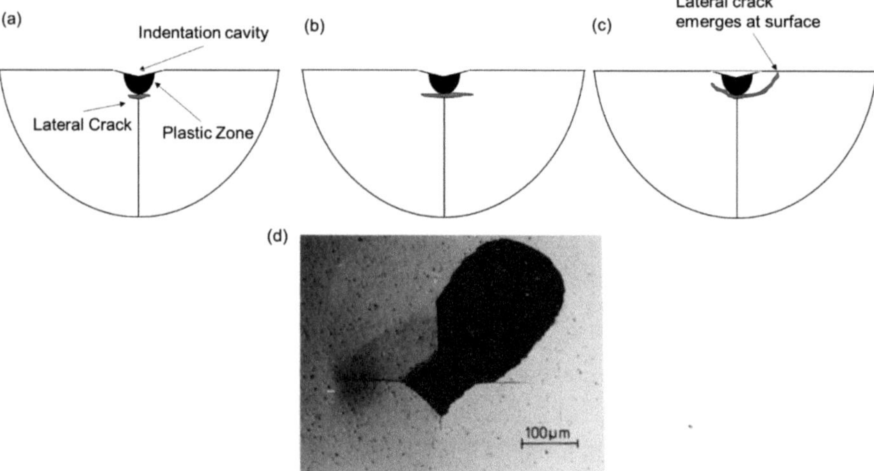

Fig. 12.17 Schematic of **a** lateral crack formation after peak load is first applied, **b** propagation of lateral crack under the influence of repeated load (of magnitude lower than the peak load), **c** lateral crack emerges at the surface causing spalling of matter, **d** micrograph of Vitox alumina sample showing spalling after a lateral crack has emerged at the surface (causing spalling which appear as large black blob to the top right of the indentation in this case) following 9 indentation cycles with a load of 125N (the original peak load to produce the lateral crack was 150N). D. A. Vaughan, F. Guiu, M. R. Dalmau, "Indentation Fatigue of alumina, Journal of Materials Science Letters, 6 (1987), 689–691, (**d**) reproduced with permission from Springer-Nature

in general a compressive stress, the presence of small ceramic debris inside the lateral crack and close to its tip can generate tensile stresses at the crack tip which aids fatigue.

12.7 Wear of Ceramics

Wear is complex. While most mechanical behaviors depend solely on the response of one material to an applied load, wear involves two surfaces/materials contacting. Hence, it is not a material property. The wear behavior will change depending on which type of surfaces are contacting each other. The magnitude of the contact force, whether it is uniaxial or more complex, among other factors, affects wear behavior. Wear occurs in metals, polymers, ceramics, intermetallics, and their composites.

12.7.1 Surfaces

However, since material wear occurs at the surface of a component, it is important to talk first about surfaces. We have often visually examined artifacts, and if they appear smooth and somewhat shiny, we say it has a good surface finish. We may even assist our visual examination by touching the artifact and running our finger up and down its surface to get a feel. In engineering, however, we want some quantification. Even on a polished surface, when we zoom in at higher magnifications, we will always observe some surface irregularities or extrusions called **asperities**. We now define two important parameters for surface evaluation: **Surface texture** and **surface integrity**. As presented in Fig. 12.18, surface *texture* includes surface roughness, waviness, flaws (including holes, scratches, or inclusions), and lay (which is the physical texture of the surface, for example, emerging from the direction of machining).

Surface integrity, however, refers to the subsurface material and its condition in terms of composition, grain structure, impurities, etc., which may be different

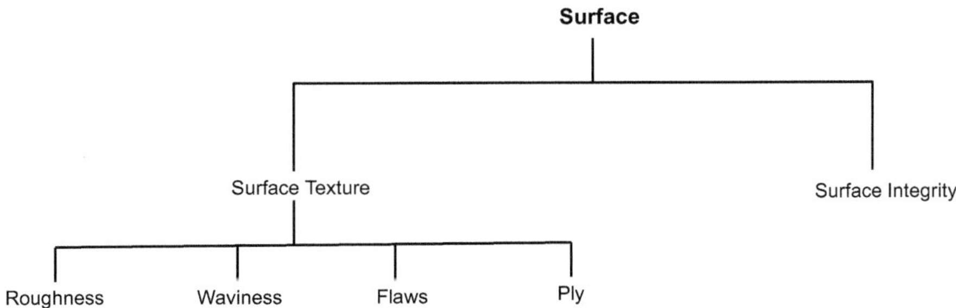

Fig. 12.18 Surface texture and integrity

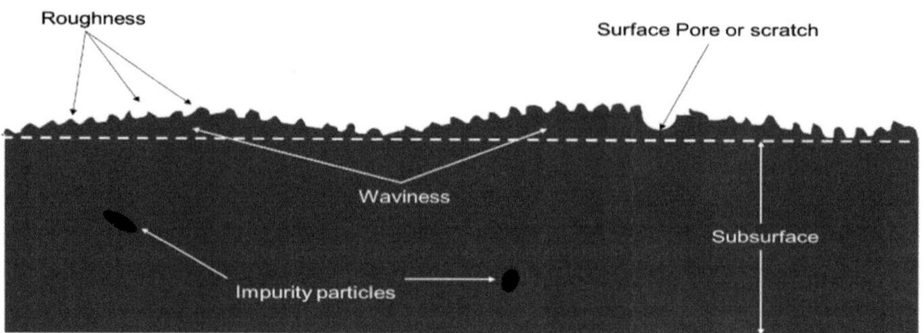

Fig. 12.19 Surface and subsurface features

from the bulk material. Figure 12.19 shows a simplified schematic of some of these features.

One of the most fundamental surface properties is roughness. A common way of measuring it is through the **mean roughness (R_a)**, which is calculated with the help of a profilometer, where a stylus moves over a surface, measuring the peaks and valleys of the surface. Non-contacting optical profilometers are also used. After an arithmetic mean between the peaks and valleys of the surface is calculated, the R_a is simply the average of the deviation of the peaks and valleys from that mean. Note that the distance between the mean and the lowest point in the valley is also taken as positive in the calculation. Historically, the root mean square of these measurements was used to depict RMS roughness, but that has since lost popularity.

Roughness is typically represented in units of micrometers or micro-inches. Surface roughness is primarily dependent on the manufacturing method and process parameters used. Table 12.7 lists some examples of surface roughness values resulting from different manufacturing processes, including some for metals (sand casting and die casting) and ceramic components, including electro-discharge machining and surface finishing operations conducted, such as surface grinding, polishing, and lapping.

Table 12.7 Surface roughness values for some processes

Process	Finish	Roughness (μm)
Die casting	Good	1–2
Sand casting	Poor	12–25
Grinding	Very good	0.1–2
Lapping	Excellent	0.05–0.5
Polishing	Excellent	0.1–0.5
Superfinish	Excellent	0.02–0.3
Electric discharge	Medium	1.5–15

Source M. P. Groover, Fundamentals of Modern Manufacturing: Materials, Processes, and Systems, 5th edition, Wiley (2013)

12.7.2 Friction

Friction is one of these phenomena that can be viewed positively or negatively. On the positive side, we need friction to walk on a floor; otherwise, we would always slip and fall. The same goes for writing with a pencil; without friction, graphite would not have been deposited on the paper to produce the writing. Welding/bonding processes such as friction stir welding or friction bonding rely on friction and its accompanying heat generation to weld metals with better weld properties than conventional welding. On the negative side, friction is responsible for a substantial cost to any nation. In the manufacturing industry, for example, in metalworking processes such as extrusion, friction between the workpiece and the inner surface of the extrusion container must first be overcome for the deformation to be successful. This increases the magnitude of the required extrusion force (and hence energy) to carry out the extrusion operation. In powder metallurgy, friction is also responsible for the inhomogeneous density distribution within a powder compact during powder compaction, eventually leading to product distortion upon sintering. Over time, friction also results in the wear of the working dies and containers, also leading to dimensional inaccuracies of the product and, eventually, tool replacement, which is also quite costly. There are instances where friction is needed, but must also be controlled. For example, when driving, pressing on the brakes too much will result in high levels of friction, which could cause a sudden, scary stop, while too little friction may not be enough to stop the car from hitting another car parked ahead of it. Hence, friction control is critical in automobiles and any moving vehicle. Lubricants are used in solid or liquid form to reduce the negative impacts of friction in manufacturing.

Consider friction from a block sliding on a surface, as shown in Fig. 12.20. A normal force is applied to the block (which could also be the block's weight), and a friction force acts opposite to the block's motion. If the block were initially at rest, the friction force would represent the force we must apply to move it. In other words, we must first overcome the frictional force before moving the block.

$$F_f = \mu F_N \tag{12.11}$$

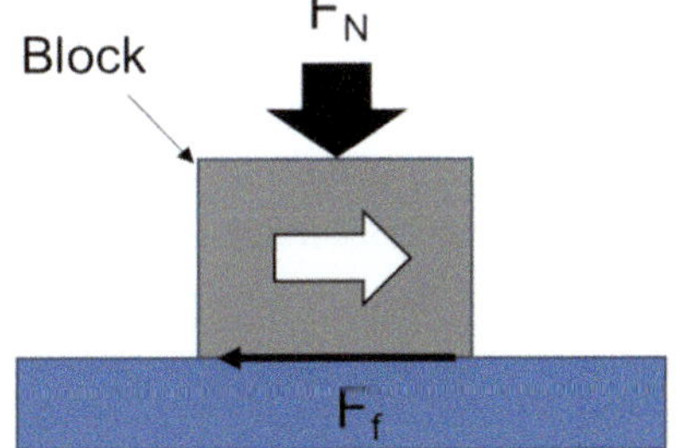

Fig. 12.20 Block moving on surface, showing normal force (F_N) and frictional force (F_f)

The friction force (F_f) is related to the normal force (F_N) by Eq. 12.11), where μ is the coefficient of friction. A high value of μ represents high friction conditions. Here, we have to differentiate between the value of μ when the block is initially at rest (coefficient of static friction, μ_s), and that when the block is sliding (coefficient of kinetic friction, μ_k). Typically, $\mu_s > \mu_k$. Interestingly, to avoid slipping while walking at a normal speed (~ 3 mph), the kinetic friction coefficient would need to be greater than 0.25. For trains with steel wheels on steel rails, the coefficient of friction in dry conditions is ~ 0.2, while in wet conditions it is ~ 0.1. This has a direct effect on the braking distances for trains. Typically, trains require three times the braking distance as cars because they have approximately a third of the friction coefficient values of cars with rubber tires.

The coefficient of friction (COF) typically does not depend on the area of the object; hence, generally, the friction coefficient of large objects and small objects of the same material have approximately the same coefficient of friction. It is very difficult to quote friction coefficients with a high degree of accuracy, as they are highly dependent on the type of contacting (mating) surfaces, among other factors. However, general ballpark values for some ceramics are listed in Table 12.8. Lubrication can be very effective in reducing the coefficient of friction. This data is for ceramics; however, when other dissimilar materials slide over each other without lubrication, we can still reach low values of the coefficient of friction; for example, the coefficient of friction of PTFE on steel could be around 0.04. It is also important to note that the COF values for lubrication surfaces can be much lower (down to 0.01 for boundary lubrication and 0.001 for hydrodynamic lubrication) than those shown in Table 12.8.

Table 12.8 Values for coefficient of friction for some mating ceramics

Mating surfaces	Lubrication	Coefficient of friction
Alumina on Alumina	No lubrication	0.50
Alumina on Alumina	Poor lubrication	0.32
Alumina on Alumina	Good lubrication	0.15
Alumina on Alumina	Excellent lubrication	0.12
Alumina on Zirconia	No lubrication	0.34
Alumina on Zirconia	Poor lubrication	0.21
Alumina on Zirconia	Good lubrication	0.11
Alumina on Zirconia	Excellent lubrication	0.10
Alumina on Silicon Carbide	No lubrication	0.24
Alumina on Silicon Carbide	Poor lubrication	0.14
Alumina on Silicon Carbide	Good lubrication	0.09
Alumina on Silicon Carbide	Excellent lubrication	0.08

Source Ernest Rabinowicz, Friction and Wear of Materials, 2nd edition, Wiley Interscience, p. 118 (1995)

12.7.3   Wear

Material wear is a significant economic problem for nations. It is not unusual that the cost of wear could be as high as 5% of a nation's Gross National Product (GDP). Typically, it does not take much wear to render an engineering component unusable, especially in applications with critical tolerances and dimensions. Wear is a major problem in orthopedic implants, affecting an implant's longevity. The ball and acetabular cup involve an articulating motion of the ball within the cup, in which wear ultimately becomes a problem. Typically, cobalt chrome balls are placed in ultrahigh molecular weight polyethylene cups, and wear debris produced over time finds its way to surrounding tissues, resulting in their inflammation and implant loosening and, ultimately, implant failure. At this stage, the patient will require repeat surgery. Implants, in this respect, have a life expectancy of about 20–25 years, making repeat surgery inevitable for younger patients. An important area of research is increasing the longevity of implants. Ceramic-on-ceramic hip implants have also been used to combat the problem of wear. Wear could also lead to costly tool replacement for extrusion dies. The wear of cutting tools is another problem that could lead to their blunting, hence requiring re-sharpening. In other applications, wear is required. The example of a pencil writing on paper is relevant here, as the adhesive wear of graphite allows for its deposition on the paper. All grinding and polishing operations used to prepare materials for microstructural examination, by design, result in abrasive wear. The different types of wear will be discussed later.

When considering wear, one must consider the types of mating surfaces, for example, metal-on-metal, ceramic-on-ceramic, polymer-on-polymer, ceramic-on-polymer, metal-on-ceramic, etc. In all cases, wear involves the removal of material from the surface. When two surfaces slide over each other, one would expect material to be removed from both surfaces. The wear of metals is somewhat different than that of ceramics or non-metals. The surface of a metal typically has a protective oxide layer; however, the action of another surface sliding over it may break the oxide layer, thus exposing the virgin metal to further oxidation, making the situation untenable. Most ceramics, however, don't suffer from this problem, especially oxide ceramics. The strong bonding in ceramics (ionic or covalent) forms the basis for the superior wear resistance of ceramics as opposed to metals and polymers.

We follow specific standards again to measure the wear properties of materials. For example, the ASTM G99 is a standard for sliding wear of ceramics, which refers to a pin-on-disk experimental setup and procedures. Understandably, many experimental configurations can be used to measure the wear of materials. However, we will now consider the **pin-on-disk** configuration, which measures sliding wear. A typical application where sliding wear resistance is needed would be a ceramic used as a thread guide.

We can quantify wear by measuring the wear rate, *W*, as presented in Eq. 12.12.

$$W = \frac{V_\mathrm{m}}{X} \tag{12.12}$$

where V_m is the volume of material removed due to the sliding motion, and X is the distance traveled by the sliding surface. W has units of m^2. The specific wear rate is another property that happens to be dimensionless, as it is defined in accordance with Eq. 12.13.

$$\Omega = \frac{W}{A_n} \tag{12.13}$$

where A_n is the nominal area.

In wear studies, the wear rate constant, k_a, is an important parameter that measures how high or low the wear is. The wear rate constant is defined by Eq. 12.14:

$$k_a = \frac{\Omega}{P} \tag{12.14}$$

P is the bearing pressure, the applied force divided by the nominal area (i.e., the assumed measured area). Although we understand that the actual contact area is much smaller. k_a has dimensions of 1/MPa. It is found that k_a is inversely proportional to Hardness, giving rise to the general understanding that materials with high hardness typically have low wear rates and vice versa. This is generally true as ceramics wear much less than metals, for example. It is wrong, however, to assume that Hardness is the only controlling property. For example, a ceramic can have a lower hardness than another ceramic. Still, because it has a higher fracture toughness, it does not necessarily suffer a higher wear rate. Suppose we view the hardness test in its simplest form. In that case, it involves the indentation of the surface with a harder material (e.g., a diamond indenter), so, understandably, a material that suffers less damage (indentation size) under an indenter can also present good wear properties. However, other factors that influence hardness should be considered, including fracture toughness, thermal conductivity, thermal expansion coefficient, and even corrosion. If material is going to be removed from the surface of a ceramic, then crack formation and propagation will ultimately need to occur. This brings fracture toughness into the picture, so materials with higher fracture toughness should help improve wear resistance. The sliding of surfaces against each other can also give rise to heat from friction, which can give rise to thermal stresses.

Finally, another dimensionless wear rate coefficient is obtained by simply multiplying k_a by the Hardness of the softer material out of the two mating surfaces, as shown in Eq. 12.15:

$$K = k_a \cdot H \tag{12.15}$$

K is typically referred to as the wear coefficient.

We now consider the types of sliding wear.

Abrasive Wear

This typically occurs on the softer of the two sliding mating surfaces by the rougher and harder surface sliding over it. The rough surface basically scratches the softer one and removes wear particles. Such a mechanism is found in grinding operations for example.

Adhesive Wear

This is where adhesive bonding between two sliding surfaces (typically smooth) occurs in specific locations, resulting in pieces of material being removed from one surface and deposited onto the other. Such fragments may later be removed and redeposited on the original surface or remain loose between the two sliding surfaces.

Corrosive Wear

This is when the sliding of two surfaces occurs within a corrosive environment. The sliding action forms corrosive products and then removes them.

Fatigue Wear

This occurs at the surface and subsurface in metals. When rolling or sliding is repeated over the same area, cracks can form and grow, releasing fractured fragments that leave a cavity on one of the surfaces. This effect is exaggerated when the sliding or rolling action reciprocates (back-and-forth).

Another type of "non-sliding" wear is **Impingement wear (erosion)**.

In impingement wear, particles in a fluid (gas or liquid) impact the surface of the ceramic and ultimately remove material over time. A typical application where this type of wear is important involves a ceramic used as a nozzle in sandblasting.

Problems

12.1. What name is given to a ceramic used in engineering applications?

12.2. When a ceramic was heated from 25 to 1025 °C, it produced a strain of 0.0001. Calculate the thermal expansion coefficient of this ceramic.

12.3. Table 12.2 lists Young's moduli and Poisson's ratios of a number of ceramics. What is the stress developed at the surface of Pyrex after it had been quenched from 200 to 25 °C by dipping in water?

12.4. What would be the stress level if instead SiC (α) was quenched from the same temperature (given in problem 12.3)? Which of these materials would you prefer under these conditions, and why?

12.5. What are the main factors that decide the structure/class of a ceramic?

12.6. Why do many ceramics have a range of strength values rather than a single value?

12.7. A ceramic was tested for 4-point strength, and a force of 95 N was required to fracture it. What is its modulus of rupture? Note: $a - 3.7$ mm, $b = 2.9$ mm, and $X = 10$ mm.

12.8. If the same ceramic is tested in 3-point, 4-point, and uniaxial tensile testing, which type of test will show the lowest strength and why?

12.9.   In bend/flexural strength testing of ceramics, what kind of specimen preparation is needed and why?

12.10. Discuss three toughening mechanisms used to improve the fracture toughness of ceramics.

12.11. What is an R-curve? And what microstructural features can cause it?

12.12. A double cantilever fracture toughness test was conducted on a ceramic. If the fracture toughness was found to be 3.5 MPam$^{1/2}$, what was the load at fracture? Note: $a = 19$ mm, $h = 10$ mm, $b = 3$ mm.

12.13. What does surface texture refer to?

12.14. What is surface integrity?

12.15. Is friction good or bad?

12.16. What are the coefficients of static and dynamic friction? Which one is typically higher?

12.17. Discuss the difference between adhesive wear and abrasive wear.

12.18. Can a ceramic have a higher hardness than another ceramic but still have a lower wear resistance? If so, how can this occur?

12.19. What is corrosive wear?

12.20. What is the difference between fatigue wear and impingement wear?

Chapter 13
Electrical and Magnetic Properties of Ceramics

13.1 Introduction

Ceramics are used in a wide range of applications, from permanent magnets to diodes. This chapter introduces the electrical and magnetic properties of ceramics.

13.2 Introduction to Electrical Properties of Ceramics

This book classifies materials as metals, ceramics, polymers, or composites. We also mentioned that other classifications can be used. One of these is classifying materials in terms of their electrical properties. Following this classification, we can divide materials into *conductors*, *semiconductors*, and *insulators*. Although this section discusses the electrical properties of ceramics, it is essential to first delve into the electrical properties of metals to gain a cumulative insight into how materials behave with respect to electricity. Let us first define a few terms that will be helpful down the line.

Electrical conductivity is based on the conduction of *charge* carriers within a material. These **charge carriers** can be electrons, electron holes, or even ions. When electrons are the charge carriers, we refer to the conductivity as **electronic conductivity**, while **ionic conductivity** refers to ions as charge carriers.

The charge of an electron is negative, and to be specific, $- 1.6 \times 10^{-19}$ C. The charge of ions depends on the type of ion. Cations are positively charged, and anions are negatively charged; the magnitude of the ionic charge itself can be equal to or multiples of 1.6×10^{-19} C, depending on the valence of the ion. For example, the charge on a Na^+ cation is $+ 1.6 \times 10^{-19}$C while that on an oxygen anion O^{2-} is $- 2 \times 1.6 \times 10^{-19}$ C $= - 3.3 \times 10^{-19}$ C. So, ions can be positively or negatively charged. We also have a similar situation with electrons. An electron is negatively charged, but when it travels, it leaves behind an **electron hole**, whether the electron

© The Author(s), under exclusive license to Springer Nature Switzerland AG 2026   433
K. Morsi, *An Engaging Approach to the Science and Engineering of Materials*,
https://doi.org/10.1007/978-3-032-06231-4_13

was originally within a covalent bond or even an electron cloud. This electron hole is regarded as being positive with the same charge magnitude as the electron, i.e., $+ 1.6 \times 10^{-19}$ C. **Electric current** (also called electric current intensity (I)) is the *amount of charge flowing per unit time*. In other words, it is the number of charge carriers traveling multiplied by the charge (e) on one carrier divided by the time interval (Δt) during which the charge traveled. This is presented in Eq. 13.1.

$$I = \frac{ne}{\Delta t} \tag{13.1}$$

The charge can be that of an electron–hole or that of an ion.

Electric current, by definition, should be noted in relation to the travel of positive charge carriers and not negative ones. Since the direction of motion of an electron hole (regarded as having a positive charge) is opposite to that of an electron, electric current flows in the opposite direction to electrons.

Metals are known to be electrically conductive with electrons as charge carriers, but why? Well, it all depends on the ability of electrons to travel. Electrons typically reside in what we call the valence band, and it all depends on how much energy is needed to push these electrons into the **conduction band**, where they can move freely. This energy is insignificant for metals as the conduction and valence bands can overlap over a range of energies, making electrons great travelers. For semiconductors, there is an energy gap between the valence band and the conduction band, called the **band gap**. The valence electrons must acquire enough energy to bridge this gap and move into the conduction band. This band gap is relatively small (< 2 eV) for semiconductors compared with insulators. Figure 13.1 shows the difference between the three types of materials. For metals, as seen from the figure, if the valence band is half-filled, it means there is an empty energy state that the electron can very easily go to and travel. We must discuss the **Fermi energy (μ)**. If an electron acquires energy above the Fermi energy, it can travel, and the material will conduct electricity. Depending on the type of material, the Fermi energy will be different. For example, with reference to Fig. 13.1 for metals, two scenarios are depicted. First, an S orbital contains only one electron and is half full. The Fermi energy is at the top of the filled state. Hence, only slight energy is needed to allow a certain number of electrons to become mobile. Such energy can come from thermal energy due to the temperature. Another scenario could occur if the energy of the valence electrons overlaps the next energy orbital, for example, S orbital electrons overlap the next P orbital. In this case, the Fermi energy is located at the bottom of the overlapping region. Again, a small amount of energy is needed to release a fraction of electrons, allowing them to participate in electrical conduction. Only a fraction of electrons would gain enough energy within an electron cloud to travel and conduct electricity.

Some materials have an **energy band gap**. This gap acts as an energy barrier that needs to be overcome before electrons can conduct. This energy is quite large for insulators, but for semiconductors, the energy gap is typically less than 2 eV.

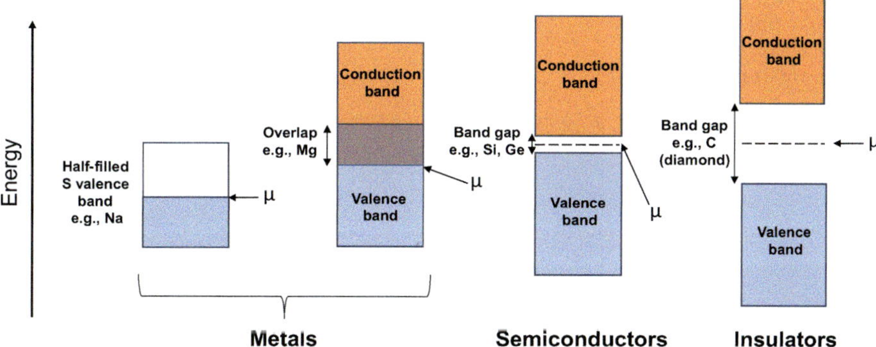

Fig. 13.1 Valence and conduction bands for various materials. Band gaps and fermi energies are also indicated. (Multiple sources including R. A. Higgins, Properties of Engineering Materials, Hodder and Stoughton, 1986)

Ohm's Law

Ohm's law outlines the relationship between voltage (V), current (I), and electrical resistance (R) as shown in Eq. 13.2:

$$V = IR \tag{13.2}$$

According to Eq. 13.3, the electrical resistance of a metal wire is related to its length (l), cross-sectional area (A), and the resistivity (ρ) of the material from which the wire is made.

$$R = \frac{\rho l}{A} \tag{13.3}$$

Resistivity is purely a material property. Figure 13.2 shows the general range of electrical resistivities of metallic solids, semiconductors, and insulators. However, as Eq. 13.3 shows, the resistance depends not only on the material's resistivity but also on the dimensions of the wire. Electrical resistance increases with an increase in the length of the wire and a reduction in its cross-sectional area. It is also noteworthy that **electrical resistance** (in ohms) is the reciprocal of **electrical conductance (G)** (in Siemens).

Electrical *conductivity* is the reciprocal of electrical *resistivity*. In simple terms, electrical conductivity is the ability of a material to conduct electricity. The higher it is, the better the material conducts electric current.

Example Problem 13.1

Determine the resistance of an aluminum wire of length 1 m and a diameter of 0.5 mm. The electrical resistivity of annealed aluminum 1100 alloy is 2.9×10^{-8} Ω m. Also calculate the conductance.

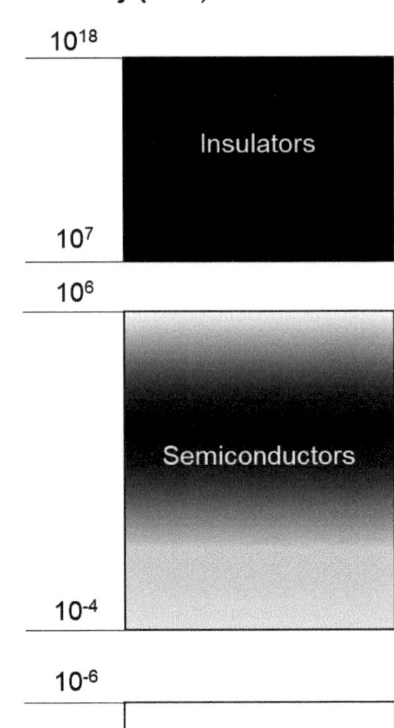

Fig. 13.2 General electrical resistivity ranges of metallic solids, semiconductors and insulators (data from Z. D. Jastrzebski, Nature and Properties of Engineering Materials, John Wiley and Sons, 1959)

Solution

$$R = \frac{\rho l}{A} = \frac{4\rho l}{D^2} = \frac{4 \times \left(2.9 \times 10^{-8} \times 1\right)}{\left(5 \times 10^{-4}\right)^2} = \frac{11.6}{78.5} =\sim 0.15a$$

Therefore, the resistance is 0.15 Ω. Electrical conductance is the reciprocal of the electrical resistance, which is 6.769 Siemens.

Electric conductivity (and resistivity) is affected by several things. For example, alloying or the presence of crystallographic defects typically decreases electrical conductivity and increases electrical resistivity. For metals, electrical conductivity is governed by the flow of electrons. Lattice vibration due to thermal energy can make it difficult for electrons to flow. Hence, we expect to see an increase in resistivity with an increase in temperature. Typically, this takes on a linear relationship for metals between $-$ 200 and 200 °C. Equation 13.4 depicts such a relation.

$$\rho_T = \rho_0(1 + X_T T) \tag{13.4}$$

where ρ_0 is the resistivity at 0 °C, and X_T is the temperature resistivity coefficient. For pure metals, X_T typically has a value of 0.004/°C; however, it can be significantly lower for alloys.

Example Problem 13.2

The electrical resistivity of pure magnesium at 0 °C is 42×10^{-9} Ω.m. Estimate the resistivity at 100 °C.

Solution

The electrical resistivity at 100 °C can be obtained using Eq. 13.4.

$$\rho_T = \rho_0(1 + X_T T)$$

Substituting $X_T = 0.004/°C$, $\rho_0 = 42 \times 10^{-9}$ Ω m and $T = 100$ °C in Eq. 13.4, gives:

$$\rho_{100} = 42 \times 10^{-9}(1 + 0.004(100)) = 58.8 \times 10^{-9} \Omega \text{ m}$$

Electrical resistivity is also affected by crystallographic defects, including impurity atoms. Hence, the total resistivity depends on both the thermal vibration contribution as well as the crystallographic defect contribution, as shown in Eq. 13.5

$$\rho_{\text{Total}} = \rho_T + \rho_d \tag{13.5}$$

ρ_{Total} is the total resistivity of the material, which is the sum of the resistivity due to lattice thermal vibrations (ρ_T) and that due to defects (ρ_d), including point defects such as vacancies and impurity atoms, line defects such as dislocations, etc. It is typically assumed that at 0 K, there are no lattice vibrations, and hence, any resistance to electron flow will come only from crystallographic defects. Hence ρ_d controls resistivity at 0 K. Obviously, for metals, the presence of crystallographic defects, such as excess dislocations, point defects, etc., is typically the result of deformation. It is also possible for material to retain their high-temperature excess vacancy concentrations if rapidly quenched to room temperature.

For solid solutions, the effect of impurity atoms as point defects can be considerable. Equation 13.6 shows this.

$$\rho_x = \rho_o + Y_s.x.(1 - x) \tag{13.6}$$

ρ_o is the resistivity of the unalloyed metal, ρ_x is the resistivity due to the addition of solutes at an atomic fraction of x, and Y_s is the solution resistivity coefficient (also called the **Nordheim Coefficient**). Table 13.1 lists solution resistivity coefficient values for copper (as the solvent) at 20 °C for solute atomic fractions less than 0.1.

Table 13.1 Solution resistivity coefficient for copper (for solute atomic fractions less than 0.1 and at 20 °C)

Solute	Solution Resistivity Coefficient (Ω.m)
Zn	0.2×10^{-6}
Al	0.8×10^{-6}
Ni	1.2×10^{-6}
Sn	2.5×10^{-6}
Si	2.0×10^{-6}
Ag	0.2×10^{-6}

Data from L.H. Van Vlack, Elements of Materials Science and Engineering, 6th edition, Addison Wesley, 1989

Example Problem 13.3

The electrical resistivity of pure copper (> 99.9%) is 17×10^{-9} Ω m at 20 °C. Three elements (Sn, Al, and Ag) are being considered to be added as solute atoms at a content level of 0.05 atomic fraction. Which of these elements would result in the lowest effect on the resistivity of copper?

Solution

A quick look at Eq. 13.6 and Table 13.1 shows that the elements will be ranked in accordance with their solution resistivity coefficient, with silver being the element that has the least effect on copper's resistivity.

Table 13.2 lists electrical resistivity data for various materials, including ceramics and glasses, semiconductors, metals and alloys, and polymers. It is clear that ceramics, glasses, and polymers typically have high resistivities, making them insulators. Metals have significantly lower resistivities, making them conductors. Semiconductors typically have intermediate electrical resistivities between metals and insulators.

13.2.1 Electronic and Ionic Conduction in Ceramics

It should be noted that metals are not the only materials that rely on electronic conduction; some ceramics also do. These are transition metal oxides, such as TiO, NiO, Cr_2O_3, CrO_2, Fe_2O_3, FeO, and Fe_3O_4, where the charge carriers are electrons.

In other ionic ceramics, electrical conduction occurs through the diffusion of ions, which are the charge carriers. Here, unlike electrons, where the charge is fixed at 1.6×10^{-19} C, the charge of an ion will depend on its valency. While an ion with a valency of $+ 1$ will only carry a charge of $+ 1.6 \times 10^{-19}$ C, an ion with a valency of two will carry twice the charge, and so on. Unlike electrons, ions are fixed in specific locations within the lattice and are surrounded by other ions. In ionic compounds, negatively charged ions (anions) are typically surrounded by positively charged ions (cations) to keep the overall charge of the compound neutral. For ions to diffuse and electrically conduct, they must break away from their bonds with surrounding ions

Table 13.2 Electrical resistivity data for various materials (*data from W.D. Callister & D. G. Rethwisch, Materials Science and Engineering, An Introduction, 9th edition, Wiley, 2014*)

Material	Electrical resistivity (Ω.m)	Material	Electrical resistivity (Ω.m)
Metals		Alumina (99.9%) pure)	$> 10^{13}$
Aluminum 1100 (annealed) alloy	2.9×10^{-8}	Silicon Nitride (sintered)	$> 10^{12}$
CP-Titanium (grade 1)	4.2×10^{-7}-5.2×10^{-7}	Zirconia (3 mol % Y_2O_3)	10^{10}
Cartridge brass (C26000)	6.2×10^{-8}	Silicon Carbide (sintered)	1.0–10^9
Bearing Bronze (C93200)	14.4×10^{-8}	Graphite (isostatically molded)	10×10^{-6}-18×10^{-6}
1020 steel (annealed)	1.6×10^{-7}	Pyrex (borosilicate glass)	$\sim 10^{13}$
Lead–tin solder (60Sn-40Pb)	1.5×10^{-7}	Soda-lime glass	10^{10}–10^{11}
Semiconductors		**Polymers**	
Silicon (Intrinsic)	2500	Silicone	10^{13}
Gallium Arsenide (intrinsic)	10^6	Nylon 6,6	10^{12}–10^{13}
Ceramics and Glasses		Epoxy	10^{10}–10^{13}
Concrete (dry)	10^9	LDPE	10^{15}–5×10^{16}

and find a place to diffuse to. Unlike diffusion in metals, where a vacancy or an interstitial site is needed for metal ions to diffuse, charge neutrality in ionic ceramics must also be maintained. This affects the type of defects that can be accommodated. For example, it is possible to control the number of vacancies in ionic ceramics by simply doping the ceramic with ions that have a different valency than the parent ions belonging to the compound. This can be explained by the fact that if we consider ZrO_2, zircon (Zr) can be substituted with calcium (Ca). Ca (Ca^{2+}) has a valency of + 2, while zircon (Zr^{4+}) has a valency of + 4. So, an oxygen vacancy must be created for every zircon ion substituted with a calcium ion to keep the charge neutral. This process allows us to control the number of vacancies in an ionic ceramic. Although both vacancies and interstitial sites can be available in ionic ceramics for electrical conduction, two more crystallographic defects can also be available. These are the **Frenkel** and **Schottky** defects. A *Frenkel defect* forms when an ion within the ionic lattice occupies an interstitial site instead of its normal position on the main lattice, hence leaving a vacancy in the position where it was supposed to be. Typically, cations do that, and since all ions remain in the compound, charge neutrality is still maintained. A *Schottky defect* maintains charge neutrality as both an anion and

Fig. 13.3 Schottky defect in NaCl. Since both a negatively charged Cl ion and a positively charged sodium ion are missing; charge neutrality is maintained

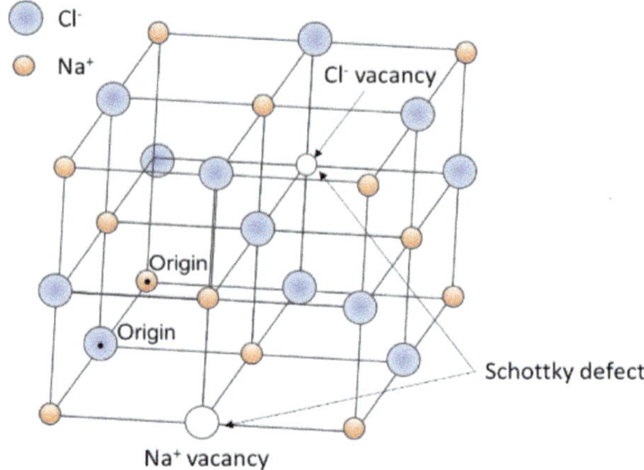

cation become missing, leaving two vacancies. If only one of them were missing, then charge neutrality would not be possible. Figure 13.3 shows a Schottky defect.

Going back to **semiconductors**, these materials have unique electrical characteristics; as mentioned, they typically have a band gap < 2 eV. Here, we divide semiconductors into two types: **intrinsic** and **extrinsic semiconductors**.

13.2.2   Intrinsic semiconductors

Intrinsic semiconductors have an electrical conductivity that increases with an increase in temperature. This is because an increase in temperature increases the energy of electrons, allowing more and more of them to cross the band gap from the valence band to the conduction band. These materials can either be simple elements such as silicon and germanium (both covalently bonded) or compounds of elements from groups III and V or groups II and VI in the periodic table. The band gaps and intrinsic electrical conductivities of such materials are presented in Table 13.3.

Professor, you said that semiconductors have a band gap < 2.0 eV, but I see several materials with band gaps greater than 2.0 eV in Table 13.3; why is that?

Table 13.3 Band gaps and electrical conductivities for various semiconductors at room temperature

Material	Band gap (eV)	Intrinsic electrical conductivity $(\Omega\,m)^{-1}$
Si	1.11	3.4×10^{-4}
Ge	0.67	2.2
Group III-V compounds		
GaAs	1.42	3×10^{-7}
InP	1.35	2.5×10^{-6}
AlP	2.42	–
InSb	0.17	2×10^{4}
Group II-VI compounds		
CdS	2.4	–
CdTe	1.56	–
ZnS	3.66	–
Other ceramics		
α-SiC	2.8–3	10
CdO	2.1	–
Fe_2O_3	3.1	–
Cu_2O	2.1	–
TiO_2	3.05–3.80	–

From W. D. Callister D. Rethwisch, Materials Science and Engineering and Introduction, 9E edition, 2014, Wiley & Sons and W. D. Kingery, H. K. Bowen, and D. R. Uhlmann, Introduction to Ceramics, 2nd Edition, Wiley, 1976

*Don't you hate it when you think you know something and then someone changes the rules on you? The 2.0 eV limit is arbitrary, but it helps to provide a general rule regarding conventional semiconductors. However, there will still be some exceptions. For example, a ceramic like SiC has a band gap of 2.8–3.0 eV. To overcome this problem, a **wide band gap (WBG) semiconductor** is defined. WBG semiconductors typically have band gaps > 2.0 eV.*

In intrinsic semiconductors, both electron holes and electrons carry charge, the holes are positive, and the electrons are negative. The electrical conductivity (σ) of such materials depends on the mobilities of the charge carriers for electrons and holes (μ_e and μ_h), and the number of charge carriers, as shown in Eq. 13.7.

$$\sigma = N_e|e|\mu_e + N_h|h|\mu_h \tag{13.7}$$

N_e and N_h are the number of electrons and holes per meter cubed, respectively, also referred to as the **charge carrier density**. N_e and N_h are typically equal for intrinsic semiconductors, but not so for extrinsic ones.

The charge mobility is related to the applied electric field (voltage across a material divided by the distance between the material's ends) and the **drift velocity** (v_d) according to Eq. 13.8:

$$\mu_e = \frac{v_d}{E} \tag{13.8}$$

Note that the mobility of holes is typically lower than that of electrons. Table 13.4 shows values for some semiconductors at room temperature.

Example Problem 13.4

Calculate the electrical conductivities of silicon and germanium.

Solution

Table 13.4 shows charge carrier density, hole, and electron mobility values for both Si and Ge. Using the data and Eq. 13.7, we have:

$$\sigma = N_e|e|\mu_e + N_h|h|\mu_h$$

Since Si and Ge are both intrinsic semiconductors, then $N_e = N_h = N$, and since $|e| = |h| = 1.9 \times 10^{-19}$ C Eq. 13.7 simplifies to:

$$\sigma = N \times 1.6 \times 10^{-19}(\mu_e + \mu_h)$$

From Table 13.4, for Si, we have $N = 14 \times 10^{15}$ m^{-3}, $\mu_e = 0.14$ and $\mu_h = 0.038$

$$\sigma = 14 \times 10^{15} \times 1.6 \times 10^{-19}(0.14 + 0.038)$$

$$\sigma = 3.98 \times 10^{-4}(\Omega \text{ m})^{-1}(\text{or S/m})$$

Table 13.4 Carrier densities, electron mobilities, and hole mobilities of some semiconductors at room temperature. Note electron mobility is always greater than hole mobility

Semiconductor	Carrier density $(N_e$ or $N_h)$ (m^{-3})	Electron mobility, μ_e (m^2/(V.s))	Hole mobility, μ_h (m^2/(V.s))
Ge	23×10^{18}	0.364	0.190
Si	14×10^{15}	0.140	0.038
InSb	13.5×10^{21}	8.000	0.045
GaAs	1.4×10^{12}	0.720	0.020

Data from: J. Shackelford, Introduction to Materials Science for Engineers, 7th edition, Pearson, Prentice Hall, (2009), C.A. Harper, Ed., Handbook of Materials and Processes for Electronics, McGraw-Hill Books Company, NY, 1970.

Likewise, for Ge:

$$\sigma = 23 \times 10^{18} \times 1.6 \times 10^{-19}(0.364 + 0.190)$$

$$\sigma = 2.039(\Omega \text{ m})^{-1}(\text{or S/m})$$

13.2.3 Extrinsic Semiconductors

On the other hand, extrinsic semiconductors are formed by doping very high-purity intrinsic semiconductors to enhance their electrical conductivity. Here, we will take the example of silicon. In its pure form (with limited impurities of around 10^{-7} at.%) Si possesses atoms with 4 electrons in their outermost shells. As such, each silicon atom bonds covalently with four other silicon atoms in a tetrahedra-type structure called a giant covalent structure, the same structure as diamond. The same could be said for germanium. The only difference is that germanium has a lower melting point (938.25 °C) than silicon (1414 °C). This has prompted many early researchers to use germanium when conducting semiconductor research. The interest in Si, however, later grew due to several factors, including the instability of germanium at high temperatures and silicon having stable oxides.

Imagine that we dope Si with atoms of an element with more than four electrons in its outermost shell. As mentioned, Si has 4 electrons in its outermost shell. Still, suppose we dope it with a pentavalent element (five electrons in its outermost shell) such that atoms of this element substitute the position of a Si atom in the crystal lattice. In that case, they will covalently bond with surrounding Si atoms using only four of their outermost electrons, leaving one electron free to move, for example, under an applied field. In this case, such a semiconductor will be called an **N-type semiconductor**. Pentavalent elements (from group V of the period table) that can be used for this purpose include phosphorus (P), antimony (Sb), and Arsenic (As). In N-type semiconductors, it is evident that the charge carriers are negatively charged electrons. Alternatively, if we dope the intrinsic semiconductor, for example, Si, with atoms from elements from group III in the periodic table, which are trivalent (three electrons in the outermost shell), we produce a **P-type semiconductor**. Again, the incoming dopant atoms substitute silicon atoms in the crystal structure and covalently bond with the surrounding silicon atoms. However, this time, only the available three outermost electrons can bond when, in fact, four electrons are needed. This leaves an electron hole, which is a positive charge carrier. P-type dopants include aluminum (Al), boron (B), and gallium (Ga).

Figure 13.4 shows schematics of silicon as an intrinsic semiconductor, when it is doped with pentavalent phosphorus to produce an extrinsic N-type semiconductor, and when it is doped with boron to produce an extrinsic P-type semiconductor.

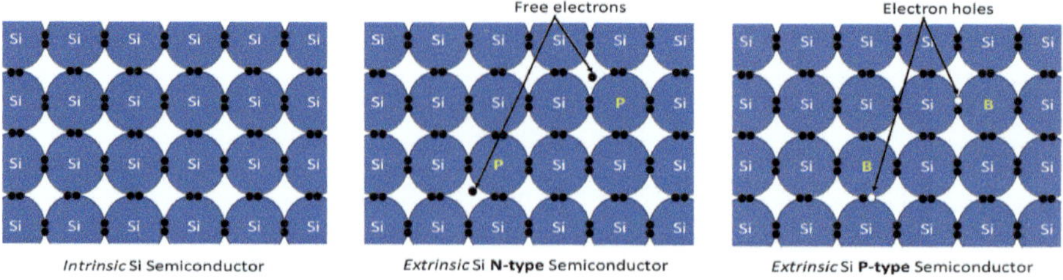

Fig. 13.4 Schematic showing intrinsic Si semiconductor, extrinsic Si N-type semiconductor and extrinsic P-type semiconductor and charge carriers (electrons and electron holes)

Figure 13.5a shows two blocks of P-type and N-type semiconductors; within the P-type block, the holes (h) are greater than the free electrons (e), while in the N-type block, the electrons are greater than the holes. When N-type and P-type semiconductors are brought together, some interesting things happen, which have become the basis for the semiconductor technologies we see today. As soon as N-type and P-type regions are joined (Fig. 13.4b), naturally, the electrons from the N-type region will flow across the boundary between the N- and P-type regions (called the P–N junction) and fill the holes on the opposite side of the junction. This will generate negatively charged ions in that region. Likewise, the holes from the P-type region will flow across the P–N junction into the N-junction and generate positively charged ions. These can be seen in Fig. 13.4c. Hence, a region called **the depletion layer** develops with no mobile charge carriers. Interestingly, the formation of negative and positive ions within this depletion layer repels any more electrons trying to flow from the N-type region toward the P-type region or positive holes trying to flow toward the N-type region. For example, if an electron tries to flow towards the + ve hole (P-type) region, it will find negative ions in front of it that will act to repel it back. Likewise, for a hole, as it approaches the negative (N-type region), it will meet the positive ions and will, too, be repelled back. This means that the depletion layer presents some **potential barrier** that would need to be overcome by electrons or holes for them to continue flowing. This potential barrier is typically around 0.6–0.7 V for Si and about 0.3 V for Ge

One of the outcomes of the P–N junction is the diode. Figure 13.6 schematically shows how a diode works. In the forward bias operation (Fig. 13.6a), the negative terminal of the voltage source (battery in this case) is facing the N-type region. In contrast, the positive terminal of the voltage source faces the P-type region. The voltage source must exceed 0.7 V for a Si P–N junction to overcome the potential barrier. When that is the case, current flows through the diode. In the reverse bias case (Fig. 13.6b), the battery's negative terminal is connected to the p-type region; this will cause the holes in the P-type region to move to the left and the electrons in the N-type region to move to the right. Reverse bias will lead to an increase in the width of the depletion layer, thus posing a higher potential barrier. In other words, the diode acts as an insulator to a large extent. Hence, the diode can act as a switch by reversing the applied voltage from forward bias to reverse bias.

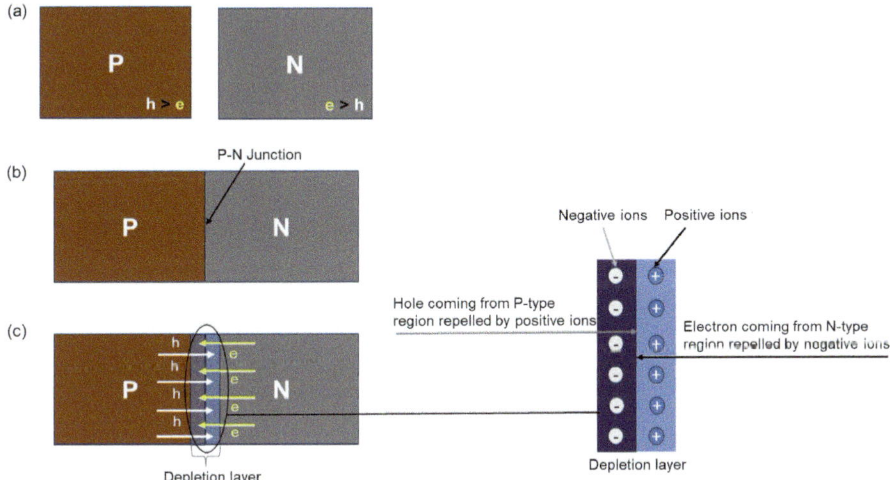

Fig. 13.5 Schematics of **a** P and N type semiconductors separated, **b** P and N type semiconductors connected, **c** Development of depletion layer due to interdiffusion of holes and electrons into N-type and P-type regions respectively

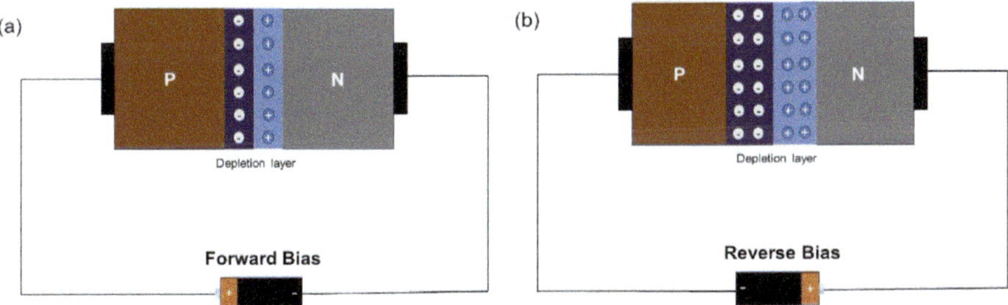

Fig. 13.6 Schematics of **a** diode in forward bias operation where electrons flow anticlockwise in the circuit and the through the P–N junction, **b** diode in reverse bias operation where no current flows. Note the increase in the width of the depletion layer which poses a larger potential barrier

Semiconducting ceramics also include SiC, B_4C, and ZnO. SiC and ZnO with band gaps of ~ 3.3 eV and 3.4 eV, respectively, are termed wide-bandgap semiconductors since their band gaps exceed 2 eV. Wide-bandgap semiconductors can be useful at higher temperatures than conventional semiconductors and operate at higher voltages.

Table 13.5 Examples of ceramic insulators (from: D. W. Richerson, Modern Ceramic Engineering: Third edition, Taylor and Francis, (CRC Press 2006)

Ceramic insulators
Silicon dioxide (SiO_2)
Aluminum oxide (Al_2O_3)
Magnesium oxide (MgO)
Silicon nitride (Si_3N_4)
Aluminum nitride (AlN)
Mullite ($Al_6Si_2O_{13}$)
Some Spinels (e.g. $MgAl_2O_4$)
Porcelains
Most glass ceramics and silicate glasses

Example Problem 13.5

The potential barrier of a silicon P-N diode is 0.7 V. What voltage is needed to overcome this barrier? And how can the diode then act as a switch?

Solution

Any voltage greater than 0.7 V with forward bias. The diode can act as an on–off switch by reversing the polarity of the voltage supply, i.e. reverse bias.

Insulators

These are materials with strongly held electrons in covalent or ionic bonding. Insulators include ceramics and polymers. Their band gap is large. For example, the band gap of a ceramic-like alumina is 8.8 eV. Table 13.5 lists ceramics that are considered insulators. Insulating ceramics find many applications, including thermocouple protection tubes, components for microwave tubes, filament support for lights (incandescent), insulators for power lines, components for X-ray tubes, spark plug housing, and substrates for integrated circuits.

As insulators, many ceramics and polymers fall under the definition of a **dielectric**. This means they can be placed between two conductors without allowing any current to pass. They are polarized, can store charge, and can be used as capacitors. There are several types of polarization mechanisms, two of which are shown in Fig. 13.7. In Fig. 13.7a, **electronic polarization** is shown, where, under an applied electric field, the charges within the atom are redistributed, with electrons becoming closer to the positive electrode and the nucleus carrying the protons brought closer to the negative electrode. The electrons are spatially distributed around the nucleus when there is no field. Although this mechanism occurs in all materials, the polarization effect caused by electron polarization is weak. **Atomic** or **ionic polarization**, on the other hand, can be seen in Fig. 13.7b. We see here that the positive ions within a lattice are attracted and brought closer to the negative electrode while the negative ions are repelled. Likewise, the negative ions are attracted and brought closer to the positive electrodes while the positive ions are repelled. All atoms/ions return to their positions

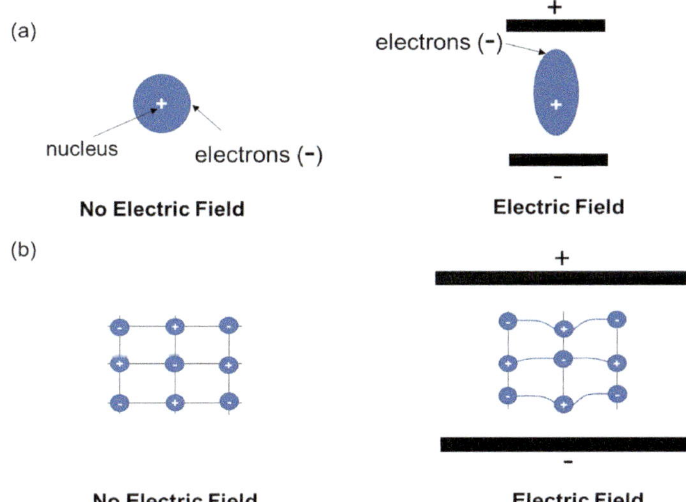

Fig. 13.7 **a** Electronic Polarization, **b** atomic/ionic Polarization (*modified from A.R. von Hippel, Dielectrics and Waves, Wiley, 1954*)

within the lattice when the field is removed. There are other polarization mechanisms, including orientation polarization for polar asymmetric molecules (with permanent electric dipoles) and space charge polarization, which will not be discussed here.

A couple of parameters are relevant to dielectric materials: the dielectric constant and the dielectric strength. Whenever we refer to the field (E), we mean the voltage (V) gradient across a thickness (t), as shown in Eq. 13.9.

$$E = \frac{V}{t} \tag{13.9}$$

If we imagine two parallel capacitor plates separated by vacuum, the charge density (ρ_o) in C/m^2. on each of the plates is given by Eq. 13.10:

$$\rho_o = \varepsilon_o \times E \tag{13.10}$$

ε_o is a constant $= 8.85 \times 10^{-12}$ C/V m.

Now, if, instead, a dielectric material is placed between the plates, one would expect that the charge density would increase due to the polarization within the dielectric material. Hence, the charge density (ρ_m) in this case, would be given by Eq. 13.11:

$$\rho_m = \kappa \times \varepsilon_o \times E = \kappa \times \rho_o$$

Hence,

Table 13.6 Dielectric constants and dielectric strengths for various materials (Data from multiple sources including *L. H. Van Vlack, Elements of Materials Science and Engineering, 6th edition, Addison Wesley (1989), W. D. Kingery, H. K. Bowen, D. R. Uhlmann, Introduction to Ceramics, 2nd edition, Wiley Interscience, (1976), and D. W. Richerson, Modern Ceramic Engineering, 3rd Edition, CRC Taylor and Francis, (2006)*

Material	Dielectric constant (at 60 Hz)	Dielectric strength (MV/m)
Steatite	6	6–11
Soda-lime glass	7	10
Pyrex glass	4.3	14
Fused silica	4	10
Barium titanate	1600*	2000
MgO	9.65	
Diamond	5.68	1000
NaCl	5.9	
Porcelain	6	12
Polyethylene	2.3	20

*Frequency unknown

$$\kappa = \frac{\rho_m}{\rho_o} \tag{13.11}$$

where κ is the relative dielectric constant, in other words, the relative dielectric constant is equal to the charge density in the presence of the dielectric material divided by the charge density in vacuum. Hence, it is unitless. A material with a low dielectric constant is suitable in applications where it is an electrical insulator. Examples include Al_2O_3, BeO, rubber (insulating electric wires), porcelain, and steatite. On the other hand, materials with high dielectric constants are suitable for use in capacitors. Barium titanate has a dielectric constant of 1600, while a dielectric constant of 9500 has been reported for $BaTiO_3 + 10\%\ CaZrO_3 + 10\%\ SrTiO_3$. The dependence of the dielectric constant on temperature is contingent on the type of polarization in the material. For atomic/ionic polarization, an increase in temperature increases the dielectric constant, while materials with electronic polarization have dielectric constants independent of temperature. The frequency of the applied electric field or an imposed electromagnetic field can also affect the dielectric constant.

Although dielectrics are insulators, at some critical applied voltage, they can break down and allow electrons to be ripped away from bonds and participate in electric conduction. In other words, the **dielectric strength** is the potential gradient across the thickness of the dielectric sample above which electrical breakdown can occur. It is the maximum field that can be tolerated without electrical conduction. Dielectric strength is measured in MV/m or V/mil. Table 13.6 lists dielectric constants and dielectric strengths for several substances. Dielectric strength can vary with thickness, grain size, composition, frequency of applied field, length of time voltage is applied, and other factors. Composite dielectrics have also been investigated, for example,

$0.92\ MgO - 0.08\ Al_2O_3$ composite, which had a dielectric strength of 126.4 MV/m. Having a high dielectric strength is beneficial in insulation. For example, a material with a dielectric strength of 2000 MV/m, is equivalent to 2000 V/μm. This means you can have a 1 μm coat that can electrically insulate as long as the electric potential across it does not exceed 2000 V.

Thermoelectric Materials

These are materials that play a role in energy conversion. Imagine a material that can convert heat into electricity. The heat used can be waste heat or solar heat, with energy conversion efficiencies of around 5–20%. Here, a temperature gradient across a material generates a voltage gradient, and the opposite is also true. A voltage gradient will generate a temperature gradient. Potential applications for these materials include wearable devices where heat can be converted to electric energy, utilizing waste heat from combustion engines, and air conditioning in vehicles, just to name a few. The $Ca_3Co_4O_9$ ceramic is an example of a thermoelectric material.

Piezoelectric Ceramics

Piezoelectricity was reported in 1880 by Jacques and Pierre Curie through investigations of single crystals of quartz, sugar, and zinc blend (sphalerite), among others. The word *piezo* comes from the Greek word for press. In essence, it refers to the ability of certain dielectric materials to build up static electric charge when mechanically deformed to small mechanical strains. The effect is also reversible, meaning applying an electric field to such materials results in mechanical strains. As one can imagine, this kind of behavior opened the door to so many applications, including transducers, microphones, actuators, and sensors.

Piezoelectricity occurs in anisotropic ceramic crystals with no symmetry center to their unit cells. For example, a cubic unit cell with a center of symmetry at the center of the cube does not qualify to be a piezoelectric material. If we account for symmetry variations of the Bravais lattices, we find that there are 32 possible classes of crystals in terms of symmetry. Only 12 of those have a center of symmetry, meaning they cannot be piezoelectric. This leaves 20 that can be piezoelectric; hundreds of dielectric materials have been formed, but relatively few have been optimized to a level where they can be used in practical applications. Barium titanate, lead zirconate titanate (PZT), lead titanate, zinc oxide, zinc sulfide are examples of piezoelectric ceramics.

Pyroelectric Ceramics

These are a subclass of piezoelectric crystals that exhibit a spontaneous polarization in one crystallographic direction at a minimum. Pyro comes from the Greek word "fire". In other words, pyroelectricity is electricity produced through heat. This should not be confused with the thermoelectric effect, as the latter concerns a temperature "gradient". The thermal expansion upon heating of pyroelectric materials results in mechanical strain, changing the extent of polarization. 10 crystal types exhibit pyroelectricity out of the 20 piezoelectric types or classes of crystals mentioned above. Lithium sulfate, hexagonal zinc sulfate, and barium titanate are pyroelectric ceramics. The pyroelectric effect is typically limited to relatively low temperatures up to a few hundred degrees centigrade, with certain exceptions such as $LiTaO_3$, which is pyrolytic just under 610 °C. Uses include self-powered thermal sensors, gas and fire alarms, imaging, and laser detectors.

Ferroelectric Ceramics

Ferroelectric materials are yet again a subclass of pyroelectric materials. In pyroelectric materials, we talked about ceramics that demonstrate spontaneous polarization. With ferroelectric materials, it is also possible to reverse the polarization by applying an electric field. Barium and lead titanate are examples of ferroelectric ceramics. Applications include ultrasonic motors, electro-optical light valves, and transducers for medical diagnosis.

Superconductivity

As mentioned earlier, the electrical resistivity of metals tends to increase with temperature. This also means that as the temperature is reduced, so will the resistivity. However, **Kammerlingh Onnes** in 1911 discovered that for certain metals below a critical temperature, T_c (also called the transition temperature), the resistivity suddenly drops to zero. The material becomes superconducting, transmitting electric current without resistance or energy losses. That critical temperature was found to be ~ 4 K for solid mercury, and indeed, later, in 1913, Onnes received the Nobel Prize for his contributions. He was also the first to liquefy helium. Another breakthrough was forwarding a mechanism that explains superconductivity in 1957 by John Bardeen, Leon Cooper, and John Robert Schrieffer, who later received the Nobel Prize in 1972. The mechanism involved pairing electrons in what has been termed **Cooper pairs**, which find it much easier to travel within the material with no resistance. Metals that display superconductivity are typically from the transition metal groups. Moreover, all elements that display superconductivity have critical temperatures of less than 10 K. One needs to cool and maintain the material below the critical temperature to use such an extraordinary phenomenon. Initially, liquid helium was needed, which was quite expensive. Ideally, one would want to be able to increase the critical temperature to room temperature or above. This would allow easy utilization of such an effect in many applications. Historically, many materials have been developed, including intermetallics and ceramics, that have indeed achieved such welcomed increases in T_c. For example, Nb_3Sn and Nb_3Ge intermetallics have critical temperatures of 18.1 K and 22.5 K, respectively. In 1986, the development of the first superconducting ceramic, $La_{1.8}Sr_{0.2}CuO_4$, increased T_c to 40 K. A few years later, $YBa_2Cu_3O_7$ was developed with a T_C of 93 K. Hence, it was the first superconductor to have a T_c value above the boiling point of liquid nitrogen (77 K). This meant achieving superconductivity no longer required liquid nitrogen. $Tl_2Ba_2Ca_2Cu_3O_{10}$ followed with a T_c of 112 K. In 2004, Neil Ashcroft put forward the idea of producing hydrides with even higher T_C values. However, high pressures were needed. Super hydrides have hence been researched heavily for this purpose. H_3S has since exhibited superconductivity up to 203 K at high pressure. As of the writing of this book, the best superconductor in terms of T_c is the superhydride lanthanum decahydride (LaH_{10}) with a T_c of 250 K but again at very high pressure (170 GPa). Although recent studies reported superconductors with T_c values above room temperature, these studies were later retracted. Judging by the progress over the past 100 years, it would not be surprising if room-temperature superconductivity is eventually realized. Its impact on technology would be immense. It is noteworthy that superconductivity can be destroyed at high current densities, strong magnetic fields, and, of course, if the superconductor is heated to temperatures greater than T_c.

**The Sphere
Layer 6**

Professor, I carried out an XRD scan on sample 6, and all the peaks point to silicon.

Alex, I'm starting to get a little suspicious about the identity of sample 6, which represents layer 6 in the Sphere. Of course, silicon is very important, and all the other properties you have already measured match perfectly, which makes this silicon. But I have a feeling that there is something more about this sample. I would like you to learn about the SIMS (Secondary ion mass spectrometry) analysis technique and use it to obtain a compositional profile of sample 6. You will, however, need to send the sample to a specialized company that provides this analysis service. Don't worry, I'll pay for it; I have funds in one of my discretionary accounts. I believe SIMS is discussed in the next section you are about to read in As a Matter of Fact. Do that, and we can talk more later. I feel we are very close!

Sure thing, Prof.

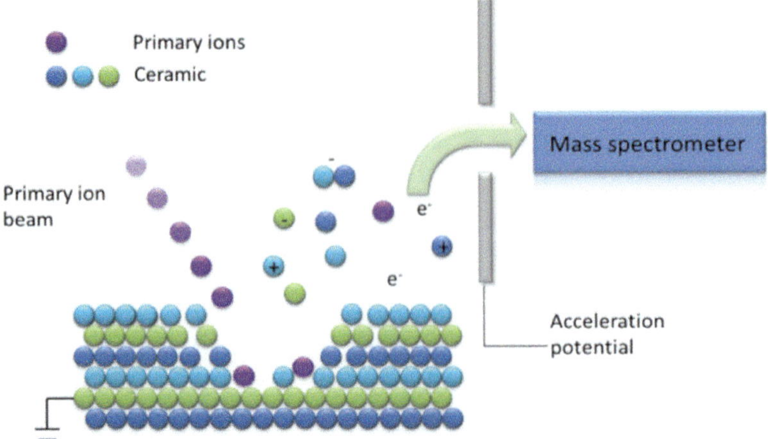

Fig. 13.8 Schematic of SIMS technique showing primary and secondary ions. (Fig. 1a in Zonghao Shen, Sarah Fearn, Back-to-basics tutorial: Secondary ion mass spectrometry (SIMS) in ceramics, Journal of Electroceramics https://doi.org/10.1007/s10832-024-00375-9), reproduced under terms and conditions of Creative Commons Attribution 4.0 International License http://creativecommons. org/licenses/by/4.0/

Secondary ion mass spectrometry (SIMS)

We discussed diodes and the silicon P–N junction. These materials rely heavily on the dopants added to silicon. However, how can we confirm the level of doping? For example, N-type semiconductors may include extremely small amounts of phosphorous atoms, while N-type semiconductors will instead contain boron at very low levels, for example, 1×10^{14} atoms/cm^3. SIMS has been used to measure such minute levels. The process involved the bombardment of the surface of a material with what are referred to as primary ions, which deplete the surface. However, secondary ions are expelled from the material surface during the process, which are picked up by a mass spectrometer and analyzed (Fig. 13.8). The process has extreme precision and can measure concentrations of dopants down to parts per million or parts per billion. One can develop a concentration profile by measuring the impurity concentrations as more and more of the material is depleted or sputtered. i.e., a through-thickness analysis. The process was first used to analyze moon rocks, but now it is extensively used to measure dopant concentrations in semiconductors, among other applications.

13.3 Introduction to Magnetic Properties of Ceramics

Magnetite (Fe_3O_4) is a naturally occurring mineral that is attracted to a magnet. Due to its combination of Fe^{++} and Fe^{+++}, it is often designated as iron (II, III) oxide. A rare form of magnetite, known as **Lodestones**, is a permanent magnet and was the basis for the first compass. Magnetite was introduced due to its ancient discovery in Magnesia (a region in ancient Greece, now modern Turkey).

This section presents the magnetic properties, but first, it is essential to lay down some foundations regarding magnetism, in which metals will also be considered.

It is often mentioned that passing an electric current (or electrons) in a conducting wire generates a magnetic field. The higher the electric current intensity, the higher the magnetic field. If we now direct our attention to atoms, we recall that no more than two electrons occupy an orbital in an atom. When there are two electrons in an orbital, they will have opposite magnetic spins. The motion of an electron in an orbital generates a mini magnetic field. When we have two electrons of opposite spins, they tend to cancel each other's magnetic field. Hence, materials with electronic configurations with only paired (2) electrons in the orbitals will not be magnetic. However, some elements have what are referred to as unpaired electrons in one or more orbitals within each of their atoms. This means that only one electron is present in each of the one or more orbitals. A magnetic field is generated as they spin but would not be canceled out; and magnetic dipole moments are generated. Examples of materials with unpaired electrons include aluminum, iron, nickel, and cobalt. One would not ordinarily acquaint aluminum with magnetism, and rightly so, but let us explain this a little more. When it comes to magnetism, we can classify materials into the following categories.

13.3.1 Ferromagnetic Materials

These materials are attracted to magnets and can indeed be magnetized by exposure to a magnetic field. Their magnetization can be permanent, meaning the material remains magnetized after removing the external magnetic field. In this case, the ferromagnetic material will be called a **hard magnet**. Other ferromagnetic materials are what we call **soft magnets**. The difference is that the material loses its magnetization when the external magnetic field is removed. We have so far stated that each atom will have magnetic dipole moments; normally, the directions of these moments will be randomly oriented on the atomic scale, as in the case of **paramagnetic** and **diamagnetic** materials (discussed below). In ferromagnetic materials, a collection of atoms forms a **magnetic domain** where all the atoms' magnetic dipole moments inside the domain are aligned in a specific direction. The size of these domains is typically less than 50 μm. In a ferromagnetic material, millions of these domains have random orientations of magnetic dipole moment. Figure 13.9 clarifies this further. As can be seen initially, a ferromagnetic material will consist of domains with random magnetic dipole moments, even without the presence of a magnet. However, when a magnet is used, its magnetic field will influence the domains within the ferromagnetic materials, and the magnetic dipole moments in each domain will align similarly. In essence, domain boundaries will disappear, and the domain size will grow. When the magnet is removed, the domains stay aligned. In this case, the ferromagnetic material behaves as a **hard magnet**. If the material were a soft magnet, then after removal of the magnet, the magnetic domains would revert to being mostly randomly oriented concerning their magnetic dipole moments, and only slight magnetization remains. However, they remain magnetized in an externally applied magnetic field.

Examples of ferromagnetic materials include transition metals, iron, cobalt, nickel, and gadolinium. It turns out that only when the atomic radius lies between 1.5 and 2.0 times greater than the radius of the shell in which the unpaired electrons reside is magnetism possible (nickel, cobalt, iron, and gadolinium satisfy this requirement, but not other transition metals such as titanium and magnesium). The interatomic space sensitivity is because when atoms are too far away from each other, the thermal

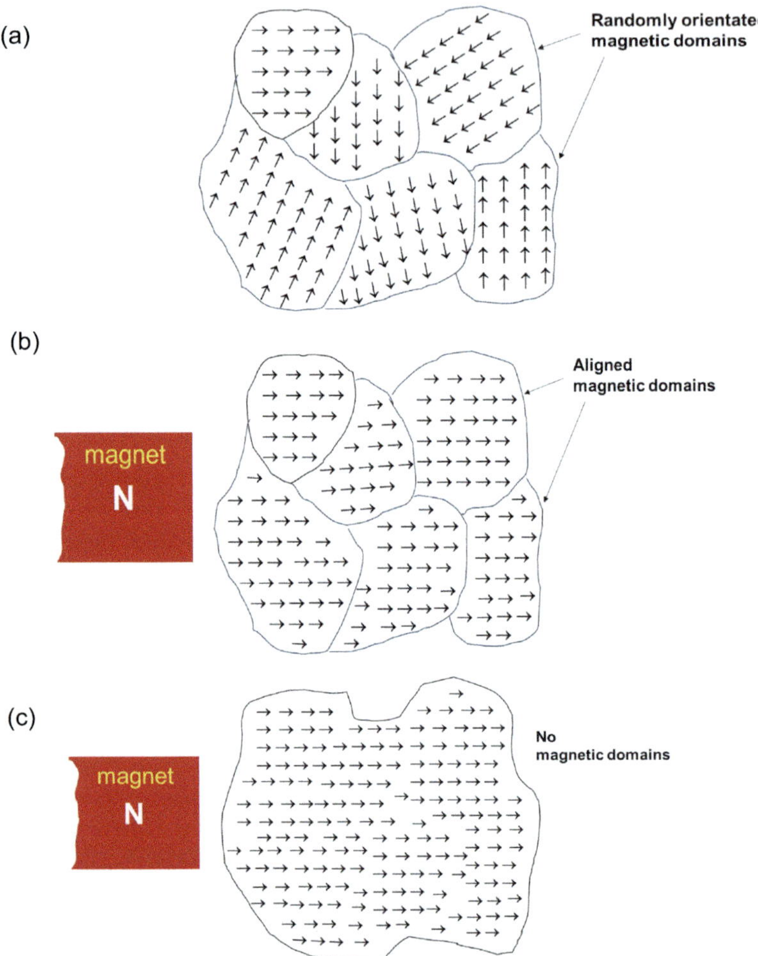

Fig. 13.9 **a** Ferromagnetic material with randomly oriented magnetic domains, **b** application of magnetic field aligns magnetic domains, **c** For aligned magnetic domains, domain boundaries are eliminated leaving only on large domain (multiple sources)

atomic vibration is enough to break away any magnetic dipole alignments and, hence, produce random orientations. Likewise, when atoms are too close together, the large interatomic forces make it impossible for magnetic dipole moments to align in the first place. In fact, to show how this rule is valid, consider magnesium, which is not ferromagnetic, and note its ratio lies outside the 1.5–2.0 times range. When nitrogen is interstitially introduced in magnesium, suddenly the Mg interatomic spacing is modified, making magnesium lie within the 1.5–2.0 times range, and hence renders it ferromagnetic. Even ceramics such as $BaFe_{12}O_{19}$ satisfy this criterion. Generally, hard ferromagnetic materials can be used as permanent magnets.

Here, we need to define three important terms for permanent magnets (after the ferromagnetic material has been magnetized). The first is *coercive force*; a magnet (for example, a hard magnet) with a high coercive force means it is difficult to be demagnetized electrically or magnetically. The opposite is true for a *soft magnet*, as it does not need as much to demagnetize it. The second is **remanence**. A ferromagnetic material initially has zero internally induced magnetism (i.e., induced magnetic flux density) since all the random domains cancel each other out. Once the external magnetic field (H) is applied to the material and removed, the ferromagnetic material becomes permanently magnetized with an induced magnetic flux density (B). This is the remanence. In soft magnets, B is small in magnitude. However, a hard magnet will have a significantly larger B value. Finally, the **energy product value** represents the stored magnetic energy in the ferromagnetic material (i.e. the magnet) following magnetization. An interesting fact: Hard magnets typically have higher mechanical hardness values than soft magnets. For example, hardened steel behaves as a hard magnet and can be used as a permanent magnet. Soft steel, however, is also a soft magnet. Permanent magnets find applications in loudspeakers, microphones, magnetic bearings, refrigerator door seals, contactless switches, starter motors, camera apertures, refrigerator magnets, disc store devices, and stepping motors. Soft magnets are used in recording heads, PCs, electric generators, smartphones, and transformer cores. Did you know that soft magnets are used in what are called electromagnets? The magnet core is encircled by wire windings. The passage of a current generates an external magnetic field, which induces a strong magnetic field inside the core, making it a very strong magnet. This strong magnet can lift heavy steel scrap, which help identify and move scrap. Once the scrap is moved using a crane, the electric current is switched off, making the magnet lose its magnetism, and the steel scrap falls. The alignment of magnetic domains is critical to the characteristics and operation of a permanent magnet. However, this alignment can be disrupted by thermal atomic vibrations. Hence, a ferromagnetic material may retain its magnetism only up to a certain temperature called the **Curie temperature**, above which randomness of magnetic dipole moments prevails, which signals the loss of magnetism. After cooling below the Curie temperature, the ferromagnetic material must be magnetized again. Above the Curie temperature, the material will not be called ferromagnetic but rather paramagnetic, as explained in the next section. Table 13.7 shows examples of soft and hard magnets.

Table 13.7 Examples of soft and hard magnets

Soft magnets	Hard magnets
Pure α-iron	Alnico V
Metallic glass ($Fe_{81}B_{13.5}Si_{3.5}C_2$)	RE-Co
Permalloy, Ni–Fe	Plain carbon steel
Ferroxcube A, $(Mn, Zn)Fe_2O_4$	$Nd_2Fe_{14}B$
Ferroxcube B, $(Ni, Zn)Fe_2O_4$	$BaFe_{12}O_{19}$

Source L. H. Van Vlack, Elements of Materials Science and Engineering, 6th edition, Addison Wesley, 1989

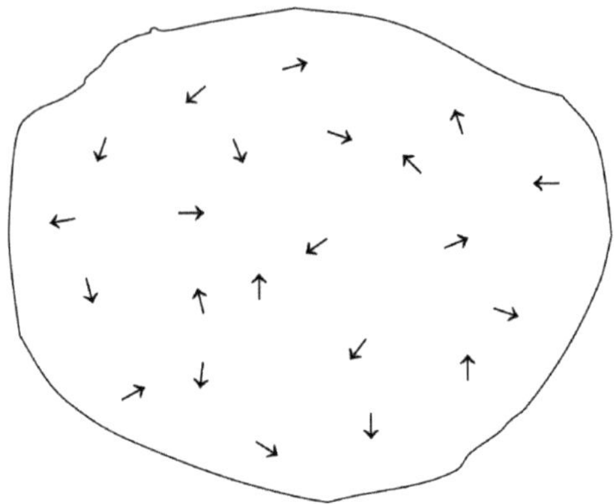

Fig. 13.10 Random orientations of magnetic dipole moments inside paramagnetic material

13.3.2   Paramagnetic Materials

Paramagnetic materials also have unpaired electrons but no magnetic domains. Rather, individual atoms will have randomly oriented magnetic dipole moments, leading to zero net magnetic moment (Fig. 13.10). The observed randomness in magnetic dipole moments is due to thermal atomic vibrations. Most metals, including aluminum, are classified as paramagnetic. These materials are only slightly attracted to magnets, and when exposed to a magnetic field, only small and slight alignment of magnetic dipole moments are induced. The material will not retain any magnetism when the magnetic field is removed. An increase in temperature will increase the randomness of the magnetic dipole moments.

13.3.3   Diamagnetic Materials

Unlike paramagnetic and ferromagnetic materials, diamagnetic materials have paired electrons. Normally, one would expect the opposite spin of the two electrons to cancel any potential net magnetic dipole moment. However, when brought close to a strong magnet, the induced magnetic field strengthens the magnetic spin of one electron over the other. This leads to the material generating a weak net magnetic dipole (opposite to that of the magnet), and hence, it gets weakly repelled by the magnet. It is important to realize that this diamagnetic phenomenon occurs in all materials to some extent since all materials will still have paired electrons in their subshells. However, its influence is weak within paramagnetic and ferromagnetic materials and hence cannot significantly change the overall behavior of these materials. Some metals and all nonmetals are diamagnetic.

13.3.4 Ferrimagnetic Materials

Ferrimagnetic materials will have magnetic dipole moments only in two or three directions (hence not completely random), the moments in these materials do not completely cancel each other out, leaving ferrimagnetic materials with a net magnetic moment that is lower than that of ferromagnetic materials. Popular ferrimagnetic ceramics are Ferrites, based on iron oxide. The formula of ferrites is $[RFe_2O_4]_8$ where R represents Ni^{2+} or Co^+ or Mn^{2+} or Zn^{2+} or Fe^{2+} etc. Nickel Ferrite is an example of ferrites with chemical formula $[Ni^{2+}Fe_2^{3+}O_4]_8$. Ferrites are used a lot as ceramic soft magnets.

13.4 Examples of Ceramics

This section discusses some of the notable and technologically relevant ceramic materials we have today.

Alumina

Probably unknown to much of the public, sapphire and ruby are nothing other than alumina (Al_2O_3), with the color difference only brought about by the presence of certain impurity elements; for example, chromium makes rubies red. It can be claimed that α-alumina is the most successful of all oxides. It is obtained from bauxite, and it is the oxide from which aluminum is obtained. Its unique properties lend it to some unique applications in addition to conventional ones; its low cost is an advantage. It has excellent oxidation properties as the material is already an oxide; it is biocompatible and hence can be used as a bioimplant, for example, as a load-bearing dental implant or a hip prosthesis, also taking advantage of its good wear resistance. It has a density of ~ 3.97 g/cm^3, and a hardness on the Mohs scale of around 9, i.e., below diamond with a Mohs hardness of 10. Since the Mohs test is somewhat of a scratch test, a ceramic with a high Mohs hardness will be scratch-resistant to a lower hardness material. Hence, alumina finds applications as a scratch-resistant transparent material, e.g., scratch-resistant optics. The mechanical properties of polycrystalline alumina are highly sensitive to grain size, typically below a grain size of 4 μm; the material has good flexural strength and outstanding compressive strength. Above 7 μm grain size is a notable decline in mechanical properties. This ceramic has been shown to display a fracture toughness that increases with crack growth, leading to R-curve behavior. Hence, larger-grained alumina will have a higher fracture toughness than fine-grained alumina. Polycrystalline alumina has a melting point just above 2000 °C; this and its translucent optical properties prompted its use in high-pressure sodium vapor lamps, which efficiently light our streets. One of its uses is as armor plates in military applications. Its high electrical resistivity enabled its use in spark plugs as the encasing body. Other metastable phases of alumina can be formed; these are called transition alumina phases, which are γ-alumina, η-alumina, δ-alumina, θ-alumina, and κ-alumina.

Zirconia (ZrO$_2$)

Zirconia is another important ceramic. This ceramic has three polymorphs, i.e., it exists in three different crystal structures: monoclinic, tetragonal, and cubic. The control of these poly morphic transformations through various heat treatments forms the basis of some unique properties of zirconia. The material can be used as a high-temperature refractory up to 2500 °C, and like alumina, zirconia is also biocompatible. Moreover, zirconia boasts fracture toughness and bend strength values generally twice those of alumina.

Silicon Carbide

Silicon carbide (SiC) is a covalent compound with a relatively high melting point (2500 °C) and very high hardness just below diamond and boron carbide. It is known for its high hardness, good thermal shock resistance, high-temperature strength, and good oxidation resistance. It is also a polymorphic compound, with the most commonly used α-SiC (hexagonal crystal structure) appearing as equiaxed grains within the SiC microstructure, whereas another type of morphology is β-SiC (cubic, Zinc Blende) which has a plate-like morphology. The different morphologies have a direct effect on properties such as fracture toughness. While an increase in the grain size of α-SiC results in an increase in fracture toughness. A significant effect on fracture toughness is observed in β-SiC where the plate-like morphology results in the activation of potent toughening mechanisms such as crack bridging and crack deflection. SiC finds applications as grinding media and cutting tools. It's very good wear properties resulted in SiC being added as a reinforcement in aluminum to produce aluminum-SiC brake discs, crucibles, and in armor. SiC also displays semi-conducting properties and can be doped to produce n- and p- type semiconductors, suitable for high temperature and high voltage usage. The high thermal conductivity of SiC helps its use in power electronics since it makes thermal management much easier. The recently accepted thermal conductivity value for –SiC (6H–SiC) is ~ 320 W/m K, compared that to the values for aluminum (~ 200 W/m/K) and copper (~ 400 W/m.K). Recent research has shown that obtaining wafer size -SiC with isotropic thermal conductivities greater than 500 W/m k at room temperature is possible. This is exceptional, making it the second highest after diamond for large crystals. As thin films, they even exceed diamond.

Silicon Nitride (Si$_3$N$_4$)

Silicon nitride (a covalent compound) was first used in the 1950s as a material for rocket nozzles and tubes that host thermocouples in addition to crucibles for the processing of molten metals and as masks in the fabrication of integrated circuits. This material displays a very good combination of properties, including low thermal expansion coefficient, high fracture toughness, high creep resistance (with proper grain boundary engineering), good wear resistance, good thermal shock resistance, high resistance to erosion, high strength, and hardness. Silicon nitride has also found application as a biomaterial due to its chemical inertness; for example, it has been used in spacers for spinal fusion and acetabular cups in hip-joint replacement. The material's melting point is often stated as 1910 °C. However, it does not turn into a liquid but rather sublimes. Silicon nitride is polymorphic, β-Si$_3$N$_4$ and α-Si$_3$N$_4$ are hexagonal, and a cubic spinel phase is only produced under very high pressure (typically above 15 GPa).

A notable method for its production is nitriding or introducing nitrogen gas to a hot porous silicon powder compact. The process is known as **reaction bonding**, and the silicon nitride produced using this approach is called **reaction-bonded silicon nitride**. Here, the nitrogen reacts on the surface of silicon powder and forms silicon nitride, typically at temperatures between 1100 and 1450 °C. The biggest advantage of the technique is that limited or no compact shrinkage is observed. This means one can shape the silicon powder into the desired simple or complex shape using powder metallurgy techniques and allow the nitrogen to react with the compact and form the silicon nitride without the compact changing size, or if it does, it only does so slightly. This results in a product that requires very little post-machining. The problem, however, is that the final product is typically highly porous (15–25% porosity), which limits its bend strengths to ~ 200–350 MPa (although it can increase with an increase in temperature).

The material needs to be produced at high densities to realize the high strengths expected from its covalent bonding. Usually, one would produce silicon nitride in powder form and use powder-based methods to shape the powder into a green powder compact or body. The green body or compact is then heated in a high-temperature furnace, allowing atomic diffusion and the sintering process to proceed and densify the powder compact into a high-density final product. A problem, however, arises in that for sintering to take place, significantly high temperatures are required, which can exceed the dissociation temperature of silicon nitride (> 1850 °C). This problem is avoided by the addition of sintering aids such as magnesia (MgO), alumina (Al$_2$O$_3$), and yttria (Y$_2$O$_3$), which produce a liquid phase during sintering by reacting with the silicon dioxide surface layer on Si$_3$N$_4$ powders and enable sintering at lower temperatures. The product is typically β-Si$_3$N$_4$ produced through the first dissolving of the α-Si$_3$N$_4$ particles in the liquid phase and the subsequent precipitation of β-Si$_3$N$_4$. The microstructure hence consists of β-Si$_3$N$_4$ grains and an intergranular glass phase. Although this phase helps achieve high densities, it can limit high-temperature performance. Sintered Si$_3$N$_4$ has been shown to possess bend strengths of 80–220 MPa. Hot pressing techniques (including sintering aids) have been applied to produce Si_3N_4 with limited residual porosity ($< 1\%$). The nature of the technique (powder placed in a container and pressure applied using an upper and lower punch at high temperatures) results in simple product geometries, which are suited for limited products such as cutting tools. However, bend strengths of 620–965 MPa have been reported. Other processes, such as spark plasma sintering, have also produced high-density silicon nitride.

SIALONs

In the 1970s it was discovered that silicon nitride 'solid solutions' can be obtained, which are called SIALONs. The Si is for silicon, Al for aluminum, O for oxygen, and N for nitrogen. They are called solid solutions for example, since oxygen can partially replace nitrogen, and aluminum can partially replace silicon on the β-Si_6N_8 unit cell and achieve charge neutrality and form β'-SIALON. Its composition can be expressed as:

$$Si_{6-z}Al_zO_zN_{8-z}$$

z values can range from 0 to 4.2 depending on the composition.

β'-SIALONs have elongated hexagonal grains very much like those of the β-SIALON, except that they have less intergranular glass phase.

It is also possible to form what is referred to as α'-SIALONs. Here, not only oxygen and aluminum substitute nitrogen and silicon, this time on a α-$Si_{12}N_{16}$ unit cell, but to achieve charge neutrality other metal ions (Me) are also added for example, Li^+, Y^{+++}, Mg^{++}, Ca^{++}. This leads to a composition for α'-SIALONs as:

$$Me_xSi_{12-(m+n)}Al_{(m+n)}O_nN_{(16-n)}$$

$x < 2$, m and n represent, respectively, the number of Al–N and Al–O bonds that replace the $(m + n)$ Si–N bonds. For electroneutrality to be in place, assuming Me has a valency of v, then $x = m/v$.

Whereas β-Si_3N_4 is typically the product of sintered or hot-pressed α-silicon nitride powder, SIALONs are different. They too, require sintering aids; however, α'-SIALONs and β'-SIALONs phases can both exist together within the microstructure. In fact, their proportions can be adjusted, making it possible to obtain 0–100% β'-SIALONs for example. β'-SIALONs and α'-SIALONs display different properties. α'-SIALONs displays high hardness (1900–2100 kg/mm^2, Vickers hardness) and good wear resistance, however it has relatively low fracture toughness values (3–4 MPam$^{1/2}$) and thermal conductivity (8.16–8.22 W/m K). Flexural strength values typically range from 350 to 500 MPa. β'-SIALONs on the other hand have higher strength values ranging from 700 to 1100 MPa, higher thermal conductivity (13.5–19.7 W/m.K) and fracture toughness values of ~ 7–8 MPa.m$^{1/2}$. Vickers hardness has been reported to range from ~ 1500–1700 kg/mm^2). By controlling the relative proportions of α'-SIALONs and β'-SIALONs we can have significant control over the properties of the final product. For example, if strength and hardness is what are desired then the volume fraction of α'-SIALONs is increased in the microstructure,

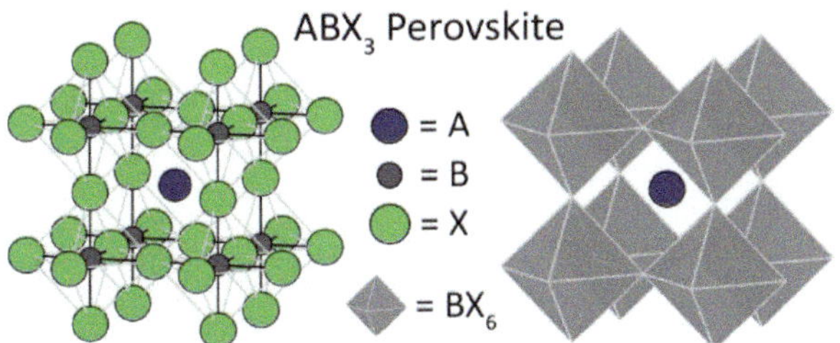

Fig. 13.11 Structure of Perovskite. (Reproduced from Quinten A. Akkerman and Liberato Manna, What Defines a Halide Perovskite? ACS Energy Letters 2020 5 (2), 604–610. https://doi.org/10.1021/acsenergylett.0c00039, under Creative Commons Attribution (CC-BY) License.http://pubs.acs.org/page/policy/authorchoice_ccby_termsofuse.html

while β'-SIALONs is increased if improved toughness is desired. Moreover, the glassy intergranular phase is greatly reduced compared with silicon nitride, which has a bearing on the high-temperature properties.

Perovskites

These are compounds with the structure of $CaTiO_3$, in other words, the formula ABX_3, as stated in Table 12.4. The name perovskite was introduced in 1939 by Gustav Rose after **Count Lev Alexevich von Perovski**, a Russian mineralogist. Perovskites are found in nature, predominantly as oxides, largely silicates, and others. However, they are still limited in comparison to the much wider array of perovskites synthetically produced, which involve a large scope of elements. Perovskites can be halides, carbides, hydrides, and, not least, oxides. What distinguishes each of them is the X in the formula (ABX_3); for example, for carbides, X=C (carbon); for oxides, X=O (oxygen); for halides, X can be F (fluorine); for hydrides, X=H (hydrogen). Figure 13.11 shows the general structure of perovskites. Perovskites can also have other formulas; for example, ordered perovskites can have the formula A_3BX_6.

Perovskites are highly sought after because of their optical, magnetic, catalytic, and electrical properties. They are also of significant technological importance and interest in solar cells (halide perovskites with over 25% efficiencies), fuel cells, electrochemical sensing, optoelectronics, and catalysts.

**The Sphere
Layer 5**

Professor, I did some SEM work and sent off the sample as you requested for SIMS analysis. The SIMS results are back. The SEM shows the sample to consist of silicon regions separated by much thinner aluminum films. The Si occupied most of the sample, thus confirming the previous XRD results. Since aluminum is present at much smaller levels, I guess XRD did not pick it up due to XRD's resolution limit. The SIMS analysis, however, revealed something very unusual. Within each silicon region, what could be described as three layers emerge. The middle layer was found to be doped with 1×10^{16} boron atoms/ cm^3. That layer is sandwiched between two outer layers, each instead doped to a level of 1×10^{16} phosphorus atoms/cm^3.

*Alex, you hit the jackpot! This is what I suspected. Remember the N-P junction diodes? What you have just found, my friend, is an NPN bipolar junction transistor (BJT). It all makes sense now why it was included in the Sphere. This is probably the most important discovery in history, and certainly in the 20th Century. As mentioned, the NPN junction forms the basis of a transistor. It could have been a PNP junction, and I would have been equally impressed. Transistors are used as switches or signal amplifiers. Unlike the PNP or the NPN bipolar junction transistor, a less energy-consuming transistor with the advantage of miniaturization and ease of manufacturing is the MOSFET (Metal Oxide Field Effect Transistor). You can now place billions of transistors on a chip. The impact of transistors on mankind cannot be overstated: computers, cellphones, sensors, TVs, communication, microprocessors, computer memory, and so on. Try to imagine a world without it. It would be what the world was like before 1947, when you had computers the size of a room, and large vacuum tubes took the role of transistors and were also very prone to breakdown. **John Bardeen, Walter Brattain, and William Shockley** invented the first transistor at Bell Labs in 1947. The transistor was called the point-contact transistor. They received the Nobel Prize in 1956 for their incredible invention. Soon after 1947, Shockley invented the bipolar junction transistor. However, the MOSFET invented by Mohamed Atalla and Dawon Kahng allowed the extraordinary ability to miniaturize transistors Well done, Alex. Now, let's see the next layer.*

Problems

13.1. Discuss the difference between the band gap of a semiconductor and that of an insulator.

13.2. What is the difference between an intrinsic and an extrinsic semiconductor? Give an example of each.

13.3. What are wide band gap semiconductors? Give an example of one.

13.4. A pure metal has an electrical resistivity of 1×10^{-7} Ω.m at 100 °C. Determine its electrical resistivity at 0 °C.

13.5. Describe the Frenkel defect with the aid of diagram(s).

13.6. Calculate the electrical conductivity of GaAs from first principles. Assume it has an electron density of $N = 1.82 \times 10^{12}$ m^{-3}.

13.7. Calculate for Si the drift velocity of electrons under an applied field of 20 V/m.

13.8. Calculate the drift velocity of electron holes under an applied field of 20 V/m for Si.

13.9. What is the depletion layer as it relates to diodes? Discuss its size under forward and reverse bias conditions.

13.10. What is a dielectric?

13.11. Describe the difference between electronic and ionic polarization.

13.12. What would a high dielectric strength material be used for?

13.13. A soda-lime glass coating is to be placed on the surface of a metal to insulate it electrically. If a voltage drop of 200 V is expected to be established across the thickness, what thickness should the coat be for the coat to be an effective electrical insulator?

13.14. Describe the difference between a thermoelectric and a piezoelectric material.

13.15. What is superconductivity?

13.16. What is the difference between a hard and a soft magnet? What applications can each of them be used for? Give one example of a soft magnet and one for a hard magnet.

13.17. Explain the difference between a paramagnetic and a diamagnetic material.

13.18. What is a ferromagnetic material? Give an example of a material considered to be one.

13.19. What are sialons and perovskites?

13.20. What is a ferrimagnetic material?

Part VI

Layer 1: Looks can be Deceiving

The Sphere

Layer 1
"Looks can be deceiving."

Professor, I'm so sorry. I know you left Sample 1 for later, but I couldn't help myself. I tried grinding it, and it turns out that the sample was coated. After removing part of the coat, I found the sample highly transparent. I also measured the sample's density, which was 2.5 g/cm^3. Its hardness was 480 HV. It also appeared brittle. I know I shouldn't jump to conclusions, as many transparent materials are around.

Maybe the coat was there to allow the sample number to be written on its surface. Who knows? The density and hardness are also good data to have. Let's keep investigating. I would advise you to carry out X-ray microanalysis on the sample, and let's talk again.

Chapter 14
Introduction to Nanomaterials

14.1 Introduction

This chapter deals with a very technologically important subject: nanomaterials. The discussion builds on the knowledge gained in subsequent chapters to highlight the importance of this relatively new field, which is gaining momentum. The chapter introduces the definition of nanomaterials and presents how their behavior differs from conventional materials and what new properties they offer.

14.2 Nanomaterials: History and Applications

So far, we have discussed different materials, from metals and alloys to polymers and ceramics, and different material properties, including mechanical, electrical, and magnetic. In this section, we define a nanomaterial and how the nanoscale affects physical and other properties.

The word nano comes from the Greek dwarf; a nanometer is one billionth of a meter in length scale. To give you some perspective, the diameter of a human hair is about 50–100 μm, which roughly represents the limit of the human eye's resolution. A nanometer is 50,000–100,000 times smaller, so you can forget trying to see something of that size. But what exactly qualifies a material to be called a nanomaterial? There are a couple of general definitions. The first definition is that a **nanomaterial** is *a material in which at least one of its dimensions is 100 nm or less.* For example, this could be a nano-coat covering a large component surface, but the coat has a thickness of less than 100 nm. This thickness alone qualifies it as a nanomaterial. Another example is a carbon nanotube, a tube of length 10–15 μm and a diameter of ~ 50 nm. The diameter here qualifies it as a nanomaterial. Other nanomaterials can have all three dimensions below 100 nm; an example is fullerenes. If the grain size of a bulk polycrystalline material is less than 100 nm, it also qualifies

K. Morsi, *An Engaging Approach to the Science and Engineering of Materials*,
https://doi.org/10.1007/978-3-032-06231-4_14

as a nanostructured material. A composite material where a matrix is reinforced with nano-scale reinforcements or particles would also be called a nanocomposite. Even porous materials can be classified as nanomaterials if their pore sizes lie at or below 100 nm.

The second definition seems to be a little more targeted. Here, a nanomaterial is the one where new properties or substantial improvements in properties are only explicitly attained because one or more of its dimensions are on the nanoscale. This is because some materials could have all dimensions below 100 nm, yet no new property is attained, and no significant improvements in properties are realized. So why should we reward this material with the famous title of "Nanomaterial"? In this section, we will only use the first definition for simplicity. If you think you haven't already come across nanomaterials, think again. Have you ever enjoyed ice cream, used toothpaste, or even put on sunscreen? Well, the likelihood is that you would have encountered TiO_2 nanoparticles. The material has antibacterial properties and is also used in food packaging films. It has a high refractive index (and hence can reflect light), and its ability to absorb UV light (while remaining transparent to the eye) prompted its use in sunscreens and cosmetics. It is also used as a food additive in milk, candy, and bakery foods and had previously been approved by the U.S. Food and Drug Administration (FDA) for such an application, as long as titanium does not exceed 1% of the weight of the food. As of 2023, though, the FDA has been reviewing a petition to stop using TiO_2 in foods due to potential health risks. Europe has already banned TiO_2 as a food additive in 2022. As of the writing of this book, the jury is still out on this topic in the US.

You are mistaken if you think that nanomaterials are something that humans have only been exposed to recently. Nanomaterials were found to have been used in ancient times, of course, without our ancestors identifying them as nanomaterials, since they would have also needed to have invented relatively recent instruments, like the electron microscope or the X-ray diffractometer, for example. When analyzed, lusters on glaze from the 9th century CE in Iraq were recently found to contain large amounts of silver nanoparticles. Luster from Spain and Italy from the Middle Ages and Renaissance revealed not only a luster "almost"-nanofilm (200–500 nm thick), but the film also contained nanocrystals, separated by nano-scale distances, and an outer glass nanolayer (10–20 nm) that contained no metal. The quality and reproducibility of these artifacts point to these craftsmen's high level of empirical material knowledge. Analysis of a 7th Century Damascus steel sword that exhibited superior properties to European swords at the time was found to contain carbon nanotubes and cementite nanowires. In 2020, single-wall carbon nanotubes with average diameters of 0.6 nm $\pm$ 0.05 nm were found in 6th-century BCE Indian pottery. This is incredible, given the fact that carbon nanotubes had been center stage for the first time only recently.

Nanomaterials can be naturally/biologically occurring (e.g., nano clays, DNA) or synthetically produced. Figure 14.1 shows the DNA double-helix structure with a diameter of 2 nm. DNA is now being considered in molecular-scale electronic devices, capitalizing on its stability and programmability.

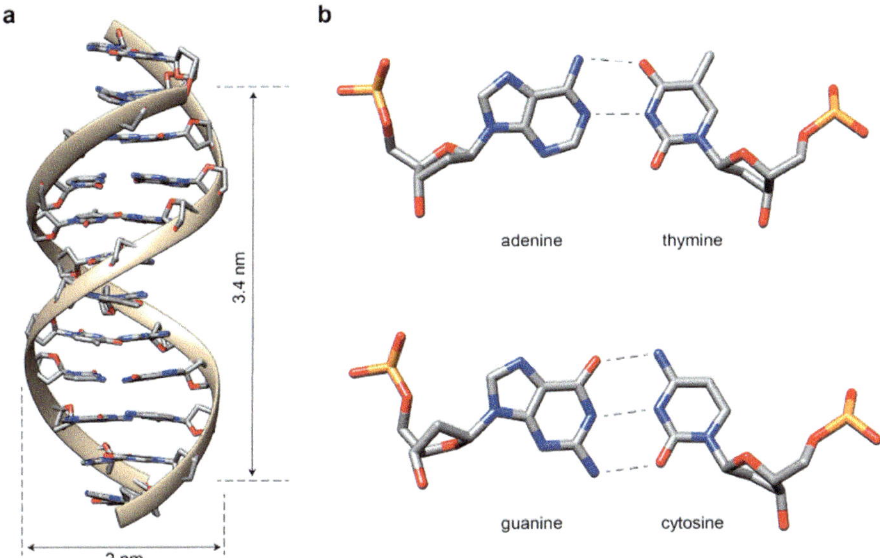

Fig. 14.1 Part of DNA structure and its four nucleobases. (*From Andreas Jaekel, Pascal Lill, Stephen Whitelam, Barbara Saccà, Molecules 2020, 25(23), 5466;* https://doi.org/10.3390/molecu les25235466, *the figure was originally modified from open sources (Protein Data Bank (PDB): 1DUF). Reproduced under Creative Commons Attribution (CC BY) license (*)http://creativecomm ons.org/licenses/by/4.0/)

The applications of nanotechnology are currently skyrocketing, as outlined in the National Nanotechnology Initiative (NNI) (www.nano.gov), and numerous products have benefited so far from nanotechnology. Many current and future nanomaterials applications span many science and technology sectors. Transistors, which are the cornerstone and possibly one of the greatest inventions of all time, have shrunk in size over the years to the extent that in May 2021, IBM unveiled a 2 nm transistor, where ~ 50 "billion" of them can be fitted on something as small as one of your fingernails. This remarkable achievement would result in 45% greater performance and energy efficiency (75% less energy) than the current IBM 7 nm nano chips. It would also translate to extremely fast internet access and a longer battery life. Nanoporous thin film MoS_2 membranes have been developed to desalinate water energy-efficiently, filtering 2–5 times the water that current filters can do.

Imagine nanoparticles delivering vaccines without requiring a prick of a needle. Imagine nanoparticles encapsulating cancer cells and directly targeting the cells with drugs. A major hope is that the harmful effects of chemotherapy can be reduced. But nanomaterials are already on the market. Nanomaterials have been incorporated into fabrics to give them anti-crease, anti-bacterial, and anti-staining properties. They have allowed fabrics to be made with the potential of harvesting energy and providing information on your vital signs. Their incorporation in sports equipment/gear, such as bicycles, baseball bats, and tennis rackets, has improved their properties and performance. Even nanofilms on eyeglasses have provided properties such as scratch resistance, anti-reflection, and self-cleaning, and they can even render glasses electrically conductive. Imagine what can come out of that.

Flexible electronics benefit tremendously from nanomaterials. Imagine foldable, stretchable, and rollable electronics. Even now, we have TVs based on quantum dot technology. Nanomaterials can be used as flexible nanosensors, referred to as tattoo sensors. Also, imagine harnessing solar energy on something you wear. Finally, nanomaterials are used in waste treatment and catalysis.

We are truly in exciting times, with a future that could have been described as sci-fi only a few decades ago. OK, enough about how great nanomaterials are. We will now delve into the inner workings of the nanoworld and explore what new laws and properties emerge at the nanoscale.

14.3 Nanomaterial Nomenclature

Natural and synthetic nanomaterials can be classified into 0D, 1D, 2D, and 3D. An example of a 0D nanomaterial is a spherical nanoparticle, a fullerene or a quantum dot. Figure 14.2 shows examples of lipid, protein, and TiO_2 (agglomerated) nanoparticles. A 1D nanomaterial could be a nanofiber, nanorod, or nanotube (Fig. 14.3 shows gold nanorods). A 2D nanomaterial would be a nanosheet, nanoplate, or nano coat/film; this is seen in Fig. 14.4 with gold nanoplates. See how these nanoplates can be controlled in terms of their size. Bulk materials with nanofeatures (nanopores, nano-scale grain size, etc.) can be considered 3D nanomaterials; Fig. 14.5 shows nano-scale pores in a ceramic filter. The external dimensions of the filter are greater than 100 nm, yet pores are seen with sizes less than 100 nm. Zeolites are an excellent example of nanoporous ceramics, with typical pore sizes of 1 nm and less, and have been referred to as molecular sieves because of these small pore sizes. Nanomaterials have different dimensions and can be seen in many materials, including ceramics, polymers, metals, intermetallic, composites, and biological materials.

There are several abbreviations that we need to familiarize ourselves with; these are listed in Table 14.1.

Prof., how can researchers and scientists produce nanoscale gold not only as rods but also as plates? How can they control that?

Great question, Alex. Although 'As a Matter of Fact' does not get too much into synthesis techniques, it does cover fundamental reasons that answer the exact question you are asking. Keep reading, my friend. For now, let's say it depends on surface energy considerations. So why don't you start reading about surfaces? Believe me, you can learn a lot about the behavior of nanomaterials from learning about their surfaces.

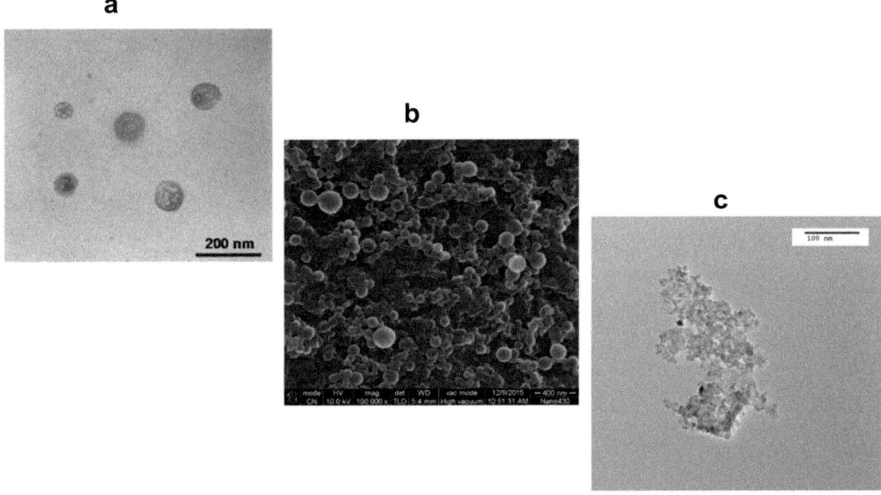

Fig. 14.2 0D nanomaterials **a** lipid nanoparticles, **b** protein nanoparticles, **c** titanium dioxide (TiO$_2$) agglomerated nanoparticles, scale bar in Fig. 14.2c is 100 nm (*reproduced from McClements, D.J., Xiao, H. Is nano safe in foods? Establishing the factors impacting the gastrointestinal fate and toxicity of organic and inorganic food-grade nanoparticles. NPJ Sci Food 1, 6 (2017).* https://doi.org/10.1038/s41538-017-0005-1 *(under license:* http://creativecommons.org/licenses/by/4.0/.*)*

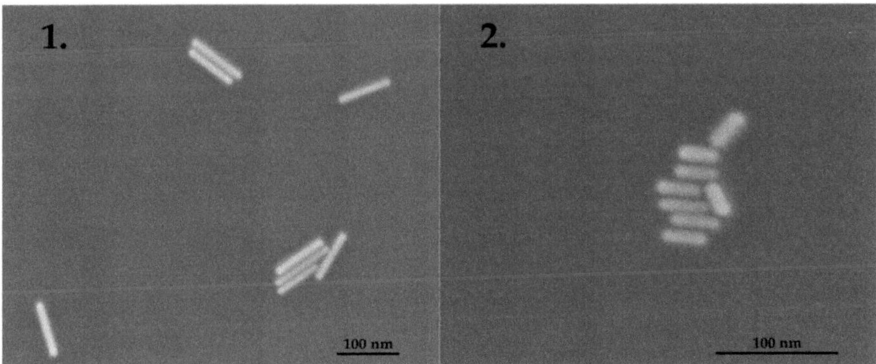

Fig. 14.3 Gold nanorods, note gold nanorods in image 1 are longer than those in image 2. *(Figure from M. Sidorova and A. Popov, Synthesis of Gold Nanorods for Multifaceted Applications, Chem. Proc. 2023, 14(1), 11;* https://doi.org/10.3390/ecsoc-27-16057, *MDPI, Reproduced under Creative Commons Attribution (CC BY) license (*https://creativecommons.org/licenses/by/4.0/*))*

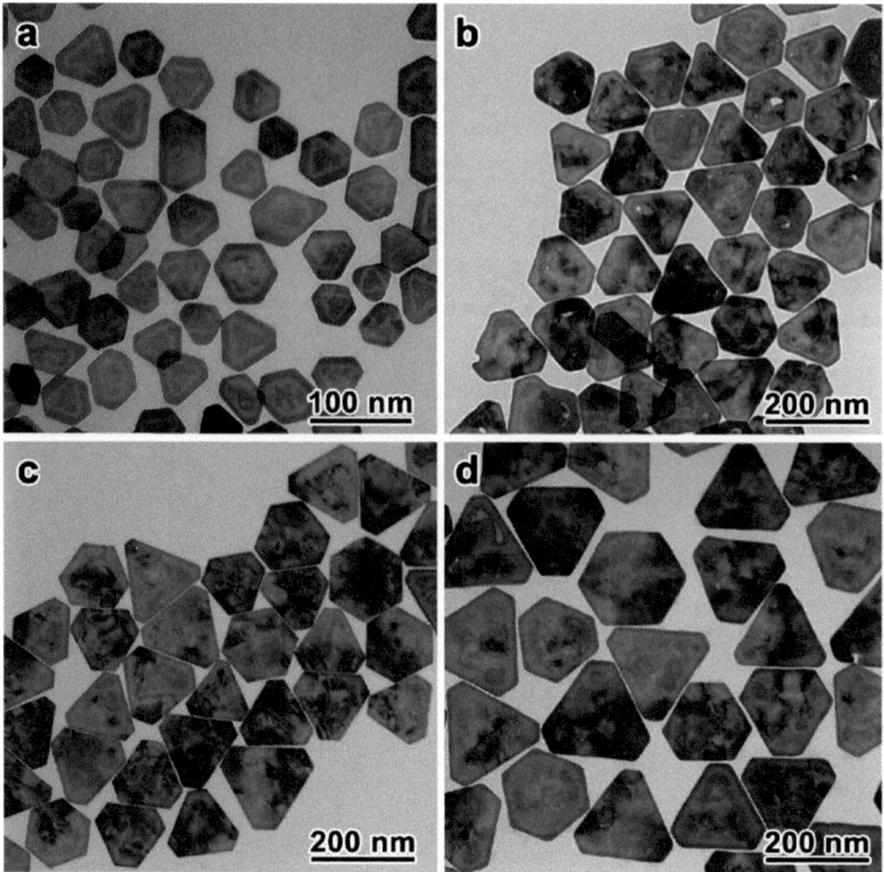

Fig. 14.4 2D nanomaterials-TEM micrographs of gold nanoplates with varying edge lengths. *(Image from Qiao, Z.; Wei, X.; Liu, H.; Liu, K.; Gao, C. Seed-Mediated Synthesis of Thin Gold Nanoplates with Tunable Edge Lengths and Optical Properties. Nanomaterials 2023, 13, 711.* https://doi.org/https://doi.org/10.3390/nano13040711, *MDPI, Reproduced under Creative Commons Attribution (CC BY) license (*https://creativecommons.org/licenses/by/4.0/)*)*

14.4   Surfaces in Nanomaterials

One of the main differences between a material in bulk form and the same material with dimensions on the nanoscale is the surface area to volume ratio. Imagine a block of aluminum; the block will contain billions and billions of aluminum atoms within its volume; however, at its surface, the number of surface aluminum atoms will be relatively insignificant. However, if we talk about a 10 nm aluminum particle, the relative amount of surface atoms will be significant compared to interior atoms. Figure 14.6 is a schematic of a spherical nanoparticle of diameter (D_o). In nanomaterials, a surface is regarded as having a thickness (t); the surface influences atoms

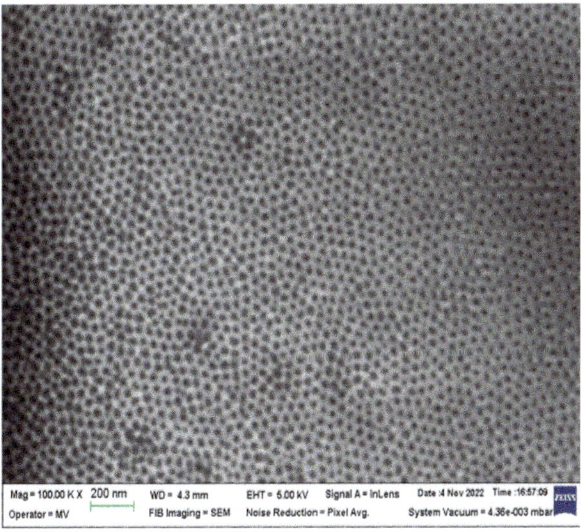

Fig. 14.5 3D nanomaterial—a nanoporous ceramic protein filter, with possible use in treating Alzheimer's. *(reproduced with modification from T. G. Schreiner, M. Menéndez-González, M. Adam, B. O. Popescu, A. Szilagyi, G. Dumitrita Stanciu, B. I. Tamba and R. C. Ciobanu, A Nanostructured Protein Filtration Device for Possible Use in the Treatment of Alzheimer's Disease—Concept and Feasibility after* In Vivo *Tests, Bioengineering, 2023, 10(11), 1303;* https://doi.org/10.3390/bioeng ineering10111303. *Reproduced with modification under Creative Commons Attribution (CC BY) license (*https://creativecommons.org/licenses/by/4.0/*))*

Table 14.1 Abbreviations for some important nanomaterials (multiple sources)

Nanomaterial	Abbreviation
Single layer graphene	1LG
Bilayer graphene	2LG
Few layer graphene	FLG
Carbon nanotube	CNT
Single wall carbon nanotube	SWCNT
Double-wall carbon nanotube	DWCNT
Multiwall carbon nanotube	MWCNT
Nanoparticle	NP
Fullerenes	C_n (where n is the number of carbon atoms)

within this thickness. These atoms would behave differently than atoms residing in the particle's interior. This thickness is typically between 0.5 and 1.5 nm. D_i represents the diameter of an inner sphere where all the atoms can be regarded as interior atoms (unaffected by the surface).

It is now possible to calculate the volume of atoms within this surface-influenced thickness for a spherical nanoparticle, by subtracting the inner sphere (V_i) volume

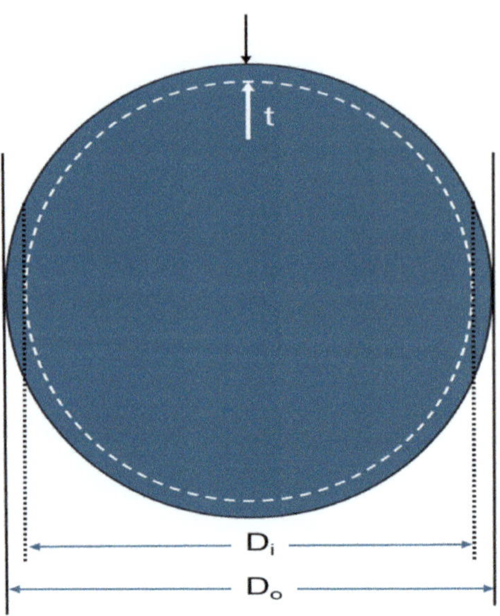

Fig. 14.6 A schematic of a nanoparticle, showing thickness (t) of surface influenced atoms/ions

from the total volume of the nanoparticle (V_o). If we then divide it by the total volume of the nanoparticle, we obtain the fraction of atoms within the volume of the nanoparticle that can be considered as surface-influenced atoms (X_s). This is given by Eq. 14.1.

$$X_s = \frac{V_s}{V_o} = \frac{V_o - V_i}{V_o} = \frac{D_o^3 - D_i^3}{D_o^3} = \frac{D_o^3 - (D_o - 2t)^3}{D_o^3} = 1 - \left(\frac{D_o - 2t}{D_o}\right)^3 \quad (14.1)$$

Equation 14.1 is plotted in Fig. 14.7 for a t value of 1 nm and particle diameters between 1 and 50 nm. It is clear from the figure that you only need a particle size slightly smaller than 9 nm for 50% of the atoms to be surface-influenced atoms, and for a 2 nm particle, the whole volume becomes surface-influenced. This means the whole concept of a particle consisting of interior atoms arranged in a crystal lattice ceases to exist. Particles at this point will tend to behave in a way between a solid and a liquid, and more so on the liquid side. **Landau's ordering parameter** has values between 0 and 1, where one refers to ideal crystals, while zero refers to a molten or liquid phase. When applied to nanoparticles, it becomes clear that as particle size decreases within the nano-range, the nanoparticle starts to take on a more liquid-type structure. For example, Landau's parameter for a 2 nm diameter tin nanoparticle is about 0.15 at the surface and about 0.275 in the center of the particle.

Remember, atoms at the surface will behave differently than atoms within the volume of the material. We discussed previously that many other atoms will surround an atom that exists within the interior of a material. For example, for an FCC material

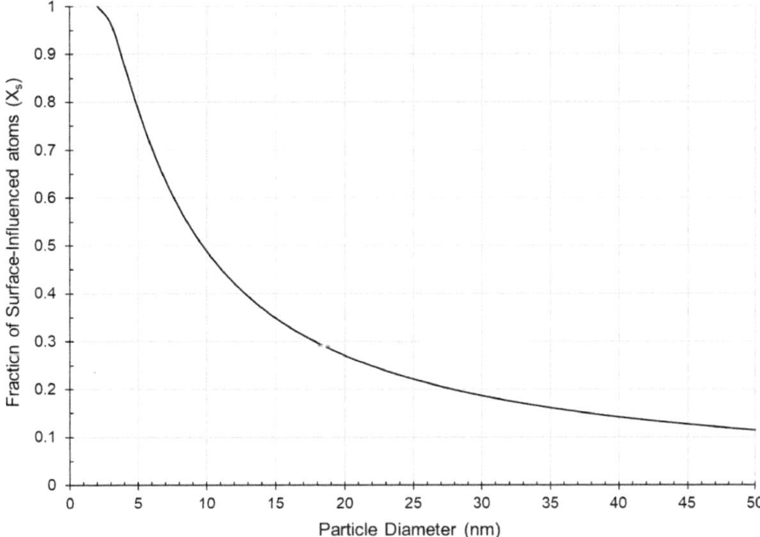

Fig. 14.7 Effect of nanoparticle diameter (1–50 nm) on the fraction of surface influenced atoms, assuming a surface-influenced thickness of 1 nm

like aluminum, the inner aluminum atoms will be surrounded by 12 other nearest-neighbor touching atoms (coordination number $= 12$), and its bonding will be satisfied. The same cannot be valid for surface-bound aluminum atoms since no aluminum atoms exist above surface atoms; there are only ones surrounding them on the particle surface and beneath them in the particle itself. This leaves surface atoms possessing more energy than interior ones, thus making them highly reactive. Hence, it is unsurprising that while an aluminum block is stable, a nano-aluminum particle can be explosive or catch fire; aluminum nanopowder is used as rocket fuel.

Professor, I understand that nanoparticles of a metal like aluminum are more energetic and reactive (because of their excess surfaces) than, for example, micro-scale aluminum powder particles, but how exactly can nano-aluminum powder be explosive?

Well, we need to ask the question as to what these highly reactive materials react with. Did you know that all metal powders have a passivating surface oxide layer? You must have guessed by now that these powders react with oxygen to form the oxide layer. That's why coarse microscale aluminum would not simply catch fire without external influences. The aluminum oxide layer protects it from further oxidation. If you, however, break the oxide surface layer and expose virgin aluminum metal to the atmosphere, then good luck. The difference between microscale and nanoscale particles of the same metal will be the surface area-to-volume ratio. Take, for example, a microscale spherical aluminum particle of a particular size, mass, and volume (Fig. 14.8a in As a Matter of Fact). Within the same volume of the micro-scale particle, we can fit a significantly greater number of smaller spherical aluminum particles (Fig. 14.8b in As a Matter of Fact). The total surface area of all these smaller particles is much greater than that of a single large aluminum particle. Now, imagine if these smaller particles were nanoscale, their surface area would be astronomically large. Remember, in all cases, you still have the same volume and mass since it's the same material and its density is the same. When a large collection of particles is exposed to an oxygen-rich environment, the metal can react with oxygen to form aluminum oxide. This reaction is highly exothermic, so heat is generated. The powder will catch fire or explode if this heat cannot escape. This is called pyrophoricity. So, be very careful when dealing with metal powders, especially nanoscale. Typically, they are processed in inert environments to avoid oxidation and its disastrous consequences.

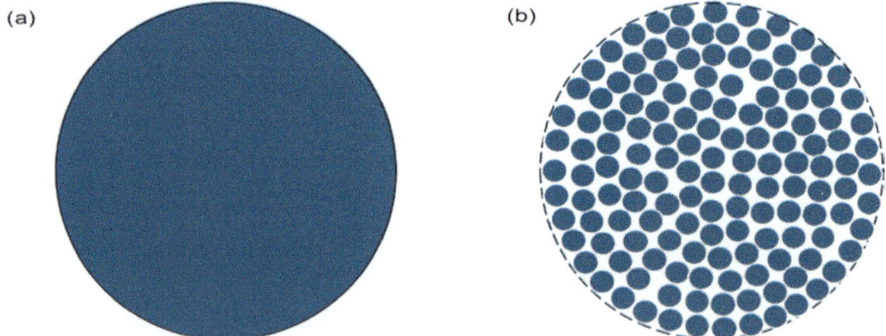

Fig. 14.8 **a** One large spherical particle of a metal, **b** roughly fitting within the same volume boundary (and hence for the same mass) many more smaller particles of the same metal. These numerous particles also come with much more surfaces, and their associated surface energies

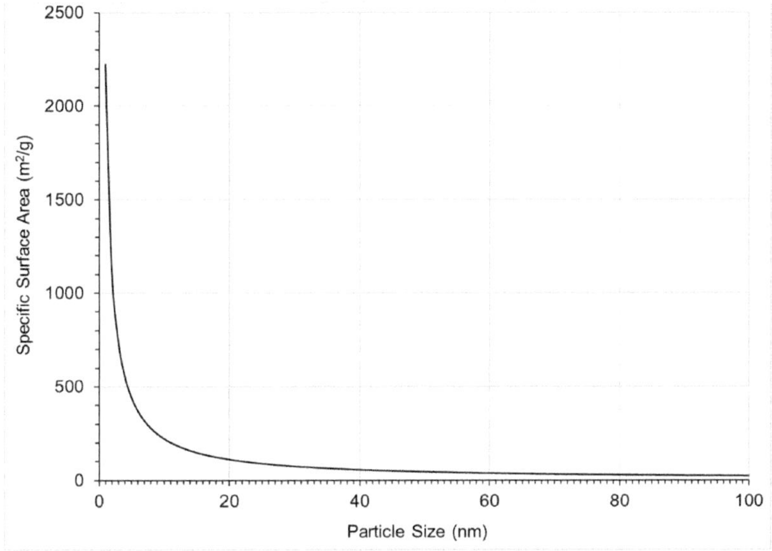

Fig. 14.9 Effect of particle size on the specific surface area of nanoparticles

The specific surface area (S) of powders is typically given in units of m^2/g and can be calculated using Eq. 14.2.

$$S = \frac{6}{\rho D} \tag{14.2}$$

where ρ is the density in g/cm^3 and the average particle diameter, D, is in micrometers. If we input in the equation D in units of micrometers and ρ in units of g/cm^3, S will come out in m^2/g, and all unit conversions will cancel out. Figure 14.9 is a plot showing the effect of the particle size of a spherical aluminum particle (with a density of 2.7 g/cm^3) on the specific surface area. As the particle size decreases significantly below 10 nm, the specific surface area rapidly increases to very high values.

Example Problem 14.1

What is the specific surface area of a gold spherical nanoparticle with a diameter of 5 nm? What would happen to the specific surface area if the particles were aluminum and not gold?

Solution

Inputting the density of gold as 19.3 g/cm^3. $D = 5 \times 10^{-3}$ μm, into Eq. 14.2, gives

$$S = \frac{6}{\rho D} = \frac{6}{19.3 \times 5 \times 10^{-3}} = 62.2 \ m^2/g$$

If it was a 5 nm aluminum nanoparticle, we replaced $19.3/cm^3$ for the density of gold with 2.7 g/cm^3 for the density of aluminum and obtained a specific surface area of 444.4 m^2/g. This means that for only one gram of 5 nm aluminum particles, the surface area would be 444.4 m^2, which is equivalent to the square footage of a 4779 SQF house.

A significant consequence of the excessive surface area of nanopowders is excessive surface energy. Metals' surface energies are typically in the ballpark of 0.2–3.0 J/m^2. Table 14.2 lists experimental surface energy values for some metals.

For nanoparticles, the surface energy is not only regarded as the energy coming from broken bonds but also arises from surface stress (a discussion of surface stresses will not be covered here). These surface stresses result in the generation of hydrostatic pressure (compressive stress) on the nanoparticles, the magnitude of which increases with a decline in particle size and can reach the GPa range. This increasing pressure can lead to a decline in lattice parameters as the particle size decreases. Indeed, this is the case for metallic nanoparticles, as atoms are brought closer together due to this imposed hydrostatic pressure. Oxide ceramics, on the other hand, may show a different behavior; the lattice can expand as the particle size is decreased. An example would be γ-Fe$_2$O, where negatively charged oxygen ions are the terminating ions at the surface; the repulsion between these negatively charged oxygen ions results in expansion.

Example Problem 14.2

What is the total surface energy of 10 g of spherical aluminum powder with a diameter of 5 nm? Assume all powder particles are monosized, which means they are all 5 nm in diameter. The density of aluminum is 2.7 g/cm^3.

Solution

Table 14.2 shows the surface energy of aluminum is 1.43 Jm^{-2}. We also have

Table 14.2 Experimental surface energy values for some metals (from data previously compiled by L. Vitos, A. V. Ruban, H. L. Skriver, J. Kollar, The surface energy of metals, Surface Science 411 (1998) 186–202)

Material	Surface energy (Jm^{-2})	Material	Surface energy (Jm^{-2})
Magnesium (Mg)	0.785	Silver (Ag)	1.246
Aluminum (Al)	1.43	Gold (Au)	1.507
Titanium (Ti)	1.989	Platinum (Pt)	2.489
Iron (BCC)	2.417	Tungsten (W)	3.265
Nickel (Ni)	2.380	Palladium (Pd)	2.03
Copper (Cu)	1.790	Molybdenum (Mo)	2.907
Ruthenium (Ru)	3.043	Niobium (Nb)	2.655
Sodium (Na)	0.261	Potassium (K)	0.145

$$S = \frac{6}{\rho D} = \frac{6}{2.7 \times 5 \times 10^{-3}} = 444.4 m^2/g$$

Hence, to obtain the total surface area within 10 g of aluminum nanopowder, we multiply $444.4 m^2/g \times 10g = 4444\ m^2$.

Since the surface energy of aluminum is 1.43 Jm^{-2}, then the total surface energy of 10 g of aluminum nanopowder of 5 nm diameter is:

Total surface energy $= 1.43 \times 4444 = 6,355$ J.

Recall that for powders, the total energy is the sum of the surface energy and the internal energy (inside the particle's interior, which includes the energy of crystallographic defects, etc.).

Alex, do you see how particle size on the nanoscale can dramatically affect something as basic as surface energy? Believe me, there is more. Let's take vapor pressure, for example. For solids, the vapor in equilibrium with the solid surface of the same material has a certain pressure. This is the vapor pressure, and it is typically low. But did you know that as the particle size of a material decreases, its vapor pressure increases?

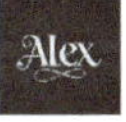

Ok, but how could that matter in the greater scheme of things?

Go to the section on the vapor pressure of nanomaterials in As a Matter of Fact, my friend.

14.5   Vapor Pressure of Nanomaterials

Another consequence of a reduction in particle size is an increase in vapor pressure, per the Kelvin (or Thomson) Eq. 14.3. The equation provides the vapor pressure for the curved surface of a particle (P) in relation to the vapor pressure of the same material but as a flat surface (P_∞). The equation includes the molar volume (V), the surface energy (γ), the gas constant (R), temperature (T) in Kelvins and the particle diameter (D):

$$\frac{P}{P_\infty} = e^{\left(\frac{4\gamma V}{RTD}\right)} \tag{14.3}$$

Since the vapor pressure depends exponentially on the particle diameter at a fixed temperature, it increases exponentially as the particle diameter decreases. Thus, the vapor pressure for a nanoparticle can be one hundred times that of the same vapor pressure for a flat surface.

Professor, you mentioned that the equation is called the Kelvin and Thomson equations. Who came up with it, Kelvin or Thomson?

William Thomson is Baron Kelvin. He chose that latter name when, in 1892, he was awarded the title Baron of Largs in Scotland. So, we are really talking about the same person. He is quite a famous guy; the unit Kelvin is named after him.

The increase in vapor pressure as particle size decreases significantly affects the shape of nanoparticles. The high vapor pressure allows the particle edges and irregularities (which tend to increase the particle's surface area to volume ratio, and hence the surface energy for that particle volume) to evaporate and recondense on other parts of the particle. This is done in a way that produces a spherical particle. Recall, the spherical shape has the lowest surface area-to-volume ratio (hence the lowest surface energy for that volume). It is not surprising that most nanoparticles are spherical. There are, of course, other nanoparticles that remain faceted; an example is ceria (CeO_2), which has a very low vapor pressure even at the nanoscale. The typically higher vapor pressures of nanoparticles can also affect sintering and nanoparticle synthesis.

14.6   Phase Transformations

So far, we have discussed binary phase diagrams in which the X-axis is composition, and the Y-axis is temperature. For nanomaterials, particle size can additionally influence the phase formed. Hence, phase diagrams can be drawn with particle size on the X-axis. We have always heard that thermodynamics determines one phase's stability over another (of course, kinetics can also have a significant influence). The BCC iron phase, for example, is said to be stable at room temperature up to a temperature of around 912 °C, above which the BCC structure is substituted with an FCC structure. Whenever we asked why? The answer was always that it is thermodynamically favorable or between room temperature and 912 °C the BCC structure provides a lower energy configuration than the FCC structure. However, above 912 °C, the FCC structure becomes the more energy-efficient crystal structure. What energy are they talking about? This happens to be **Gibbs free energy (G)**, given by Eq. 14.4:

$$G = H - TS \tag{14.4}$$

where G, H, and S are the Gibbs free energy, enthalpy, and entropy of one mol of coarse particles. T is the temperature in Kelvin. Gibbs free energy can be regarded as the energy available to do work. Entropy can be defined as the degree of randomness.

The Enthalpy (H) is given by Eq. 14.5:

$$H = E + PV \tag{14.5}$$

where E is the internal energy of 1 mol of coarse particles.

Within the context of phase transformations, the Gibbs free energy is extremely important in determining whether a particular phase transformation will occur. Let's take, for instance, BCC iron transforming into FCC iron by heating. Let's call BCC iron phase 1 and FCC iron phase 2. In Fig. 14.10, the Gibbs free energy of the two phases is roughly and schematically drawn with respect to temperature. At low temperatures, phase 1 (BCC iron or ferrite) has a lower Gibbs free energy than phase 2 (FCC iron or austenite), making it more stable. Thus, at room temperature, we see ferrite. At T_{trans} (~ 912 °C), both phases are equally stable and can coexist at that temperature. Let's look at a more familiar scenario to understand what this means. When a metal is heated to its melting point, initially, all the material is solid metal. Still, as time passes at the melting temperature, more and more of the solid metal will convert to liquid until all the solid metal converts to liquid. During this period, both solid metal and liquid metal (two separate phases) coexist. Similarly, when a solid of a particular crystal structure transforms and changes crystal structure from phase 1 to phase 2, at the transformation temperature, both phase 1 and phase 2 can coexist. Above T_{trans} phase 2 (FCC iron) now has the lower Gibbs free energy and hence becomes the more stable phase (austenite, γ). Figure 14.10 illustrates these dependencies.

Figure 14.10 clearly shows that at the transformation temperature (T_{trans}), the Gibbs free energy of phase 1 is equal to phase 2, making $\Delta G = 0$.

Note that under normal circumstances for conventional materials, Eq. 14.6 is valid:

$$\Delta G = \Delta H - T \Delta S \tag{14.6}$$

Hence at the transformation temperature, T_{trans} for coarse (non-nanometric) particles (now called T_{coarse}) the following is true, leading to Eq. 14.7:

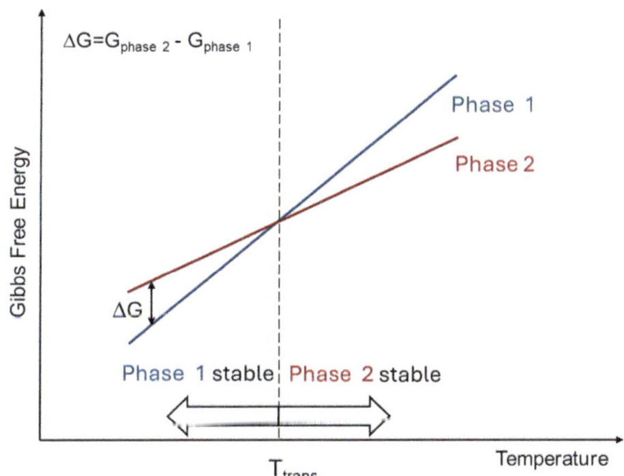

Fig. 14.10 Gibbs free energy change with temperature for two polymorphs of the same material. The stable phase at a particular temperature is the one with the lower Gibbs free energy

$$\Delta G = 0 = \Delta H - T_{\text{coarse}} \Delta S$$

$$T_{\text{coarse}} = \frac{\Delta H}{\Delta S} \tag{14.7}$$

This result shows that the transformation temperature for conventional "non-nanometric" materials is independent of particle size and only depends on the change in enthalpy and entropy between phase 1 and 2.

The analysis so far is valid for conventional materials (as mentioned, non-nanometric). However, what about nanomaterials? When considering nanoparticles for example, the surface area and hence surface energy term can become quite significant. So far Gibbs free energy has been determined from enthalpy and entropy values given the temperature. For nanomaterials, we need to include a surface energy term, hence Eq. 14.4 needs to be modified to include surface energy as seen in Eq. 14.8.

$$G = H - TS + \gamma A \tag{14.8}$$

where γ is the surface energy in J/m^2 for one mol of particles, and A is the surface area per mol (for spherical particles $= 6\, M/\rho D$, where D is the particle diameter, ρ is the density, and M is the molar mass.

$$G = H - TS + \gamma \frac{6M}{\rho D} \tag{14.9}$$

Suddenly, particle size now plays a role in the Gibbs free energy. This only happens on the nanoscale and has major consequences on the phase transformation temperature. Let's examine this a little further. Imagine a nanoparticle initially of a particular crystal structure (we will call phase 1), let's say BCC, which at a higher temperature transforms to an FCC crystal structure (phase 2). The difference between the Gibbs free energy of the two phases will be given by ΔG which is $G_2 - G_1$. Equation 14.9 can now be rewritten as Eq. 14.10 as:

$$\Delta G_{\text{nano}} = \Delta H - T \Delta S + \left(\gamma_2 \frac{6M}{\rho_2 D_2} - \gamma_1 \frac{6M}{\rho_1 D_1} \right) \tag{14.10}$$

where $\Delta H = H_2 - H_1$, and $\Delta S = S_2 - S_1$

At the transformation temperature, $T = T_{nano}$ and $\Delta G = 0$. One cannot assume that the diameter of a nanoparticle will not change when the crystal structure changes from phase 1 to phase 2. The diameter and the density may change. However, what remains constant will be the mass of the particle since we are not adding or subtracting atoms from the particle (assuming no evaporation or oxidation taking place).

Hence,

$$m_1 = m_2$$

$$\rho_1 V_1 = \rho_2 V_2$$

Using the volume of a sphere for the spherical nanoparticles:

$$\rho_1 \frac{\pi D_1^3}{6} = \rho_2 \frac{\pi D_2^3}{6}$$

Hence,

$$D_1 = \left(\frac{\rho_2}{\rho_1}\right)^{1/3} D_2$$

By substituting D_1 for $\left(\frac{\rho_2}{\rho_1}\right)^{1/3} D_2$ we obtain:

$$\Delta G = \Delta H - T_{\text{nano}} \Delta S + \left(\gamma_2 \frac{6M}{\rho_2 D_2} - \frac{\gamma_1 6M}{\rho_1 \left(\frac{\rho_2}{\rho_1}\right)^{1/3} D_2} \right)$$

$$\Delta G = \Delta H - T_{\text{nano}} \Delta S + \gamma_2 \frac{6M}{\rho_2 D_2} \left(1 - \frac{\gamma_1}{\gamma_2} \left(\frac{\rho_2}{\rho_1}\right)^{2/3} \right)$$

At the transformation temperature (T_{nano}), we have $\Delta G = 0$ and hence:

$$0 = \Delta H - T_{\text{nano}} \Delta S + \gamma_2 \frac{6M}{\rho_2 D_2} \left(1 - \frac{\gamma_1}{\gamma_2} \left(\frac{\rho_2}{\rho_1}\right)^{2/3} \right)$$

$$T_{\text{nano}} = \frac{\Delta H}{\Delta S} + \frac{\gamma_2 6M}{\rho_2 D_2 \Delta S} \left(1 - \left(\frac{\gamma_1}{\gamma_2}\right) \left(\frac{\rho_2}{\rho_1}\right)^{2/3} \right)$$

If we assume that the change in enthalpy and entropy from phase 1 to phase 2 is the same for coarse particles and nanoparticles. Then, we can substitute $\frac{\Delta H}{\Delta S}$ with T_{coarse} and use $\Delta S = \Delta H / T_{\text{coarse}}$, and eventually come to Eq. 14.11.

$$T_{\text{coarse}} - T_{\text{nano}} = -\frac{\gamma_2 6M T_{\text{coarse}}}{\rho_2 D_2 \Delta H} \left(1 - \left(\frac{\gamma_1}{\gamma_2}\right) \left(\frac{\rho_2}{\rho_1}\right)^{2/3} \right)$$

$$T_{\text{coarse}} - T_{\text{nano}} = -\frac{\gamma_2 6M T_{\text{coarse}}}{\rho_2 D_2 \Delta H} \left(1 - \left(\frac{\gamma_1}{\gamma_2}\right) \left(\frac{\rho_2}{\rho_1}\right)^{2/3} \right) \tag{14.11}$$

Let's consider some surface energy and density values, as seen in Table 14.3.

It is clear from Table 14.3 that the factor $1 - \left(\frac{\gamma_1}{\gamma_2}\right) \left(\frac{\rho_2}{\rho_1}\right)^{2/3}$ will be negative, which will result in a positive value for $T_{\text{coarse}} - T_{\text{nano}}$. This means that the melting point of a nanoparticle is lower than that of the same material that is not on the nanoscale

Table 14.3 Some material data for three metals. (*Based on: D. Vollath, Nanomaterials: Synthesis, properties and applications, Wiley, (2013) with slightly different presentation*)

Metal	$\gamma_{solid}/\gamma_{liquid}$	$(\gamma_{liquid}/\rho_{solid})$	$(\rho_{liquid}/\rho_{solid})^{2/3}$	$\gamma_{solid}/\gamma_{liquid}\,(\rho_{liquid}/\rho_{solid})^{2/3}$
Copper	1.11	0.90	0.932	1.035
Silver	1.15	0.90	0.932	1.072
Gold	1.22	0.89	0.925	1.129

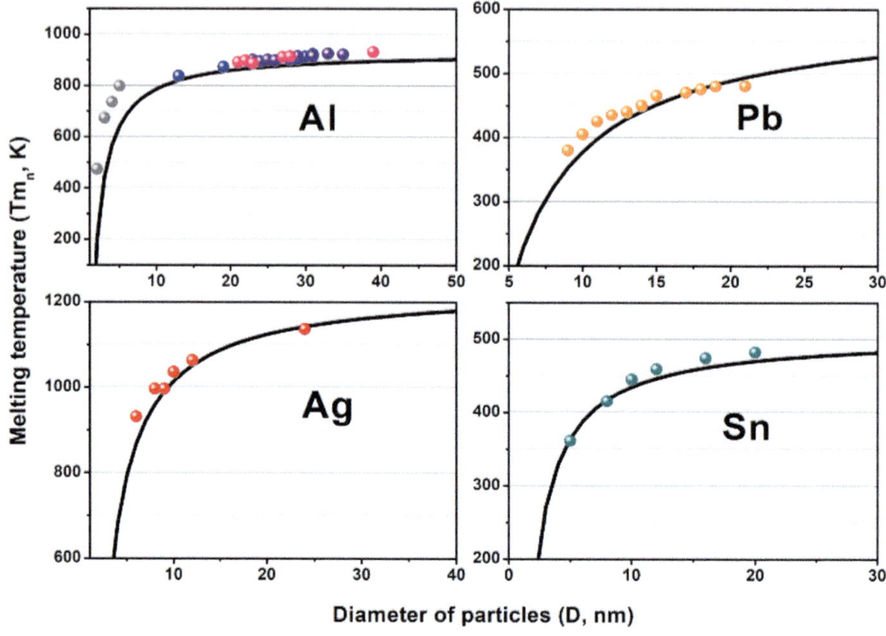

Fig. 14.11 Model showing the effect of particle size on the melting points of Al, Pb, Ag and Sn. Points in the figure are experimental and simulation data from previously published sources included for comparison by Ansari (from *Manauwar Ali Ansari, Modelling of size-dependent thermodynamic properties of metallic nanocrystals based on modified Gibbs–Thomson equation, Applied Physics A (2021) 127:385. image reproduced under license,* http://creativecommons.org/licenses/by/4.0/.)

(i.e., the melting point of bulk material). The difference between the two melting points is given by $T_{coarse} - T_{nano} = \Delta T$. Note the dependence on particle size, D_2. It is clear that as particle size decreases, so will the melting point, or, in general, the phase transformation temperature decreases.

Hence, this result applies not only to the melting of nanoparticles in comparison to the melting of bulk materials but also to all phase transformations, such as the monoclinic-tetragonal transformation in ZrO_2 or the ferrite-austenite transformation in iron and steel. We should, therefore, expect phase transformation temperatures to decline with a decline in particle size within the nano-regime. This has major implications for phase diagrams. The equation can then lead to **Thomson's law**, as

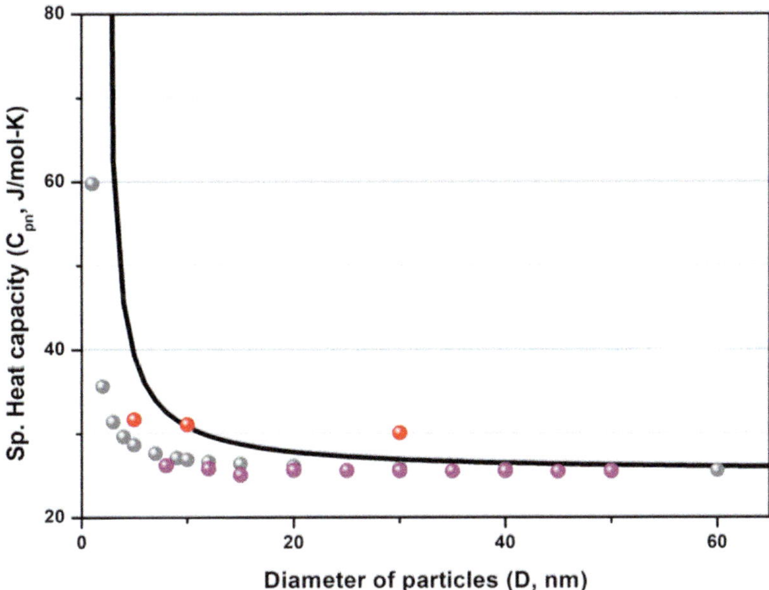

Fig. 14.12 Effect of silver particle size on specific heat capacity. Points in the figure are experimental and other simulation data from previously published sources included for comparison by Ansari (from *Manauwar Ali Ansari, Modelling of size-dependent thermodynamic properties of metallic nanocrystals based on modified Gibbs–Thomson equation, Applied Physics A (2021) 127:385. image reproduced under license,* http://creativecommons.org/licenses/by/4.0/.)

shown in Eq. 14.12, assuming the particle size does not change significantly through the transformation.

$$\Delta T = \alpha \frac{\gamma T_{\text{coarse}}}{D \Delta H} \tag{14.12}$$

where $\alpha = 1 - \beta = 1 - \left(\frac{\gamma_1}{\gamma_2}\right)\left(\frac{\rho_2}{\rho_1}\right)^{2/3}$

Because of this thermodynamics exercise, we find that the melting point of aluminum (660 °C) can be reduced to room temperature for nanoscale aluminum particles. Depending on the nanoscale particle size, the phase transformation temperature can be reduced by hundreds of degrees. Figure 14.11 shows published simulation results for four metals, showing how the melting point decreases with particle size. Also included by the authors is a comparison of the model with experimental and other simulation results (shown as points).

Another important outcome of reduced particle size is a change in the crystal structure. A particle of a material tends to revert to its high-temperature phase when the particle size is small enough. Examples are alumina (Al_2O_3) and Iron Oxide (Fe_2O_3), which occupy the hexagonal crystal structure at room temperature. Both revert to γ-Al_2O_3 and γ-Fe_2O_3 at high temperatures. The reduction in particle size within the nano-regime acts similarly to increasing the temperature in conventional materials and hence the particles undergo phase transformations. For both oxides at particle sizes equal to and lower than 20 nm the materials switch to the respective high-temperature γ-phase. Similar effects also occur in zirconia. By now, you would have realized that phase diagrams, which are normally constructed using composition on the x-axis and temperature on the y-axis, would now have to include particle size as a variable.

Other properties are also affected by a reduction in particle size within the nano-regime. These include heat capacity, which is the amount of heat needed to raise the temperature of a material of unit mass by 1 °C. The specific heat capacity of nanoparticles is larger than that of the same bulk-form material. Figure 14.12 shows the effect of particle size on specific heat capacity for silver nanoparticles. The solid curve is from a model, and the points are from other simulation results and experiments, which are also included in the figure.

Even thermal conductivity is affected by a reduction in particle size within the nano-range. Figure 14.13 suggests a reduction in thermal conductivity with a decline in nanometric particle size for silver.

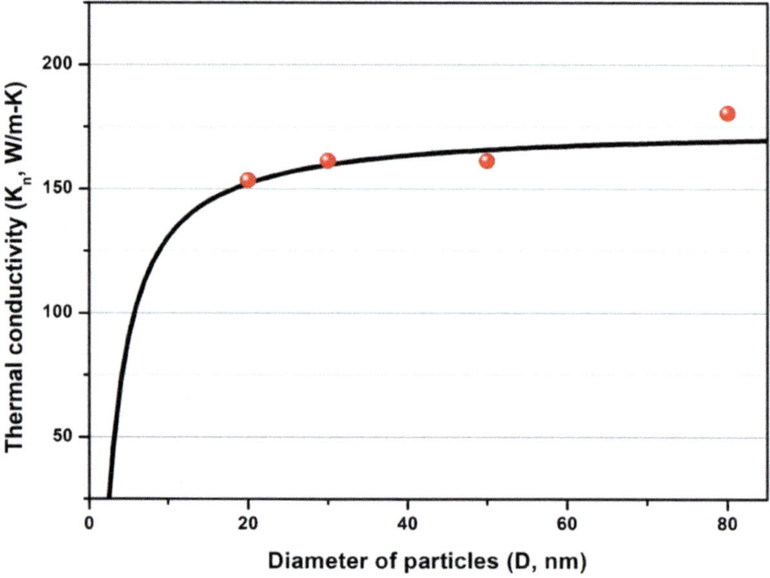

Fig. 14.13 Effect of silver particle size on thermal conductivity. Points in the figure are experimental data from previously published sources included for comparison by Ansari (from *Manauwar Ali Ansari, Modelling of size-dependent thermodynamic properties of metallic nanocrystals based on modified Gibbs–Thomson equation, Applied Physics A (2021) 127:385. image reproduced under license,* http://creativecommons.org/licenses/by/4.0/.)

Professor, I took the sample to the SEM and found that it charged a lot, making me think it was an electrical insulator. So, I coated the sample with a very thin coat of gold, which allowed an electron beam to pass over the sample surface and generate an image. I performed an X-ray microanalysis study over an area of the sample surface. I found the following elements: silicon and oxygen in significant quantities, followed by sodium, then calcium.

I also carried out an XRD scan on it, and the material is amorphous, as no clear peaks were observed; only a wide hump between 20 and 30 degrees (2θ) was observed in the scan.

You have done great work, Alex! You are almost a Pro!

14.7 Electrical Properties of Nanomaterials

Electrical conductivity in nanomaterials is affected by multiple factors and can be complicated. For one, electrical conductivity decreases due to the scattering of electrons by the extensive surfaces found on the nanoscale, which affects their conduction. Hence, if that mechanism is dominant, then electrical conductivity decreases with a decline in particle size. It is interesting to note that the reduction in particle size within the nano-regime can also influence the electronic structure, which can change the electrical characteristics of a nanomaterial and make materials less conductive with a decline in particle size. For example, below a critical diameter, a material can change from being a conductor to being a semiconductor, as shown for metal nanowires (i.e. undergoing a metal-to-semiconductor transition). An example is Bi nanowires, which go through such a transition below a diameter of 52 nm. Additionally, semiconducting nanowires can even become insulators below a critical diameter. Also, a semiconductor can become an insulator; for example, Si nanowires at or below 15 nm become insulating.

The electrical conductivity of MWCNTs has been reported to range from 10^3–10^6 S/m, and hence their addition to enhancing the electrical properties of materials has been investigated extensively, especially for non-conducting matrix materials such as polymers and ceramics (but naturally to a lesser extent metals). Notably, the electrical conductivity of metals can be comparable to or even better than CNTs. However, the application of CNTs to metals can enhance another critical property which is **Ampacity**, defined as the maximum current density that can be sustained before failure, which can be important for thin wires or thick and thin films. The ampacity of CNTs can be as high as 1×10^9 A/cm^2, which is orders of magnitude higher than that for metals, and hence, the addition of CNTs to metals can significantly increase the ampacity of the metal (by even 100 times).

The electrical conductivity of nanomaterials can also be affected by the degree of disorder. For crystalline nanoparticles, the degree of disorder is expected to increase with a reduction in particle size, as depicted by the Landau ordering parameter. Figure 14.14 shows such a trend for copper nanoparticles.

For polymers the situation can be somewhat different. For example, when polymer fibrils are reduced in diameter within the nano-regime, the individual molecules become more and more aligned as the fibrils' diameter is reduced, resulting in an

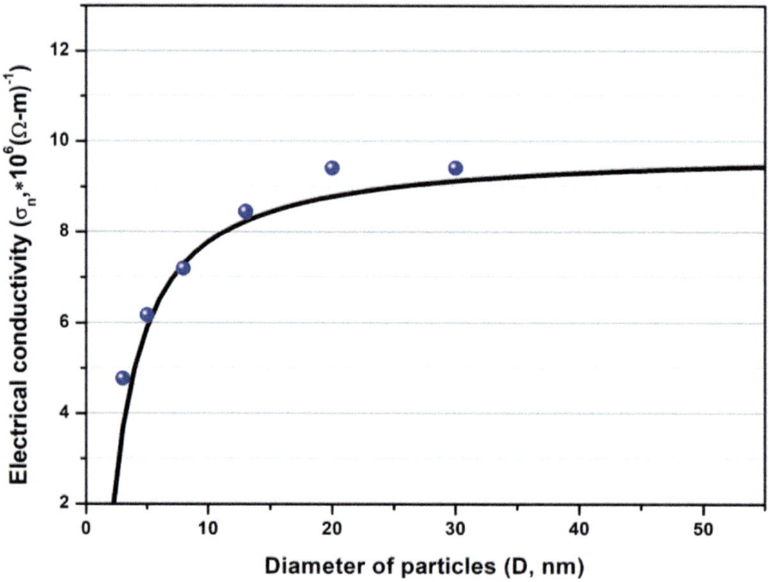

Fig. 14.14 Effect of copper particle size on electrical conductivity. Points in the figure are experimental data from previously published sources included for comparison by Ansari (from *Manauwar Ali Ansari, Modelling of size-dependent thermodynamic properties of metallic nanocrystals based on modified Gibbs–Thomson equation, Applied Physics A (2021) 127:385. image reproduced under license,* http://creativecommons.org/licenses/by/4.0/.)

increase in order. This leads to considerable gains in electrical conductivity. For example, the electrical conductivity of polyheterocyclic fibrils dramatically increased below a diameter of 500 nm. Going from around 200 Scm^{-1} to above 2000 Scm^{-1} at much smaller nanoscale fibril diameters.

All these facts mean that we need to be cautious when dealing with nanomaterials and not assume bulk properties for these materials.

14.8   0D, 1D, and 2D Nanomaterials

Here we discuss some notable examples of 0D, 1D, and 2D nanomaterials.

0D Nanoparticles

There are many examples of 0D nanoparticles; any spherical nanoparticle would certainly qualify, whether nanoparticles of gold, silver, ceramic, or even polymer. However, here we will direct our attention to carbon-based 0D nanoparticles, since there will be carbon-based 1D and 2D particles that we will discuss later. An example of 0D carbon-based nanoparticles is what we call **fullerenes**, which were discovered relatively recently in 1985. Specifically, the Buckminster*fullerene* C_{60} can be viewed as a hollow ball made from 20 hexagons and 12 pentagons of carbon. The subscript 60 refers to the 60 atoms of carbon from which the fullerene is made. The name Buckminster*fullerene* is after the American architect **Richard Buckminster Fuller**, the designer of Geodesic domes. This fullerene has also been named *Buckyball*. It is essentially a ~ 1 nm diameter soccer ball-shaped molecule. Figure 14.15 shows a schematic of several fullerenes including the C_{60} (which is the most abundant). C_{70} is the second most abundant fullerene, which can be viewed as a C_{60} molecule but with an extra 10 atoms added around the waist of the cage. In fact, it is, therefore, somewhat elongated.

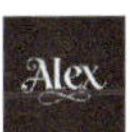

Professor, discovering a molecule should be worthy of a Nobel prize, don't you think?

You've got the right idea, Alex. Indeed, the discoverers of fullerenes (Smalley, Curl, and Kroto) received the Nobel Prize in Chemistry in 1996

All fullerenes consist of 12 pentagons but vary in the number of hexagons. As one can imagine the more hexagons there are, the greater the distance between the pentagons. Indeed, the **"isolated pentagon rule"** says that for fullerenes to be stable, the pentagon in the structure needs to be isolated. This happens when you have more and more hexagons separating them. It should follow that the greater the distance between pentagons (i.e., more hexagons), the more stable the fullerene is. The International Union of Pure and Applied Chemistry (*E. W. Godly and R. Taylor, Nomenclature and terminology of fullerenes: A preliminary study, Pure &Appl. Chem., Vol.*

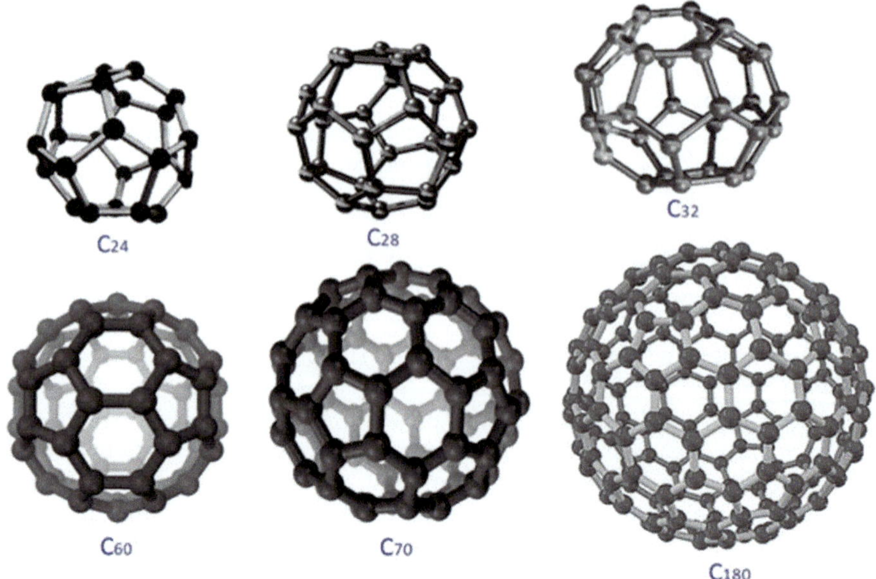

Fig. 14.15 Different types of fullerenes, from Kausar, A.; Ahmad, I.; Maaza, M.; Eisa, M. H. State-of-the-Art of Polymer/Fullerene C_{60} Nanocomposite Membranes for Water Treatment: Conceptions, Structural Diversity and Topographies. *Membranes* **2023**, *13*, 27. https://doi.org/10.3390/membranes13010027 (reproduced under Creative Commons Attribution (CC BY) license (https://creativecommons.org/licenses/by/4.0/)

69, No. 7, pp. 1411–1434, 1997.) put forward a definition for fullerenes and pseudo fullerenes. The definition is stated below:

> Fullerenes are defined as polyhedral closed cages made up entirely of n three-coordinate carbon atoms and having 12 pentagonal and (n/2-10) hexagonal faces, where n $\geq$ 20.
>
> Other polyhedral closed cages made up entirely of n three-coordinate carbon atoms shall be known as quasi-fullerenes

Hence, fullerenes are designated as C_n, where n $\geq$ 20, which means that C_{20} is the smallest possible fullerene. Although C_{20}, in accordance with the isolated pentagon rule, should be unstable (since hexagons do not isolate pentagons), however, researchers have managed to use approaches to stabilize it. Different types of fullerenes are shown in Fig. 14.15.

Alex, you should know by now that C_{20} is the smallest fullerene. Can you confirm the number of hexagons in C_{20} using the definition of fullerenes?

Since this fullerene has 20 atoms, it means n = 20. The number of hexagons is n/2–10 = 20/2–10 = 0. So, it has no hexagons, only 12 pentagons

Spot on, my friend. While the C_{60} has a diameter of ~ 1 nm, the diameter of C_{20} is approximately 0.5 nm. Its size and structure make it a potential candidate for many technological applications. Imagine drugs being encapsulated within its cage; it could then be used for targeted drug delivery. Don't forget, carbon is also biocompatible. This molecule is highly reactive and, hence, can easily bind to proteins. Due to its electronic properties, it can also be a great candidate for use in molecular electronics and quantum computing. Its reactive properties can be useful in wastewater treatments.

There are other types of fullerenes out there. For example, **nested fullerenes** are multilayered fullerenes with multiple layers where fullerenes are consecutively encapsulated in other fullerenes, like an onion. These can have applications in lubricants, which can reduce friction by the process of fullerene rolling. By the way, **Giant fullerenes** are large fullerenes containing 100 or more carbon atoms.

Solid/Crystalline Fullerenes

Individual fullerenes can bond together via van der Waals bonding into crystalline structures. For example, C_{60} fullerenes can bond together, forming an FCC structure. The properties of this crystalline phase are very interesting. The bulk modulus is about 18 GPa, and many other properties of the solid fullerenes resemble those of individual fullerene molecules. Table 14.4 shows some of these properties for C_{60} solid fullerene. Compared to diamond, which has a thermal conductivity of around 2000 W/m. K, the thermal conductivity of C_{60} solid fullerene is extremely low.

1D nanoparticles

Under this classification, we find carbon nanofibers, nanorods, nanowires, and nanotubes. Here, we will consider carbon nanotubes due to the extensive interest in such materials. Our starting point will be a graphene layer found in graphite.

Table 14.4 Some properties of C_{60} solid (*Compiled from different sources by D. R. Huffman in Synthesis, Structure, and Properties of Fullerenes, in Nanomaterials: Synthesis, Properties, and applications, edited by A. S. Edelstein and R. C. Cammarata, Institute of Physics Publishing, 1996.*)

Property	Value
Density	1.68 g/cm^3
Crystal structure	FCC
Lattice constant	1.417 nm
Bulk modulus	18 GPa
Electron band gap	1.85 eV
Thermal conductivity (at 300K)	0.4 W/m.K
Thermal expansion coefficient	6.2X10^{-5}/K

Figure 14.16 shows the structures of 3D graphite, consisting of parallel graphene layers. The carbon atoms within the graphene layers are covalently bonded; however, the bonding between graphene layers is weak secondary Van der Waals bonding. Imagine one 2D graphene layer as our starting point. Now, imagine this graphene layer being rolled into a tube. This would be referred to as a carbon nanotube, as seen in Fig. 14.16. Of course, carbon nanotubes are not produced in this manner, but this simple description is helpful in visualization.

The direction in which these graphene layers are rolled results in different carbon nanotubes with different chirality and properties. Carbon nanotubes can be classified according to their chirality. The chirality vector $c = ne_1 + me_2$ describes a carbon nanotube's chirality, with n representing the column number and m representing the row number. For example, for the blue vector c (4, 4) in Fig. 14.17, when the graphene layer is rolled along the c (4, 4) vector, a nanotube is formed **with** a chirality of (4, 4). Likewise, a CNT with chirality (9, 0), would have been rolled along the c

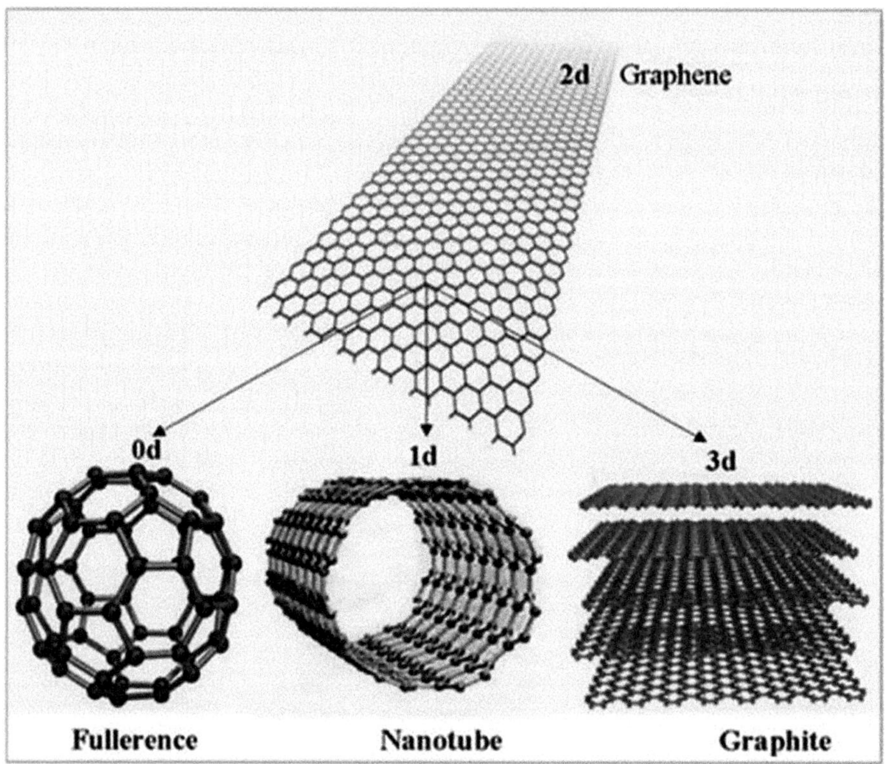

Fig. 14.16 Four forms of carbon 3D graphite (consisting of layers of graphene), 2D graphene, 1D carbon nanotube and 0D fullerene. (*Ibrahim, A., Klopocinska, A., Horvat, K., Abdel Hamid, Z. Graphene-Based Nanocomposites: Synthesis, Mechanical Properties, and Characterizations. Polymers 2021, 13, 2869.* https://doi.org/10.3390/polym13172869, *reproduced with modification under Creative Commons Attribution (CC BY) license* (https://creativecommons.org/licenses/by/4.0/))

(9, 0) vector. Note that the actual longitudinal axis of the formed carbon nanotube would be perpendicular to the chirality vector, as shown in Fig. 14.17. As can be imagined, many chiralities are possible; two well-known chiralities are the **zigzag** and the **armchair** carbon nanotubes. The zigzag CNT chirality is (n,0), and the zigzag shape is seen in the figure. For armchair CNTs, the chirality is (n, n), i.e., $n = m$; the shape is also red in the figure. From simple geometric considerations, it is possible to arrive at an expression for the diameter (d) of the carbon nanotube based on the chirality vector. This is presented in Eq. 14.13.

$$d = \frac{\sqrt{3}}{\pi} L_{c-c} \times \sqrt{(n^2 + m^2 + nm)} = 0.0783 \times \sqrt{(n^2 + m^2 + nm)} \qquad (14.13)$$

where, L_{c-c} is the distance between two adjacently bonded carbon atoms in the graphene layer, which is 0.14 nm. n and m are the column and row number respectively.

Example Problem 14.4
You are provided with a CNT having a chirality of (10, 10). Based on this chirality, what type of CNT is it? And what is its diameter?

Solution

Since the chirality is (10, 10), that means $n = m$. Hence it is an armchair CNT. We can calculate the diameter of the CNT using Eq. 14.13:

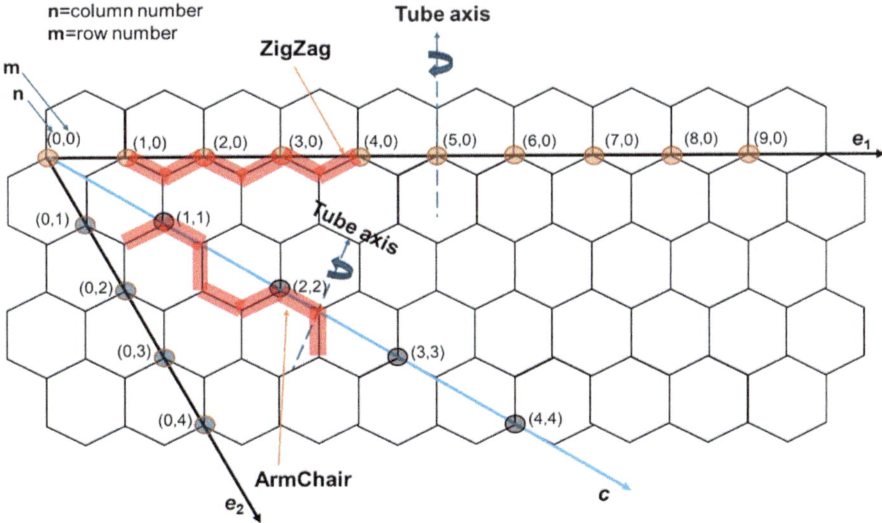

Fig. 14.17 Graphene layer with hexagonal structure showing chiral vectors, including armchair and zigzag carbon nanotubes (modified *based on D. Vollath, Nanomaterials: Synthesis, Properties and Applications, Wiley (2013)*)

$$d = 0.0783 \times \sqrt{\left(10^2 + 10^2 + 10 \times 10\right)} = 0.0783 \times \sqrt{300} = 1.36 \text{ nm}$$

The **chiral angle** (ϕ) is also used to describe nanotubes. This angle is the angle between the axis e_1, and the chiral vector c, and is given by:

$$\phi = \arctan\left[\sqrt{3}\,\frac{3}{2n + m}\right] \tag{14.14}$$

If the proper n and m values are inputted into Eq. 14.14, it would reveal that the chiral angle of zigzag CNTs is 0, while that of armchair CNTs is 30°.

The example presented is for a nanotube produced by the hypothetical rolling of one graphene layer. Such a nanotube is called a **single-wall carbon nanotube** and can have diameters between 0.6 and 2.0 nm. A **double-wall carbon nanotube** is one with two concentric tubes (a second graphene layer rolled around the first nanotube). A **multi-wall carbon nanotube** consists of more than two concentric nanotubes and can have a diameter typically between 5 and 100 nm. These nanotubes can also be capped at their ends with half a fullerene. Properties can vary depending on which kind of nanotube is being tested. Figure 14.18 shows a transmission electron micrograph of a multi-wall carbon nanotube.

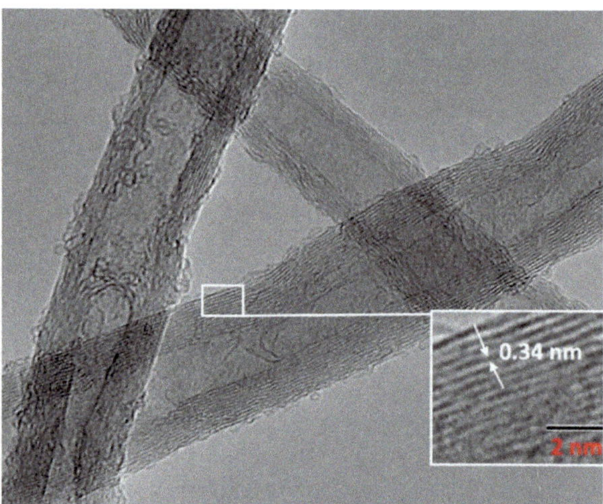

Fig. 14.18 Transmission electron microscopy of multiwall carbon nanotubes showing wall separations of 0.34 nm. Some amorphous regions are also observed. The scale bar in the inset is 2 nm. *The figure is reproduced and modified from Korusenko, P.; Kharisova, K.; Knyazev, E.; Levin, O.; Vinogradov, A.; Alekseeva, E. Surface Engineering of Multi-Walled Carbon Nanotubes via Ion-Beam Doping: Pyridinic and Pyrrolic Nitrogen Defect Formation. Appl. Sci. 2023, 13, 11,057.* https://doi.org/10.3390/app131911057. *Under Creative Commons Attribution (CC BY) license* (https://creativecommons.org/licenses/by/4.0/)

2D nanoparticles

Graphene is a notable example of a 2D nanomaterial. The graphene nanosheet consists of hexagons with each carbon atom covalently bonded with three other carbon atoms. This leaves one free electron, making the graphene nanosheet or nanolayer electrically conductive. These materials have been used to reinforce other materials and produce nanocomposites

14.9 Mechanical Properties of Carbon Nanotubes and Bulk Nanomaterials

14.9.1 Mechanical Properties of CNTs

One of the advantages of carbon nanotubes is their exceptional mechanical properties. For example, their Young's modulus is ~ 1TPa, compared to steel (210 GPa). Tensile strengths can be as high as 150GPa, orders of magnitudes greater than steel. Such properties are highly sensitive, though, to the presence of defects in the structure of the CNTs. Moreover, recent research has shown that tensile strength increases with a decline in the diameter of SWCNT and an increase in chiral angle, becoming the highest close to armchair designation. The Poisson's ratio of single-wall carbon nanotubes ranges from 0.149 to 0.34; a value of 0.27 has also been reported for double-wall and multi-wall carbon nanotubes (from 3 to 10 walls), which are all within the range of conventional materials. Table 14.5 lists and compares Young's moduli and tensile strengths of CNTs with those of metals and polymers. With such overall impressive mechanical properties, it is of no surprise that CNTs have been considered as reinforcements for many materials (including metals, polymers, ceramics, and intermetallics) to produce nanocomposites with better properties than the material receiving the nanotubes.

Mechanical Properties of Bulk Nanostructured Materials.

Consider nanostructured bulk materials with grain sizes less than 100 nm. If we take the Hall–Petch equation at face value, we will initially think that hardness and strength will keep increasing as the grain size is reduced below 100 nm. However, the Hall–Petch equation breaks down at a particular grain size within the nano-regime. The Hall–Petch equation relies on mobile dislocations (that can move under the influence of a critical resolved shear stress) interacting with grain boundaries. If a condition is reached where no mobile dislocations are present, fundamentally, the Hall–Petch equation should break down. This is indeed the case on the nanoscale. We need to understand that mobile dislocations can either be initially present in a material or generated (i.e., via the Frank-Read Source), for example, during cold working. Generating new dislocations below a critical grain size within the nano-regime is impossible. If the Hall–Petch equation were indeed valid throughout the

Table 14.5 Young's modulus and tensile strength data for carbon nanotubes and other materials. (data from *multiple sources, including F. Li, H-M Cheng, S. Bai, G. Su, M. S. Dresselhaus, Tensile strength of single wall carbon nanotubes directly measured from their macroscopic ropes, Applied Physics Letters, 77 (20), pp.3161–3163 (2000), W. D. Callister, D. G. Rethwisch, Materials Science and Engineering, An Introduction, 10th edition, Wiley, 2018, S. Thomas, et. al, Composite Materials, in Ullmann's Encyclopedia of Industrial Chemistry, 2016, Wiley–VCH.*)

Material	Young's modulus (TPa)	Tensile strength (GPa)
SWCNT	0.79–3.6	22
MWCNT	1.0	11–63
1020 steel	0.207	0.38
Aluminum	0.07	0.09
Titanium	0.11	0.52
Kevlar 49 improved	0.124	4.1 (impregnated strand)

nanoscale, the yield strength would increase 30 times from a 10 μm grain size to a 10 nm grain size. What happens, however, is that nanomaterials fall short of this expectation, meaning that there can be, in many cases, an increase in yield strength with a reduction in grain size (below a critical grain size), but not at the same rate as predicted by the Hall–Petch equation. This critical grain size is different for different materials. Nanostructured nickel and TiO_2 are examples of materials that exhibit such behavior. The failure of nanostructured materials to meet the expectations of Hall–Petch is due to softening processes taking place on the nanoscale, primarily due to the presence of an extremely large volume fraction of grain boundaries. Hence, grain boundary diffusion/creep plays a role. In some cases, the rate of hardening with a decline in grain size can reach a value of zero. Also, in some cases, such as for nanostructured copper and palladium, a reduction in yield strength is observed with a decline in grain size below a critical grain size. This is referred to as the **inverse Hall–Petch relation**. Figure 14.19 describes all these different dependencies. It is important to note that when nanomaterials first came on the stage, they were very expensive to produce and even took considerable time to produce. As such, most of the early investigations on the mechanical behavior of nanomaterials, or more specifically, nanostructured materials, used hardness testing instead of tensile testing (which was impractical at the time). Although the Hall–Petch equation was initially formulated as a relation between the yield strength of a polycrystalline material and its grain size, a similar relation between hardness and grain size is also valid. These are presented in Eqs. 14.15a and 14.15b.

$$\sigma_y = \sigma_o + \frac{k}{d^{1/2}} \tag{14.15a}$$

$$H = H_o + \frac{C}{d^{1/2}} \tag{14.15b}$$

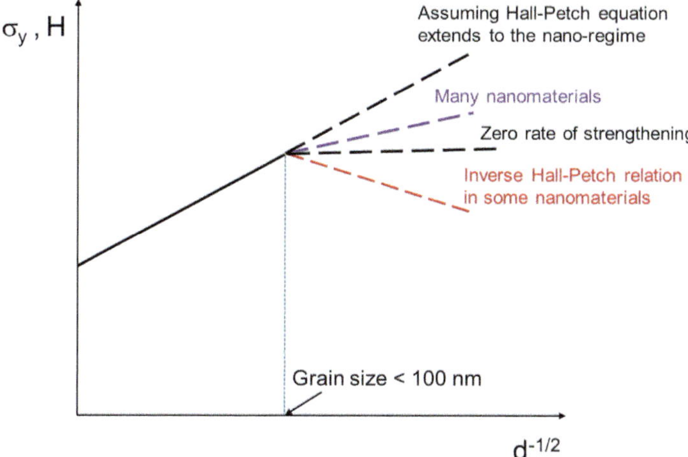

Fig. 14.19 Hall–Petch relation for conventional materials and nanomaterials

σ_y and H are the yield strength and hardness respectively. σ_o, H_o, k and C are constants.

14.9.2 Nanoindentation

With the advent of nanotechnology and the production of smaller and smaller components or structures, for example, thin films (typically less than 100 nm in thickness), it becomes essential to apply some sort of mechanical test to gauge their mechanical response and measure properties such as hardness and modulus of elasticity. Macro- and micro-hardness tests are out of the question due to the relatively enormous size of the indentation they leave behind. As such, nanoindentation involves indenting the surface with forces on the order of micronewtons, resulting in nano/micro-scale indentation sizes. These machines have shown great precision with position-sensing resolutions of 1 nm. Some machines today can measure loads from 500 pN (500 10^{-12} N) to 500mN (10^{-3}N), and displacements from < 0.1 nm to 500 μm. One of the outputs we can obtain is a map of hardness or modulus values over a particular area of a material surface, say 20 μm × 20 μm, which can show locations of high or low values. Depending on the options and modes, nanoindentation machines can measure the properties of materials that vary significantly in their mechanical behaviors, such as metals, ceramics, and polymers.

Figure 14.20 shows a schematic of the subsurface impression produced by a nano-indenter, one of which is a triangular diamond pyramid indenter called the **Berkovich indenter**. The figure shows the contours of the indenter itself in red, and several dimensions are also highlighted. $X_{\max}$ is the maximum depth the indenter reaches during loading, and X_c is the length over which the indenter is in contact with the material. X_f is the final depth of the permanent impression left behind after the load has been applied and removed. X_s is the distance over which the indenter is not contacting the material beneath the surface.

Figure 14.21a shows an electron micrograph of a Berkovich indenter, while Fig. 14.21b shows an impression left behind after the indenter is applied and removed. Figure 14.21c is a schematic of a force-indentation depth plot showing a typical loading and unloading curve during nanoindentation. It is assumed that the material during the loading curve experiences elastic and plastic deformation, whereas only elastic deformation is experienced during the unloading curve. This, therefore, provides a reasonable basis for measuring the stiffness (S) of the material using the unloading curve.

In the case of a Berkovich indenter, the unloading curve is not a straight line, and in fact the relationship between the force and depth has been shown to follow Eq. (14.16).

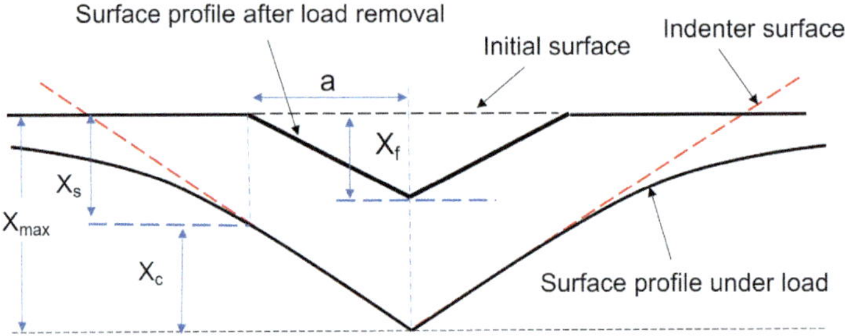

Fig. 14.20 Schematic of subsurface indentation impression under load and after the load has been applied and removed. (Based on: *W. C. Oliver, G. M. Pharr, An Improved Technique for determining hardness and elastic modulus using load and displacement sensing indentation experiments, Journal of Materials Research vol.7, no.6. 1992*)

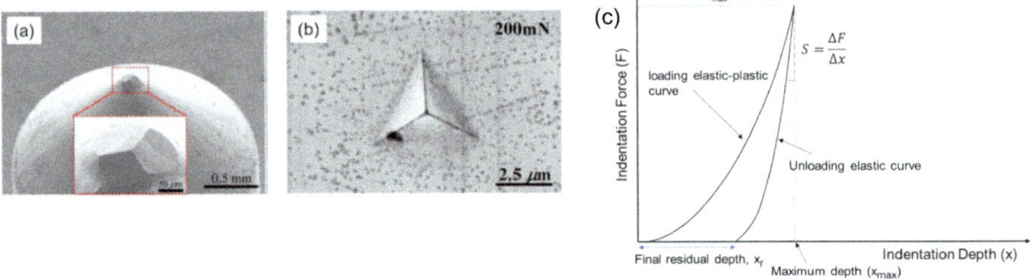

Fig. 14.21 **a** Electron micrograph of Berkovich indent, **b** impression left behind after indentation, **c** a schematic of loading and unloading curves produced during nanoindentation. (*Figures a and b slightly modified from Lixia Xu, Lingqi Kong, Hongwei Zhao, Shunbo Wang, Sihan Liu, and Long Qian, Materials, Mechanical Behavior of Undoped n-Type GaAs under the Indentation of Berkovich and Flat-Tip Indenters, Materials, 2019, 12, 1192;* https://doi.org/10.3390/ma12071192, *published by MDPI.*) Reproduced under creative commons license: http://creativecommons.org/ licenses/by/ 4.0/

$$F = \alpha(X - X_r)^m \tag{14.16}$$

where α and m are power–law fitting constants (Table 14.6 summarizes values for such constants for a few materials). X is the distance below the surface (X = 0 at the surface). It follows that the stiffness can be expressed using Eq. 14.17.

$$S = \frac{\mathrm{d}F}{\mathrm{d}x} = m\alpha(x - x_r)^{m-1} \tag{14.17}$$

Graphically, the stiffness (S) is obtained by taking the slope close to the start of the unloading curve, as seen in Fig. 14.21c. This stiffness has also been called **contact stiffness**. The maximum depth (X_{max}) and load (F_{max}) experienced during the test are recorded so is the final residual depth (X_r) after the load has been removed. An important parameter is also the contact depth (X_c), which is the distance over which the indenter is in contact with the material. This is important since it allows us to determine the contact area considering the geometry of the indenter. This depth can be calculated using Eq. 14.18.

$$X_c = X_{max} - \varepsilon \frac{F_{max}}{S} \tag{14.18}$$

For a Berkovich indenter, $\varepsilon = 0.75$.

The contact area A_c is calculated from X_c per Eq. 14.19.

$$A_c = 24.5 X_c^2 \tag{14.19}$$

The **hardness** is then simply calculated using Eq. 14.20:

$$H = \frac{F_{max}}{A_c} \tag{14.20}$$

Additionally, the effective elastic modulus is given by Eq. 14.21:

Table 14.6 α and m values for different materials (Source *W. C. Oliver, G. M. Pharr, Measurement of hardness and elastic modulus by instrumented indentation: Advances in understanding and refinements to methodology, Journal of Materials Research vol.19, no.1. 2004*, pages 3–20)

Material	α (mN/mm^m)	m
Aluminum	0.265	1.38
Soda lime glass	0.0279	1.37
Sapphire	0.0435	1.47
Fused silica	0.0500	1.25
Tungsten	0.141	1.51
Silica	0.021	1.43

$$E_{\text{eff}} = \frac{\sqrt{\pi}}{2} \frac{\beta S}{\sqrt{A_c}} \qquad (14.21)$$

Note that the elastic modulus is related to stiffness.

For a Berkovich indenter, $\beta = 1.034$, E_{eff} is the effective elastic modulus. However, the value of this measured modulus is affected by the modulus of both the material being tested and the indenter itself. The equation has been found applicable to a range of indenter geometries.

To obtain the actual Young's modulus of the material being tested, we must use Eq. 14.22.

$$E_{\text{eff}} = \left[\frac{1 - v^2}{E} + \frac{1 - v_{\text{indenter}}^2}{E_{\text{indenter}}} \right] \qquad (14.22)$$

where E and v are the Young's modulus and Poisson's ratio of the material being tested, and E_{indenter} and v_{indenter} are the Young's modulus and Poisson's ratio of the material from which the Berkovich indenter is made, which is diamond, hence E_{indener} = 1140 GPa and v_{indenter} = 0.07 respectively.

The main advantage of these analyses is that they allow the measurement of the Hardness and Young's modulus without the need to measure the area of the indentation left behind experimentally. Instead, the calculated contact area A_c is used, which is derived from the indentation contact depth, X_c. This is useful when the indentation size is small or when investigating the mechanical response of micro- and nano-features and components, such as thin films.

One of the very useful features of nanoindentation is the ability to produce a property map of a targeted region of the material's microstructure. Figure 14.22a shows such a map for an aluminum alloy 2024 in a region of the microstructure where a hard intermetallic particle is present. Note that the blue region is the hardness of the aluminum alloy, which surrounds the region where the particle is (which has higher hardness values). Figure 14.22b shows the same, but this time, the modulus within the region is being mapped.

14.9.3   Superplasticity

Regarding ductility, typical values for engineering metals and alloys regarded as being ductile can generally range from 20 to 70% (percent elongation to fracture). Imagine, however, if we can obtain ductility values of 100's of percent or even 1000%; the name **superplastic** suddenly seems appropriate to describe such materials. Super-plastic behavior has been observed for nanostructured materials, including ceramics; one notable example is the superplastic forming of nano-structured zirconia, which occurs at high temperatures under load and low strain rates.

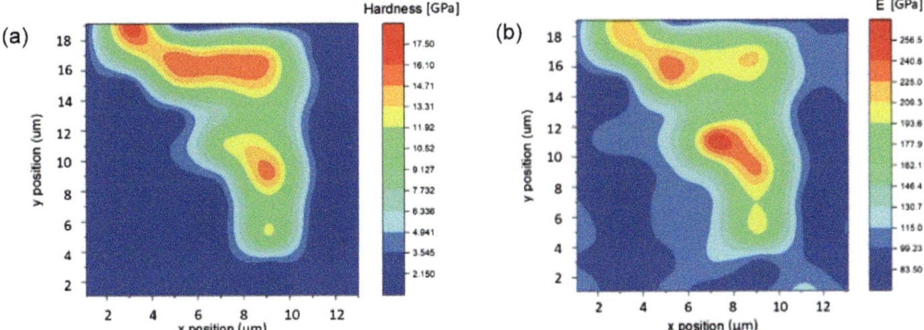

Fig. 14.22 Nanoindentation mapping of **a** hardness and **b** young's modulus across a 13 μm × 19 μm area of the microstructure of a 2024 Aluminum Alloy reinforced with an intermetallic particle. The high hardness and modulus values are in the vicinity of the intermetallic particle. (*Anna Staszczyk, Jacek Sawicki, Łukasz Kołodziejczyk and Sebastian Lipa, Nanoindentation Study of Intermetallic Particles in 2024 Aluminium Alloy, Coatings, 10, 846, (2020)* https://doi.org/10. 3390/coatings10090846). Reproduced under creative commons license: http://creativecommons. org/licenses/by/4.0/

14.9.4   Superparamagnetism

Due to the increased surface area and hence surface energy with a reduction in particle size in the nano-regime, ferromagnetic materials become paramagnetic, allowing spontaneous switching of directions of polarization. What is interesting is that while microscale ferromagnetic particles contain multiple magnetic domains with their accompanying domain walls, as the particle size is reduced, the number of domains decreases until, at some point, the particle contains only a single magnetic domain. This can happen at a particle size typically between 50 and 100 nm. The coercivity at this stage is large (representing the magnitude of the magnetic field needed to bring the material's magnetization to zero). However, as the particle size decreases further, the coercivity keeps decreasing with particle size until some critical particle size is reached, where the coercivity becomes zero. The particles at this critical size and below are called **superparamagnetic.** They have many applications in medicine for treating cancer and, in contrast dyes used in MRI.

Professor, I carried out the resonance frequency test to measure the modulus of this sample, and it is 70 GPa

In all, indications my friend points to soda-lime glass as the identity of Layer 1. Don't be surprised that it qualified to be included in the Sphere. Glass was always available to mankind, even during the Stone Age. Remember Obsidian rock? To be more specific, it is nothing but dirty, volcanic glass. Over time, glass makers produced extremely colorful glass by adding certain additives, to produce vases, and other beautiful artifacts. Surprisingly, it was only in the fifteenth century that transparent glass was invented. Later, glass was used in so many essential things. Could Galileo have looked at the stars and planets without glass lenses in his telescope? To comprehend how an incredible material glass is, look around you. What would our homes look like without glass windows that protect us from external elements while providing light and the ability to see outside our house? What about people with eyesight problems? How would some of them be able to read? Or function? Of course, now there are many types of glass. My friend, optical fibers are extremely important, and glass is all around us. It certainly qualifies, in my opinion. But why isn't there a glass layer 1? That was the question we wondered about early on in our journey. Well, while you were busy with your investigations. I took half of the sphere to a good friend specializing in surface coatings. He determined that the inner surface of the Sphere cavity was coated with a 50 nm glass coat. This means that layer 1 not only identifies glass as a significant material in our history. However, it also alludes to thin film technology, which has had a significant impact on present-day technology. You know, Alex, as Professors, our job is to teach students and provide them with the skills for lifelong learning. I would highly advise you to research this important topic. Remember, thick and thin film technology!

I got it, Prof. This was so fascinating. I also realized that glass is present in the doors of microwaves and ovens. They must have a low thermal expansion coefficient to resist heat and thermal shock, something I learned from As a Matter of Fact! I intend to learn all about thin and thick films! Now, I have one more layer to identify!

Problems

14.1. What is the general definition of a nanomaterial.

14.2. Give examples of 0D, 1D, 2D and 3D nanomaterials.

14.3. Consider a 5 nm diameter nanoparticle. If the surface thickness is considered to be 0.5 nm, what volume fraction of the particle is occupied by surface atoms.

14.4. What is the specific surface area (in m^2/g) of a spherical silver nanopowder with an average diameter of 5 nm? (Hint: you can search for the density of silver in reliable sources.)

14.5. Consider 5 g of Ni (density 8.9 g/cm^3) nanopowder 25 nm in diameter. What is the specific surface area of this powder and the total surface area of 5 g of powder?

14.6. If the specific surface energy of the powder in problem 14.5. is 2.38 J/m^2, what is the total surface energy content in the 5 g of Ni nanopowder?

14.7. How many hexagons does the C_{60} fullerene have?

14.8. What is the difference between a single-walled carbon nanotube and a multi-walled carbon nanotube?

14.9. Describe the difference between a zigzag and an armchair carbon nanotube.

14.10. Draw the chirality vector of a (5, 5) armchair carbon nanotube.

14.11. What type of CNT is a CNT with a chirality of (20, 20)? Explain.

14.12. What is the diameter of a CNT with a chirality of (35, 35)?

14.13. Determine the chirality angle of a CNT having a chirality of (0, 15).

14.14. What is the inverse Hall–Petch relation, and why does it occur?

14.15. A material having a modulus of 60GPa and Poisson's ratio of 0.32 is nanoindented with a diamond Berkovich indenter. If the modulus of the indenter is 1140 GPa and its Poisson's ratio is 0.07, determine the effective modulus.

14.16. What is meant by superplasticity?

14.17. Explain superparamagnetism as it relates to nanoparticles.

14.18. What is the relation between particle size and melting point of nanoparticles within the nano size range?

14.19. What is the relation between heat capacity and particle size in the nanorange?

14.20. As particle size decreases within the nanorange, what happens to the electrical conductivity of polymer fibrils and copper nanoparticles? Discuss the underlying mechanisms for each behavior.

Part VII

Layer 7: This is Cutting Edge Technology

The Sphere

Layer 7
"This is Cutting-edge Technology."

I believe I have learned a lot, Prof. In fact, I feel I'm ready to reveal the identity of Layer 7 all by myself ☺. I carried out many tests and did some research, and I believe I can say what this material is.

Hit me with it, Alex!

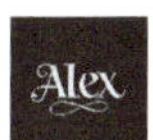

The following are some of the properties of the layer 7 sample.
Color: silver-grey.
Density: 14.8 g/cm^3
$K_{IC} = 9.4$ MPa.m$^{1/2}$

*I also observed the microstructure using the SEM. Unfortunately, the image capture feature of the SEM was down. However, I did sketch the microstructure; here it is in **Fig. 1**.*

Fig. 1 Sketch of
microstructure of sample 7

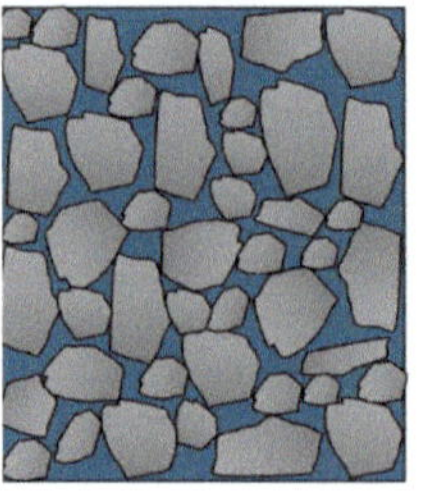

Moreover, I also carried out point X-ray micro-analysis. I discovered that the blue region in the sketch was cobalt, and the grey particles had a lot of tungsten and carbon. So, I took the sample and carried out an X-ray diffraction scan. The peaks that came out were for WC and cobalt. Professor, I can report that sample 7 is none other than a WC-Co cermet, i.e., a cemented carbide. Although it does appear that the cobalt is a little more than usual.

Alex, I can't tell you how proud I am of you. You are talking like a true scientist. Indeed, you are correct. It is a tungsten carbide-cobalt composite. The material's hardness, among other properties, has prompted its use as a cutting tool material. They are also used as drill bits for oil drilling. But did you know this material is regarded as the most important triumph of the powder metallurgy field? Wow, I guess we are all done!

Professor, it would not be right to leave you without summarizing the results of our adventure.
I made a table that summarizes our findings.

Layer	Material
1	Soda-Lime Glass and thin film (*without glass no telescopes, eyeglasses, windows in houses, planes, cars,…etc. and thin film technology is now critical in modern electronics and more*)
2	Cu-As alloy from the Chalcolithic period (*the start of smelting, which influenced the Bronze and Iron Ages*)
3	Tennalum® 7068 (*the strongest aluminum on Earth*)
4	Steel from the Hull of the Titanic (*Also, a reference to Steel, one of the most impactful materials of all time.*)

(continued)

(continued)

Layer	Material
5	Kevlar *(Saving human lives)*
6	Bipolar junction transistor (Si) *(Transistors are probably the most important invention of our time)*
7	WC-Co Cermet *(The pride and joy of Powder Metallurgy)*

You concluded the adventure like a Pro. It was quite an adventure, Alex. Thank you for making me part of it. I hope you are a better scientist for this experience. Indeed, these are fantastic materials that have had significant impacts on our humanity. I'm sure others may argue that other materials are worthy of inclusion in the Sphere, but it has undoubtedly been one of the most enjoyable periods of my career.

The most important thing, Alex, is to always seek knowledge and truth in everything through evidence and not hearsay. Only when you acquire true knowledge can you help other seekers of knowledge and benefit mankind. Stay humble and patient; your journey is not over yet.

Since layer 7 turned out to be a composite, why don't you increase your knowledge of composites by reading Chap. 15: Introduction to Composite Materials?

Let's meet in my research lab on Monday and tell the rest of the research group and faculty colleagues all about our adventure. You can prepare a PowerPoint Presentation; I'll bring coffee and sweets! I wish you all the best for your career, Alex. Hopefully, you and your work will be a beacon of light for future generations!

By the way, I also teach a powder metallurgy class. I would advise you to register for it and learn how very important engineering components are made from powder. Bye for now!

Chapter 15
Introduction to Composite Materials

15.1 Introduction

Composites have been around for thousands of years. Today, they are found in planes, automobiles, sports equipment, boats, and many other products. This chapter discusses composites, their classification, and their properties. Finally, nanocomposites are discussed.

15.2 What is a Composite, and What is Its Importance?

We had previously discussed the concept of alloying, which involves the atomic mixing of elements. However, it is also possible to produce a material with a microstructure composed of two or more different types or classes of materials (not atomically mixed). We have seen how every material class has advantages and disadvantages. This goes for metals, ceramics, and polymers. It is, however, possible to construct a material with a microstructure that contains both a metal and a ceramic constituent, or a polymer and ceramic, or a polymer and a metal, and so on. This way, we can combine the advantages of different classes of materials while reducing or eliminating their disadvantages. For example, a metal can be great in conducting electricity and heat but poor in resisting wear. Here, adding ceramic particles to the metal's microstructure provides the advantage of wear resistance that ceramics offer. The material produced is then called a **composite**. The material making up most of the composite is called the **matrix**, while the minor constituent is called the **reinforcement**. Composites are identified as follows. A **metal matrix composite** has most of its microstructure made out of a metal, a **polymer matrix composite** has most of its microstructure made out of a polymer. Finally, a **ceramic matrix composite** is one where most of the microstructure is made of a ceramic. Reinforced concrete is an example of a ceramic matrix composite, in which concrete is reinforced

© The Author(s), under exclusive license to Springer Nature Switzerland AG 2026
K. Morsi, *An Engaging Approach to the Science and Engineering of Materials*,
https://doi.org/10.1007/978-3-032-06231-4_15

with steel bars, rods, or meshes to produce materials with much-improved properties over just concrete. It is also a major building material today. Yet the concept of a composite material was present thousands of years ago, with mud mixed with stones and straws to produce house bricks to construct dwellings. Steel with a dual or multiphase microstructure may also be considered a composite. Pearlite is also a layered composite structure of alternating ferrite and cementite layers. An alumina-zirconia ceramic is also a composite, and so on.

Professor, it does seem that the potential of composite materials is enormous. This must undoubtedly be one of the greatest triumphs of mankind.

Believe me, Alex, if I could speak to the people who first produced and used composites, I would say to them, "What took you so long?" You see, composites have always been around since the beginning of life as we know it. Did you know that there are 'natural' composites all around us? And you don't need to look far. Your bones are natural composites made of protein, collagen, and mineral apatite. Wood from trees is another prime example of a natural composite of cellulose fibers and lignin. As we discussed earlier, there are almost endless natural materials to ponder and learn from within our world. Not only composite materials, but also other materials, including foams and even nanomaterials. The field of biomimicking capitalizes on this fact.

Composite design allows the engineer to tailor the properties of engineering materials in a versatile way unmatched by any conventional material, irrespective of the class of that material. These properties include thermal properties (such as thermal conductivity, specific heat capacity, and thermal expansion coefficients) and electrical properties such as electrical conductivity. Mechanical properties are also greatly influenced and controlled through composite design. Such properties include strength, fracture toughness, modulus of elasticity, wear resistance, and more. The engineer is left with a plethora of options at his or her disposal. Which matrix and reinforcement material to use? In terms of reinforcements, we have many options; for example, what size reinforcements do we add to the matrix? Macro-scale reinforcements, as in steel-reinforced concrete? Micro-scale reinforcements, as in SiC particle-reinforced aluminum composites, used as vehicle brake discs? Or nanoscale reinforcement, such as in the case of carbon nanotube reinforced aluminum nanocomposites? What about the shape of these reinforcements? They can be spherical particles, irregular-shaped particles, discs, short fibers, continuous fibers, meshes, etc. Each one of these choices will generate a whole set of properties. We have not even talked about the volume fraction (content) of the added reinforcements; do we add 5% reinforcement or 40%? I think by now, the vast scope of properties that can be achieved through composite design is clear. These properties lead to a broad scope of applications, some of which are listed in Table 15.1. For example, carbon–carbon composites are known for their very high thermal conductivities, which reduce temperature build-up on exiting and re-entering the Earth's atmosphere by space shuttles. This property makes them suitable for applications such as nose cones in space shuttles. **Borsic fiber** reinforced aluminum finds applications as fan blades in engines. Borsic fibers are boron fibers coated with SiC for protection. The low densities of polymers when stiffened with low-density reinforcements have made possible applications that include the shafts in golf clubs. When the weight of such shafts is reduced by using these polymer composites, it allows the actual club head to be heavier. Polymer composites are still, however, limited to low-temperature applications due to the vulnerability of the polymer matrix to degradation at high temperatures.

Figure 15.1 shows examples of matrices that can benefit from adding reinforcements to produce composites. Examples of metal matrices include aluminum, nickel,

Table 15.1 Examples of some applications for composites (*Multiple Sources, including D. R. Askeland and W. J. Wright, Essentials of Materials Science and Engineering, 3rd edition (SI edition), Cengage learning, (2014), S. Dhanasekar, Arul Thayammal Ganesan, Taneti Lilly Rani, Venkata Kamesh Vinjamuri, Medikondu Nageswara Rao, E. Shankar, Dharamvir, P. Suresh Kumar, Wondalem Misganaw Golie, A Comprehensive Study of Ceramic Matrix Composites for Space Applications Advances in Materials Science and Engineering, Vol. 2022, Article ID 6160591, 9 pages* https://doi.org/10.1155/2022/6160591)

Composite	Applications
Polymer matrix composite	
Kevlar reinforced epoxy or polyester	Aviation and aerospace applications, golf club shafts, fishing rods, tennis rackets
Graphite reinforced polymer	Automotive, aviation and aerospace applications in addition to sporting goods
Glass fiber reinforced polymer	Automotive, marine and water applications, sporting equipment
Metal matrix composite	
Borsic fiber reinforced aluminum	Fan blades in engines, aviation and aerospace applications, struts used in space shuttle
Carbon fiber reinforced aluminum	Hubble telescope antenna mast
SiC fiber and whisker reinforced aluminum	Missile fins, Stiffeners
SiC fiber reinforced copper alloys	Ship propellers
Alumina fiber reinforced aluminum	Pistons in diesel engines
SiC particle reinforced aluminum	Brake discs
Pb-PbO	Battery grids
Ceramic matrix composite	
Carbon–Carbon composites	Nose cones in space shuttle, brake discs in in commercial jet planes and racing cars,
Carbon-SiC composites	Brake disks for aircrafts and top-end cars
Zirconia-alumina composites	Cutting tools, thread guides, nozzles, coatings

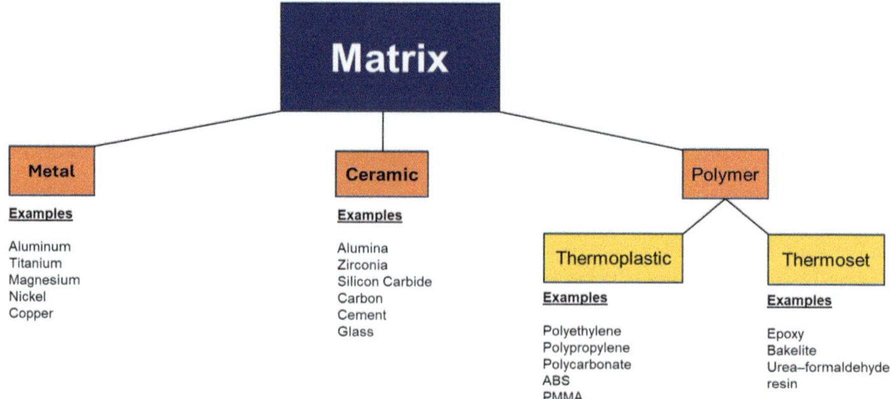

Fig. 15.1 Three basic matrices found in composites. (*Partly based on Sivasubramanian Palanisamy, K. Mayandi, V. Arumugaprabu, N. Rajini, Rajesh Shanmugavel, History of Composites and Polymers, Polymer-Based Composites, CRC Press, (pp.1–21) July 2021,* https://doi.org/10.1201/9781003126300-1)

magnesium, titanium, and their alloys. The properties of the matrix can be controlled through heat treatment, mechanical deformation, and compositional variations, thus providing the material designer with an added degree of tailorability of properties. Examples of ceramic matrices include alumina, zirconia, silicon nitride, carbon, and silicon carbide, to name only a few. As far as polymers are concerned, matrices can be thermoplastic (e.g., Nylon, polyethylene, ABS, and PMMA) or thermosetting (e.g., Epoxy and Bakelite).

15.3   Mechanical Properties of Composites

Examples of properties for several matrix materials are given in Table 15.2.

As mentioned, to produce a composite, we must reinforce the matrix material with a reinforcing material (*reinforcement*). Reinforcements are typically divided into different types, as seen in Fig. 15.2.

A material can be reinforced by adding whiskers, particles, or fibers. In addition, structural composites can also be produced, as will be explained later. It is important to note that reinforcements that can be used successfully to reinforce the matrix material must be chemically stable with the matrix during service and manufacturing. An example of unsuitable reinforcement is SiC when it is added to the Ni_3Al intermetallic compound. During high-temperature processing to produce the composites, SiC can dissociate, leaving a product free of SiC. So, selecting the correct reinforcement in terms of chemical stability with the matrix and other factors, such as their thermal expansion coefficient, is important. Let us now discuss each of the reinforcement types presented in Fig. 15.2.

Table 15.2 Mechanical properties of potential matrix Materials. *(Source J.F. Shackelford, Introduction to Materials Science for Engineers, 7th edition, Pearson-Prentice Hall, (2009))*

Matrix	Young's modulus(GPa)	Tensile strength (MPa)	Fracture strength (MPa)
Polymer			
Epoxy	6.9	69	–
Polyester (thermoset)	6.9	28	–
HDPE	0.83	28	–
LDPE	0.17	14	–
Polycarbonate	2.4	62	–
Metal			
Al	69	76	–
Cu	115	170	–
Ti–6Al–4 V	110	895	–
AZ31B magnesium	45	290	–
Ceramic			
Al_2O_3 (5% porosity)	380	–	210–340
SiC (~ 5% porosity)	470	–	170
Reaction bonded Si_3N_4	–	–	260
Boron carbide (~ 5% porosity)	290	–	340

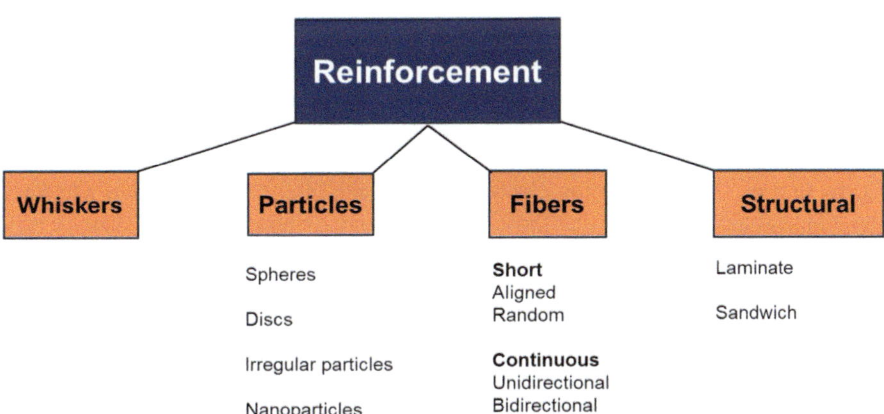

Fig. 15.2 Classification of the different types of reinforcements *(Partly based on Sivasubramanian Palanisamy, K. Mayandi, V. Arumugaprabu, N. Rajini, Rajesh Shanmugavel, History of Composites and Polymers, Polymer-Based Composites, CRC Press, (pp. 1–21) July 2021,* https://doi.org/10.1201/9781003126300-1

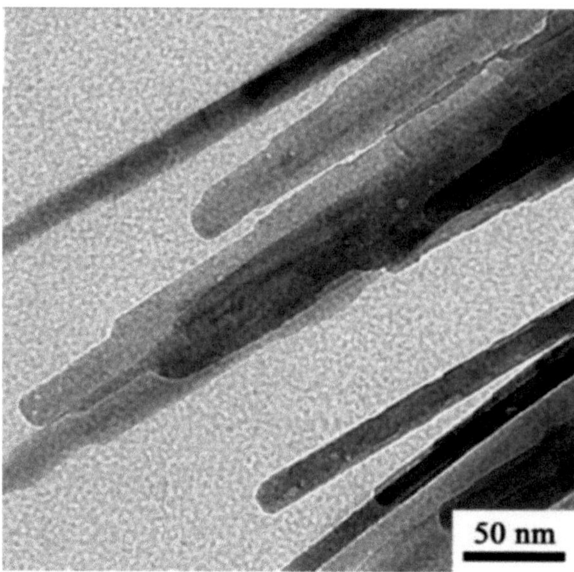

Fig. 15.3 High resolution transmission electron micrograph of θ-Alumina whiskers, note the 50 nm scale bar in the image. (*Image reproduced and modified from Liao, N.; Su, X.; Zhang, H.; Feng, Q.; Grasso, S.; Hu, C. Synthesis of θ-Al₂O₃ Whiskers with Twins. Metals 2021, 11, 895.* https://doi.org/ 10.3390/met11060895, *under the terms and conditions of* the Creative Commons Attribution (CC BY) license (https://creativecommons.org/licenses/by/4.0/)

Ceramic whiskers are typically single crystals grown in a specific crystallographic direction. Their size can vary between 50 nm and 10 μm in diameter. Their lengths can also vary between a few micrometers and a few centimeters. What is so special about these reinforcements is that their inherent flaw sizes are quite small due to how they were synthesized, giving rise to excellent strength and mechanical properties. Figure 15.3 is a high-resolution electron micrograph image of θ-Alumina whiskers. The 50 nm scale bar shows that these whiskers have a diameter of less than 50 nm.

Table 15.3 shows examples of some ceramic whiskers and their properties. Note the extraordinarily high strength values even of graphite whiskers. Now, if added, one can only assume that such whiskers can significantly enhance the mechanical properties of metals, resulting in high-performance whisker-reinforced metal matrix composites. Even when used to reinforce ceramics, fracture strength and toughness improvements have been realized. For example, adding 20 vol% SiC whiskers to alumina (Al₂O₃) can double the fracture toughness of alumina and significantly increase its fracture strength. However, dealing with fine whiskers can be a **serious** health hazard as they have been known to lodge in the lungs if inhaled. Hence, dealing with them in the manufacturing of composites requires significant precautions. The mechanical properties of fibers are far inferior, as also shown in the table. This is due to their larger inherent flaw sizes.

Table 15.3 Properties of some ceramic Whiskers and Fibers. (*Source R. A. Lowden, Characterization and Control of the Fiber-Matrix Interface in Ceramic Matrix Composites, ONRL, report no. TM-11039, March 1989, and W. F. Smith, J. Hashemi, Foundations of Materials Science and Engineering, 5th edition, McGraw Hill (2010)*)

Whisker	Density (g/cm^3)	Tensile strength (GPa)	Young's modulus (GPa)
Silicon Carbide (SiC)	3.2	21	840
Graphite	1.7	21	700
Silicon nitride (Si$_3$N$_4$)	3.2	14	380
Alumina (Al$_2$O$_3$)	3.9	21	430
Boron carbide (B$_4$C)	2.5	14	480
Fiber			
Alumina	3.9	1.4	385
Boron carbide	2.4	2.275	90
E-Glass	2.54	3.1	76
CVD SiC monofilaments	3.2	3.45	415
High strength carbon	1.8	2.76	275
High modulus carbon	1.8	1.9	530
Boron nitride	1.9	1.38	90
Aramid (kevlar 49)	1.44	3.6	131

Particles can provide another alternative to whiskers; such particles can have various shapes (spheres, discs, or other irregular equiaxed shapes), sizes (even down to the nanoscale), and compositions (different types of ceramics, for example). Figure 15.4a shows a schematic of a particle-reinforced composite microstructure showing the matrix and spherical reinforcements. Figure 15.3b, c shows what can be varied. In Fig. 15.4b, the volume fraction of the reinforcements is changed, and in Fig. 15.4c, the size of the reinforcements is changed. The reinforcements' size, shape, and composition can all play major roles in determining the properties of the final composite as the matrix does. One of the advantages of particles as reinforcements is their ease of incorporation in materials as opposed to whiskers; hence, simpler processing and composite manufacturing methods suffice.

If we had to classify these particles, we could call them 0D particles (excluding discs). Typically, these particles have improved properties over the matrix material. In terms of mechanical properties, we can improve the Young's modulus of a material by reinforcing it with higher Young's modulus reinforcements to produce a composite material with an improved modulus compared with the matrix material. Here, we can use the **Rule of Mixtures** to estimate the Young's modulus of the composite in terms of the volume fraction of the reinforcements. We can calculate an upper bound (UB) and a lower bound (LB) for Young's modulus, and the experimental results should lie in between these two values. The upper and lower bounds can be determined using Eqs. 15.1 and 15.2, respectively.

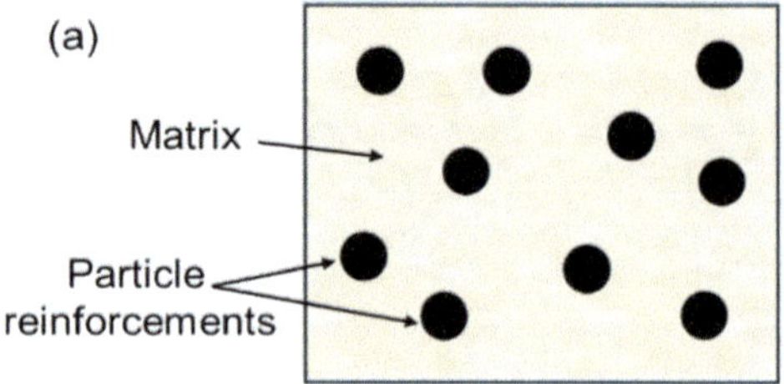

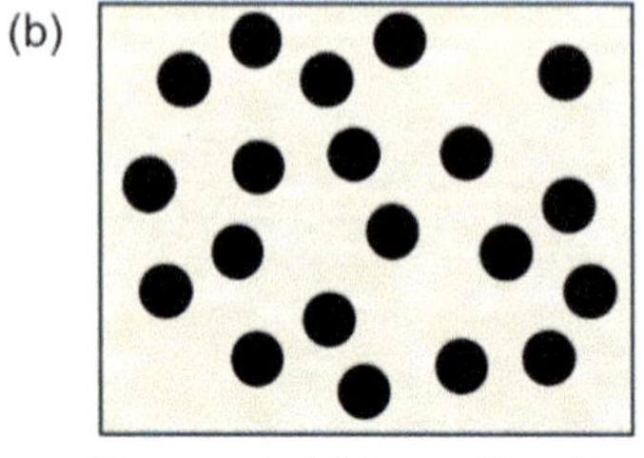

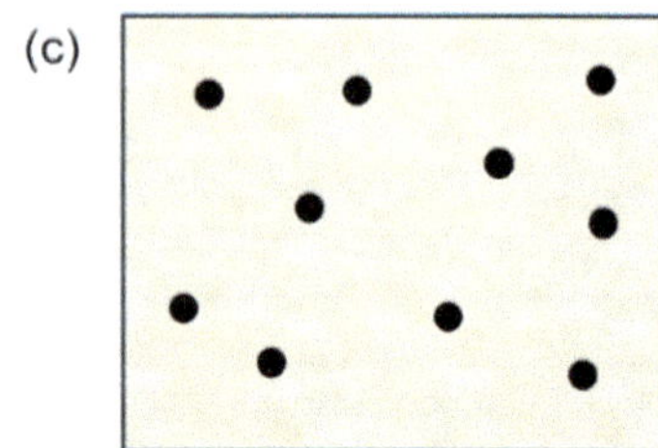

Fig. 15.4 a Particle-reinforced composite with matrix and particle reinforcement, **b** change in volume fraction of reinforcements or **c** change in size of the reinforcements can lead to significant changes in properties

$$E_{c(UB)} = E_p V_p + E_m V_m \tag{15.1}$$

$$E_{c(LB)} = \frac{E_p E_m}{E_p V_m + E_m V_p} \tag{15.2}$$

Note that $V_p + V_m = 1$.

Example Problem 15.1

A lightweight component is required with a Young's modulus of 120 GPa. Aluminum-SiC has been suggested as a possible solution. Is this possible, and if so, what minimum volume fraction of SiC is needed to achieve this? Note that Young's modulus of aluminum is 70 GPa, and that of SiC is 410 GPa.

Solution

Using Eq. 15.2 for the lower bound Young's modulus, we have:

$$E_{c\ (LB)} = 120\ \text{GPa} = \frac{E_p E_m}{E_p V_m + E_m V_p} = \frac{E_p E_m}{E_p\left(1 - V_p\right) + E_m V_p}$$

$$= \frac{410 \times 10^9 \times 70 \times 10^9}{410 \times 10^9 \times (1 - V_p) + 70 \times 10^9 \times V_p}$$

Solving for V_p gives:

$$V_p = 0.5.$$

Since we used the lower bound rule of mixtures, adding 0.5 volume fraction of SiC to aluminum would result in a Young's modulus of <u>at least</u> 120GPa for the Al–SiC composite. So, the answer is yes; it is possible to produce a composite with Young's modulus of 120 GPa. This would happen if we added 0.5 volume fraction of SiC. Of course, we did not use the upper bound rule of mixtures, so it may be possible that a SiC content of less than 0.5 could still satisfy the requirement. However, the solution we did was more of a conservative approach.

It should be noted that the rule of mixtures can also be used to estimate other properties for composites, including strength, thermal conductivity, electrical conductivity, density, and thermal expansion coefficient.

Professor, sometimes I come across composites where the amount of reinforcement is given in weight fraction or weight percent, but you seem to be referring to volume fraction or volume percent all the time. Why is that?

Great question, Alex: Do you see if I gave you two composite samples, one with 10 wt% carbon fibers and the other with 10 wt% glass fibers? We really wouldn't be able to compare appropriately their properties. Since carbon is typically lighter than glass, the number of carbon fibers weighing 10% of the composite's weight would be much more than the glass fibers. Each glass fiber would weigh much more than carbon fiber, and hence, you would be able to achieve the 10% (i.e., 0.1 weight content) with much fewer glass fibers than carbon fibers. In other words, less composite volume would be occupied by glass fibers instead of carbon ones. Using weight percent to describe composites may only, I guess, be applicable when you are trying to weigh the different constituents to manufacture the composite in the first place. Still, volume percentage shows you how much of the polymer volume is occupied by reinforcements (e.g., fibers) and is, hence more useful in calculating the overall density of the composite, among other properties. This would be the same when adding whiskers or even just particles. As I remember, there is an example in As a Matter of Fact that deals with this, if I recall, it was Example Problem 15.2. Why don't you look at it?

It is often required to convert the weight fraction of reinforcements (X_R) to the volume fraction (V_R) and vice versa. Equations 15.3 and 15.4 can be used to make these conversions:

Converting from reinforcement mass fraction to reinforcement volume fraction (V_R).

$$V_R = \frac{\dfrac{X_R}{\rho_R}}{\dfrac{X_R}{\rho_R} + \dfrac{X_M}{\rho_M}} \tag{15.3}$$

Converting from reinforcement volume fraction to mass fraction (X_R).

$$X_R = \frac{V_R \rho_R}{V_R \rho_R + V_M \rho_M} \tag{15.4}$$

where V_M, X_M, and ρ_M are the matrix volume fraction, weight fraction, and density, respectively.

Example Problem 15.2

Consider two fiber-reinforced epoxy composites, one reinforced with 0.1 volume fraction of carbon fibers and the other with 0.1 volume fraction of E-glass fiber. Determine the corresponding reinforcement mass fraction for both composites. Note that E-glass, carbon fiber, and epoxy densities are 2.55 g/cm^3, 1.77 g/cm^3, and 1.2 g/cm^3, respectively.

Solution

To convert volume fraction to mass fraction, we use Eq. 15.4:

$$X_R = \frac{V_R \rho_R}{V_R \rho_R + V_M \rho_M}$$

For carbon fibers:

$$X_{R\_C} = \frac{0.1 \times 1.77}{0.1 \times 1.77 + 0.9 \times 1.2} = 0.14$$

For E glass fibers

$$X_{R\_G} = \frac{0.1 \times 2.55}{0.1 \times 2.55 + 0.9 \times 1.2} = 0.19$$

As can be seen, since glass fibers are denser than carbon fibers, they take up more of the weight of the final composites, even though they are added at the same volume fraction. In the case of carbon fibers, the mass fraction of the epoxy will be 1.0–0.14 = 0.86. Also, for the E-glass fiber reinforced epoxy composite the mass fraction of the epoxy would be 1.0–0.19 = 0.81.

In contrast to 0D reinforcement particles, ***fibers*** are an example of 1D reinforcements, they can be short (or chopped) fibers or long continuous ones (Fig. 15.5). Due to the 1D nature of short fibers, it is possible to introduce them to the matrix randomly or align them in a particular direction within the matrix (Fig. 15.5a, b. This provides added options for the material designer. Adding short fibers in a random way to the matrix has some advantages, including the isotropic properties of the final composite. The alignment of short fibers can have the advantage of improving properties in a particular direction (the direction of fiber alignment). The same applies to the continuous fibers (Fig. 15.5c). Let's take mechanical properties as an example; for fibers to be able to reinforce a matrix mechanically, load transfer between the matrix and the reinforcement needs to be efficient. This is achieved when there is strong bonding with high interfacial strength between the matrix and the fiber reinforcement (whether short or continuous). Assuming good interfacial strength, the fibers must also have a length above a minimum critical length (L_c), given by Eq. 15.5, to effectively strengthen and stiffen the composite.

$$L_c = \frac{d\,\sigma_f}{2\tau_c} \tag{15.5}$$

d diameter of fiber

σ_f fracture or tensile strength of fiber.

τ_c interfacial bond strength between matrix and fiber, or the failure/yield strength of the matrix. The one with the lower value is used here.

With this critical length in mind, it is obvious that the fibers used would need to be longer than this critical length. Typically, fibers of length greater than **15L_c** are referred to as **continuous fibers** and those less than that will be termed **short fibers**.

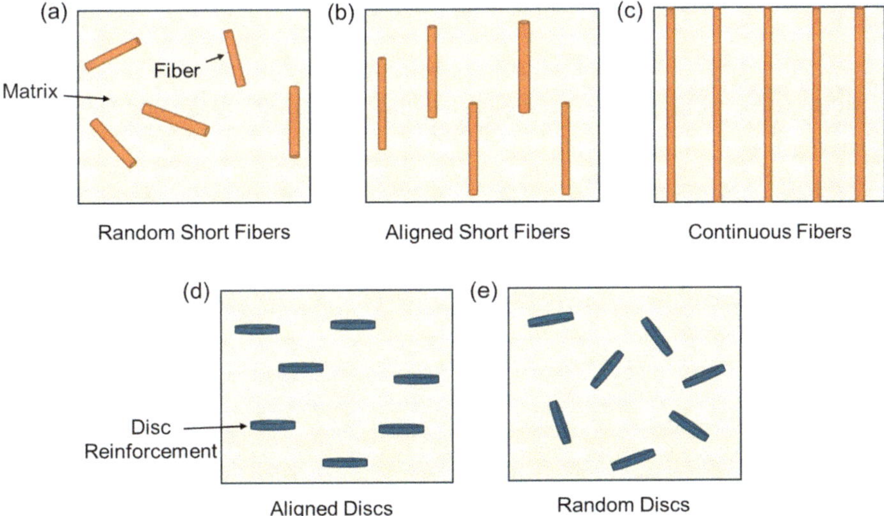

Fig. 15.5 **a** Fiber-reinforced composite with randomly oriented short fiber reinforcements **b** Short fibers are aligned, **c** fiber-reinforced composite with continuous fiber reinforcements **d** Particle-reinforced composite showing matrix and aligned disc (particle) reinforcements **e** discs are randomly oriented

One of the most widely used fibers for polymer matrix composites are **glass fibers**. These fibers can have several advantages, including ease of forming, chemical inertness in corrosive environments, are relatively cheap compared to carbon fibers, and can add significant stiffness to polymer composites. However, the surface sensitivity of glass to flaw formations can drastically reduce their strength. This is also true for ceramic fibers, and if they are to be incorporated into metallic or ceramic matrices, great care will be needed to protect these fibers from damage. **Carbon fibers** have a higher modulus, strength, and typically lower density than glass fibers. They are, however, more expensive and may be vulnerable to high-temperature oxidation. Other types of fibers include **Aramid** fibers (poly paraphenylene terephthalamide), which are polymeric fibers, that came to the scene at the beginning of the 1970s. One of their greatest advantages is their high strength-to-weight ratios, which outperform metals. **Kevlar** is an example of an aramid fiber. These fibers have impressive properties, including high toughness, very high tensile strength and modulus, good fatigue and creep properties, and good impact resistance. They, however, leave a lot to be desired when it comes to compressive strength. Acids and bases have been known to attack these materials if they are strong acids and bases.

The mechanical behavior of fiber-reinforced composites will be highly dependent on the orientation and obviously on the content of the reinforcing phase.

Equation 15.6 determines the Young's modulus for a composite containing **randomly oriented short fibers (ROSF)**.

$$E_{c(ROSF)} = CV_f E_f + V_m E_m = CV_f E_f + \left(1 - V_f\right)E_m \tag{15.6}$$

$E_{c(ROSF)}$ is the composite Young's modulus, and C is a parameter that accounts for fiber efficiency (generally having values ranging from 0.1 to 0.6 and always less than 1).

Example Problem 15.3

The Young's modulus of an aluminum component is 69 GPa; what would be the modulus if we reinforce the aluminum with randomly oriented short fibers of SiC at a volume fraction of 0.4? Assume $C = 0.5$, and the Young's modulus of SiC short fibers is 400GPa.

Solution

Following Eq. 15.6

$$E_{c(ROSF)} = CV_f E_f + V_m E_m = CV_f E_f + \left(1 - V_f\right)E_m$$

$$E_{c(ROSF)} = 0.5 \times (0.4) \times \left(400 \times 10^9\right) + (0.6)(69 \times 10^9)$$

$$E_{c(ROSF)} = 80 \times 10^9 + 41.4 \times 10^9 = 121.4 \times 10^9 \, \text{Pa}$$

Hence, the Young's modulus of the composite is 121.4 GPa.

If the short fibers are not randomly oriented but instead aligned, we cannot assume isotropic properties. The properties of the composite along the direction of fiber alignment (longitudinal direction) will be different than the transverse direction (90° to the direction of fiber alignment). Imagine reinforcing a material of a low Young's modulus with fibers of higher Young's modulus along a specific direction; let's call it the longitudinal direction. Along this direction, it is expected that the fibers will be able to exert stiff resistance to the applied force and, in turn, increase the overall Young's modulus of the material. If we, however, try to apply a tensile force in the transverse direction, the fibers will be inadequate at providing the same type of stiffening as in the longitudinal direction. This will lead to a lower Young's modulus of the composite in the transverse direction compared to the longitudinal direction. Other properties can also be examined further. Equation 15.7 shows how the strength of an aligned short fiber (of length > L_c) reinforced composite along the longitudinal direction (i.e., parallel direction).

$$\sigma_{C(L)} = \sigma_{fF} V_f \left(1 - \frac{L_c}{2L} \right) + \sigma_{mF} \left(1 - V_f \right) \tag{15.7}$$

$\sigma_{C(L)}$ is the strength of the composite in the longitudinal direction, L is the length of the fiber, σ_{fF} is the failure stress of the fiber, σ_{mF} is the failure stress of the matrix.

On the other hand, if the fibers have a length shorter than L_c, then Eq. 15.8 applies.

$$\sigma_{C(L)} = \frac{\tau_c L}{d} V_f + \sigma_{mF} \left(1 - V_f \right) \tag{15.8}$$

d is the fiber diameter, τ_c is the strength of the fiber/matrix interfacial bond. If the matrix's shear yield strength is lower than the interfacial bond strength, the equation uses that value instead. This means the lower strength is the one used in the equation.

Composites reinforced with 2D reinforcements

Reinforcements in the form of discs or micro-scale platelets can also be added to matrix materials. Discs and platelets are considered 2D reinforcements and can effectively deflect propagating cracks, promoting toughening mechanisms such as crack deflection and crack bridging. Such mechanisms have been observed in ceramic matrix composites, among other systems. These 2D reinforcements are also considered particle reinforcements, and they can be aligned or randomly distributed within a composite, as seen in Fig. 15.4.

Certain manufacturing processes can produce aligned short fibers and platelets within a composite. These include extrusion for polymer matrix composites and metal matrix composites, where reinforcements align along the extrusion direction. Extrusion can also be used to align fibers and platelets in ceramic matrices when the ceramics are initially in powder form and mixed with a polymeric binder. Extrusion of this powder mix (including the short fibers/platelets) above the glass transition temperature/melting point of the polymer would allow flow, and the shear strains experienced can align the reinforcements in the extrusion direction. In processes such as tape casting, ceramic particles are suspended in a solvent, including polymeric plasticizers and binders, and passed through a thin slit in a doctor blade setup which allows the production of a tape. After drying, the tape can be sintered at high temperatures to produce the final composite. Here too, the flow of the solvent containing the ceramic particles and fibers/platelets through the slip will experience shearing, which also aligns the reinforcement in the tape casting direction.

Effect of the Volume Fraction of Reinforcements

It is important to note that one of the obvious ways to change the properties of composites is to increase the volume fraction of reinforcements. Indeed, when you have a soft matrix material, adding more of the harder and stiffer reinforcements increases the stiffness and strength of the resulting composite. Assuming the matrix is a ductile material, and the reinforcement is a brittle and hard one, increasing the reinforcement's volume fraction may increase the composite's strength, stiffness, and wear resistance. Still, it may come at the expense of ductility. Within this context, imagine continuously adding hard reinforcements; at some critical volume fraction, these hard and brittle reinforcements will be so numerous that they may start to touch each other within the microstructure. At this point, a 3D skeleton of a brittle interconnected reinforcement phase would have formed, which can result in a further sudden decline in ductility and fracture toughness. When particles of a reinforcement phase become interconnected within the composite microstructure, they are said to be percolated. There is typically a critical volume fraction of particles at which point this occurs, called the **percolation threshold**. The properties of the composite can significantly change at reinforcement volume fractions above the percolation threshold. For example, imagine an electrically non-conducting material (e.g., Bakelite), which is essentially an insulator. We could envision adding electrically conducting reinforcements (fillers) such as copper particles. Bakelite will only be able to conduct electricity when the copper particles have become percolated within the Bakelite-Copper composite. Indeed, such a product exists; Bakelite is used to mount metallurgical samples (just like epoxy) for later characterization using electron microscopy. Electron microscopy uses an electron beam that hits the surface of the sample. A metallic sample that is placed in the middle of a non-conducting Bakelite sample mount would interact with the electron beam. Still, since the electrically insulating Bakelitesurrounds the metallic sample, it leads to charging, making the process of obtaining any proper image impossible. Scientists and researchers coat the Bakelite's surface (containing the sample to be viewed) with a thin layer of conductive carbon or gold, allowing the electrons from the electron beam a pathway out and thus relieving the charging issue. This means an additional coating process is needed to observe the mounted sample under an electron microscope. Examples of such coating processes include sputtering or evaporation techniques. However, the same effect can be achieved without needing a thin electrically conducting layer by using conductive Bakelite, which contains electrically conductive copper particles at a volume content above their percolation threshold.

One of the significant successes of powder metallurgy or particulate materials is the tungsten carbide-cobalt (WC–Co) **cermet**, formed through powder metallurgical processes, including liquid phase sintering. The *Cer* in cermet is for ceramic and *Met* for metal. This composite has cobalt content typically between 3 and 12 wt%, where the cobalt essentially coats the WC particles. In other words, the cobalt cements the WC particles together; this material is also called a **cemented carbide**. Adding cobalt to the hard, brittle, and wear-resistant WC adds toughness and improved thermal conductivity to the resulting composite. The material is a well-known cutting tool used as drill bits in oil drilling. The high level of friction during oil drilling generates excessive heat, and cobalt provides a way to dissipate some of that heat. The cobalt also improves the fracture toughness of the WC, thus combating brittle fracture. Recently, a new kind of composite was developed called double cemented carbides, where the WC is not distributed homogeneously within the composite but in-homogeneously, leaving regions within the microstructure free of WC. This microstructural composite design uniquely combines properties, including very good wear and fracture toughness.

Reinforcement Particle Size

If, for the same reinforcement volume fraction, the particle size is decreased, the composite can gain strength. This can be due to the interparticle spacing, i.e., the average distance between the reinforcements, sometimes called the mean free path. Figure 15.6 schematically demonstrates how, at the same reinforcement volume fraction, if the particle size is decreased, then the interparticle (inter-reinforcement) spacing also decreases. The interparticle spacing also decreases if we simply increase the volume fraction of reinforcements.

It can be understood from Fig. 15.6 that the smaller the particle sizes, the more particles there are and the greater the matrix/reinforcement interfacial area. This is truly exemplified in dispersoid-strengthened and precipitation-hardened alloys, where nanometric-scale precipitates or dispersoids interfere with the motion of dislocations, causing a strengthening effect. For ceramic matrix composites, the spacing between reinforcements can limit the length of the inherent flaws, and hence, the smaller the interparticle spacing between reinforcements, the smaller the flaw size and, hence, the higher the fracture strength of a ceramic.

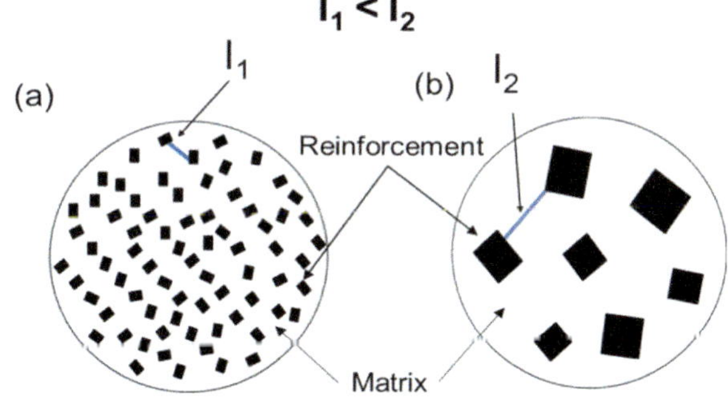

Fig. 15.6 Schematic showing how the interparticle spacing decreases with a decrease in reinforcement particle size **a**, compared with larger reinforcement particles **b**, both have the same reinforcement volume fraction

In-Situ Reinforcements

The conventional way to produce particle- or fiber-reinforced composites is to phys-ically add reinforcements to the matrix during manufacturing. However, sometimes, these reinforcements are either expensive or not even available commercially in stand-alone form. In this case, one approach is to grow the reinforcing phase in-situ during the high-temperature processing of the composite. An example of this approach is found in the production of titanium-TiB_w composites. Although titanium and its alloys possess good properties, including high specific strength, corrosion resistance, and low density, they come with disadvantages, including low specific stiffness and wear resistance. The addition of hard reinforcements to titanium has provided a means to overcome such disadvantages, and titanium boride (TiB_w) in-situ formed single crystal whiskers are currently seen as one of the most compatible reinforcements for Ti, enabling improvements in Young's modulus, strength, and wear resistance. TiB_w is, however, not commercially available in powder form and can only be produced through in-situ reactions, for example, the reaction between Ti and titanium diboride (TiB_2) particles during high-temperature processing, as shown in Eq. (15.9).

$$Ti + TiB_2 \rightarrow 2TiB_w \tag{15.9}$$

The synthesized TiB_w, however, bonds very well with the Ti matrix with very clean interfaces (i.e., no interfacial products). Some applications/potential applications for Ti matrix composites include die-casting shot sleeve liners, automotive valves, connecting rods and piston pins, spinal articulating devices, hockey skate blades, knives, tank tracks, and wheeled vehicle undercarriage components. Moreover, Ti composites have been used as exhaust valves in Toyota engines. TiB_w is much stiffer ($E_{TiB} = 371$–550 GPa) than Ti ($E = 105$ GPa), have similar densities and thermal expansion coefficients (ρ_{Ti} & $\rho_{TiB} \sim 4.5$ g/cm^3), $TEC_{TiC} = 8.2 \times 10^{-6}$/°C, $TEC_{TiB} = 6.2 \times 10^{-6}$/°C). Figure 15.7 shows grown single crystals of TiB within a titanium matrix.

15.4 Hybrid Composites

The word hybrid in **Hybrid composites** refers to the use of more than one type of reinforcement in the composite. For example, adding SiC and alumina short fibers to a matrix material would qualify the final composite as a hybrid composite. It should be noted that adding SiC and alumina "particles" would also be called a hybrid particle-reinforced composite. Hybrid composites further demonstrate the significant versatility in composite design, i.e. yet another variable can be controlled. These materials can display enhanced properties over single-type fiber or particle-reinforced composites. Table 15.4 presents properties of some composites.

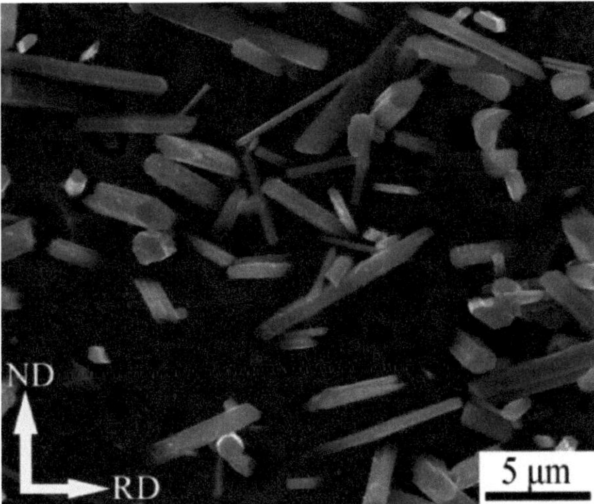

Fig. 15.7 TiB$_w$ whiskers formed from a reaction between titanium matrix and TiB$_2$ particle reinforcements. Note the whiskers are seen protruding from the surface of the composite only because some of the titanium matrix was selectively etched out. (Image from Zherebtsov, *S.; Ozerov, M.; Povolyaeva, E.; Sokolovsky, V.; Stepanov, N.; Moskovskikh, D.; Salishchev, G. Effect of Hot Rolling on the Microstructure and Mechanical Properties of a Ti-15Mo/TiB Metal-Matrix Composite. Metals **2020**, 10, 40.* https://doi.org/10.3390/met10010040. Reproduced under the terms and conditions of the Creative Commons Attribution (CC BY) license (http://creativecommons.org/licenses/by/4.0/)

15.5 Laminar/Layered Composites

Yet another example of the versatility of composite design is that of laminates/layered composites. We have understood so far that aligning a short fiber or a continuous fiber along one direction can improve mechanical properties along that direction, however, the material will be anisotropic. Stacking and bonding multiple thin lamellae or plies of aligned fiber-reinforced composites along different orientations is possible to improve the in-plane anisotropy. Figure 15.8a shows an example of three plies stacked on each other, each with a different orientation. The 0° orientation is parallel to the loading direction of the composite. 90° is aligned perpendicular to the loading direction, and the 45° is aligned at 45° to the loading direction. As one can imagine, other directions of orientation can also be used, including − 45°, which would be a mirror reflection of the 45° along the vertical axis. If we subdivide the laminates produced from this stacking process, we can end up with:

1. Unidirectional laminates, in which all the plies have the same direction.
2. Cross-ply, which is an alternating 0° and 90° stacking sequence.
3. Angle ply, in which 45° and − 45° plies are stacked in an alternating sequence.

Table 15.4 Properties of some composites (*Source James F. Shackelford, Introduction to Materials Science for Engineers, 6th edition, Pearson Prentice Hall, 2005 and Essentials of Materials Science and Engineering, SI edition, D. R. Askeland, W. J. Wright, 3rd edition, Cengage Learning (2014), W. F. Smith, J. Hashemi, Foundations of Materials Science and Engineering, 5th edition, McGraw Hill (2010), T. W. Clyne, P. J. Withers, An Introduction to Metal Matrix Composites, Cabridge University Press (1995)*

Matrix	Reinforcement	E (GPa)	Tensile strength (MPa)	Fracture Stress (MPa)	Ductility (%)	K_{IC} (MPam$^{1/2}$)
Polymer matrix composite						
Epoxy	73.3 vol% E-glass continuous fibers (axial loading)	56	1640	–	2.9	42–60
Epoxy	14 vol% Al_2O_3 whiskers	41	779	–	–	–
Epoxy	70 vol% B continuous filaments (axial loading)	210–280	1400–2100	–	–	46
Epoxy	62 vol% carbon fibers (axial)	145	1860	–	–	–
Epoxy	62 vol% carbon fibers (transverse)	9.4	65	–	–	–
Metal matrix composite						
Aluminum (Al)	10 vol% Al_2O_3 dispersions	–	330		–	–
Al 6061-T6	51% vol% continuous boron fibers (axial loading)	231	1417	–	0.735	–
Copper	50 vol% W continuous filaments (parallel loading)	260	1100	–	–	–
Copper	50 vol% W particles	190	380	–	-	–
Al 6061	20% SiC particles	103	496	–	5.5	–
Al 6061	20% SiC whiskers	–	545	–	-	–
Al 2024	20% SiC Whiskers	–	532	–	-	–

(continued)

Table 15.4 (continued)

Matrix	Reinforcement	E (GPa)	Tensile strength (MPa)	Fracture Stress (MPa)	Ductility (%)	K_{IC} (MPam$^{1/2}$)
Ceramic matrix composite						
Al_2O_3	SiC whiskers	–	–	800	–	8.7
SiC	SiC fibers	–	–	750	–	25
Si_3N_4	SiC whiskers	–	–	900	–	20
ZrO_2	SiC	–	–	450	–	22.2

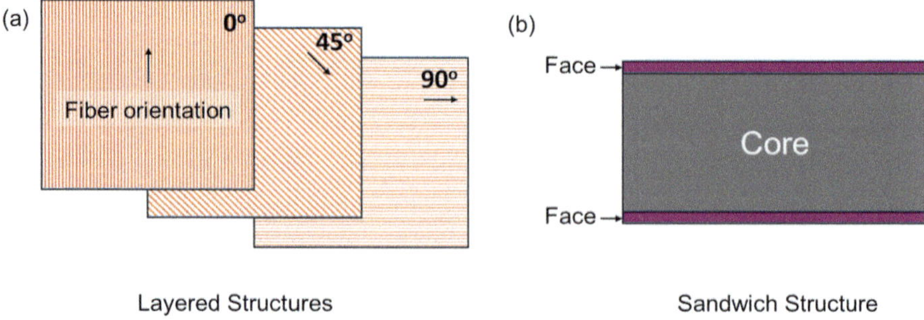

Fig. 15.8 **a** Layered composite of continuous fiber reinforced polymer, each layer with different orientation of fibers, **b** a sandwich-structured composite

4. Multi-directional laminates, in which more than two orientations are stacked. When stacking these plies, we need to ensure that the plies are symmetric above and below the central plane of the laminate.

Layered composites can also include layers of different materials bonded to each other, again to produce a laminate. For example, alternating layers of ceramic and metals, or metals and polymers, etc. All these combinations are to produce materials with properties that cannot be achieved by any of the individual layers making up the composite. A notable example of a laminate structure is safety glass; glass sheets are bonded with a polymeric adhesive, where the adhesive layer is sandwiched between two glass sheets. This way, when the glass shatters, the adhesive keeps the broken glass from shattering and flying everywhere. The laminate composite design is found in Formica countertops and even circuit boards. A simple cladding of a material surface can qualify the material to be a laminate structure. Let's take the US silver coin. Here, a Cu–20Ni alloy is sandwiched between two Cu-80 alloy layers, which are silvery in color.

Professor, why bother with this process since they are both Cu–Ni alloys?

Often, my friend, cost plays a significant role. In this case, Cu–20Ni is cheaper than Cu–80Ni.

Imagine a laminar composite structure with multiple layers of different materials (Fig. 15.9). Using the rule of mixtures, we can determine average properties along the

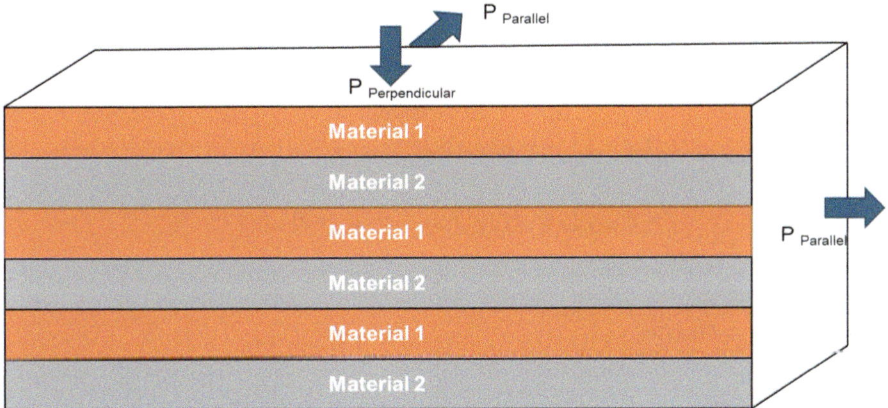

Fig. 15.9 Schematic of laminar composite structure consisting of alternating laminates of two different types of materials (Material 1 and Material 2). The figure also shows the parallel and perpendicular directions in relation to the calculated parallel and perpendicular average properties

laminar structure's layer length and width (also identified as parallel to the laminates) and those perpendicular (or through-thickness) to that structure.

The properties can be deduced from mechanical to thermal and even electrical. Let's consider first the average property (P) parallel to the laminates.

Average properties parallel to the laminate direction are given by Eq. 15.10:

$$P_{\text{Parallel}} = \sum v_i P_i \qquad (15.10)$$

P_{Parallel} is the average property along the laminates, v_i is the volume fraction of the i laminates (e.g., aluminum), P_i is the property of the laminate material (this property could be density, thermal conductivity, electrical conductivity or modulus of elasticity.

Average properties perpendicular to laminates (i.e., through-thickness) are given by Eq. 15.11.

$$\frac{1}{P_{\text{Perpendicular}}} = \sum \frac{v_i}{P_i} \qquad (15.11)$$

Example Problem 15.4

Consider a laminar composite of the type shown in Fig. 15.9, where Material 1 is aluminum and Material 2 is steel. What would be the average modulus of this composite structure along the parallel directions and the perpendicular direction? Assume perfect bonding between the laminates. Also note that the modulus of steel is 207 GPa, and aluminum's is 70 GPa. The volume fractions of steel and aluminum laminates are 0.5 each.

Solution

$$P_{\text{Parallel}} = \sum v_i P_i$$

The modulus along the parallel direction would be

$$P_{\text{Parallel}} = v_{\text{Al}} \times P_{\text{Al}} + v_{\text{steel}} \times P_{\text{steel}}$$

$v_{\text{Al}} = v_{\text{steel}} = 0.5$
$P_{\text{Al}} = E_{\text{Al}} = 70\,\text{GPa.}$
$P_{\text{steel}} = E_{\text{steel}} = 207\,\text{GPa.}$

Hence,

$$E_{\text{Parallel}} = 0.5 \times 70 \times 10^9 + 0.5 \times 207 \times 10^9 = 138.5\,\text{GPa}$$

15.6 Sandwich Composite Structures

This structure consists of a material core sandwiched between two sheets or faces glued to it using an adhesive layer (Fig. 15.8b). A notable example of this type of composite structure is lightweight honeycomb sandwich structures, which are honeycomb structures sandwiched between two sheets of metal, e.g., aluminum, titanium, steel, or even wood or a fiber-reinforced polymer. This structure adds strength and stiffness to the honeycomb structure or the core. The core could also be a polymeric foam or even wood. Metallic foams containing a high content of large pores (also called cells) are also used as cores in sandwich structures. Such structures can have applications in aerospace structures and even in building construction, where they can be used in roofs and walls.

15.7 Natural Composites

Natural Fiber Polymer Composites (NFPCs), which rely on natural fibers, find applications within the automotive, medical, building, sports, packaging, and furniture industries. Natural fibers provide an environmentally friendly and cheaper alternative to synthetic fibers used in fiber-reinforced polymer composites. If you have ever wondered how strong or stiff natural fibers are, the tensile strength of Flax can range from 340 MPa to 1.6 GPa, while its modulus of elasticity ranges between 25 and 81 GPa. Even cotton fibers and banana fibers boast strengths of 287–800 MPa and 711–789 MPa, respectively. Hemp boasts an even higher tensile strength of 1.7

Table 15.5 Properties of some natural composites (*Source Data presented are some of the data compiled from multiple sources by: Kamarudin, S. H.; Mohd Basri, M. S.; Rayung, M.; Abu, F.; Ahmad, S.; Norizan, M. N.; Osman, S.; Sarifuddin, N.; Desa, M. S. Z. M.; Abdullah, U. H.; et al. A Review on Natural Fiber Reinforced Polymer Composites (NFRPC) for Sustainable Industrial Applications. Polymers 2022, 14, 3698.* https://doi.org/10.3390/polym14173698)

Fiber	Density (g/cm^3)	Tensile Strength (MPa)	Young's Modulus (GPa)
Hemp	1.1–1.6	285–1735	14.4–44.0
Jute	1.3–1.5	393–773	13–26.5
Flax	1.3–1.5	340–1600	25–81
Bamboo	1.2–1.5	500–575	27.0–40.0
Cotton	1.5–1.6	287–800	5.5–12.6
Banana	0.5–1.5	711–789	4.0–32.7
Sugarcane	1.1–1.6	170–350	5.1–6.2

GPa. So mechanically speaking, they seem viable; Table 15.5 lists some properties of certain natural fibers. One must, however, consider other factors, such as fiber/matrix bonding, which can be a challenge. Table 15.5 lists properties of some natural composites. Note the high strengths of flax and hemp, not to mention banana.

15.8 Nanocomposites

All the knowledge we have gained about composites is very relevant when discussing nanocomposites. When composites are formed by reinforcing a matrix material with nanoscale reinforcements, then the resulting composite is called a **nanocomposite**.

At the outset, one must understand that the inherent difference between conventional materials and nanomaterials is the excessive surface area found in nanomaterials. Whether we reinforce materials with nanoparticles, nanoplates, nanofibers, or nanotubes to produce a nanocomposite, we will always be faced with the problem of agglomeration. These nano-reinforcements will possess excessive surface areas and hence will tend to stick together with weak bonds into what would be referred to as **agglomerates**. Dry mud is an example of agglomerates; when we try to break it down, it easily crumbles under the pressure of our fingers. That is because the forces holding the particles together are weak. Figure 15.10a shows agglomerated TiO_2 nanoparticles, and Fig. 15.10b shows agglomerated carbon nanotubes. The tendency for nano-reinforcements to agglomerate poses a problem for nanocomposites. For the reinforcements to positively impact the properties of the matrix material, they usually need to be homogenously distributed and dispersed within the matrix of the composite. Many manufacturing approaches have been developed to allow nano-reinforcement dispersion within polymeric, metallic, and ceramic matrices. One method is to produce the nanoparticles already coated with a layer

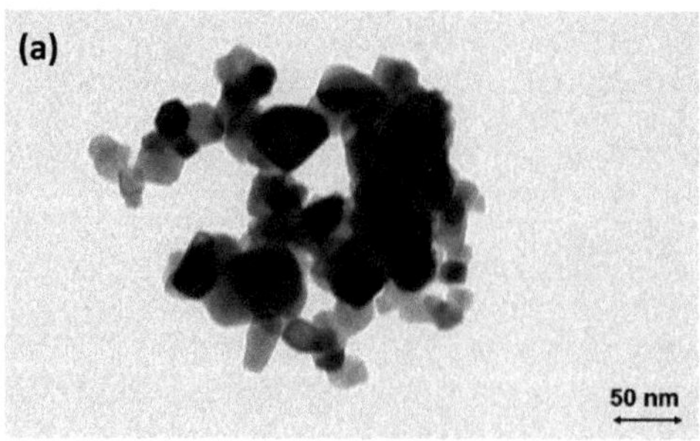

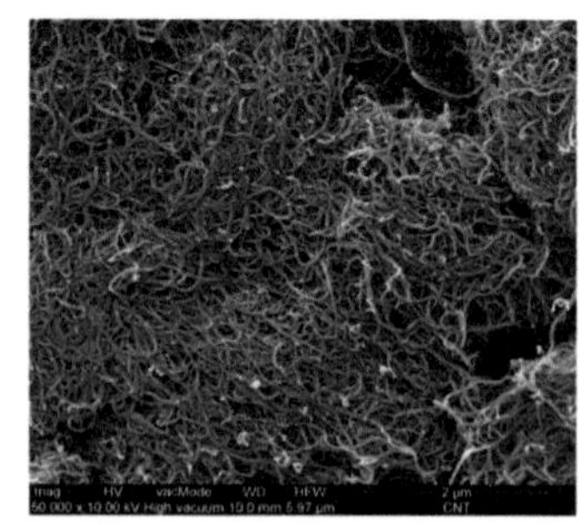

Fig. 15.10 **a** Transmission electron micrograph of agglomerated TiO_2 nanoparticles (Image from *May, N.; Baumann, W.; Hauser, M.; Yin, Z.; Geigle, K. P.; Stapf, D. Degradation and Recondensation of Metal Oxide Nanoparticles in Laminar Premixed Flames. Nanomaterials 2024, 14, 1047.* https://doi.org/10.3390/nano14121047, **b** agglomerated carbon nanotubes (image from *Szeląg, M. Properties of Cracking Patterns of Multi-Walled Carbon Nanotube-Reinforced Cement Matrix. Materials 2019, 12, 2942.* https://doi.org/10.3390/ma12182942. *Both images reproduced under terms and conditions of the Creative Commons Attribution (CC BY) license (*https://creativecomm ons.org/licenses/by/4.0/)

of the matrix material. This guarantees that the matrix will always separate the nano-reinforcements. Other CNT dispersion methods have been reported, including wet processing approaches that involve sonication, surface modification coupled with mechanical stirring, molecular-level mixing, and the nanoscale dispersion method.

For ultrahigh strength and stiffness, reinforcements such as carbon nanotubes need to bond well with the matrix material to improve the strength and stiffness of carbon nanotube-reinforced composites. An example of such bonding occurs during the high-temperature processing (above 450 °C) of aluminum-carbon nanotube composites, where aluminum carbide (Al_4C_3) forms at the interface between carbon nanotubes and aluminum, strengthening the interfacial bond. The thermal expansion mismatch between CNTs and Al can also improve contact and adhesion.

The Percolation Threshold

As mentioned earlier, at a critical volume fraction of added reinforcements (V_c), called the percolation threshold, the reinforcements become interconnected (touching each other) within the composite microstructure. For 1D reinforcements, this can occur at relatively high reinforcement volume fractions. However, for 1D reinforcements, percolation can occur at much lower reinforcement contents. Equation 15.12 shows an expression for the percolation threshold for equal length (L) and diameter (d) 1D reinforcements with a matrix, assuming random orientation.

$$V_c = 0.7\frac{d}{L} \tag{15.12}$$

As can be seen, this equation is highly dependent on the aspect ratio (L/d). A high L/d gives rise to a very low percolation threshold (V_c), which can be useful when rendering an insulating material conductive. Flexible solar cells can significantly benefit from including high aspect ratio carbon nanotubes in a transparent polymer. Adding very small amounts of carbon nanotubes can make a normally insulating transparent polymer electrically conductive. Example 15.5 exemplifies this.

Example Problem 15.5
The traditional brittle indium tin oxide (ITO) transparent electrode used in solar cells is proposed to be substituted with a flexible alternative. A transparent and flexible polymer is sought, namely Poly (2,7–9,9-(di(oxy-2,5,8-trioxadecane)) fluorene known as PFO. However, the electrical conductivity of PFO is only 10^{-13} S/cm. A proposed solution is to add single-wall carbon nanotubes to significantly improve the electrical conductivity of the polymer to an acceptable level. What volume fraction of SWCNT needs to be added to enable this? Assume the single-wall carbon nanotubes all have the same diameter of 2 nm and length of 5 μm.

Solution

Equation 15.12 can be used to find the percolation threshold that would render the polymer electrically conductive, assuming a random distribution of the SWCNTs within PFO.

$$V_c = 0.7\frac{d}{L} = 0.7 \times \frac{2 \times 10^{-9}}{5 \times 10^{-6}} = 2.8 \times 10^{-4}$$

This is a truly low critical volume fraction, and you hardly need any SWCNT (which are expensive) to make it happen. Not only will the polymer be electrically conductive, but it will also remain optically transparent.

Assuming that the electrical conductivity of the 1D reinforcement is known and that of the matrix is almost zero, Eq. 15.13 can give the electrical conductivity of the reinforced composite above the percolation threshold.

$$\sigma_c = \sigma_o(V - V_c)^a \tag{15.13}$$

σ_c is the electrical conductivity of the composite, σ_o is the electrical conductivity of the conducting reinforcement material, V is the reinforcement volume fraction, V_c is the percolation threshold, and 'a' is an exponent representing the reinforcement network dimensionality, typically a non-integer with experimental values between 1.3 and 3.0.

Example Problem 15.6
The percolation threshold of short multiwall carbon nanotubes (40 nm diameter and 5 μm length) in an alumina ceramic is 2.8×10^{-3}. What would be the electrical

conductivity of an alumina ceramic reinforced with 5 vol.% short MWCNT? Note that the electrical conductivity of alumina and MWCNT are 1×10^{-16} S/cm and 10^5 S/cm, respectively, and a = 1.5.

Solution

The percolation threshold seems so low that we can give it a value of zero to facilitate the calculations. The same is true for the electrical conductivity of alumina, which will also be taken as zero, since it is known to be an insulator.

Hence, Eq. 15.13 can then be used as follows, using a = 1.5 and V = 0.05:

$$\sigma_c = \sigma_o(V - V_c)^a = 1 \times 10^5(0.05 - 0)^{1.5} = 1118 \, \text{S/cm}$$

Professor, I heard that carbon nanotubes can be quite expensive. It seems like having a percolation threshold that is so low is also beneficial from the cost perspective, since we don't have to use much of the expensive nanotubes to achieve our aim. I have a few questions for you. Aren't carbon nanotubes very long? Wouldn't they get entangled when being introduced to the material? Would that have an effect on the percolation threshold and, for that matter, properties in general?

Well, Alex, you have encountered one of the most important issues with carbon nanotubes. Did you know some carbon nanotubes can have aspect ratios exceeding 1,000,000 to 1? Imagine that! You are correct to think that this poses a significant problem for adding CNTs to matrices (metallic, polymeric, and even ceramic). The examples we covered so far assume that the CNTs are well dispersed within the matrix material. However, CNT agglomeration (clumping up together of CNTs) can occur, meaning that the percolation threshold can shift to much larger values since the CNTS are now not well dispersed within the matrix material. An interesting finding by researchers is that adding CNTs to materials typically positively affects mechanical properties. Still, at higher additions of CNTs, the mechanical properties can deteriorate. This is usually the result of the vulnerability of the CNTs to agglomeration at higher concentrations.

The nano-structural design of nanocomposites can yield substantial benefits in terms of mechanical behavior and damage tolerance. An example of that is the brick-and-mortar nanostructure of nacre (mother of pearl) found in sea snails. Figure 15.11a

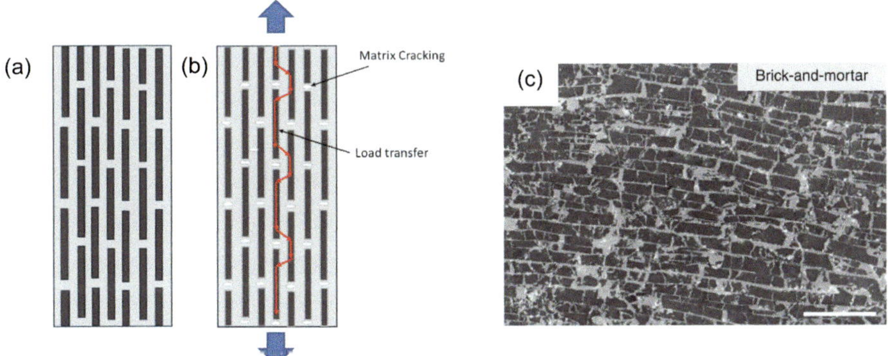

Fig. 15.11 Schematic of internal structure of nacre (mother of pearl) **a** showing aligned aragonite platelets (dark grey) separated by lustrin biopolymer (light grey). **b** effectiveness of the structure in carrying load even after matrix failure in certain locations (**a** and **b** based on *D. Vollath, Introduction to Nanomaterials, Wiley, (2013)*, **c** example of the brick-and-mortar design found in nacre implemented in alumina platelet reinforced metallic glass, scale bar is 100 μm (image modified from *A. Wat, Je In Lee, C. W. Ryu, B. Gludovatz, J. Kim, A. P. Tomsia, T. Ishikawa, J. Schmitz, A. Meyer, M. Alfreider, D. Kiener, E. Soo Park & R. O. Ritchie, Bioinspired nacre-like alumina with a bulk-metallic glass-forming alloy as a compliant phase, Nature Communications, (2019) 10:961* | https://doi.org/10.1038/s41467-019-08753-6. *(Image reproduced with modification under terms and condition of* http://creativecommons.org/licenses/by/4.0/

shows the structure of nacre, where plates of aragonite (a crystalline form of calcium carbonate) are bonded together with a biopolymer. This brick-and-mortar structure has remarkable benefits. For example, in Fig. 15.11b, it can be seen that even after the matrix material fails through the generation of cracks, the load can still be transferred, and the material does not fall to pieces. If weak, the interfacial bond between the platelets and the matrix can result in the crack traveling through the interface, which can promote platelets bridging the faces of a propagating crack. Figure 15.11c shows an example of the brick-and-mortar design implemented in alumina platelet-reinforced metallic glass composites. The composite has very high strength values.

One of the recent uses of nanocomposites is in the game of tennis. Some tennis tournaments have adopted new tennis balls, called double-core tennis balls. Here, an inner nanocoat (10–50 μm thick) hinders the diffusion of molecules of the pressurized air out of the ball and improves the ball's performance. For example, the ball remains pressurized much longer than regular balls and keeps its bouncing for twice as long as regular balls. This inner nanocoat core is termed 'nano' because it consists of

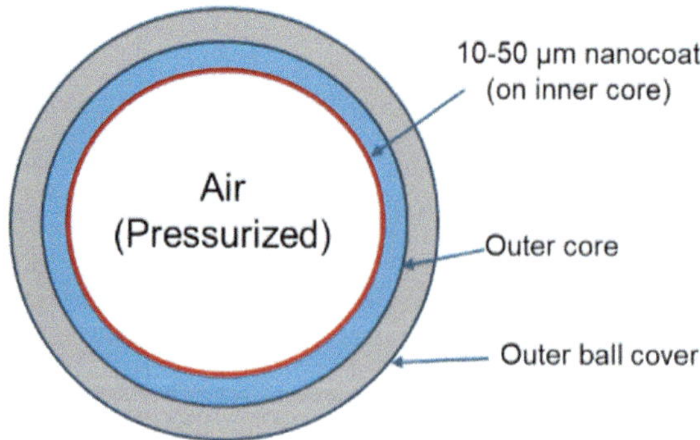

Fig. 15.12 Schematic of double core tennis ball with inner nanocomposite coat. (*after: W. D. Callister, D. G. Rethwisch, Materials Science and Engineering-An Introduction, Eighth edition, Wiley (2010)*)

naturally occurring clay mineral nanoplatelets, with very high aspect ratios of ~ 10,000:1, within a butyl rubber matrix. They are aligned along the lining and provide a great diffusion barrier for air trying to escape; they also help maintain the mechanical response of the ball (Fig. 15.12).

Problems

15.1. What is the difference between a composite and an alloy?

15.2. In the field of composites, what is meant by a matrix and reinforcement?

15.3. What is meant by a ceramic matrix composite?

15.4. What is the difference between a whisker and a fiber as reinforcements?

15.5. Give one disadvantage of using whiskers as reinforcements

15.6. Why is the strength of whiskers superior to that of fibers of the same material?

15.7. Determine the upper and lower bound modulus of a copper-10 vol.% Boron nitride particle composite. (Hint: Property data is found in the chapter.)

15.8. Your company has decided that an aluminum-20 vol.% boron nitride composite should be manufactured. Determine the weight percent of reinforcement needed to make this composite.

15.9. Consider aluminum reinforced with 0.3 volume content of aligned short fibers of SiC (50 μm, diameter and length 100 μm). The fibers' length is smaller than the critical length for effective load transfer. Calculate the expected strength of the composite along the fiber direction. The strength of SiC is 3 GPa, and the failure stress of the aluminum matrix is 110 MPa.

15.10. Consider the same problem 15.9., but assume the aligned fibers have the same diameter but a length of 1 mm. Assume the critical fiber length for load transfer is 350 μm.

15.11. Give examples of three natural fibers used in composites.

15.12. What are in-situ reinforcements? Give an example.

15.13. Consider a lamellar composite structure consisting of five layers of aluminum and five layers of copper. If the thermal conductivity of aluminum is 200 W/m.K, and that of copper is 400 W/m.K, determine the thermal conductivity of the structure parallel to the layer direction and the perpendicular direction. All individual layers have the same dimension.

15.14. Consider a lamellar composite structure consisting of five layers of aluminum, five layers of copper, and five layers of steel. If the thermal conductivity of aluminum is 200 W/m.K, and those of copper and steel are 400 W/m.K and 50 W/m.K., respectively. Determine the thermal conductivity of the structure parallel to the layer direction and the perpendicular direction. All individual layers have the same dimension.

15.15. What is a nanocomposite?

15.16. Determine the percolation threshold of carbon nanofiber in epoxy. The nanofiber dimensions are 50 nm in diameter and 10 μm in length.

15.17. Determine the electrical conductivity of PFO reinforced with 0.1 volume percent MWCNT (50 nm diameter and 5 μm long). Note that the electrical conductivity of PFO is 10^{-13} S/cm, and MWCNTs is 10^5 S/cm.

15.18. Discuss general ways by which you could improve the strength of a composite.

15.19. Discuss general ways in which you can improve the fracture toughness of composites.

15.20. How does adding SiC to aluminum affect its Young's modulus?

15.21. Give an example of a composite used in tall buildings.

Bibliography[1]

Books

R.A. Higgins, Properties of Engineering Materials, Hodder and Stoughton, (1986).

K.G. Budinski, M.K. Budinski, Engineering Materials: Properties and Selection, 8th edition, Pearson Prentice Hall, (2005).

W.D. Callister and D. G. Rethwisch, Materials Science and Engineering, 10th edition (and other editions), Wiley, (2018).

D. R. Askeland, W. J. Wright, Essentials of Materials Science and Engineering, SI-edition (and other editions), Cengage Learning, (2014).

W. F. Smith, J. Hashemi, Foundations of Materials Science and Engineering, 5th edition, McGraw-Hill, (2010) (and other editions).

J. F. Shackelford, Introduction to Materials Science for Engineers. Pearson Education Limited, (2016).

D. Vollath, Nanomaterials: An introduction to Synthesis, Properties, and Applications, Wiley-VCH, second edition, (2013).

Nanomaterials: Synthesis, Properties and Applications (edited by A.S. Edelstein and R.C. Cammarata), IoP, (1998).

B.D. Cullity, Elements of X-Ray Diffraction, Addison-Wesley Pub. Co. (1956)

C.Suryanarayana, M. Grant Norton, X-Ray Diffraction: A Practical Approach, Plenum, (Press (1998)

J.E. Gordon, The New Science of Strong Materials: Or Why You Don't Fall Through the Floor, Penguin Group, (1991).

S. M. Lindsay, Introduction to Nanoscience, Oxford University Press (2010)

G. Cao, Y. Wang, Nanostructures and Nanomaterials: Synthesis, Properties and Applications, 2nd edition, World Scientific, (2011).

C. P. Poole Jr., F. J. Owen, Introduction to Nanotechnology, Wiley, (2003).

Eliade, Mircea. The Forge and the Crucible. Second ed., University of Chicago, (1978).

W.F. Hosford, and R. M. Caddell. Metal Forming: Mechanics and Metallurgy. Cambridge University Press, (2014).

G. E. Dieter, Mechanical Metallurgy, McGraw Hill, SI metric edition, (1988).

---

[1] The writing of this book would not have been possible without the valuable past works, insights and contributions of scholars, researchers, institutions and organizations in the field The following are examples of the works that impacted the author and his present contribution.

540   Bibliography

M.P. Groover, Fundamentals of Modern Manufacturing: Materials, Processes and Systems, 5th edition, (2013).

W. Hume-Rothery, Electrons, Atoms, Metals, and Alloys, 3rd edition, Dover Publications Inc., (1963).

W.T. Read, Dislocations in Crystals, McGraw-Hill, (1953).

S. L. Sass, The Substance of Civilization, Arcade Publishing, New York (2011)

D.A. Porter, K.E. Eastering, Phase Transformations in Metals and Alloys, V. Nostrand Reinhold (International), (1981).

Z. D. Jastrzebski, Nature and Properties of Engineering Materials, John Wiley and Sons, (1959).

W. Cai, W. D. Nix, Imperfections in Crystalline Solids, MRS Cambridge Materials Fundamentals, Cambridge University Press, United Kingdom (2016).

Mary Elvira Weeks and Henry M. Leicester, Discovery of the Elements, 7th Edition, Journal of Chemical Education, Easton, PA, USA (1968).

J. Weertman, J.R. Weertman "Elementary Dislocation Theory", Oxford University Press, Oxford (1992).

M. F. Ashby and D.R. Jones, Engineering Materials 1, An introduction to Properties, applications and Design, 4th edition, Butterworth-Heinemann, (2009).

M.F. Ashby and D.R. Jones, Engineering Materials I-An introduction to their properties and applications, second edition, Butterworth and Heinmann (1996).

M. F. Ashby, D.R. Jones, Engineering Materials 2, An Introduction to Microstructures, Processing and Design (2nd edition), Butterworth and Heinmann (2001).

Engineering Materials Handbook, volume 4: Ceramics and Glasses, ASM International, (1991).

Handbook of Materials for Medical Devices, edited by J.R. Davis, ASM International, (2003).

M. P. Groover, Fundamentals of Modern Manufacturing: Materials, Processes, and Systems, 5th edition, Wiley (2013)

Sir Alan Cottrell, An Introduction to Metallurgy, 2nd edition, The Institute of Materials, (1995).

R.E. Reed-Hill, R. Abbaschian, Physical Metallurgy Principles, 3rd Edition, PWS Publishing Company, (1994).

W.D. Kingery, H.K. Bowen, D.R. Uhlmann, Introduction to Ceramics, 2nd edition, Wiley Interscience, (1976).

D. W. Richerson, Modern Ceramic Engineering, 3rd Edition, CRC Taylor and Francis, (2006).

J. Gerald Byrne, Recover, Recrystallization and Grain Growth, Macmillam series in Materials Science (1965).

R.J. Young, Introduction to Polymers, Chapman and Hall, (1983).

N.G. McCrum, C.P. Buckley, and C.B. Bucknall, Principles of Polymer Engineering, 2nd edition, Oxford Science Publications, Oxford University Press, (1997).

P. C. Powell, Engineering with polymers, Chapman and Hall Ltd (1983).

Ernest Rabinowiczm Friction and Wear of Materials, 2nd edition, Wiley Interscience, (1995).

A. Ul-Hamid, A Beginners' Guide to Scanning Electron Microscopy, Springer Nature, (2018).

S. Bell, K. Morris, An Introduction to Microscopy, CRC Press, (2010).

David B. Williams, C. Barry Carter, Transmission Electron Microscopy, 2nd edition, Springer (2009).

M. G. Fontana, N. D. Greene, Corrosion Engineering, McGraw-Hill, (1978).

S.B. Roy, A Short History of Magnetism and Magnetic Materials. In: Experimental Techniques in Magnetism and Magnetic Materials. Cambridge University Press; (2023):3–12.

Encyclopedia of Materials: Science and Technology, Editors-in-Chief: K.H. Jürgen Buschow, Robert W. Cahn, Merton C. Flemings, Bernhard Ilschner, Edward J. Kramer, Subhash Mahajan. Patrick Veyssière, Elsevier Ltd (2001).

Modern Plastics Encyclopedia, McGraw-Hill, 1985–1986.

M. Bhavisha, K. Anjali, S. Aswani, A. Sakthivel, 2 - Catalytic applications of perovskites, Editor(s): Manju Kurian, Smitha Thankachan, Swapna S. Nair, In Elsevier Series in Advanced Ceramic Materials, Ceramic Catalysts, Elsevier, (2023), Pages 19-55, ISBN 9780323857468, https://doi.org/10.1016/B978-0-323-85746-8.00005-9.

S. Hampshire, Chapter 2: Fundamental Aspects of Hard Ceramics, in Comprehensive Hard Materials: Ceramics, vol. 2, editors: Vinod K. Sarin, Luis Llanes, Daniele Mari, Elsevier (2014), pp. 3–28. https://doi.org/10.1016/B978-0-08-096527-7.00020-9
Humphry Davy, Elements of Chemical Philosophy, Bradford and Inskeep, (1812).

Journal Papers

M. Kokarneswaran, P. Selvaraj, T. Ashokan, S. Perumal, P. Sellappan, K. D. Murugan, S. Ramalingam, N. Mohan & V. Chandrasekaran, Discovery of carbon nanotubes in sixth century BC potteries from Keeladi, India, *Scientific reports, 10*, 19786 (2020). https://doi.org/10.1038/s41598-020-76720-7
J. Pérez-Arantegui et al. Luster pottery from the thirteenth century to the sixteenth century: a nanostructured thin metallic film. *Journal of the American Ceramic Society, 84*, 442–446, (2004)
R. Chakraborty, A. Dey, A.K. Mukhopadhyay, Loading Rate Effect on Nanohardness of Soda-Lime-Silica Glass. *Metallurgical and Materials Transactions A, 41*, 1301–1312, (2010). https://doi.org/10.1007/s11661-010-0176-8
M. Reibold, et al. Carbon nanotubes in an ancient Damascus sabre. *Nature, 444*, 286–286, (2006)
T. Pradell, J. Molera, A.D. Smith, M.S. Tite, The invention of lustre: Iraq 9th and 10th centuries AD, *Journal of Archaeological Science, 35*, 1201–1215, (2008).
E. Lucon, Experimental Assessment of the Equivalent Strain Rate for an Instrumented Charpy Test, *Journal of Research of the National Institute of Standards and Technology, 121*, 165–179, (2016).https://doi.org/10.6028/jres.121.007
K. Felkins, H.P. Leighly, Jr., and A. Jankovic, "The Royal Mail Ship Titanic: Did a Metallurgical Failure Cause a Night to Remember? *JOM, 50*(1), 12–18, (1998).
H.E. Cleaves, J.M. Hiegel, Properties of high purity iron, Research paper RP1472., *Part of Journal of Research of the National Bureau of Standards, 28*(May), 643–667, (1942).
J. Zhai, X. Song, A. Xu, Y. Chen, Q. Han, Dislocation Damping and Defect Friction Damping in Magnesium: Molecular Dynamics Study, *Metals and Materials International, 27*, 1458–1468, (2021)
C. Yu, Q. Xie, Y. Bao, G. Shan, P. Pan, Crystalline and Spherulitic Morphology of Polymers Crystallized in Confined Systems. *Crystals, 7*, 147, (2017).https://doi.org/10.3390/cryst7050147.
J. K. Mackenzie, Proc. Phys. Soc. London 63B (1950)
W. Pabst, E. Gregorová. New relation for the porosity dependence of the effective tensile modulus of brittle materials. *Journal of Materials Science, 39*(10) 3501–3503, (2004).
E.A. Friis, R.S. Lakes, J.B. Park, Negative Poisson's ratio polymeric and metallic foams, *Journal of Materials Science, 23*, 4406–4414, (1988).
K.A. Hart, J.J. Rimoli, Generation of statistically representative microstructures with direct grain geometry control, *Computer Methods in Applied Mechanics and Engineering, 370*, 113242. (BibTeX) (DOI), (2020).
K.A. Hart, J.J. Rimoli, MicroStructPy: A statistical microstructure mesh generator in Python, *SoftwareX, 12*, 100595. (BibTeX) (DOI), (2020).
V. Krishnan, T. Lakshmi, Bioglass: A novel biocompatible innovation, *Journal of advanced pharmaceutical technology & research, 4*(2), 78–83, Apr–Jun, (2013).https://doi.org/10.4103/2231-4040.111523
J.T. Pandayil, N.G. Boetti, D. Janner, Advancements in Biomedical Applications of Calcium Phosphate Glass and Glass-Based Devices—A Review, *Journal of Functional Biomaterials, 15*(3), 79, (2024). https://doi.org/10.3390/jfb15030079
R.L. Fullman, American Society for Metals. Seminar Metal Interfaces. 179, 54, (1952).
R.B. Heimann, Silicon Nitride Ceramics: Structure, Synthesis, Properties, and Biomedical Applications, *Materials*, 2023, *16*(14), 5142.

A.A.M. El-Amir, A.A. El-Maddah, E.M.M. Ewais, S.M. El-Sheikh, I.M.I. Bayoumi, Y.M.Z. Ahmed, Sialon from synthesis to applications: an overview, *Journal of Asian Ceramic Societies, 9*(4), 1390–1418, (2021).

Z. Cheng, J. Liang, K. Kawamura, H. Zhou, H. Asamura, H. Uratani, J. Tiwari, S. Graham, Y. Ohno, Y. Nagai, T. Feng, N. Shigekawa, D.G. Cahill , High thermal conductivity in wafer-scale cubic silicon carbide crystals, *Nature Communications, 13*, 7201 (2022).

J. Cañas, J.C. Piñero, F. Lloret, M. Gutierrez, T. Pham, J. Pernot, D. Araujo, Determination of alumina bandgap and dielectric functions of diamond MOS by STEM-VEELS, *Applied Surface Science, 461*, 93–97, 15 December 2018.

C. Zhang, Y. Chen, M. Zhou, X. Li, L. Wang, L. Xia, Y. Shenc, X. Dong, Achieving ultrahigh dielectric breakdown strength in MgO-based ceramics by composite structure design, *Journal of Materials Chemistry C, 7*(26), 8120–8130, (2019).

M.N. Yoder, Chapter 1: The Vision of Diamond as an Engineered Material, in Synthetic Diamond: Emerging CVD Science and Technology (edited by K.A. Spear, J.P. Dismukes, Wiley-Interscience Publication, John Wiley and Sons, Inc. (page 4, 1994).

D.J. McClements, H. Xiao, Is nano safe in foods? Establishing the factors impacting the gastrointestinal fate and toxicity of organic and inorganic food-grade nanoparticles. *npj Science of Food, 1*, 6, (2017). https://doi.org/10.1038/s41538-017-0005-1

M.M. El-Kadya, I. Ansarib, C. Arorac, N. Raic, S. Sonic, D.K. Vermad, P. Singhd, A.E.D. Mahmoud, Nanomaterials: A comprehensive review of applications, toxicity, impact, and fate to environment. *Journal of Molecular Liquids, 370*, 121046, (2023).

S. Bayda, M. Adeel, T. Tuccinardi, M. Cordani, F. Rizzolio, The History of Nanoscience and Nanotechnology: From Chemical–Physical Applications to Nanomedicine, *Molecules, 25*(1), 112, (2020). https://doi.org/10.3390/molecules25010112.

F. Li, H-M Cheng, S. Bai, G. Su, M.S. Dresselhaus, Tensile strength of single wall carbon nanotubes directly measured from their macroscopic ropes, *Applied Physics Letters, 77*(20), 3161–3163, (2000).

E.W. Godly, R. Taylor, Nomenclature and terminology of fullerenes: A preliminary study, *Pure and applied chemistry, 69*(7), 1411–1434, (1997).

T.G. Schreiner, M. Menéndez-González, M. Adam, B.O. Popescu, A. Szilagyi, G. Dumitrita Stanciu, B.I. Tamba, R.C. Ciobanu, A Nanostructured Protein Filtration Device for Possible Use in the Treatment of Alzheimer's Disease—Concept and Feasibility after In Vivo Tests *Bioengineering, 10*(11), 1303, (2023).

K. Wang, DNA-Based Single-Molecule Electronics: From Concept to Function. *Journal of functional biomaterials, 9*(1), 8, 2018 Jan 17. https://doi.org/10.3390/jfb9010008. PMID: 29342091; PMCID: PMC5872094.

K. Kulkarni, Y. Moghe, A. Tangadpalliwar, J. Kaur, R. Vohra, A review on the smallest carbon fullerene C20: Applications and device formation, Materials Today: Proceedings (in press, 2024, https://doi.org/10.1016/j.matpr.2024.05.147)

A. Takakura, K. Beppu, T. Nishihara, et al. Strength of carbon nanotubes depends on their chemical structures. *Nature communications, 10*, 3040, (2019). https://doi.org/10.1038/s41467-019-109 59-7.

M. Sakai, R. Bradt, Fracture toughness testing of brittle materials", *International Materials Reviews, 38*(2), 66, (1993).

M. Miao, J. McDonnell, L. Vuckovic, S.C. Hawkins, Poisson's ratio and porosity of carbon nanotube dry-spun yarns, *Carbon, 48*,(10), 2802–2811, August (2010).

C. Ponton, R. Rawlings, "Vickers Indentation fracture toughness test: part 1, *Materials Science and Technology, 5*, 865–872, (1989).

D.A. Vaughan, F. Guiu, M.R. Dalmau, "Indentation Fatigue of alumina, *Journal of Materials Science Letters, 6*, 689–691, (1987).

Adina Luican, Guohong Li, Eva Y. Andrei, Scanning tunneling microscopy and spectroscopy of graphene layers on graphite, *Solid State Communications, 149*, 1151–1156, (2009). https://doi.org/10.1016/j.ssc.2009.02.059

E.A. Friis, R.S. Lakes, J.B. Park, Negative Poisson's ratio polymeric and metallic foams, *Journal of Materials Science, 23*, 4406–4414, (1988).

J.K. Mackenzie, Proc. Phys. Soc. London 63B (1950).

W. Pabst, E. Gregorová. New relation for the porosity dependence of the effective tensile modulus of brittle materials. *Journal of Materials Science, 39*(10), 3501–3503, (2004).

L.B. Kong, T.S. Zhang, J. Ma, F. Boey, Progress in synthesis of ferroelectric ceramic materials via high-energy mechanochemical technique, *Progress in Materials Science, 53*(2), 207-322, February (2008).

I. Mayo, The Metal With A Memory: A combination of accident, luck, and hard work produced nitinol, an advanced "intelligent" metal, *American Heritage's Invention and Technology magazine, 9*(2), (Fall 1993).

J.T. Pandayil, N.G. Boetti, D. Janner, Advancements in Biomedical Applications of Calcium Phosphate Glass and Glass-Based Devices-A Review. *Journal of Functional Biomaterials, 15*(3), 79, 2024 Mar 21. https://doi.org/10.3390/jfb15030079. PMID: 38535272; PMCID: PMC10970746.

Deubener et. al., "Updated definition of glass ceramics", *Journal of Non-Crystalline Solids, 501*, 3-10, (2018).

Reports

R.A. Lowden, Characterization and Control of the Fiber-Matrix Interface in Ceramic Matrix Composites, Office of Naval Research Laboratory (ONRL), Report no. TM-11039, March 1989

Standards

ASTM standard E18 (1984)
ASTM Standard E-8
ASTM C1361-10 (2019)
ASTM G99

Websites

National Nanotechnology Initiative (NNI). www.nano.gov (accessed on 11 August 2024).
https://research.ibm.com/projects/advanced-logic-technology-at-2nm-node (accessed on 11 August 2024).
www.Britannica.com
www.Dupont.com (latest access 23 January 2025).
https://www.aluminum.org/ (The Aluminum Association) (latest access April 2025).

Index

A

Abrasive wear, 431
Activation energy, 133, 159–164, 286, 317
Adhesive wear, 429, 431
Allotropic, 238
Alternating copolymer, 373
Alumina, 20, 21, 54, 55, 211, 212, 230, 231,
    296, 298, 401, 402, 405, 408, 412,
    419, 424, 439, 446, 457–459, 485,
    510–512, 514, 515, 524, 533–535
Aluminum alloy(s), 56, 57, 136, 150, 151,
    185, 186, 194, 228, 270, 272, 306,
    500, 501
Amorphous materials, 19, 67, 68, 80, 81,
    89, 328, 349
Anelastic behavior, 212
Aqueous corrosion, 294
Atactic isomer, 361
Atomic diffusion, 134, 135, 142, 155, 156,
    159, 241, 244–246, 283, 286, 313,
    319, 320, 396, 398, 459
Atomic force microscope, 67, 89
Atomic mass, 29, 31, 32, 106, 150, 298
Atomic packing, 97, 107, 108, 162
Atomic packing factor, 107, 110, 239
Atomic weight, 31
Auger electrons, 63, 89
Auger electron spectroscopy, 63
Austempering, 251, 252
Austenite, 90, 170, 238, 239, 241–255, 481,
    484
Avrami equation, 285

B

Backscattered electrons, 61, 62

Bainite, 246–248, 250, 252, 257
Band gap, 434, 435, 440, 441, 445, 446, 491
Beryllium alloy(s), 153
Bifunctional polymers, 352
Binding energy, 36
Block copolymer, 373, 375
Bloomery iron, 12, 237
Body-centered cubic unit cell, 97, 127
Bond energy, 36–38
Bragg's law, 125, 126
Branched polymer, 367, 368, 384
Bravais lattices, 101, 127, 449
Brinell hardness, 195, 199, 260, 261

C

Carbonitriding, 257
Carbonization, 157, 256
Carbonizing, 164, 256
Carbon nanotube, 15, 16, 22, 126, 206, 225,
    226, 231, 467, 468, 473, 491–496,
    510, 531–534
Carrier density, 441, 442
Case hardening, 256
Casting, 17, 75, 76, 81, 82, 84, 89, 151,
    153, 426, 522, 524
Cast iron, 194, 195, 213, 237, 241, 257–261
Cementite, 238, 241–243, 246, 250, 251,
    254, 257, 259, 260, 302, 468, 510
Ceramic glass, 401, 402
Ceramic matrix composite, 21, 417, 509,
    511, 515, 521, 523, 527
Ceramics, 18–22, 34, 40, 43, 54, 68, 69, 95,
    131, 141, 149, 205, 207, 208,
    211–213, 215, 230, 233, 277, 312,
    313, 323, 332–334, 349, 370, 385,